D1548671

VIROLOGY

VIROLOGY

Directory & Dictionary of Animal, Bacterial and Plant Viruses

Roger Hull
Fred Brown
Chris Payne

M
STOCKTON
PRESS

Published in the United States and Canada by
STOCKTON PRESS, 1989
15 East 26th Street
NEW YORK, N.Y. 10010, USA

Library of Congress Cataloguing in Publication Data

Hull, Roger, 1937-
    Virology: directory & dictionary of animal, bacterial, and plant viruses/Roger Hull, Fred Brown, Chris Payne.
    p. cm.
    ISBN 0-935859-59-4 $79.00

    1. Viruses — Dictionaries.
    I. Brown, Fred, 1925- .
    II. Payne, Chris, 1946- .
    III. Title.
    [DNLM:. 1. Viruses — dictionaries. QW 13 H913v]
QR358.H85 1989
576'.64'0321 — dc19
DNLM/DLC
for Library of Congress 88-38137 CIP

Published in the United Kingdom by
THE MACMILLAN PRESS LTD, 1989
London and Basingstoke

Associated companies in Auckland, Delhi, Dublin, Gaborone, Hamburg, Harare, Hong Kong, Johannesburg, Kuala Lumpur, Lagos, Manzini, Melbourne, Mexico City, Nairobi, New York, Singapore, Tokyo.

ISBN 0-333-39063-0

Printed in Great Britain

Contents

Introduction   vii

Abbreviations   viii

References   ix

A Note on Use   x

THE DICTIONARY   1

Appendix A *A list of insect species in which infections caused by occluded baculovirus (nuclear polyhedrosis (NPV) and granulosis (GV)) infections have been recorded*   253

Appendix B *A list of insect species in which Cytoplasmic Polyhedrosis Virus infections have been recorded*   266

Appendix C *A list of insect species in which Densovirus infections have been recorded*   272

Appendix D *A list of insect species in which Entomopoxvirus infections have been recorded*   274

Appendix E *A list of insect species in which Iridescent virus infections have been recorded*   277

Appendix F *Properties of Phage Isolates*   282

Introduction

Virology brings together many branches of science ranging from human pathology to plant pathology, entomology to biochemistry, immunology to biophysics, and molecular biology to evolution. This confluence of subjects has resulted in a diversity of terminology which can confuse even leading workers. The aim of this dictionary is to provide a ready reference for all scientists working in or teaching virology and also for people interested in the subject. The dictionary comprises names of viruses and their higher order taxa as well as terms which are commonly used in the virological literature. In many cases references are given to recent reviews and papers so that readers have access to yet further information.

The virus classification is based on that approved by the International Committee on the Taxonomy of Viruses (ICTV) which is the regulatory body in this area. The most recent definitive listing published by R.E.F. Matthews (1982, Intervirology **17**, 1) has been updated using the minutes from subsequent meetings of the ICTV. Various groupings of viruses, not yet approved by the ICTV, are listed and are identified as being unofficial.

The ICTV classification groups viruses into families and subfamilies, genera, subgenera or groups and subgroups, and species or individual viruses. There is not yet full concordance amongst virologists on the formal classification and this is reflected in the plant virus entries which are under groups and individual viruses. We have attempted to list most, if not all, viruses of vertebrates, invertebrates, plants and bacteria. In some cases we have omitted older names of viruses; in other cases, with less well-characterised viruses, it is likely that the same virus has two or more names. For each virus we record the family or group to which it belongs and it is at this level of classification that the characteristic features of the virus will be found.

The general terms include those relating to techniques and laboratory reagents frequently used in virology as well as some historical facts and organisations closely associated with the subject. It is in these general terms that the overlap with other disciplines becomes obvious and we hope that this dictionary will also be of use to scientists other than virologists. We have included some useful formulae and facts so that the "bench virologist" can have easy access to this information.

We extend our most sincere thanks to those who have helped us in this venture: Dr. H.W. Ackermann, Université Laval, Quebec, Canada; Professor K.W. Buck, Imperial College, University of London; Professor D.A. Ritchie, University of Liverpool; Dr. N. E. Crook, AFRC Institute of Horticultural Research, Littlehampton, and to the numerous colleagues who have been pestered with questions about definitions. These acknowledgements are not a delegation of responsibility and we take full blame for any mistakes. We are indebted to Mrs. Patricia Thomas (KITES Project, University of Surrey) for assembling and collating many of the entries, to Mr. John Hodder, Computing Unit, University of Surrey, and to our respective wives for their continuous support.

Abbreviations

The following abbreviations are used throughout this dictionary. Other abbreviations are identified as entries.

(+)-sense	plus-sense (nucleic acid)
(–)-sense	minus-sense (nucleic acid)
CNS	central nervous system
DNA	deoxyribonucleic acid
ds	double-stranded (nucleic acid)
E. coli	*Escherichia coli*
kb	kilobases
kbp	kilobase pairs
mRNA	messenger ribonucleic acid
mw.	molecular weight
RBC	red blood cells
RF	replicative form
RNA	ribonucleic acid
r.p.m.	revolutions per minute
ss	single-stranded (nucleic acid)
tRNA	transfer ribonucleic acid
UV	ultraviolet (radiation)

References

1. Ackermann, H.W. (1978) **In** Handbook of Microbiology, pp.639, 643 and 673, ed. A.I. Laskin and H.A. Lechevalier. CRC Press: Boca Raton, Florida.
2. Ackermann, H.W. *et al.* (1984) Intervirology **22**, 61.
3. Ackermann, H.W. *et al.* (1984) Intervirology **22**, 181.
4. Ackermann, H.W. *et al.* (1985) Intervirology **23**, 121.
5. Ackermann, H.W. *et al.* (1985) Annls Virol. **136**, 175.
6. Cole, R.M. (1978) **In** Handbook of Microbiology, p.683, ed. A.I. Laskin and H.A. Lechevalier. CRC Press: Boca Raton, Florida.
7. Fiers, W. (1978) **In** Comprehensive Virology vol. 13, p.69, ed. H. Fraenkel-Conrat and R.R. Wagner. Plenum Press: New York.
8. Liss, A. *et al.* (1981) Intervirology **15**, 71.
9. Maniloff, J. *et al.* (1982) Intervirology **18**, 177.
10. Matthews, R.E.F. (1982) Intervirology **17**, 1.
11. Reanney, D.C. and Ackermann, H.W. (1981) Intervirology **15**, 190.
12. Safferman, R.S. *et al.* (1983) Intervirology **19**, 61.
13. Ackermann, H.W. *et al.* (1981) Intervirology **16**, 1.
14. Denhardt, D.T. (1977), **In** Comprehensive Virology vol. 7, p.1, ed. H. Fraenkel-Conrat and R.R. Wagner, Plenum Press: New York.
15. Rocourt, J. *et al.* (1983) Annls Virol. **134**, 245.
16. Ackermann, H.W. *et al.* (1981) Annls Virol. **132**, 371.
17. Ackermann, H.W. *et al.* (1978) Path. Biol. **26**, 507.
18. Lomovskaya, N.D. *et al.* (1980) Microbiol. Rev. **44**, 206.
19. Ackermann, H.W. and Du Bow, M.S. (1987) Viruses of Prokaryotes. CRC Press: Boca Raton, Florida.
20. D.A. Ritchie, personal communication.
21. Ackermann, H.W. (1987) Abstr. VII Internat. Congr. Virol. p.195.
22. Chowdhury, R. *et al.* (1987) J. Virol. **61**, 3999
23. Trautwetter, A. *et al.* (1987) J. Virol. **61**, 1540.
24. Liao, Y-D. *et al.* (1987) J. Virol. **61**, 1695.
25. Santos, M.A. *et al.* (1986) J. Virol. **60**, 702.
26. Poon, A.P.W. and Dhillon, T.S. (1986) J. Virol. **60**, 317.
27. Trautwetter, A. *et al.* (1986) J. Virol. **59**, 551.
28. Ito, S-I. *et al.* (1986) J. Virol. **59**, 103.
29. Chowdhury, R. and Das, J. (1986) J. Virol. **57**, 960.
30. Donohue, T.J. *et al.* (1985) J. Virol. **55**, 147.
31. Bess, U.H. and Birge, E.A. (1987) Virology **156**, 122.
32. Iida, S. *et al.* (1985) Virology **143**, 347.
33. Moynet, D.J. *et al.* (1985) Virology **142**, 263.
34. Bancroft, I. and Smith, R.J. (1988) J. gen. Virol. **69**, 739.
35. Pretorius, G.H.J. and Coetzee, W.F. (1979) J. gen. Virol. **45**, 389.
36. Reakes, C.F.L. *et al.* (1987) J. gen. Virol, **68**, 263.
37. Reddy, A.B. and Gopinathan, K.P. (1987) J. gen. Virol. **68**, 949.
38. Poon, A.P.W. and Dhillon, T.S. (1986) J. gen. Virol. **67**, 2781.
39. Karsten, K.H. *et al.* (1979) J. gen. Virol. **43**, 57.
40. Dietz, A. *et al.* (1986) J. gen. Virol. **67**, 831.
41. Powell, I.B. and Davidson, B.E. (1985) J. gen. Virol. **66**, 2737.
42. Poon, A.P.W. and Dhillon, T.S. (1986) J. gen. Virol. **67**, 789.
43. Hug, H. *et al.* (1986) J. gen. Virol. **67**, 333.

A Note on Use

The head words defined in the dictionary are placed in alphabetical order. This applies to the complete term regardless of spaces and hyphens. For example, **bee virus Y** appears before **beet cryptic virus 1**.

Where numerals occur they count first in the alpha ordering, unless they are placed before the first letter. Thus **H3 virus** appears before **Haden virus**, but **734B B virus** appears under 'B' before **Babahoyo virus**.

Cross references appear in SMALL CAPITALS.

A

A-type inclusion body. Inclusion bodies produced late in infection in the cytoplasm of cells infected with certain pox viruses (e.g. fowlpox, ectromelia, cowpox, canarypox). Some contain virus particles. In cowpox virus infections the inclusions consist almost entirely of a single protein species (mw. 160 x 10^3), abundantly synthesised late in infection. Cytoplasmic A-type inclusions of some vertebrate poxviruses (e.g. fowlpox) resemble POLYHEDRA in several properties as they contain virus particles, are infective, can be isolated intact, are resistant to digestion by some proteases and are extremely resistant to prolonged storage.
Patel, D.D. *et al.* (1986) Virology **149**, 174.

A-type virus particle. A term used originally for a morphologically defined group of RNA virus particles, often found in tumour cells. They are double-shelled spherical particles with a diameter of 65-75 nm. for the outer and 50 nm. for the inner shell; the inner ring appears denser. Various members of RETROVIRIDAE have A-type particles. *See* B-, C-, D-TYPE VIRUS PARTICLE.
Dalton, A.J. (1972) J. Natl. Canc. Inst. **49**, 323.

AAB. *See* CMI/AAB.

Abadina virus. Family *Reoviridae*, genus *Orbivirus*. Isolated from *Culicoides*.

Abelson leukaemia virus. Family *Retroviridae*, subfamily *Oncovirinae*, genus *Type C Oncovirus*, sub-genus *Mouse type C Oncovirus*. A virus isolated from Balb/c mice inoculated with Moloney leukaemia virus. Produces lymphoid leukaemia and can transform mouse cells *in vitro*. Requires a helper virus for virus replication. Risser, R. *et al.* (1978) J. exp. Med. **148**, 714.

abortive infection. An infection which does not produce infectious progeny. The cell may not allow the expression of all the viral genes or the virus may be defective.

abortive transformation. A TRANSFORMATION of cells which is unstable so the cells revert to normal after a few generations.

abras virus. Family *Bunyaviridae*, genus *Bunyavirus*.

Abraxas grossulariata Cytoplasmic Polyhedrosis Virus. CYTOPLASMIC POLYHEADROSIS VIRUS (CPV) isolated in the United Kingdom from larvae of the magpie moth, *Abraxas grossulariata* (Geometridae, Lepidoptera). The virus is the type member of 'type 8' CPVs. Unrelated to *Bombyx mori* (type 1) CPV on the basis of RNA electropherotype. Viruses of similar electropherotype have been observed in other Lepidoptera (*see* APPENDIX B).
Payne, C.C. and Mertens, P.P.C. (1983) **In** The *Reoviridae*. p. 425. ed. W.K. Joklik. Plenum Press: New York.

Absettarov virus. Family *Flaviviridae*, genus *Flavivirus*. Isolated from a boy with biphasic fever and signs of meningitis. Tick-borne. Found in many countries. Pathogenic in Rhesus monkeys.

absorbance. Amount of light absorbed by a substance at a particular wavelength. It has no official units although the term optical density units is often used. *See* SPECIFIC ABSORBANCE.

absorption spectrum. Graphical representation of ABSORBANCE of a substance at different wavelengths. Valuable in obtaining an approximate estimate of the percentage of nucleic acid in a virus from the ratio of absorbances at 254 and 280 nm. Spectral characteristics of nucleotides can be found in Sober, H.A. ed. (1968) Handbook of Biochemistry: Selected Data for Molecular Biology. CRC Press: Cleveland, Ohio.

Abu Hammad virus. Family *Bunyaviridae*,

1

genus *Nairovirus*. Isolated from a tick *Argas hermanni* in Egypt.

Abu Mina virus. Family *Bunyaviridae*, genus *Nairovirus*.

abutilon mosaic virus. A possible *Geminivirus*, subgroup B.
Jaske, H. and Schuchalter-Eicke, G. (1984) Phytopath. Z. **109**, 353.

acado virus. Family *Reoviridae*, genus *Orbivirus*. Isolated from *Culex antennatus* and *Culex univittatus neavi* in Ethiopia.

acara virus. Family *Bunyaviridae*, genus *Bunyavirus*. Isolated from sentinel mice *Culex* sp. and *Nectomys squamipes* in Brazil and Panama.

acholeplasmavirus. Virus isolated from *Acholeplasma* sp. e.g. phages L1, L2 and L3.

AcMNPV. Abbreviation for *Autographa californica* nuclear polyhedrosis virus.

AcNPV. Abbreviation for *Autographa californica* nuclear polyhedrosis virus.

acquired immunodeficiency syndrome (AIDS). A disease of man caused by HUMAN IMMUNODEFICIENCY VIRUS (HIV). AIDS is primarily a disease of the immune system so the infection usually results in a wide range of adverse immunological and clinical conditions. The opportunistic infections (i.e. those caused by micro-organisms that seldom cause· disease in persons with normal defence mechanisms) and cancers resulting from immune deficiency are generally the most severe but neurological problems such as dementia resulting from infection of the brain can also occur. It is now recognised that the disease is likely to be fatal. The disease is generally transmitted sexually but is also prevalent among drug addicts and occurs occasionally in children from infected mothers and in patients receiving blood transfusions. It can be controlled by AZT (AZIDOTHYMIDINE) but the side effects of the drug are not negligible.

acridine orange. A compound used for determining the nature of the nucleic acid in virus particles or cells. It binds to the nucleic acid and, when exposed to UV-light, fluoresces green if bound to ds nucleic acid and orange if bound to ss nucleic acid.

acriflavine. A dye which inactivates viruses in the presence of light (photodynamic inactivation) by binding to nucleic acid. Also used as an ANTIBIOTIC.

acronym. A word created from the initial letters of the principal words in a compound term, e.g. enteric cytopathic human orphan virus = ECHO virus.

acrylamide. A chemical which is polymerised using a cross-linking agent to give polyacrylamide, one of the most commonly used supports for GEL ELECTROPHORESIS.

Actias selene Cytoplasmic Polyhedrosis Virus. Cytoplasmic polyhedrosis virus (CPV) isolated in the UK from laboratory-reared larvae of *Actias selene* bred from pupae obtained from a site in the Himalayas. The virus is the type member of 'type 4' CPVs. Unrelated to *Bombyx mori* (type 1) CPV on the basis of RNA electropherotype. Viruses of similar electropherotype have been observed in other Lepidoptera (*see* APPENDIX B).
Payne, C.C. and Mertens, P.P.C. (1983) In The *Reoviridae*. p. 425. ed. W.K. Joklik. Plenum Press: New York.

actinomycin D. A compound which inhibits transcription by interacting with the guanine residues of helical DNA. Replication of DNA-containing viruses and those RNA viruses which require DNA to RNA transcription (e.g. RETROVIRIDAE, INFLUENZA VIRUS) is inhibited, while replication of other RNA-containing viruses is unaffected.

actinophages. Viruses infecting members of the order Actinomycetales (filamentous, branching bacteria; e.g. *Streptomyces, Nocardia, Mycobacterium*) and related organisms including *Corynebacterium, Arthrobacter* and *Kurthia*. All of the virus isolates of known morphology from these hosts are TAILED PHAGES (*see* PHAGE).

activator. In molecular biology an activator is a protein which binds to DNA upstream of a gene and activates transcription of that gene.

active immunity. Immunity induced by injection of a virus or virus subunit.

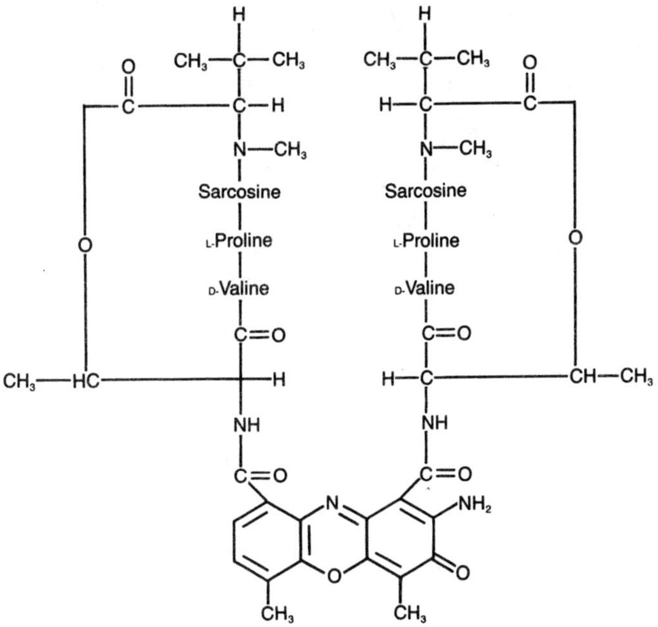

Actinomycin

acute bee-paralysis virus. *See* BEE ACUTE PARALYSIS VIRUS.

acute epidemic gastroenteritis virus of man. Unclassified. Consists of a group of viruses, Norwalk, Hawaii and Wollan. Viruses are *c*.30 nm. in diameter and are acid- and ether-stable. They are found in faeces by electron microscopy. Flewett, T.H. (1977) Rec. Adv. clin. Virol. **1,** 151.

acute haemorrhagic conjunctivitis virus. Family *Picornaviridae,* genus *Enterovirus.* Designated enterovirus 70. Causes acute haemorrhagic conjunctivitis in man in many parts of the world. Grows in several cell lines.

acute laryngo-tracheo-bronchitis virus. *See* PARAINFLUENZA VIRUS type 2.

acyclovir. (Synonym: acycloguanosine; Zovirax). An antiviral agent with a potent and specific action against HERPES VIRUS 1 and 2, both *in vitro* and *in vivo.* The compound is selectively PHOSPHORYLATED by herpes virus-induced THYMIDINE KINASE. The phosphorylated derivative inhibits herpes virus-induced DNA polymerase.

adenine. A constituent purine base of DNA and RNA. *See* NUCLEIC ACID.

adenine arabinoside. *See* ARABINOSYL ADENINE.

adeno-associated virus. (Latin 'dependere' = to depend.) A genus in the family *Parvoviridae.* Replication is dependent on a helper adenovirus for complete virus replication. Multiplies in cells which support adenovirus replication. The virus particles contain either + or − DNA which are complementary and anneal to form dsRNA on extraction. Occurs in many species. Now known as the *Dependovirus* genus.

adenosine. A nucleoside of adenine and ribose. *See* NUCLEIC ACID.

adenosine 5'-monophosphate. Monophosphate of the nucleoside ADENOSINE.

Acyclovir

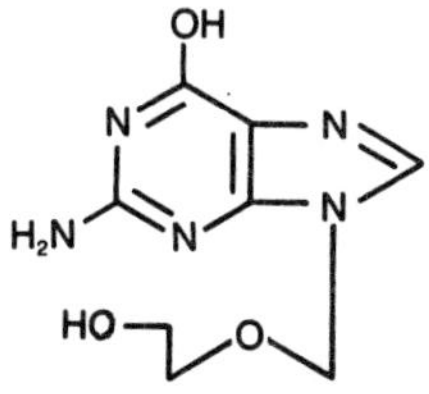

adenosine 5'-triphosphate. Triphosphate of the nucleoside ADENOSINE. *See* NUCLEIC ACID.

adenosine triphosphatase (ATPase). An enzyme which catalyses the conversion of ATP to ADP with the release of Pi. Some viruses (e.g. VACCINIA) possess an ATPase activity.

Adenoviridae. (Greek 'aden', 'adenos' = gland.) A family of DNA viruses with isometric particles 70-90 nm. in diameter which sediment at *c*.800S and band in CsCl at 1.33-1.35 g/cc. The capsid is

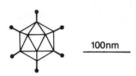

100nm

formed of 252 capsomeres, 8-9 nm. in diameter, of which the 12 pentons at the vertices have an outward projection with a knob at the end. The 240 non-vertex hexons are different from the penton bases. Antigens at the surface are mainly species-specific, the hexon for neutralisation, the fibre for haemagglutination. There are at least ten virion polypeptides, mw. ranging from 5-120 x 10^3. Each particle contains a single linear molecule of dsDNA, mw. 20-30 x 10^6 with inverted terminal repeats of *c*.100 base pairs. Genome transcription takes place in two stages, 'early' before initiation of viral DNA synthesis when non-virion polypeptides are synthesised and 'late' when viral DNA and structural polypeptides are synthesised. DNA replication is by strand-displacement mechanism. Transcription in the nucleus is followed by processing into mRNAs migrating into cytoplasm. Structural polypeptides are synthesised in cytoplasm but assembly is nuclear. The natural host range is mostly confined to one host or closely related animal species. Several species cause tumours in new-born hosts of heterologous species. Transmission is direct or indirect from throat, eye, faeces or urine. There are two genera, *MASTADEN-OVIRUS* and *AVIADENOVIRUS*.
Matthews, R.E.F. (1982) Intervirology **17**, 59.

adjuvant. Substance added to antigens to enhance immune response. Salts of aluminium (e.g. hydroxide or phosphate) acceptable for use in man. Saponin or FREUND'S ADJUVANT used in experimental animals.

AdoHcy. Abbreviation for S-adenosyl homocysteine.

AdoMet. Abbreviation for S-adenosyl methionine. Synonym: SAM.

adonis mosaic virus. An unclassified plant virus, isometric particles, occurs in Japan. Doi, Y. *Personal communication.*

adsorption. The first stage of infection of a cell by a virus involves attachment at a receptor site. Often involves a specific interaction between a receptor on the cell surface and a component of the virus. It is independent of temperature.

adventitious virus. A foreign virus present in a vaccine or preparation of a different virus.

adzuki bean mosaic virus. *See* COWPEA APHID-BORNE MOSAIC VIRUS.

Aedes cells. Cell lines established from the mosquitoes *Aedes aegypti* and *Aedes albopictus* able to support the replication of a number of arboviruses. Usually grown in Mitsuhashi and Maramorosch medium.

aerobic. Requiring the presence of free oxygen or air.

aerosol. A gaseous suspension of ultramicroscopic particles or liquid droplets.

affinity chromatography. Chromatography using ligands attached to an insoluble support which interacts with the molecule of interest, retaining it and allowing unwanted molecules to be washed away. On changing the conditions the molecule of interest can be eluted. Examples are the selection of antigen by immobilised antibody and the selection of POLYADENYLATED mRNA using oligo dT sepharose.

African green monkey kidney cells. Cells used for growth of certain viruses of vertebrates, e.g. POLIOVIRUS.

African monkey cytomegalovirus. *See* CERCO-PITHECID HERPESVIRUS 2.

African swine fever virus (ASFV). Synonym: WART HOG VIRUS. Unclassified. First seen in Africa. Often causes fatal disease in domestic pigs, following high fever, cough and diarrhoea but is carried by wart hogs and bush pigs. Many strains cause chronic infection. Disease has now been spread to Western Europe, Brazil and Cuba, probably in waste food. Antibodies produced during infection do not give immunity.

AG80-24 virus. Family *Bunyaviridae*, genus *Bunyavirus*.

agar. Mixture of polysaccharides, some anionic, that forms a gel at temperatures below about 40°C. Used as a support medium when supplemented by appropriate buffers/media ingredients for electrophoresis, production of microbial cultures, overlaying tissue culture cells etc. Derived from red algae.

Agaricus bisporus virus 3 (AbV-3). An unclassified fungal virus with BACILLIFORM particles, 19 x 50 nm., a single capsid polypeptide species, mw. 24 x 10³ and a single species of ssRNA, mw. 1.4 x 10⁶.
Buck, K.W. (1986) In Fungal Virology, p. 1. ed. K.W. Buck. CRC Press: Boca Raton, Florida.

Agaricus bisporus virus 4 (AbV-4). A member of the *Partitivirus* group.

agarose. One of the constituents of AGAR. Often used in preference to agar as it does not contain inhibitors of virus growth frequently present in agar and as lower temperature gelling products of agarose are now available. Also used widely in gel electrophoresis as the pore size is more uniform than that of agar.

agarose gel electrophoresis. Technique used for separating proteins or nucleic acids by passage of an electric current through the gel.

ageing *in vitro*. Synonym: LONGEVITY *IN VITRO*.

agglutination test. Some viruses will cause clumping of cells due to attachment to more than one cell. HAEMAGGLUTINATION of red blood cells by e.g. INFLUENZA VIRUS is by far the most spectacular example. Cells other than red blood cells also exhibit the phenomenon. Also used to refer to the clumping of inert particles, e.g. latex, coated with antibody and mixed with homologous virus antigens.

agouti endogenous type C retrovirus. Family *Retroviridae*, subfamily *Oncovirinae*, genus Type C *Oncovirus*, sub-genus *Mammalian Type*

C *Oncovirus*. Isolated from kidney tissue of a New World rodent, the agouti *Dasyprocta puntata*. Has been transmitted to human and cat cells.

agropyron mosaic virus. Group: *Ryegrass mosaic virus*.
Slykhuis, J.T. (1973) CMI/AAB Descriptions of Plant Viruses No. 118.

Agrotis segetum Cytoplasmic Polyhedrosis Virus. Cytoplasmic polyhedrosis virus (CPV) isolated in the United Kingdom from larvae of the cutworm, *Agrotis segetum* (Noctuidae, Lepidoptera). The virus is the type member of 'type 9' CPVs. Unrelated to *Bombyx mori* (type 1) CPV on the basis of RNA electropherotype. Currently the sole recorded isolate of 'type 9' CPVs.
Payne, C.C. and Mertens, P.P.C. (1983) In The *Reoviridae*. p. 425. ed. W.K. Joklik. Plenum Press: New York.

AIDS. Acronym for ACQUIRED IMMUNODEFICIENCY SYNDROME.

aino virus. Family *Bunyaviridae*, genus *Bunyavirus*. Isolated from *Culex tritaeniorhynchus* in Japan. May be same as SAMFORD VIRUS.
Doherty, R.L. *et al.* (1972) Aust. vet. J. **48**, 81.

akabane virus. Family *Bunyaviridae*, genus *Bunyavirus*. Isolated from mosquitoes in Australia and Japan. *Culicoides* sp. are vectors. Sheep and goats can be infected experimentally causing disease in the foetus.
Kurogi, H. *et al.* (1975) Arch. Virol. **47**, 71.

alagoas virus. Family *Rhabdoviridae*, genus *Vesiculovirus*. Isolated from a vesicular lesion on the tongue of a mule in Alagoas, Brazil. Closely related to the classical strains of VESICULAR STOMATITIS VIRUS. Antibody present in several species in Brazil, including man. Causes fever and malaise in man.

alastrim virus. See VARIOLA MINOR VIRUS.

Aleutian disease of mink virus. Family *Parvoviridae*, genus *Parvovirus*. Causes an economically important, lethal disease in mink. The virus can cross the placenta to infect the foetus. Ferrets and skunks can be infected experimentally.
Porter, D.D. *et al.* (1977) Intervirology **8**, 129.

alfalfa cryptic virus 1. A member of the *Cryptovirus* group, subgroup A.

Boccardo, G. *et al.* (1987) Adv. Virus Res. **32**, 171.

alfalfa cryptic virus 2. Synonym: ALFALFA TEMPERATE VIRUS. A member of the *Cryptovirus* group, subgroup B.
Boccardo, G. *et al.* (1987) Adv. Virus Res. **32**, 171.

alfalfa latent virus. *See* PEA STREAK VIRUS.

alfalfa mosaic virus. Type member of the *Alfalfa mosaic virus* group.
Jaspars, E.M.J. and Bos, L. (1980) CMI/AAB Descriptions of Plant Viruses No. 229.

Alfalfa mosaic virus group. (Named after the type member ALFALFA MOSAIC VIRUS.) Monotypic genus of a MULTICOMPONENT plant virus with BACILLIFORM particles of at least four sizes. The four major RNA species are contained in separate components: bottom (B), 58 x 18 nm.; middle (M), 48 x 18 nm.; top b (Tb), 36 x 18 nm. and top a (Ta), 28x18 nm. Particles are stabilised primarily by protein-RNA interactions; they are sensitive to ribonuclease. Capsid structure considered to be based on icosahedral symmetry, the structural subunit being a single protein species of mw. 24×10^3. Genomic linear (+)-sense ssRNA comprises three species, RNA-1 (mw. 1.1×10^6; 3,644 nucleotides) in B component, RNA-2 (mw. 0.8×10^6; 2,593 nucleotides) in M component and RNA-3 (mw. 0.7×10^6; 2,037 nucleotides) in Tb component; in Ta component there is the MONO-CISTRONIC mRNA for coat protein, RNA-4 (mw. 0.3×10^6; 881 nucleotides). RNAs 1-3 plus coat protein or RNA 4 are required for infection. Replication is in the cytoplasm and, for the genomic RNAs, is via distinct ds REPLICATIVE INTERMEDIATES; RNA-4 is derived from RNA-3. RNAs -1, -2 and -4 are monocistronic messengers for proteins of mw. 125, 89 and 24 (coat protein) $x 10^3$ respectively; RNA-3 is bicistronic encoding a protein of mw. 89×10^3 at the 5' end and having the coat protein cistron at the 3' end. The natural host range is wide. The particles are found in most cell types. The virus is readily transmitted mechanically. It is transmitted by aphids in the NON-PERSISTENT TRANSMISSION manner and is seed transmitted in some species. Alfalfa mosaic virus is considered by some to be in the *ILARVIRUS* group.
Matthews, R.E.F. (1982) Intervirology **17**,77.
Jaspars, E.M.J. and Bos, L. (1980) CMI/AAB Descriptions of Plant Viruses No. 229.
Francki, R.I.B. *et al.* (1985) **In** Atlas of Plant Viruses. Vol. 2. p. 93. CRC Press.: Boca Raton, Florida.
Francki, R.I.B. (1985) **In** The Plant Viruses. Vol. 1. p. 1. ed. R.I.B. Francki. Plenum Press: New York.

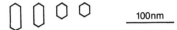

100nm

alfalfa temperate virus. *See* ALFALFA CRYPTIC VIRUS 2.

Alfuy virus. Family *Flaviviridae*, genus *Flavivirus*. Isolated from mosquitoes in Queensland, Australia. Antibodies found in human sera but no disease in man has been reported.

algophage. Synonym: CYANOPHAGE.

alkaline phosphatase. An enzyme of relevance in molecular biology as it removes the 5' terminal phosphate of linear DNA molecules.

Allerton virus. *See* BOVID (ALPHA) HERPES VIRUS 2.

Almpiwar virus. Unclassified arthropod-borne virus. Isolated from a skink *Ablepharus boutinii virgatus* in Queensland, Australia.

alpha amanitin. A cyclic peptide which selectively inhibits DNA-DEPENDENT RNA POLYMERASES II and III of eukaryotic cells at low and high concentrations respectively. It binds to the polymerase and blocks RNA synthesis after initiation, thus preventing chain elongation. Isolated from the fungus *Amanita phalloides*. Viruses such as ADENOVIRUSES, INFLUENZA VIRUS and RETRO-VIRUSES which require RNA polymerase II for their replication are inhibited.

Alphaherpesvirinae. A subfamily of the family *Herpesviridae*. Consists of two genera, *Human herpesvirus 1* and *Suid herpesvirus 1*. Replicate rapidly and latent infection is often demonstrable in nerve ganglia.

Alphavirus. (Greek letter 'a'.) A genus in the family *Togaviridae*. The type species is Sindbis virus. The members have an RNA with mw. 4×10^6, capped and polyadenylated, which forms $c.6\%$ of the particle. There is a capsid protein, mw. $30-34 \times 10^3$ and usually two envelope proteins, E1 and E2, mw. 50-59 x 10³. Sometimes there is a third envelope protein, E3. The genus contains many viruses of importance as disease agents and as tools for the study of virus replication. EASTERN, WESTERN and VENEZUELAN EQUINE ENCEPHALOMYELITIS VIRUSES and ROSS RIVER VIRUS are important disease agents. SEMLIKI FOREST VIRUS and SINDBIS VIRUS have been used extensively as laboratory models for studying replication.

alsike clover vein mosaic virus. An unclassified plant virus with isometric particles, 30 nm. in diameter, which sediments at 123S.
Gerhardson, B. and Lindsten, K. (1971) Phytopath. Z. **72**, 76.

amaas virus. See VARIOLA MINOR VIRUS.

amapari virus. Family *Arenaviridae*, genus *Arenavirus*. Isolated from forest rodents in the Ampari region of Brazil. Grows in Vero cells.

amaranthus leaf mottle virus. A *Potyvirus*. Francki, R.I.B. *et al.* (1985) **In** Atlas of Plant Viruses. Vol. 2. p. 183. CRC Press: Boca Raton, Florida.

amber codon. See STOP CODON.

amber mutant. A mutation resulting in the CODON, UAG, which prematurely terminates an open reading frame.

ambisense expression strategy. The coding of proteins on both viral-sense and viral-complementary mRNAs. Shown by one genus of the BUNYAVIRIDAE, the phleboviruses.
Bishop, D.H.L. (1986) Microbiological Sciences **3**, 183.

American haemorrhagic fever viruses. Family *Arenaviridae*, genus *Arenavirus*. A group of viruses also known as the New World arenaviruses and the Tacaribe antigenic group. Wild rodents are the natural hosts but the viruses have also been isolated from mites.

American oyster reo-like virus. Unclassified virus isolated from *Crassostrea virginica*, morphologically similar to reoviruses. Particles are isometric, 75 nm. in diameter with a double capsid shell containing at least five major proteins. Morphologically, the particles closely resemble members of the REOVIRUS genus, though the genome is composed of eleven segments of dsRNA (total mw. 15 x 10⁶). The virus replicates in certain fish cell lines and it appears to have similar properties to reo-like viruses isolated from the fish hosts, golden shiner, chum salmon and channel catfish.
Winton, J.R. *et al.* (1987) J. gen. Virol. **68**, 353.

American plum line pattern virus. A possible member of the *Ilarvirus* group.
Fulton, R.W. (1984) CMI/AAB Descriptions of Plant Viruses No. 280.

Amino acid

Some common amino acids and their properties.

Amino acid	Abbreviation 3-letter	1-letter	mw.	Partial-specific volume (g/cc)	Properties
Alanine	Ala	A	71.1	0.74	Small, hydrophobic
Arginine	Arg	R	156.2	0.70	Large, basic
Asparagine	Asn	N	132.1	0.71	Small, hydrophilic
Aspartic acid	Asp	D	115.1	0.60	Small, acidic
Cysteine	Cys	C	102.2	0.61	Small, sulphydryl
Glutamic acid	Glu	E	129.1	0.66	Small, acidic
Glutamine	Gln	Q	146.2	0.67	Small, hydrophilic
Glycine	Gly	G	57.1	0.64	Small, hydrophobic
Histidine	His	H	137.1	0.67	Large, basic
Isoleucine	Ile	I	113.2	0.90	Hydrophobic
Leucine	Leu	L	113.2	0.90	Hydrophobic
Lysine	Lys	K	128.2	0.82	Large, basic
Methionine	Met	M	131.2	0.75	Hydrophobic
Phenylalanine	Phe	F	147.2	0.77	Large, hydrophobic
Proline	Pro	P	97.1	0.76	Small, hydrophilic
Serine	Ser	S	87.1	0.63	Small, hydrophilic
Threonine	Thr	T	101.1	0.70	Small, hydrophilic
Tryptophan	Trp	W	188.2	0.74	Large, hydrophobic
Tyrosine	Tyr	Y	163.2	0.71	Large, hydrophobic
Valine	Val	V	99.1	0.86	Hydrophobic

Amino acid

Basic structure

Amino group   Carboxyl group

Peptide bond

Peptide bond

Francki, R.I.B. (1985) **In** The Plant Viruses. Vol. 1. p. 1. ed. R.I.B. Francki. Plenum Press: New York.

American Type Culture Collection (ATCC). Organisation which holds a large collection of micro-organisms and cells which are available on request on payment of a small fee. Address is Rockville, Maryland 20852, USA.

amino acid. Basic unit of proteins (*see* Figure), containing a carboxyl and an amino group and a variable side chain which determines the properties of the individual amino acid. The side chain may be simple (glycine) or a complicated ring structure (tryptophan). There are 20 commonly occurring amino acids in nature (*see* Table) and a few which occur much less frequently.

aminoacyl-tRNA. An amino acid linked via its carboxyl group to the hydroxyl group (2′ or 3′) of the ribose residue at the 3′ end of the tRNA. Amino acids are linked in this form during protein synthesis.

aminoacyl-tRNA synthetase. (Synonym: AMINOACYL-tRNA LIGASE). The enzyme which ligates the amino acid to its specific tRNA. There are 20 such enzymes, specific for each amino acid and its tRNA.

ammonium sulphate. Salt commonly used to precipitate enzymes, proteins and viruses without denaturation. Used frequently in the initial stages of protein purification as proteins precipitate at different concentrations of the salt.

AMP. Abbreviation for adenosine 5′-phosphate.

amphotericin B. An ANTIBIOTIC produced by *Streptomyces nodosus* which operates by affecting the permeability of the cytoplasmic membrane.

Amsacta moorei entomopoxvirus. Type species of probable subgenus B, *Entomopoxvirinae (Poxviridae)*, isolated from *A. moorei* (Lepidoptera). Ovoid virions 350 x 250 nm. with a sleeve-shaped LATERAL BODY and cylindrical core. Particle surface has globular units 40 nm. in diameter. Virions have a buoyant density of 1.26-1.28 g/cc. and contain *c*.36 polypeptides and a single linear molecule of dsDNA (mw. 135×10^6). During replication, virions are occluded in large occlusion bodies (SPHEROIDS), 1-4 μm. in diameter. Smaller inclusions (SPINDLES) devoid of virions, are occasionally produced. Virus will infect some lepidopteran cell lines. Morphologically similar viruses have been found in other insects, specifically *Choristoneura, Euxoa, Oreopsyche, Operophtera, Oncopera* spp. (Lepidoptera) and *Melanoplus* sp. (Orthoptera).
Arif, B.M. (1984) Adv. Virus Res. **29**, 195.

Amyelois chronic stunt virus. A possible member of the *Caliciviridae* isolated from the navel orangeworm, *Amyelois transitella* (Lepidoptera). Virions are icosahedral, 38 nm. in diameter with characteristic cup-shaped surface depressions. Intact particles sediment at 185S and have a buoyant density of 1.32 g/cc. in CsCl. Particles contain ssRNA (mw. 2.5×10^6) and several proteins with two major species 29×10^3 and 55×10^3.
Hillman, B. *et al.* (1982) J. gen. Virol. **60**, 115.

analytical ultracentrifuge. Instrument used to sediment macromolecules at high centrifugal force with appropriate optics so that the rates of sedimentation and diffusion of macromolecules

can be measured with great accuracy. *See* DIFFU-
SION COEFFICIENT, SEDIMENTATION COEFFICIENT.
Markham, R. (1962) Adv. Virus Res. **9**, 241.

Ananindena virus. Family *Bunyaviridae*, genus
Bunyavirus.

Andean potato latent virus. A *Tymovirus*.
Francki, R.I.B. *et al.* (1985) **In** Atlas of Plant
Viruses. Vol. 1. p. 117. CRC Press: Boca Raton,
Florida.
Hirth, L. and Girard, L. (1988) **In** The Plant
Viruses. Vol. 3. p. 163. ed. R. Koenig. Plenum
Press: New York.

Andean potato mottle virus. A *Comovirus*.
Fribourg, C.E. (1979) CMI/AAB Descriptions of
Plant Viruses No. 203.
Francki, R.I.B. *et al.* (1985) **In** Atlas of Plant
Viruses. Vol. 2. p. 1. CRC Press: Boca Raton,
Florida.

Anhanga virus. Family *Bunyaviridae*, genus
Phlebovirus. Isolated from the sloth *Choloepus
brasiliensis* in Brazil.

Anhembi virus. Family *Bunyaviridae*, genus
Bunyavirus. Isolated from the rodent *Proechimys
iheringi* and an arthropod *Phoniomyia pilicauda*
in Brazil.

anionic detergent. A detergent having a negatively charged surface ion, e.g. sodium DE-
OXYCHOLATE, SODIUM DODECYL SULPHATE.

anisometric. Adjective to describe virus particles which are not isometric e.g. rod-shaped
particles.

anisotropy of flow. Difference in the flow properties of a solution of macromolecules in different
directions. Shown by a solution of a virus with
rod-shaped particles e.g. TOBACCO MOSAIC VIRUS.
On stirring the solution the particles become
orientated along the line of flow and rotate polarised light.

annealing. The formation of ds nucleic acid
molecules from ss molecules with complementary base sequences. This can be done (a) in
solution, cooling the mixture of ss molecules
slowly from a high temperature to below the
MELTING TEMPERATURE or (b) with one strand
immobilised on a solid support. Applied quantita-
ively, a measure of the complementarity between

two ss nucleic acid molecules can be obtained.

Anopheles A virus group. Family *Bunyaviridae*,
genus *Bunyavirus*. Isolated from *Anopheles* sp. in
Columbia.

Anopheles B virus group. Family *Bunyaviridae*,
genus *Bunyavirus*. Isolated from mosquitoes in
South America.

anserid herpesvirus 1. Synonyms: DUCK ENTERI-
TIS VIRUS, DUCK PLAGUE VIRUS. Member of family
Herpesviridae. Causes infection of ducks and
mallards, resulting in nasal and ocular discharge
and diarrhoea. Most animals die. Ducks, geese,
swans and chicks can be infected experimentally.
The virus can be grown on the chorioallantoic
membrane, killing the embryo, or in chick fibroblast cultures.

Antheraea eucalypti Cells. The first line of
insect cells established in culture. Able to support
the growth of a number of arboviruses. Usually
grown in Grace's cell culture medium.

Antheraea eucalypti virus. Member of
Tetraviridae (formerly the NUDAURELIA β VIRUS
GROUP).

Antheraea satellite virus. Possible RNA-containing satellite virus associated with *ANTHERAEA
EUCALYPTI* VIRUS. Virions are isometric, 13 nm. in
diameter and sediment at 44S.
Longworth, J.F. (1978) Adv. Virus Res. **23**,
103.

Anthoxanthum latent blanching virus. An
unclassified plant virus with rod-shaped particles
135 nm. long, 22 nm. wide.
Catherall, P.L. and Chamberlain, J.A. (1981)
Welsh Plant Breeding Sta. Ann. Rept. 1980,
153.

Anthoxanthum mosaic virus. A possible
Potyvirus.
Francki, R.I.B. *et al.* (1985) **In** Atlas of Plant
Viruses. Vol. 2. p. 183. CRC Press: Boca Raton,
Florida.

antibiotic. Substance used to inhibit the growth
of micro-organisms including bacteria and fungi.
Many different antibiotics are now available.
Their principal application in virology is in tissue
culture media used for the cultivation of cells and
growth of viruses.

antibody. Protein molecule produced by lymphocytes following administration of a foreign protein or carbohydrate (ANTIGEN) into a vertebrate host. The induced antibody reacts specifically with the administered antigen. In virology, antibodies are frequently used as a means of discriminating between different viruses using a range of serological techniques including virus NEUTRALISATION, IMMUNODIFFUSION, COMPLEMENT FIXATION and ELISA. Antibodies can belong to several different classes, e.g. IgM, IgG and IgA (secretory antibody). *See* IMMUNOGLOBULIN.

Anticarsia gemmatalis nuclear polyhedrosis virus. A baculovirus (MNPV), from the soybean looper, *A. gemmatalis* which has been widely used in Brazil for the biological control of the pest.
Fisher, J. and Ferreira-Lima, L.C. (1988) Chemistry and Industry **6**, 184.

anticodon. A group of three bases in a tRNA molecule which recognises a codon in an mRNA molecule. It is this codon-anti-codon interaction, taking place on the ribosome, which ensures that the correct amino acid is inserted into the growing polypeptide chain. The first two bases pair A-U or G-C in an anti-parallel manner. The pairing with the third base is more flexible as one tRNA molecule can recognise several codons provided they differ only in the last base. *See* WOBBLE HYPOTHESIS.

antigen. Molecule of carbohydrate or protein which stimulates the production of an antibody, with which it reacts specifically.

antigen-antibody reaction. The specific interaction between an antigen and an antibody which recognises a specific structural feature of the antigen and binds to it. This reaction can be measured by a variety of serological methods. *See* COMPLEMENT FIXATION TEST, ELISA, IMMUNODIFFUSION, PRECIPITIN, VIRUS NEUTRALISATION.

antigenaemia. Presence of circulating viral antigen in the bloodstream as in the surface antigen levels found in patients infected with HEPATITIS B VIRUS.

antigenic drift. The appearance of a virus with slightly changed antigenic structure following passage in the natural host. This is probably due to natural selection under pressure of the immune response of the host to virus infection. Examples are restricted to virus infections of vertebrate hosts (which produce an immune response) e.g. INFLUENZA VIRUS and FOOT-AND-MOUTH DISEASE VIRUS.

antigenic shift. A major change in the antigenic structure of a virus. Occurs with viruses which have segmented genomes, e.g. INFLUENZA VIRUS, presumably as a result of gene REASSORTMENT.

antigenic site. The portion of a protein which reacts with the antibody induced in response to the entire antigen. A complex antigen (e.g. a protein) will contain several antigenic sites. Also termed EPITOPE.

antireceptor. Virion surface protein which binds specifically to a cell surface receptor. *See* HAEMAGGLUTININ,

antiserum. The serum from a vertebrate which has been exposed to an antigen and which contains antibodies that react specifically with the antigen.

Apanteles melanoscelus virus. Formerly listed as the type species of proposed subgroup D, BACULOVIRUS genus, now classified as a probable member of subgroup B, POLYDNAVIRUS. Virus particles consist of enveloped rod-shaped NUCLEOCAPSIDS, variable in length (40 nm. wide by 30-100 nm. long), with one envelope surrounding one or more nucleocapsids which usually carry a tail appendage. Particles contain at least 18 polypeptides and a polydisperse DNA genome composed of multiple supercoiled dsDNAs of variable mw. (2×10^6-25×10^6). Virus was isolated from the parasitoid *Apanteles melanoscelus* (=*Cotesia melanoscela*) (Hymenoptera: Braconidae) where it replicates in the ovary of *all* females without causing any deleterious effects. The virus appears to play a role in successful parasitism of the lepidopteran hosts, e.g. (*Lymantria dispar* and *Orgyia leucostigma*) of *A. melanoscelus*, in some way preventing encapsulation of the parasitoid egg by the defence mechanisms of the host.
Krell, P.J. and Stoltz, D.B. (1979) J. Virol. **29**, 1118.
Stoltz, D.B. and Vinson S.B. (1979) Adv. Virus Res. **24**, 125.
Stoltz, D.B. *et al.* (1984) Intervirology **21**, 1.

apeu virus. Family *Bunyavirus*, genus *Bunyavirus*.

aphid. Insect belonging to the family *Aphidae*. Members of this family are phytophagous plant pests and common vectors of many plant viruses.

aphid lethal paralysis virus. An unclassified small RNA virus (probable INSECT PICORNAVIRUS) isolated from the aphid, *Rhopalosiphum padi*, in S. Africa; infection causes paralysis and death of the host. Particles are isometric, 26 nm. in diameter, sediment at 164S and have a buoyant density in CsCl of 1.34 g/cc. The viral genome is a polyadenylated ssRNA (9.7kb). The capsid contains three major polypeptides (mw. 34.4×10^3, 32×10^3, 31.2×10^3) and one minor polypeptide (mw. 40.8×10^3). The virus is serologically unrelated to the *R. padi* virus isolated in the USA but is serologically related to Cricket Paralysis Virus. Williamson, C. *et al.* (1988) J. gen. Virol. 69, 787.

Aphis 'baculovirus'. Unclassified virus, morphologically similar to a NON-OCCLUDED BACULOVIRUS observed principally in fat body cells of two *Aphis* spp. in Brazil. Particles in section consisted of a rod-shaped nucleocapsid (180-200 x 40-50 nm.) surrounded by a single membrane. Kitajima, E.W. *et al.* (1978) J. Invertebr. Pathol. 31, 123.

Aphthovirus. (Greek 'aphtha' = vesicles in the mouth.) A genus in the family *Picornaviridae*. Consists of viruses belonging to seven distinct serotypes of foot-and-mouth disease virus. Characterised by considerable antigenic variability, making control by vaccination difficult. The particles are unstable below pH7 and have a high (1.43 g/cc) buoyant density in CsCl. Like the cardioviruses, they have a polycytidylic acid tract near the 5′ end of the genome.

aphthovirus. *See* FOOT-AND-MOUTH DISEASE VIRUS.

Apis iridescent virus. *See* BEE IRIDESCENT VIRUS.

apoi virus. Family *Flaviviridae*, genus *Flavivirus*. Isolated from the rodents *Apodemus speciosus ainu* and *Apodemus argentus hokkaidi* in Japan. Can cause encephalitis in man.

apollo virus. *See* ENTEROVIRUS 70.

Aporophyla lutulenta Cytoplasmic Polyhedrosis Virus. Cytoplasmic polyhedrosis virus (CPV) isolated in the United Kingdom from larvae of *Aporophyla lutulenta* (Noctuidae, Lepi-doptera). The virus is the type member of 'type 10' CPVs. Unrelated to *Bombyx mori* (type 1) CPV on the basis of RNA electropherotype. Currently the sole recorded isolate of 'type 10' CPVs. Payne, C.C. and Mertens, P.P.C. (1983) In The *Reoviridae*. p. 425. ed. W.K. Joklik. Plenum Press: New York.

apple chlorotic leafspot virus. Type member of the *Closterovirus* subgroup 2. Lister, R.M. (1970) CMI/AAB Descriptions of Plant Viruses No. 30. Francki, R.I.B. *et al.* (1985) In Atlas of Plant Viruses. Vol. 2. p. 219. CRC Press: Boca Raton, Florida.

apple mosaic virus. An *Ilarvirus*. Fulton, R.W. (1972) CMI/AAB Descriptions of Plant Viruses No. 83. Francki, R.I.B. (1985) In The Plant Viruses. Vol. 1. p. 1. ed. R.I.B. Francki. Plenum Press: New York.

apple necrosis virus. An *Ilarvirus*, occurs in Japan. Doi, Y. *Personal communication.*

apple stem grooving virus. Type member of the *Capillovirus* group. Lister, R.M. (1970) CMI/AAB Descriptions of Plant Viruses No. 31.

apple (Tulare) mosaic virus. An *Ilarvirus*. Fulton, R.W. (1971) CMI/AAB Descriptions of Plant Viruses No. 42. Francki, R.I.B. (1985) In The Plant Viruses. Vol. 1. p. 1. ed. R.I.B. Francki. Plenum Press: New York.

aquacate virus. Family *Bunyaviridae*, genus *Phlebovirus*. Isolated from *Lutzomyia* sp. in Panama.

aquilegia necrotic mosaic virus. A possible *Caulimovirus*; occurs in Japan. Doi, Y. *Personal communication.*

aquilegia virus. A possible *Potyvirus*. Francki, R.I.B. *et al.* (1985) In Atlas of Plant Viruses. Vol. 2. p. 183. CRC Press: Boca Raton, Florida.

ara A. *See* ARABINOSYL ADENINE.

arabinosyl adenine. (Synonym: ADENINE ARABINOSIDE; ARA A). An antiviral agent which inhibits viral DNA polymerase. Active against HERPES VIRUS and POXVIRUS infections but less so against ADENO- and PAPOVA-VIRUSES.

arabinosyl cytosine. (Synonym: CYTOSINE ARABINOSIDE; ARA C). An anti-viral agent, probably acting by inhibiting DNA polymerase.

arabis mosaic virus. A *Nepovirus*.
Murant, A.F. (1970) CMI/AAB Descriptions of Plant Viruses No. 16.
Francki, R.I.B. *et al.* (1985) **In** Atlas of Plant Viruses. Vol. 2. p. 23. CRC Press: Boca Raton, Florida.

araujia mosaic virus. A possible *Potyvirus*.
Francki, R.I.B. *et al.* (1985) **In** Atlas of Plant Viruses. Vol. 2. p. 183. CRC Press: Boca Raton, Florida.

arbovirus. A term used to describe any virus of vertebrates which is transmitted by an arthropod. The virus replicates in both the arthropod and vertebrate host. The term has no taxonomic meaning since it has no relevance to the chemistry or mode of replication of the viruses. Many virus families contain members which are arboviruses.

Arenaviridae. (Latin 'arena' = sand.) A family of RNA viruses comprising a single genus (*ARENAVIRUS*) with spherical or pleomorphic particles 50-300 nm. in diameter with club-shaped surface projections 10 nm. long surrounding a core which contains ribosome-like particles. The

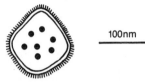

particles contain three structural proteins (two glycosylated) and two species of (−)-sense ssRNA (mw. 1.1-1.6 and 2.1-3.2 x 10^6). In addition there are three RNAs of host cell origin (28S, 18S and 4.6S). The small RNA codes for the nucleoprotein and precursor glycoprotein while the large RNA codes for the large non-glycosylated protein. The virus RNA is transcribed by virion polymerase into complementary RNA which acts as mRNA. Most of the viruses have a single restricted rodent host in which persistent infection with viraemia occurs, probably caused by a slow or insufficient immune response of the host. The viruses grow well in tissue culture. Transmission can be vertical or horizontal.
Matthews, R.E.F. (1982) Intervirology **17**, 119.

Arenavirus. (Latin 'arenosus' = sandy.) The only genus of the family *Arenaviridae*. The genus is divided into two groups on the basis of serology: (a) the Old World Arenaviruses and (b) the New World Arenaviruses. Most of the viruses have a rodent or bat host in which they cause a persistent infection. Contains several members of which the most important are the viruses causing lymphocytic choriomeningitis, Lassa fever, Argentinian haemorrhagic fever (Junin) and Bolivian haemorrhagic fever (Machupo).

Argentina virus. Family *Rhabdoviridae*, genus *Vesiculovirus*. Closely related to but distinguishable from other isolates in the Indiana serotype of vesicular stomatitis virus.

Argentinian haemorrhagic fever virus. *See* JUNIN VIRUS.

aride virus. An unclassified arthropod-borne virus.

Arkansas bee virus. Unclassified small RNA virus from the honey bee. Virions are isometric, 30 nm. in diameter, sediment at 128S and have a buoyant density in CsCl of 1.37 g/cc. Particles contain a single polypeptide (mw. 41 x 10^3). Formerly considered as a possible member of the

NODAVIRIDAE, it is now known not to have a divided genome but to contain only a single species of ssRNA (mw. 1.8×10^6).
Lommel, S.A. *et al.* (1985) Intervirology **23**, 199.

arkonam virus. An unclassified arthropod-borne virus. Isolated from *Anopheles subpictus*, *Anopheles hyrcanus* and *Culex tritaenior lynchus* mosquitoes in India.

arracacha virus A. A *Nepovirus*.
Jones, R.A.C. and Kenton, R.H. (1980) CMI/AAB Descriptions of Plant Viruses No. 216.
Francki, R.I.B. *et al.* (1985) **In** Atlas of Plant Viruses. Vol. 2. p. 23. CRC Press: Boca Raton, Florida.

arracacha virus B. An unclassified plant virus with isometric particles, 26 nm. in diameter which sediment at 126S. Capsid made of two coat protein species (mw. 26 and 20×10^3) which are in a 3:1 ratio.
Jones, R.A.C. and Kenton, R.H. (1983) CMI/AAB Descriptions of Plant Viruses No. 270.

Arrhenatherum blue dwarf virus. A *Fijivirus*.
Francki, R.I.B. *et al.* (1985) **In** Atlas of Plant Viruses. Vol. 1. p. 47. CRC Press: Boca Raton, Florida.

Arterivirus. A genus in the family *Togaviridae* which comprises only equine arteritis virus.

arthropod-borne. Virus carried by an arthropod and transmitted to its animal or plant host(s) by this route. *See* NON-PERSISTENT TRANSMISSION, PERSISTENT TRANSMISSION, SEMIPERSISTENT TRANSMISSION.

artichoke curly dwarf virus. A possible *Potexvirus*.
Francki, R.I.B. *et al.* (1985) **In** Atlas of Plant Viruses. Vol. 2. p. 159. CRC Press: Boca Raton, Florida.

artichoke Italian latent virus. A *Nepovirus*.
Martelli, G.P. *et al.* (1977) CMI/AAB Descriptions of Plant Viruses No. 176.
Francki, R.I.B. *et al.* (1985) **In** Atlas of Plant Viruses. Vol. 2. p.23. CRC Press: Boca Raton, Florida.

artichoke latent virus. A possible *Potyvirus*.
Francki, R.I.B. *et al.* (1985) **In** Atlas of Plant Viruses. Vol. 2. p. 183. CRC Press: Boca Raton,

Florida.

artichoke mottled crinkle virus. A *Tombusvirus*.
Francki, R.I.B. *et al.* (1985) **In** Atlas of Plant Viruses. Vol. 1. p. 181. CRC Press: Boca Raton, Florida.
Martelli, G.P. *et al.* (1988) **In** The Plant Viruses. Vol 3. p. 13. ed. R. Koenig. Plenum Press: New York.

artichoke vein banding virus. A possible *Nepovirus*.
Gallitelli, D. *et al.* (1984) CMI/AAB Descriptions of Plant Viruses No. 285.

artichoke yellow ringspot virus. A *Nepovirus*.
Rana, G.L. *et al.* (1983) CMI/AAB Descriptions of Plant Viruses No. 271.
Francki, R.I.B. *et al.* (1985) **In** Atlas of Plant Viruses. Vol. 2. p. 23. CRC Press: Boca Raton, Florida.

aruac virus. Unclassified arthropod-borne virus. Isolated from *Trichoprosopon theobaldi* mosquitoes in Trinidad.

Arumowot virus. Family *Bunyaviridae,* genus *Phlebovirus.* Isolated from *Culex antennatus* mosquitoes and *Thamnomys macmillani* and *Lemniscomys striatus* mice, *Crocidura* sp. shrew, *Arvicanthis niloticus* rats and *Tatera kempi* gerbils in Africa.

ascitic fluid. Fluid from the peritoneal cavity of mice. Used in raising monoclonal antibodies in large quantities following injection of hybridoma cells into the peritoneal cavity. Can also be used in the culture of ENCEPHALOMYOCARDITIS VIRUS.

Ascovirus. (Latin 'asco' = sac, bladder.) A proposed taxon of large dsDNA viruses isolated from lepidopteran larvae. Virions are enveloped (150 x

100nm

400 nm.), allantoid in shape (sausage-shaped) and composed of an inner particle (nucleoprotein core and surrounding matrix) measuring 130 x

350 nm. and surrounded by a membrane. The genome appears to be a linear dsDNA (mw. 100 x 10⁶). Virus replicates in the nuclei of several tissues (e.g. fat body and epidermis), where it causes disruption of the nuclear envelope and the formation of large numbers of virus-containing vesicles which accumulate in the haemolymph of infected larvae. Virus isolated from the cabbage looper, *Trichoplusia ni* (Noctuidae) has been transmitted experimentally to several other noctuid species.
Federici B.A. (1983) Proc. Nat. Acad. Sci. USA **80**, 7664.

asparagus stunt virus. A strain of *Tobacco Streak Virus*.
Francki, R.I.B. (1985) **In** The Plant Viruses. Vol. 1, p. 1. ed. R.I.B. Francki. Plenum Press: New York.

asparagus virus 1. A *Potyvirus*.
Howell, W.E. and Mink, G.I. (1985) Plant Disease **69**, 1044.

asparagus virus 2. An *Ilarvirus*.
Uyeda, I. and Mink, G.I. (1984) CMI/AAB Descriptions of Plant Viruses No. 288.
Francki, R.I.B. (1985) **In** The Plant Viruses. Vol. 1. p. 1. ed. R.I.B. Francki. Plenum Press: New York.

asparagus virus 3. A *Potexvirus*, occurs in Japan.
Doi, Y. *Personal communication.*

Aspergillus foetidus virus S (AfV-S). A possible member of the *Totivirus* group.

Aspergillus niger virus S (AnV-S). A possible member of the *Totivirus* group.

Aspergillus ochraceous virus (AoV). A probable member of the PARTITIVIRUS group.

astra virus. An isolate of DHORI VIRUS.

astro virus. Unclassified. Spherical virus of 28 nm. diameter found in faeces of patients with diarrhoea. Star-like appearance but can be distinguished from calicivirus particles. Has not been grown in cell culture.
Woode, G.N. & Bridger, J.C. (1978) J. med. Microbiol. **11**, 441.

asystasia mottle virus. A *Potyvirus*.

Thouvenel, J-C. (1988) Ann. Appl. Biol. **112**, 127.

ataxia of cats virus. *See* FELINE PANLEUKOPENIA VIRUS.

ATP.   Abbreviation   for   ADENOSINE 5'- TRIPHOSPHATE.

Atropa belladonna rhabdovirus. A plant *Rhabdovirus*.
Francki, R.I.B. *et al.* (1985) **In** Atlas of Plant Viruses. Vol. 1. p. 73. CRC Press: Boca Raton, Florida.

attachment. *See* ADSORPTION.

attenuated strain. A selected strain of a virulent virus which does not cause the clinical signs associated with the parent virus but still replicates well enough to induce immunity. Usually obtained by passage in cell culture or in a host different from the natural host. Many highly successful vaccines have been produced in this way, e.g. POLIO, YELLOW FEVER.

attenuation. The process of producing an attenuated virus strain.

aucuba ringspot virus. A member of the *Cacao Swollen Shoot Virus* group.
Francki, R.I.B. *et al.* (1985) **In** Atlas of Plant Viruses. Vol. 2. p. 235. CRC Press: Boca Raton, Florida.

Aujezky's disease virus. Synonyms: MAD ITCH VIRUS, PSEUDORABIES VIRUS, SUID (ALPHA) HERPESVIRUS 1. Family *Herpesviridae*, subfamily *Suid alphaherpesvirus 1*, genus *Suid herpesvirus 1 group*. An infection of pigs. Cattle, sheep, dogs, cats, foxes and mink are also susceptible. In pigs, the disease is one of the central nervous system but the animals recover. In other species the disease is usually fatal. Many species of animals can be infected experimentally. Virus grows in chorioallantoic membrane with plaque production and causes cpe in many kinds of tissue culture cell.
Basinger, D. (1979) Brit. vet. J. **135**, 215.

aura virus. Family *Togaviridae*, genus *Alphavirus*. Isolated from mosquitoes in Brazil. Antibodies found in man, rodents, marsupials and horses but no disease associated with the virus.

Australia antigen. The HEPATITIS B VIRUS surface antigen, first found in the blood of an Australian aboriginal.
Blumberg, B.S. *et al.* (1965) Jour. Am. med. Soc. **191**, 541.

Autographa californica nuclear polyhedrosis virus (AcMNPV). Type species of subgroup A of BACULOVIRUS genus (BACULOVIRIDAE) - NUCLEAR POLYHEDROSIS VIRUSES (NPV). Originally isolated from the alfalfa looper, *A californica*, AcMNPV has a wider host range than most NPVs, experimentally infecting at least 43 species of Lepidoptera from eleven families. The virus is representative of the MNPV subtype where virions contain one to many nucleocapsids within a single viral envelope. The morphological and structural properties of the virus are consistent with those of other baculoviruses (*see* BACULOVIRUS, NUCLEAR POLYHEDROSIS VIRUS). However, AcMNPV is the best-studied of all baculoviruses principally because of its wide *in vivo* host range and the ease with which it can be grown *in vitro* in several insect cell lines. Physical maps have been prepared of the 128kbp genome which is composed primarily of unique sequences. Five regions of repeated sequences (hr=homologous regions) account for about 3% of the genome and are interspersed throughout the genome. They appear to act as enhancers of gene expression. A large number of closely-related genotypic variants have been isolated both from *A. californica* as well as *Galleria mellonella* (GmMNPV), *Rachiplusia ou* (RoMNPV), *Spodoptera exempta* (SeMNPV) and *Trichoplusia* NI (TnMNPV); several have been mapped. The designated origin for these physical maps is the POLYHEDRIN gene, which in AcMNPV has a 732 base coding region and a sequence which is broadly conserved amongst occluded baculoviruses. During virus replication, virus proteins are synthesised in at least four temporal classes; α immediate early (2 h), β delayed-early (2-6 h), γ late (8-12 h) and δ delayed late (>12 h). At least some of the temporal regulation occurs at the level of transcription. Transcripts for different temporal classes of proteins arise from dispersed regions of the genome rather than a continuous block. The primary transcripts appear not to be spliced; some of the different-sized mRNAs contain overlapping sequences. At least two proteins are produced in abundance late in infection. These are the POLYHEDRIN molecule and a methionine-deficient basic polypeptide (mw. *c.*10×10^3) – the '10K' protein. The polyhedrin gene is probably expressed more abundantly than any other gene in any virus-infected cell (at death, >15% of the dry mass of an NPV infected caterpillar is polyhedrin). For this reason AcMNPV is being used with increasing frequency as a cloning vector for the high-level expression of foreign DNA sequences inserted after the polyhedrin promoter (*see* BACULOVIRUS EXPRESSION VECTOR). AcMPNV has also been produced for field trials as a biocontrol agent for the cabbage looper (*T. ni*) and Douglas fir tussock moth (*Orgyia pseudotsugata*).
Granados, R.R. and Federici, B.A. (1986) The Biology of Baculoviruses. Vols. I and II. CRC Press: Boca Raton, Florida.

autoimmune. A situation in which an organism produces an immune response to its own tissues e.g. Hashimoto's disease which is caused by production of antibodies to thyroglobulin.

autoinfection. Reinfection of a host from an original infection.

autoradiography. A method for detecting the location of a radioisotope in a tissue, cell, or molecule. The sample is placed in contact with an X-ray film, whereby the emission of β particles from the sample activates the silver halide grains in the emulsion and allows them to be reduced to metallic silver when the film is developed.

avalon virus. Family *Bunyaviridae*, genus *Nairovirus*. Isolated from mosquitoes in Newfoundland.

Aviadenovirus. (Latin 'avis' bird.) A genus in the family *Adenoviridae* consisting of those viruses isolated from birds. They share a common antigen but do not cross-react with the viruses in the *Mastadenovirus* genus.

avian encephalomyelitis virus. Family *Picornaviridae*, genus *Enterovirus*. Causes a silent infection of adult domestic fowl, but severe disease, often followed by death, occurs in young chickens. Ducks, turkeys and pigeons can be infected experimentally. The virus is present in the excreta and the egg. Virus replicates in the egg, often killing the embryo, and in chick embryo cells.

avian enteroviruses. Family *Picornaviridae*, genus *Enterovirus*. A group of viruses which includes avian encephalomyelitis virus and duck hepatitis virus.

avian erythroblastosis virus. Family *Retroviridae*, subfamily *Oncovirinae*, genus *Type C Oncovirus group*, sub-genus *Avian Type C Oncovirus*. Consists of a transforming virus and a leukaemia virus which acts as helper for the defective transforming virus. The latter causes erythroblastosis when injected i.m. Transforms fibroblasts and bone marrow cells *in vitro*.
Hayman, M.J. *et al.* (1979) Virology, **92**, 31.

avian infectious bronchitis virus. Synonym: GASPING DISEASE VIRUS. Family *Coronaviridae*, genus *Coronavirus*. Causes an acute respiratory disease of chicks. Mortality can be as high as 90%. Also causes drop in egg production. Pheasants can be infected giving a mild disease. Replicates on the chorioallantoic membrane and in chick embryo cell cultures.

avian myeloblastosis virus (AMV). Family *Retroviridae,* subfamily *Oncovirinae,* genus *Type C Oncovirus Group,* sub-genus *Type C Oncovirus.* Causes myeblastosis, osteopetrosis, lymphoid leukosis and nephroblastoma in chickens and transformation of yolk sac or bone marrow cells *in vitro.* Probably a mixture of viruses.

avian myelocytomatosis virus. Family *Retroviridae,* subfamily *Oncovirinae,* genus *Type C Oncovirus Group,* sub-genus *Type C Oncovirus.* Causes myelocytomatosis, renal and liver tumours and occasionally erythroblastosis in chickens and cell transformation *in vitro.*

avian nephrosis virus. *See* INFECTIOUS BURSAL DISEASE VIRUS.

avian pneumoencephalitis virus. *See* NEWCASTLE DISEASE VIRUS.

avian reticuloendotheliosis virus. Family *Retroviridae,* subfamily *Oncovirinae,* genus *Type C Oncovirus Group,* sub-genus *Type C Oncovirus.* Causes leukosis in turkeys in nature but many species of bird can be infected experimentally resulting in quick death with a large dose. The virus grows in chick embryo fibroblasts with cpe followed by transformation.

avian type C oncoviruses. Family *Retroviridae*, genus *Oncovirinae*. A sub-genus in the genus *Type C Oncovirus.*

avidity. Intensity of binding of e.g. an antibody molecule to the antigen which induced its formation.

Avipoxvirus. (Greek 'avis' = bird.) A genus in the family *Poxviridae*, subfamily *Chordopoxvirinae* containing those viruses which infect birds: canary pox, junco pox, pigeon pox, quail pox, sparrow pox, starling pox and turkey pox. The type species is fowlpox virus. The members are ether-resistant and antigenically related.

avirulent. Strain of virus which is lacking in virulence; *see* ATTENUATED.

avocado sunblotch viroid. A VIROID, comprising 247 nucleotides.
Dale, J.L. *et al.* (1982) CMI/AAB Descriptions of Plant Viruses No. 254.
Desjardins, P.R. (1987) **In** The Viroids. p. 299. ed. T.O.Diener. Plenum Press: New York.

Avogadro's number. The number of molecules in one gram molecular weight of any compound: 6.023×10^{23}.

axenic (culture). The growth of a single species of organism in the absence of organisms or cells of another species.

AZT (azidothymidine). An analogue of thymidine which blocks DNA replication (*see* Figure). It is used in the treatment of ACQUIRED IMMUNODEFICIENCY SYNDROME.

AZT   Thymidine

B

734B B virus. Family *Retroviridae*, subfamily *Oncovirinae*, genus *Type B Oncovirus group*. Present in MCF-7 cells, a line derived from a patient with disseminated mammary adenocarcinoma. Related to mouse mammary tumour virus but no relationship with the Type C oncovirus group.

B-type inclusion body. Inclusion bodies observed in the cytoplasm of vertebrate POXVIRUS infected cells, representing sites of virus synthesis; characteristic of all productive poxvirus infections.

B-type virus particles. A morphologically defined group of enveloped RNA virus particles, seen outside the cells in mouse mammary carcinoma. They have a dense body or core 40-60 nm. in diameter which is contained in an envelope 90-120 nm. in diameter. The particles bud through the cell membrane. The prototype member is MOUSE MAMMARY TUMOUR VIRUS. *See* A-, C-, D-TYPE VIRUS PARTICLE.
Dalton, A. J. (1972) J. Natl. Cancer Inst. **49**, 323.

Babahoyo virus. Family *Bunyaviridae*, genus *Bunyavirus*.

Baboon Type C Oncovirus. Family *Retroviridae*, subfamily *Oncovirinae*, genus *Type C Oncovirus Group*, sub-genus *Type C Oncovirus*. Isolated from a baboon *Papio cynocephalus*. Can be grown in foetal canine cells.

baby hamster kidney cells. Heteroploid fibroblastic cells from kidneys of day-old Syrian or golden hamster; susceptible to many viruses e.g. FOOT-AND-MOUTH DISEASE VIRUS, VESICULAR STOMATITIS VIRUS, RABIES VIRUS.
MacPherson, I. and Stoker, M. (1962) Virology **6**, 147.

bacilliform. Description of shape of certain virus particles which are cylindrical with two rounded ends, e.g. PLANT RHABDOVIRUSES, BACULOVIRUSES.

Bacillus 'baculovirus'. Unclassified virus, morphologically similar to a NON-OCCLUDED BACULOVIRUS, observed in mid-gut epithelial cells of the stick insect, *Bacillus rossius*. Particles in section consist of a rod-shaped nucleocapsid (210 x 50-60 nm.) surrounded by a membrane.
Scali V. *et al.* (1980) J. Invertebr. Pathol. **35**, 109.

bacteriocinogen. Genetic determinant in some bacteria (often plasmids) which specifies the production of a BACTERIOCIN. The determinants for particulate bacteriocins are specified by chromosomal genes and behave like a defective PROPHAGE. Particulate bacteriocin production (which is normally suppressed) can be induced by agents such as UV light which normally 'activate' LYSOGENIC phages.
Hardy, K. (1986) Bacterial Plasmids. 2nd edit. Thomas Nelson: Walton-upon-Thames.

bacteriocins. Proteins released by some bacteria, which are able to kill other bacterial strains. The producing strain is relatively immune to the effects of its own bacteriocin. 'Particulate bacteriocins', which resemble phage particles or subviral particles (e.g. TAILS) adsorb to and kill sensitive cells without multiplying in them. 'True' bacteriocins are agents without complex morphology (e.g. enzymes) and of low mw. Colicins (bacteriocins produced by *E. coli*) are specified by *Col* plasmids and range from mw. 12-90 x 10³.
Ackermann, H.W. and Brochu, G. (1978) In Handbook of Microbiology. Vol II. p. 691. ed. A.I. Laskin and H.A. Lechevalier. CRC Press: Boca Raton, Florida.
Hardy, K. (1986) Bacterial Plasmids. 2nd. edit. Thomas Nelson: Walton-upon-Thames.
Reanney, D.C. and Ackermann, H.W. (1982) Adv. Virus Res. **27**, 205.

bacteriophage. A virus which replicates inside a bacterium. *See* PHAGE.

Baculoviridae. (Latin 'baculum' = 'stick'-viruses; from the rod-shaped morphology of the virion.) A family of viruses containing a single genus (BACULOVIRUS) isolated from Arthropods.

occur in the cytoplasm. Virions of subgroups A and B can be occluded during infection in crystalline protein OCCLUSION BODIES which may be polyhedral in shape and contain many virus particles (subgroup A = NPV) or may be ovicylindrical and contain only one or rarely two or more particles (subgroup B = GV); *see* Figure. Size estimates for

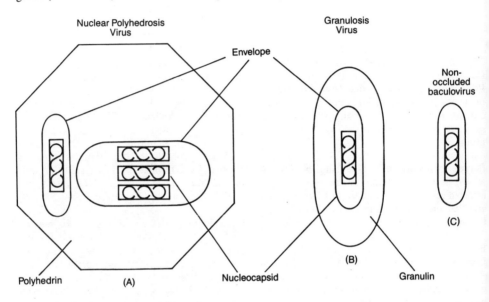

Nuclear Polyhedrosis Virus

Granulosis Virus

Envelope

Non-occluded baculovirus

(C)

(B)

Polyhedrin   (A)   Nucleocapsid   Granulin

Particles not to scale

Virions contain one or more rod-shaped nucleocapsids surrounded by an envelope. The viral genome is a single molecule of circular supercoiled dsDNA.

Baculovirus. The only genus of viruses currently classified within the *Baculoviridae*. Virions consist of one or more rod-shaped nucleocapsids (40-60 nm. x 200-400 nm.) enclosed within a single envelope, which sediment between 1230-1640S and have buoyant densities in CsCl of 1.20-1.32 g/cc. for enveloped particles and 1.47-1.48 g/cc. for nucleocapsids. Virions are structurally complex and may contain at least 25 polypeptides (mw. $10-160 \times 10^3$) of which about eleven are nucleocapsid components. The DNA genome is a single molecule of circular supercoiled dsDNA (mw. 59×10^6-154×10^6). Of the three subgroups within the genus (A = NUCLEAR POLYHEDROSIS VIRUSES (= NPVs); B = GRANULOSIS VIRUSES (= GVs) and C = NON-OCCLUDED BACULOVIRUSES), viruses of subgroups A and C replicate exclusively in the nucleus, while those of subgroup B replicate mainly in the nucleus but replication can

the matrix protein of the occlusion bodies (referred to as 'POLYHEDRIN' for NPVs and 'GRANULIN' for GVs) range from $25-33 \times 10^3$, depending on virus isolate. Virus infection in susceptible larvae follows ingestion of the virus and solubilization of the occlusion body in the gut. Virus enters the cell by fusion of the envelope with the cell membrane. Nucleocapsids uncoat at pores in the nuclear membrane and/or within the nucleus. Biochemical events in baculovirus replication have been intensively studied with the type species of subgroup A, *Autographa californica* NPV. In this virus at least, two morphological forms of the virus are known to occur; one form acquires the envelope *de novo* within the nucleus of an infected cell before occlusion; the other which is not occluded, acquires the envelope by budding through the plasma membrane and is responsible for cell-to-cell spread of infection. Baculoviruses or baculovirus-like particles have been isolated from several hundred members of the Insecta (particularly Lepidoptera and Hymenoptera but also Coleoptera, Diptera, Neuroptera and Trichoptera) as well as certain arachnids and crustaceans. A putative baculovirus with similar general morphology has been observed in a fungus (*Strongwellsea magna*). There is no formal nomenclature for each virus isolate although, conventionally, viruses have been named after the host from which they were first isolated (e.g. *A. californica* NUCLEAR POLYHEDROSIS VIRUS = AcNPV). Occasionally, Latin binomials have been used (e.g. *Baculovirus heliothis* for an NPV from *Heliothis* spp.). Relationships between the many isolates have not been clearly defined. Transmission is generally horizontal, by contamination of the food, but vertical transmission through surface contamination of the egg also occurs. The host range of individual isolates may be restricted to one or a few arthropod species. For this reason and because they often induce a lethal infection in susceptible hosts, certain baculoviruses have found favour as selective insect pest control agents.

A fourth baculovirus subgroup (D) was proposed in earlier classifications, which included non-occluded nuclear viruses with a polydisperse (multipartite) DNA genome isolated from the ovaries of parasitoid Hymenoptera. These have now been classified as a separate family (*see* POLYDNAVIRIDAE). However, certain unclassified virus isolates from braconid parasitoids do not have a polydisperse genome and most closely resemble non-occluded baculoviruses (e.g. *Mesoleius Baculovirus*).

Granados, R.R. and Federici, B.A. (1986) The Biology of Baculoviruses. Vols. I and II. CRC Press: Boca Raton, Florida.
Stoltz D.B. *et al.* (1984) Intervirology **21**, 1.

Baculovirus expression vector. A gene expression system which utilises a strong promoter found in baculoviruses to obtain high-level expression of foreign genes. The baculovirus vector uses the highly expressed and regulated polyhedrin promoter modified for the insertion of foreign genes. The baculovirus used is usually *Autographa californica* nuclear polyhedrosis virus (NPV), the type species of the NPV subgroup. One of the major advantages of this invertebrate virus expression vector over bacterial, yeast and mammalian expression systems is the abundant expression of proteins coded by the inserted gene. In addition, recombinant proteins produced in insect cells with baculovirus vectors are biologically active and, for the most part, appear to undergo post-translational processing to produce gene products with similar properties to the authentic proteins.

Lucknow, V.A. and Summers, M.D. (1988) Biotechnology **6**, 47.

Baculovirus heliothis. *See* HELIOTHIS ZEA NUCLEAR POLYHEDROSIS VIRUS.

Bagaza virus. Family *Flaviviridae*, genus *Flavivirus*. Isolated from *Culex* mosquitoes in Africa by injection into suckling mice.

Bahia Grande virus. Probable member of the family *Rhabdoviridae*.

Bahig virus. Family *Bunyaviridae*, genus *Bunyavirus*. Isolated from birds in Egypt and Italy. Antibodies found in various species in Cyprus and Israel.

Bakalasa virus. Family *Flaviviridae*, genus *Flavivirus*. Isolated from bats of *Tadarida* sp. in Uganda.

Bakau virus. Family *Bunyaviridae*, genus *Bunyavirus*. Isolated from mosquitoes in Malaya and Pakistan.

Baku virus. Family *Reoviridae*, genus *Orbivirus*. Isolated from the tick *Ornithodoros capensis*. Antibodies found in the bird *Larus argentatus* on islands in the Caspian Sea.

bal 31. An EXONUCLEASE which digests dsDNA at both the 5′ and 3′ ends. Isolated from *Alteromonas espeiiana* (formerly *Brevibacterium albidum*). Used for forming processive deletions of DNA.

balanced salt solutions. Solutions which have the composition which maintains a balance of the requirements of the cells for which they are providing nutrients. They also control the osmolarity of the nutrient solution.

Balano posthitis virus of sheep. Family *Poxviridae*, subfamily *Chordopoxvirinae*, genus *Parapoxvirus*. Causes ulcerative dermatitis, balanitis and ulcerative vulvitis. Found in Australia, S. Africa, Britain and USA.

bamble disease virus. *See* BORNHOLM DISEASE VIRUS.

bamboo mosaic virus. A possible *Potexvirus*. Francki, R.I.B. *et al.* (1985) **In** Atlas of Plant Viruses. Vol. 2. p. 159. CRC Press: Boca Raton, Florida.

banana bunchy top virus. A possible member of the *Luteovirus* group.
Francki, R.I.B. *et al.* (1985) **In** Atlas of Plant Viruses. Vol. 1. p. 137. CRC Press: Boca Raton, Florida.
Casper, R. (1988) **In** The Plant Viruses. Vol. 3. p. 235. ed. R. Koenig. Plenum Press: New York.

Bandia virus. Family *Bunyaviridae*, genus *Nairovirus*. Isolated from mice and *Ornithodoros* ticks in Bardia Forest, Senegal.

Bangor virus. Family *Paramyxoviridae*, genus *Paramyxovirus*. Isolated from a finch in Northern Ireland.
McFerran J.B. *et al.* (1974) Arch. Virusforsch. **46**, 281.

Bangoran virus. An unclassified arthropodborne virus. Isolated from *Culex perfuscus* mosquitoes and the Kurrichane thrush *Tardus iboyanus* in Africa.

Bangui virus. An unclassified arthropod-borne virus. Isolated from a man with fever in Africa.

Banzi virus. Family *Flaviviridae*, genus *Flavivirus*. Isolated from a boy with fever in S. Africa. Also found in other African countries. Mosquito-borne. Natural hosts are probably cattle and sheep.

barley B-1 virus. A possible *Potexvirus*.
Francki, R.I.B. *et al.* (1985) **In** Atlas of Plant Viruses. Vol. 2. p. 159. CRC Press: Boca Raton, Florida.

barley stripe mosaic virus. Type member of the *Hordeivirus* group.
Atabekov, J.G. and Novikov, V.K. (1971) CMI/AAB Descriptions of Plant Viruses No. 68.
Francki, R.I.B. *et al.* (1985) **In** Atlas of Plant Viruses. Vol. 2. p. 133. CRC Press: Boca Raton, Florida.

barley yellow dwarf virus. Type member of the *Luteovirus* group. Has several strains based on aphid vector specificity. Isolates RPV and RMV are considered to be strains of BEET WESTERN YELLOWS VIRUS. An important virus of temperate cereal crops.
Rochow, W.F. (1970) CMI/AAB Descriptions of Plant Viruses No. 32.
Francki, R.I.B. *et al.* (1985) **In** Atlas of Plant Viruses. Vol. 1. p. 137. CRC Press: Boca Raton, Florida.

Casper, R. (1988) **In** The Plant Viruses. Vol. 3. p. 235. ed. R. Koenig. Plenum Press: New York.

barley yellow mosaic virus. Type member of the *Barley Yellow Mosaic Virus* group.
Inouye, T. and Saito, Y. (1975) CMI/AAB Descriptions of Plant Viruses No. 143.

Barley yellow mosaic virus 'group'. (type member of group, BARLEY YELLOW MOSAIC VIRUS). Unofficial group of plant viruses with flexuous

100nm

rod-shaped particles, 550-700 nm. long and 13 nm. in diameter which sediment as two components and band in CsCl at 1.29 g/cc. The host range is narrow. Virus particles are found in most cell types. They are transmissible mechanically. In nature they are transmitted by fungi (Plasmodiophorales).

barley yellow striate mosaic virus. Synonym: CEREAL STRIATE VIRUS. A plant *Rhabdovirus*, subgroup 1; planthopper transmitted.
Milne, R.G. and Conti, M. (1986) CMI/AAB Description of Plant Viruses No. 312.
Francki, R.I.B. *et al.* (1985) **In** Atlas of Plant Viruses. Vol. 1. p. 73. CRC Press: Boca Raton, Florida.

Barmah Forest virus. Family *Bunyaviridae*, genus *Bunyavirus*.

Barur virus. Family *Rhabdoviridae*, unassigned genus. Isolated from *Rattus rattur* and *Haemaphysalis intermedia* ticks in India and Canada.

base. The purine (adenine or guanine) or pyrimidine (uridine, thymidine, or cytidine) compounds forming the structure of RNA and DNA. *See* NUCLEIC ACID.

base analogue. A compound resembling one of the natural bases occurring in RNA or DNA which is incorporated into newly synthesised nucleic acid by substituting for the 'normal' base. This can result in mutation or inhibition of growth. An example is 5-fluorouracil which substitutes for uracil.

base composition. Proportion of the four bases

within a particular RNA or DNA molecule. This is a highly specific proportion for individual viruses.

base pair. A pair of nucleotides held together by hydrogen bonding. These are found in ds nucleic acids. DNA contains the base pairs A-T and G-C and RNA the base pairs A-U and G-C. *See* NUCLEIC ACID.

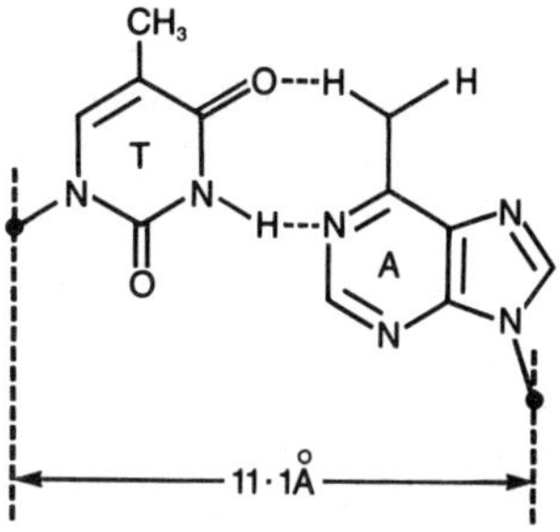

$T = A$

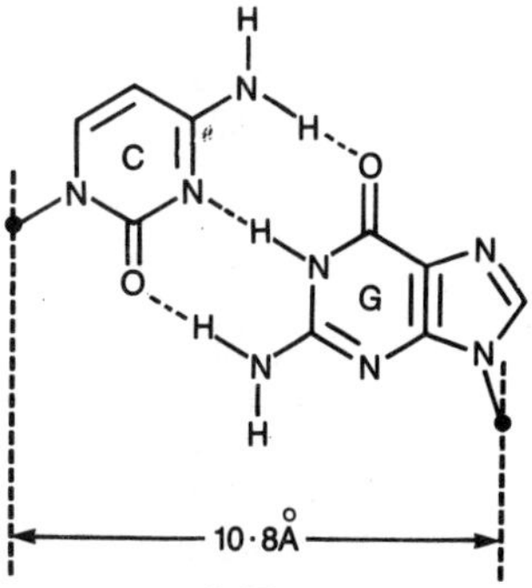

$C \equiv G$

base plate. The structure in some TAILED PHAGE particles at the distal end of the tail, to which any tail pins and tails fibres are attached (*see* T4 PHAGE and TAIL).

base sequence. The order in which the four purine and pyrimidine molecules A, G, C, U, (T) occur along the polynucleotide chain of nucleic acids.

Batai virus. Family *Bunyaviridae*, genus *Bunyavirus*. Isolated in India, Malaya, Thailand, Czechoslovakia and the Ukraine. Mosquito-borne.

Batken virus. An unclassified arthropod-borne virus. Isolated from the sheep tick *Hyalomma p. plumbeum* in the USSR.

Batu Cave virus. Family *Flaviviridae*, genus *Flavivirus*. Isolated from bats in Malaysia.

Bauline virus. Family *Reoviridae*, genus *Orbivirus*. Isolated from the tick *Ixodes uriae* in Newfoundland. Antibodies are present in puffins and petrels.

BBV. Abbreviation for BLACK BEETLE VIRUS.

BCIRL-HZ-AM3 cells. Insect cell line from *Heliothis zea* susceptible to infection by the NUCLEAR POLYHEDROSIS VIRUS of *H. zea* (HzSNPV).

Bdellovibriophage. PHAGE isolated from *Bdellovibrio* spp. (a group of intracellular parasitic bacteria which can only divide and grow within a bacterial cell).

bean common mosaic virus. A *Potyvirus*. Important as a seed-transmitted virus of Phaseolus beans.
Bos, L. (1971) CMI/AAB Descriptions of Plant Viruses No. 73.
Francki, R.I.B. *et al.* (1985) **In** Atlas of Plant Viruses. Vol. 2. p. 183. CRC Press: Boca Raton, Florida.

bean distortion dwarf virus. A possible *Geminivirus*, subgroup B.
Xi, Z. *et al.* (1982) Acta Microbiol. Sinica **22**, 293.

bean golden mosaic virus. A *Geminivirus*, subgroup B. The genome comprises ssDNA species of 2646 and 2587 nucleotides.
Goodman, R.M. and Bird, J. (1978) CMI/AAB Descriptions of Plant Viruses No. 192.
Harrison, B.D. (1985) Ann. Rev. Phytopath. **23**, 55.

bean leaf roll virus. A *Luteovirus*.
. Ashby, J.W. (1984) CMI/AAB Descriptions of Plant Viruses No. 286.
Casper, R. (1988) **In** The Plant Viruses. Vol. 3. p. 235. ed. R. Koenig. Plenum Press: New York.

bean mild mosaic virus. A possible member of the *Carmovirus* group.
Waterworth, H. (1981) CMI/AAB Descriptions of Plant Viruses No. 231.
Morris, T.J. and Carrington, J.C. (1988) **In** The Plant Viruses. Vol. 3, p. 73. ed. R. Koenig. Plenum Press: New York.

bean pod mottle virus. A *Comovirus*.
Semancik, J.S. (1972) CMI/AAB Descriptions of
Plant Viruses No. 108.
Francki, R.I.B. *et al.* (1985) **In** Atlas of Plant
Viruses. Vol. 2. p. 1. CRC Press: Boca Raton,
Florida.

bean rugose mosaic virus. A *Comovirus*.
Gamez, R. (1982) CMI/AAB Descriptions of
Plant Viruses No. 246.
Francki, R.I.B. *et al.* (1985) **In** Atlas of Plant
Viruses. Vol. 2. p. 1. CRC Press: Boca Raton,
Florida.

bean yellow mosaic virus. A *Potyvirus*. Closely
related to BEAN COMMON MOSAIC and PEA MOSAIC
VIRUSES.
Bos, L. (1970) CMI/AAB Descriptions of Plant
Viruses No. 40.
Francki, R.I.B. *et al.* (1985) **In** Atlas of Plant
Viruses. Vol. 2. p. 183. CRC Press: Boca Raton,
Florida.

bean yellow vein banding virus. Unencapsidated RNA, depending on BEAN LEAF ROLL
VIRUS for aphid transmission.
Cockbain, A.J. *et al.* (1986) Ann. Appl. Biol. **108**,
59.

bearded iris mosaic virus. A *Potyvirus*.
Barnett, O.W. and Brunt, A.A. (1975) CMI/AAB
Descriptions of Plant Viruses No. 147.
Francki, R.I.B. *et al.* (1985) **In** Atlas of Plant
Viruses. Vol. 2. p. 183. CRC Press: Boca Raton,
Florida.

Bebaru virus. Family *Togaviridae*, genus *Alphavirus*. Isolated from mosquitoes in Malaya.

bedbug reovirus. An unclassified reovirus-like
agent isolated from the bedbug, *Cimex lectu-
larius*. Particles are isometric and double-shelled,
c.56 nm. in diameter. The dsRNA genome
contains at least eleven and possibly twelve RNA
segments.
Eley, S. M. *et al.* (1987) J. gen. Virol. **68**, 195.

bee acute paralysis virus. Synonym: ACUTE BEE-
PARALYSIS VIRUS. Unclassified small RNA virus
which infects adult honey bees rendering them
flightless and killing them in 5-6 days. Virions are
isometric, 30 nm. in diameter, unenveloped, sediment at 160S and have a buoyant density in CsCl
of 1.34-1.38 g/cc. Particles contain two structural
polypeptides (23.5 x 10^3 and 31.5 x 10^3) and an

ssRNA genome. Isolated from UK, France and
Australia.
Bailey, L. and Woods R.D. (1977) J. gen. Virol.
37, 175.
Moore, N.F. *et al.* (1985) J. gen. Virol. **66**, 647.

bee chronic paralysis virus. Unclassified small
RNA virus isolated from paralysed adult honey
bees. Virions have an unusual ellipsoidal morphology (similar to *DROSOPHILA* RS VIRUS) *c*.22 nm.
wide, with average lengths of four particle sizes
of 30, 40, 55 and 65 nm. which sediment at 82,
100, 110 and 126S, respectively. All particles
have a buoyant density in CsCl of 1.34 g/cc,
contain a single structural polypeptide (23.5 x
10^3) and an ssRNA genome. Purified virions
contain five ssRNA components, two larger RNA
species (mw. 1.35 x 10^6 and 0.9 x 10^6) and three
smaller RNAs (mw. each 0.35 x 10^6). The small
RNAs are probably identical with the three RNA
components of the SATELLITE VIRUS, BEE CHRONIC
PARALYSIS VIRUS ASSOCIATE and have sequences in
common with the 0.9 x 10^6 RNA species. The
virus has a worldwide distribution.
Bailey, L. (1976) Adv. Virus Res. **20**, 271.
Overton, H.A. *et al.* (1982) J. gen. Virol.
63, 171.

bee chronic paralysis virus associate. Unclassified small RNA virus from honey bees whose
replication is absolutely dependent on BEE
CHRONIC PARALYSIS VIRUS as a helper virus. The
two viruses are serologically unrelated. Virions
are isometric, 17 nm. in diameter and sediment at
41S. The particles have a buoyant density of 1.38
g/cc. in CsCl, a single structural polypeptide (15
x 10^3) and contain three RNA molecules (each
0.35 x 10^6) which have some common oligonucleotides but are distinguishable from one
another. These RNAs are also packaged within
virions of the helper virus, bee chronic paralysis
virus, accounting for the appearance of the
associate virus in bees injected with purified bee
chronic paralysis virus.
Ball, B.V. *et al.* (1985) J. gen. Virol. **66**, 1423.
Overton, H.A. *et al.* (1982) J. gen. Virol. **63**, 171.

bee filamentous virus. An unclassified DNA
virus of the honey bee. Ovoid virus particles
measure 450 x 150 nm. and comprise a folded
NUCLEOCAPSID (3000 x 40 nm.) within a membrane. Buoyant density in CsCl is 1.28 g/cc.
Particles are structurally complex with about 12
proteins (mw. 13-70 x 10^3) and a genome of linear
dsDNA (approximately 12 x 10^6). Isolated from

bees from Europe, N. America, Asia and Australasia.
Clark, T.B. (1978) J. Invertebr. Pathol. **32**, 332.
Bailey, L. *et al.* (1981) Virology **114**, 1.

bee iridescent virus. Synonym: APIS IRIDESCENT VIRUS. Member (type 24) of small iridescent insect virus group (IRIDOVIRUS; IRIDOVIRIDAE), isolated from *Apis cerana* from Kashmir.
Bailey, L. *et al.* (1976) J. gen. Virol. **31**, 459.

bee slow paralysis virus. Unclassified small RNA virus causing paralysis in honey bees. Virions are isometric, 30 nm. in diameter, sediment at 178S and have a buoyant density in CsCl of 1.37 g/cc. Particles contain three structural proteins (mw. 27 x 10³, 29 x 10³, 46 x 10³) and an RNA genome. Considered as a possible INSECT PICORNAVIRUS.
Bailey, L. (1976) Adv. Virus Res. **20**, 271.

bee virus X. Unclassified small RNA virus of the honey bee. Virions are isometric, 35 nm. in diameter, sediment at 187S and have a buoyant density in CsCl of 1.36 g/cc. Particles contain one structural polypeptide (mw. 54.5 x 10³) and an RNA genome. Physical properties may be similar to those of NUDAURELIA B VIRUS GROUP.
Bailey, L. (1976) Adv. Virus Res. **20**, 271.

bee virus Y. Serotype of BEE VIRUS X occurring commonly in Great Britain.
Bailey, L. (1980) J. gen. Virol. **51**, 405.

beet cryptic virus 1. Synonym: BEET TEMPERATE VIRUS. A member of the *Cryptovirus* group, subgroup A.
Boccardo, G. *et al.* (1987) Adv. Virus Res. **32**, 171.

beet cryptic virus 2. A member of the *Cryptovirus* group, subgroup B.
Boccardo, G. *et al.* (1987) Adv. Virus Res. **32**, 171.

beet curly top virus. A *Geminivirus*, subgroup C. Has a wide host range. Single DNA species of 2993 nucleotides.
Thomas, P.E. and Mink, G.I. (1979) CMI/AAB Descriptions of Plant Viruses No. 210.
Harrison, B.D. (1985) Ann. Rev. Phytopath. **23**, 55.
Stanley, J. *et al.* (1986) EMBO Jour. **8**, 1761.

beet leaf curl virus. A probable plant *Rhabdo-*

$virus$, subgroup 2; transmitted by lace-bugs.
Proeseler, G. (1983) CMI/AAB Descriptions of Plant Viruses No. 268.
Francki, R.I.B. *et al.* (1985) **In** Atlas of Plant Viruses. Vol. 1. p. 73. CRC Press: Boca Raton, Florida.

beet mild yellowing virus. A *Luteovirus*; considered to be a strain of BEET WESTERN YELLOWS VIRUS. One of the two viruses, the other being BEET YELLOWS VIRUS, which causes virus yellows of sugarbeet.
Francki, R.I.B. *et al.* (1985) **In** Atlas of Plant Viruses. Vol. 1. p. 137. CRC Press: Boca Raton, Florida.
Casper, R. (1988) **In** The Plant Viruses. Vol. 3. p. 235. ed. R. Koenig. Plenum Press: New York.

beet mosaic virus. A *Potyvirus*.
Russell, G.E. (1971) CMI/AAB Descriptions of Plant Viruses No. 53.
Francki, R.I.B. *et al.* (1985) **In** Atlas of Plant Viruses. Vol. 2. p. 183. CRC Press: Boca Raton, Florida.

beet necrotic yellow vein virus. A possible member of the *Furovirus* group. Causes 'rhizomania' disease in sugarbeet, so named because of the proliferation of rootlets.
Tamada, T. (1975) CMI/AAB Descriptions of Plant Viruses No. 144.
Brunt, A.A. and Shikata, E. (1986) **In** The Plant Viruses. Vol. 2. p. 385. ed. M.H.V. van Regenmortel and H. Fraenkel-Conrat. Plenum Press: New York.

beet pseudo yellows virus. A possible *Closterovirus*, transmitted by whitefly.
Duffus, J.E. *et al.* (1986) Phytopath. **76**, 97.

beet ringspot virus. Serotype of TOMATO BLACK RING VIRUS.

beet soil-borne virus. A possible member of the *Furovirus* group.
Brunt, A.A. and Shikata, E. (1986) **In** The Plant Viruses. Vol. 2. p. 385. ed. M.H.V. van Regenmortel and H. Fraenkel-Conrat. Plenum Press: New York.

beet temperate virus. *See* BEET CRYPTIC VIRUS 1.

beet western yellows virus. A *Luteovirus*. Has a wide host range and is important in beet, lettuce and other crops.

Duffus, J.E. (1972) CMI/AAB Descriptions of Plant Viruses No. 89.
Francki, R.I.B. *et al.* (1985) In Atlas of Plant Viruses. Vol. 1. p. 137. CRC Press: Boca Raton, Florida.
Casper, R. (1988) In The Plant Viruses. Vol. 3. p. 235. ed. R. Koenig. Plenum Press: New York.

beet yellow net virus. A possible member of the *Luteoviruses*.
Francki, R.I.B. *et al.* (1985) In Atlas of Plant Viruses. Vol. 1. p. 137. CRC Press: Boca Raton, Florida.

beet yellow stunt virus. A member of the *Closterovirus* subgroup 1.
Duffus, J.E. (1979) CMI/AAB Descriptions of Plant Viruses No. 207.
Francki, R.I.B. *et al.* (1985) In Atlas of Plant Viruses. Vol. 2. p. 219. CRC Press: Boca Raton, Florida.

beet yellows virus. Type member of the *Closterovirus* group, subgroup 1. One of two viruses, the other being BEET MILD YELLOWS VIRUS which causes sugarbeet yellows disease.
Russell, G.E. (1970) CMI/AAB Descriptions of Plant Viruses No. 13.
Francki, R.I.B. *et al.* (1985) In Atlas of Plant Viruses. Vol. 2. p. 219. CRC Press: Boca Raton, Florida.

belladonna mottle virus. A *Tymovirus*.
Paul, H.L. (1971) CMI/AAB Descriptions of Plant Viruses No. 52.
Francki, R.I.B. *et al.* (1985) In Atlas of Plant Viruses. Vol. 1. p. 117. CRC Press: Boca Raton, Florida.
Hirth, L. and Girard, L. (1988) In The Plant Viruses. Vol. 3. p. 163. ed. R. Koenig. Plenum Press: New York.

Belmont virus. Family *Bunyaviridae*, genus *Bunyavirus*. Isolated from *Culex annulirostris* in Queensland, Australia. Antibodies found in cattle, wallaby and kangaroo.

Benevides virus. Family *Bunyaviridae*, genus *Bunyavirus*.

Benfica virus. Family *Bunyaviridae*, genus *Bunyavirus*.

Bergoldiavirus. Former nomenclature for GRANULOSIS VIRUS (after G.H. Bergold).

Berkeley bee picornavirus. Unclassified small RNA virus (a possible INSECT PICORNAVIRUS) isolated in mixed infections with ARKANSAS BEE VIRUS from honey bees in California, USA. Virions are isometric, sediment at 165S and contain three structural proteins (mw. 32.5×10^3, 35×10^3 and 37×10^3) and an ssRNA genome (mw. 2.8×10^6). Virus is unrelated to ARKANSAS BEE VIRUS.
Lommel, S.A. *et al.* (1985) Intervirology **23**, 199.

Bermudagrass etched-line virus. A member of the *Marafivirus* group.
Gamez, R. and Leon, P. (1988) In The Plant Viruses. Vol. 3. p. 212. ed. R.Koenig. Plenum Press: New York.

Berry-Dedrick virus. *See* RABBIT FIBROMA VIRUS.

Bertioga virus. Family *Bunyaviridae*, genus *Bunyavirus*. Isolated from sentinel mice in Brazil.

BES. N,N-bis(2-hydroxyethyl)-2-aminoethane-sulphonic acid (mw. 213.25). A biological buffer, pK_a 7.15 with a pH range 6.2-8.2.
Good, N. *et al.* (1966) Biochem. **5**, 469.

Betaherpesvirinae. Synonym: CYTOMEGALOVIRUS. A subfamily in the family *Herpesviridae*. There are two genera, *Human* and *Murid Cytomegalovirus Group*. Viruses cause latent infection in the salivary glands. Inclusion bodies are found in the nuclei and cytoplasm late in infection.

Bhanja virus. Family *Bunyaviridae*, genus *Bunyavirus*. Isolated from ticks in many countries and from cattle, sheep, hedgehog and squirrel. May cause fever in man.

BHK cells Abbreviation for BABY HAMSTER KIDNEY CELLS.

Bidens mosaic virus. A possible *Potyvirus*.
Francki, R.I.B. *et al.* (1985) In Atlas of Plant Viruses. Vol. 2. p. 183. CRC Press: Boca Raton, Florida.

Bidens mottle virus. A *Potyvirus*.
Purcifull, D.E. *et al.* (1976) CMI/AAB Descriptions of Plant Viruses No. 161.
Francki, R.I.B. *et al.* (1985) In Atlas of Plant Viruses. Vol. 2. p. 183. CRC Press: Boca Raton, Florida.

Bijou Bridge virus. Family *Togaviridae*, genus

Alphavirus.

Bimbovirus. Unclassified arthropod-borne virus. Isolated from a Golden Bishop bird *Buplectes afer* in Africa.

Bimiti virus. Family *Bunyaviridae*, genus *Bunyavirus*. Isolated from *Culex spissipes* in Trinidad, Brazil, French Guiana and Surinam.

binding site. 1. Region of a protein at which binding occurs with a compound (e.g. an enzyme and its substrate) whose structure is changed by the activity of the protein. 2. The site on a cell to which virus particles adsorb before entry.

bioassay. Determination of the amount of a virus by measuring its biological activity (e.g. infectivity for its host).

biological containment. Method for reducing or even eliminating the risk of viruses or other micro-organisms escaping from the laboratory. In the specialised sense of recombinant DNA molecules, this is a strategy which involves the use of vectors which have been genetically disabled so that they can only survive in the conditions used by the experimenter and which are unavailable outside the laboratory. Containment conditions are determined by national regulatory bodies.

biological control. Pest control using agents of biological origin including parasites, predators and pathogens. Some viruses have been used as biological control agents e.g. myxoma virus for rabbit control and baculoviruses for insect pest control.

biorational pesticides. Pest control agents of biological origin including viruses as well as bacteria, protozoa, fungi and biochemical products of natural origin or identical to a natural product.

biotin. A small water-soluble macromolecule (vitamin H) which is used as a non-radioactive reporter group for LABELLING nucleic acid PROBES. It has a very high affinity for avidin (STREPT-AVIDIN) ($k_{dis} = 10^{-15}$) which, when coupled with an indicator molecule (enzyme, fluorescent dye), is used to detect biotin.

Biotrol. Nuclear polyhedrosis virus preparations produced on a semi-commercial scale by Nutri-

lite Products Inc., USA; now discontinued. Biotrol VPO was developed for the biological control of *Prodenia* sp.; Biotrol VSE for *Spodoptera* sp.; Biotrol VHZ for *Heliothis* sp. and Biotrol VIN for *Trichoplusia ni*.

biphasic milk fever virus. *See* TICK-BORNE ENCEPHALITIS VIRUS.

birao virus. Family *Bunyaviridae*, genus *Bunyavirus*. Isolated from *Anopheles pharoensis* and *Anopheles squamosus* in Africa.

Birdiavirus. Former nomenclature for NUCLEAR POLYHEDROSIS VIRUSES isolated from Hymenoptera (Insecta) which only infect cells in the gut epithelium of susceptible hosts (after F.T. Bird).

Birnaviridae. A family of viruses roughly

   100nm

spherical in shape with a diameter of *c*.60 nm., sedimenting at 650S and banding in CsCl at 1.32 g/cc; there is no envelope. The capsid is composed of four major polypeptides and contains a genome comprising two segments of dsRNA, mw. 2.5 and 2.2 x 10^6. The family contains viruses infecting fish (infectious pancreatic necrosis viruses) and chickens (infectious bursal disease virus), both important commercially.
Dobos, P. J. Virol. (1979) **32**, 593.

Birnavirus. A genus in the family *Birnaviridae*.

Biston betularia Cytoplasmic Polyhedrosis Virus. Cytoplasmic polyhedrosis virus (CPV) isolated in the United Kingdom from field-collected larvae of the peppered moth, *Biston betularia* (Geometridae, Lepidoptera). The virus is the type member of 'type 6' CPV and is unrelated to *Bombyx mori* (type 1) CPV on the basis of RNA electropherotype. Viruses of similar electropherotype have been observed in other Lepidoptera (*see* APPENDIX B).
Payne, C.C. and Mertens, P.P.C. (1983) **In** The *Reoviridae*. p. 425. ed. W.K. Joklik. Plenum Press: New York.

BK polyoma virus. Family *Papovaviridae*, genus *Polyomavirus*. Isolated from a patient on immunosuppressive drugs after transplantation.

Antibodies in man suggest it is a common infection. Grows in Vero cells and primary human foetal kidney cells. Transforms rat and hamster cells in culture and is oncogenic in new-born hamsters.

black beetle virus (BBV). Family *Nodaviridae*, genus *Nodavirus*, isolated from *Heteronychus arator* (Coleoptera) in New Zealand and more intensively studied than other Nodaviruses. Virions are icosahedral, 30 nm. in diameter, sediment at 137S and have a buoyant density in CsCl of 1.33 g/cc. They contain 180 protein subunits arranged in T=3 quasisymmetry. The particles contain one major polypeptide (mw. 39 x 10^3) and two minor proteins (mw. 44 x 10^3 and another of $c.4.5$ x 10^3 derived by proteolytic cleavage from the 44 x 10^3 precursor). The genome is composed of two + strand ssRNA molecules (RNA1, 3106 bases and RNA2, 1399 bases) in the same particle, both of which are required for infection. The RNAs are not polyadenylated. The larger RNA codes for a polymerase (mw. 104 x 10^3), the smaller for the capsid protein precursor (mw. 44 x 10^3). The virus multiplies in the cytoplasm of infected cells. The precursor coat protein (407 aa residues) is cleaved at residue 363 within the provirion form of the particle, leading to a more stable virion structure. Unlike NODAMURA virus, BBV does not replicate in vertebrate cells or suckling mice, but replicates to high titre in *Drosophila melanogaster* line 1 cells. Extensive homology has been detected between the coat protein genes of BBV and Flock House Virus.
Moore, N.F. *et al.* (1985) J. gen. Virol. **66**, 647.

black locust true mosaic virus. A strain of PEANUT STUNT VIRUS.
Richter, J. *et al.* (1979) Acta Virol. **23**, 489.

black queen cell virus. Unclassified small RNA virus isolated from prepupae and pupae of queen honey bees in Britain. Virions are isometric, 30 nm. in diameter and sediment at 151S. They have a buoyant density in CsCl of 1.345 g/cc., contain a single structural polypeptide (mw. 30 x 10^3) and an RNA genome.
Bailey, L. and Woods, R.D. (1977) J. gen. Virol. **37**, 175.

black raspberry latent virus. An *Ilarvirus*. Lister, R.M. and Converse, R.H. (1972) CMI/AAB Descriptions of Plant Viruses No. 106.

Francki, R.I.B. (1985) **In** The Plant Viruses. Vol. 1. p. 1. ed. R.I.B. Francki. Plenum Press: New York.

blackeye cowpea mosaic virus. A *Potyvirus*. Purcifull, D. and Gonsalves, D. (1985) CMI/AAB Descriptions of Plant Viruses No. 305.
Francki, R.I.B. *et al.* (1985) **In** Atlas of Plant Viruses. Vol. 2. p. 183. CRC Press: Boca Raton, Florida.

blackgram mottle virus. A possible member of the *Carmovirus* group.
Scott, H.A. and Hoy, J.W. (1981) CMI/AAB Descriptions of Plant Viruses No. 237.
Morris, T.J. and Carrington, J.C. (1988) **In** The Plant Viruses. Vol. 3, p. 73. ed. R. Koenig. Plenum Press: New York.

blind passage. Inoculation of material (from an animal or cell culture) which shows no evidence of infection into a fresh animal or cell culture, usually with the aim of growing and identifying an infectious agent.

blotting. A technique to transfer DNA, RNA or protein to an immobilising matrix such as nitrocellulose or nylon filters or DIAZOBENZYLOXY-METHYL PAPER. *See* NORTHERN BLOTTING, SOUTHERN BLOTTING, WESTERN BLOTTING.

blueberry leaf mottle virus. A *Nepovirus*. Ramsdell, D.C. and Stace-Smith, R. (1983) CMI/AAB Descriptions of Plant Viruses No. 267.

blueberry shoestring virus. A possible *Sobemovirus*. Ramsdell, D.C. (1979) CMI/AAB Descriptions of Plant Viruses No. 204.
Francki, R.I.B. *et al.* (1985) **In** Atlas of Plant Viruses. Vol. 1. p. 153. CRC Press: Boca Raton, Florida.
Hull, R. (1988) **In** The Plant Viruses. Vol. 3. p. 113. ed. R. Koenig. Plenum Press: New York.

bluecomb disease virus. Family *Coronaviridae*, genus *Coronavirus*.

bluetongue virus. Family *Reoviridae*, genus *Orbivirus*. Causes a serious disease of sheep with mortality up to 30%. Fever, oedema of the head and neck, cyanosis, erosions round the mouth and lameness are observed. Cattle and goats develop a milder disease. Pigs have foot lesions and wild ruminants are often affected. Usually found in

Africa, Cyprus, Palestine, Turkey, Spain, Portugal, Pakistan, India, Japan and parts of the USA. Transmission is by biting *Culicoides*. Replication occurs in the insects. Virus can be grown in yolk sac of eggs and causes cpe in a variety of cells.
Gorman, B.M. (1979) J. gen. Virol. **44**, 1.
Bluetongue and Related Viruses (1985) In Progress in Clinical and Biological Research. Vol. 178. ed. T.L. Barber and M.M. Jochim. Alan R. Liss: New York.

blunt end. (Synonym: FLUSH END.) DNA fragments generated by certain RESTRICTION ENDONUCLEASES, e.g. *Hae* III, and which are perfectly base-paired along their entire length, i.e. they do not carry single-stranded regions after cleavage with the enzyme.

Bm-5 cells. Insect cell line from the silkworm, *Bombyx mori*, susceptible to infection by *B. mori* nuclear polyhedrosis virus.

BML-TC7A medium. Insect cell culture medium developed for the production of a cell line derived from *Spodoptera frugiperda* which supports the growth of *S. frugiperda* and *Autographa californica* nuclear polyhedrosis viruses. The medium also supports growth of insect cell lines from *Cydia pomonella*, *Heliothis zea* and *Trichoplusia ni*.

BML-TC10 medium. Insect cell culture medium developed for growth of a cell line derived from *Spodoptera frugiperda* (IPLB-SF-21) and for the production of its homologous nuclear polyhedrosis virus.

Bobayavirus. Family *Bunyaviridae*, genus *Bunyavirus*. Isolated from an African thrush *Turdus libonyanus* in Central Africa.

Bobia virus. Family *Bunyaviridae*, genus *Bunyavirus*.

Bocas virus. Family *Coronaviridae*, genus *Coronavirus*. Possibly the same as MOUSE HEPATITIS VIRUS.

boletus virus. A possible *Potexvirus*.
Francki, R.I.B. *et al.* (1985) In Atlas of Plant Viruses. Vol. 2. p. 159. CRC Press: Boca Raton, Florida.

Bolivian haemorrhagic fever virus. *See* MACHUPO VIRUS.

Bombyx mori cytoplasmic polyhedrosis virus.
Type species of the CYTOPHASMIC POLYHEDROSIS VIRUS group (REOVIRIDAE), isolated from the silkworm, *B. mori* in Japan. VIRIONS are icosahedral (*c*.60 nm. in diameter), with 20 nm. spikes at each of the twelve vertices. Particles sediment at 400-440S and have a buoyant density in CsCl of 1.44 g/cc. The virion contains at least five polypeptides (mw. *c*.146 x 10^3, 138 x 10^3, 125 x 10^3, 70 x 10^3 and 31 x 10^3) and a segmented dsRNA genome (ten segments; total mw. *c*.15 x 10^6). During infection, virions are occluded within large occlusion bodies (polyedra) the matrix protein of which ('polyhedrin') has a mw. of 27 x 10^3 and is coded for by the smallest genome segment. The + strands of the genome RNA segments carry a m^7G CAP. Virions contain transcriptase, nucleotide phosphohydrolase, guanylyl transferase and transmethylases necessary for the synthesis of complete viral mRNA copies of the genome + strands. The virus infects larvae after ingestion of polyhedra. Replication is confined to cells of the gut epithelium. The virus is associated with a complex of other diseases which together cause extensive losses in the silk industry. At least nine CPV isolates have been reported in the silkworm and classified according to the shape and/or intracellular location of occlusion bodies.
Payne, C.C. and Mertens, P.P.C. (1983) In The Reoviridae. p. 425. ed. W.K. Joklik. Plenum Press: New York.

Bombyx mori nuclear polyhedrosis virus.
Member of subgroup A (NUCLEAR POLYHEDROSIS VIRUS) of the BACULOVIRUS genus isolated from the silkworm, *B. mori*. It is representative of the SNPV subtype where the enveloped virions contain predominantly single nucleocapsids.

boolarra virus. A NODAVIRUS isolated from *Oncopera intricoides* (Lepidoptera) in Australia.
Reinganum, C. *et al.* (1985) Intervirology **24**, 10.

boraceiavirus. Family *Bunyaviridae*, genus *Bunyavirus*. Isolated from *Anopheles cruzii* and *Phoiomyia pilicande* in Brazil.

border disease virus. Family *Togaviridae*, genus *Pestivirus*. First described on the borders of Wales and England. Causes a congenital condition of new-born lambs, characterised by a hairy birth coat and a tremor. There is defective myelination of the CNS and virus can be isolated from this. Injection of virus into pregnant ewes causes abortion and border disease in the foetuses which survive. Disease has now been found in Germany, Australia and New Zealand. The virus replicates in primary calf and foetal lamb kidney cells. The virus is related antigenically to swine fever and mucosal disease viruses.

Barlow, R.M. *et al.* (1979). Vet. Rec. **104**, 334.

borna disease virus. Unclassified arthropod-borne virus. Found in Germany, Poland, Rumania, Russia, Syria and Egypt. Isolated from ticks and from the brains of herons and other wild birds. Causes lassitude followed by chronic spasms and paralysis in horses, sheep, cattle and deer. Rabbits infected experimentally show similar signs. No infection of man has been reported. Virus replicates in the chorioallantoic membrane and in cultures of lamb testis and monkey kidney cells.

Bornholm disease virus. Synonym: BAMBLE DISEASE VIRUS. Family *Picornaviridae*, genus *Enterovirus*. This can be any one of several Coxsackie or Echo virus strains. These viruses cause chest, abdomen and muscle pain.

Borrel bodies. Minute granules (Bollinger bodies) found in FOWLPOX VIRUS infected cells. Named after Professor Borrel (Institut Pasteur).

Borrelinavirus. Former nomenclature for NU-CLEAR POLYHEDROSIS VIRUSES from Lepidoptera (Insecta), after Professor Borrel (Institut Pasteur).

Botambi virus. Family *Bunyaviridae*, genus *Bunyavirus*. Isolated from *Culex guiarti* in Africa.

Boteke virus. Unclassified arthropod-borne virus. Isolated from *Mansonia maculipennis* in Africa.

Bothid herpesvirus 1. Synonym: TURBOT HERPES-VIRUS. Member of family *Herpesviridae*. Isolated from turbot in a fish farm.

Bouboui virus. Family *Flaviviridae*, genus *Flavivirus*. Isolated from mosquitoes and the baboon *Papio papio* in Africa.

bovid (alpha) herpesvirus 2. Synonyms: ALLER-TON VIRUS, BOVINE MAMMILLITIS VIRUS. Family *Herpesviridae*, subfamily *Alphaherpesvirinae*, genus *Human Herpesvirus 1 Group*). Causes lumpy skin disease in cattle. Probably transmitted by insects. Infection of day-old rats, mice and hamsters may lead to stunted growth, rashes and death. Grows in calf kidney cell cultures, producing syncytia.

bovid herpesvirus 1. Synonyms: INFECTIOUS BOVINE RHINOTRACHEITIS VIRUS, NECROTIC RHINITIS VIRUS. Family *Herpesviridae*, subfamily *Alpha Herpesvirinae*). Found in cattle worldwide. Can cause a silent, mild infection or an acute disease of the respiratory tract with very high mortality. Can also cause conjunctivitis and disease of the genital tract. Experimentally, goats develop fever, whereas rabbits develop an encephalitis and paralysis. Transmission is by contact. Can be grown in bovine embryo cell cultures with cytopathic effect and in pig, sheep, goat and horse kidney cells and human amnion cell cultures. Hamster cells are transformed *in vitro* by the virus.

bovid herpesvirus 2. *See* BOVID (ALPHA) HERPES-VIRUS 2.

bovid herpesvirus 3. Synonyms: MALIGNANT CATARRHAL FEVER VIRUS, WILDEBEEST HERPESVIRUS. Causes a widespread disease of cattle accompanied by fever, acute inflammation of nasal and oral membranes and involvement of pharynx and lungs. The disease is often fatal. It can be transmitted experimentally to cattle and rabbits. The virus grows in bovine thyroid and adrenal cell cultures.

bovid herpesvirus 4. Family *Herpesviridae*, unassigned subfamily and genus. Isolated from sheep with pulmonary adenomatosis.

bovid herpesvirus 5. Synonym: CAPRINE HERPES-VIRUS 1. Family *Herpesviridae*, unassigned subfamily and genus). Isolated from kids *Capra hircus* with severe febrile disease. Can be transmitted to other kids to give similar disease and to adults in which it can cause abortion. Virus grows in many cell cultures.

bovine adeno-associated virus. Family *Parvoviridae*, genus *Dependovirus*.

bovine adenovirus. Family *Adenoviridae*, genus *Mastadenovirus*. Causes respiratory infection and conjunctivitis in cattle. Serological evidence suggests high incidence of infection. Grows with cytopathic effect in bovine kidney cells. Occurs as ten serotypes, two of which are oncogenic for new-born hamsters.

bovine ephemeral fever virus. Family *Rhabdoviridae*, unassigned genus. Isolated in S. Africa, Australia and Japan from cattle with respiratory signs, increased salivation, joint pains, stiffness and tremors. The disease is of short duration. Virus grows in mice and BHK cell cultures, and in eggs. Transmitted by insects.
Della Porta, A.J. and Snowdon, W.A. (1982) **In** The Rhabdoviruses. ed. D.H.L. Bishop. CRC Press: Boca Raton, FLorida.

bovine enterovirus. *See* ECBO VIRUSES.

bovine leukosis virus. Synonyms: BOVINE LEU-KAEMIA VIRUS, BOVINE TYPE C ONCOVIRUS. Family *Retroviridae*, subfamily *Oncovirinae*, genus *Type C Oncovirus*. Causes leukaemia. Cattle, sheep and goats can be infected experimentally and produce antibo dies. Some animals develop lymphosarcoma. Cells from many species produce syncytia on infection. This can be prevented with antiserum.
Mussgay, M. and Kaaden, O.R. (1978) Current Topics in Microbiol. Immunol. **79**, 43.

bovine mammilitis virus. *See* BOVID (ALPHA) HERPESVIRUS 2.

bovine papilloma virus. Family *Papovaviridae*, genus *Papillomavirus*. Causes papillomas on the head and neck and in the mouth and oesophagus. Injection into hamsters and mice produces fibrosarcomas. Embryo cultures of mouse, hamster and bovine tissues are transformed.

bovine parvovirus. Synonym: HADEN VIRUS. Family *Parvoviridae*, genus *Parvovirus*. Can cause diarrhoea in young animals. Replicates in bovine embryo kidney cell cultures producing cytopathic effect.

bovine pustular stomatitis virus. Synonym: PAPULAR STOMATITIS OF CATTLE VIRUS. Family *Poxviridae*, subfamily *Chordopoxvirinae*, genus *Parapoxvirus*. Causes a benign non-febrile disease with ulcers in the mouth, most frequently in young animals. Some strains infect sheep and goats. May infect man.

bovine respiratory syncytial virus. Family *Paramyxoviridae*, genus *Pneumovirus*. Causes respiratory disease. Replicates in bovine kidney and lung cultures causing syncytia.

bovine rhinovirus. Family *Picornaviridae*, unassigned genus. A widespread infection causing nasal discharge. Does not infect other species.

bovine serum albumin (BSA). The major protein constituent of bovine serum. mw. 68 x 10^3.

bovine syncytial virus. Family *Retroviridae*, subfamily *Spumavirinae*. Causes lymphosarcomas. Has been isolated from normal cattle. Replicates in BHK21 cells producing syncytia.
Greig, A.S. (1978) Canad. J. comp. Med. **43**, 112.

bovine viral diarrhoea virus. *See* MUCOSAL DISEASE VIRUS.

Br1 mycoplasmavirus group. Proposed genus of phages isolated from *Mycoplasma* which have contractile tails and resemble members of the *Myoviridae*. Proposed type member is phage MVBr1.
Maniloff, J. *et al.* (1982) Intervirology **18**, 177.

Brazilian corn streak virus. *See* MAIZE RAYADO FINO VIRUS.

breakbone fever virus. *See* DENGUE VIRUS.

broad bean mottle virus. A *Bromovirus*.
Gibbs, A.J. (1972) CMI/AAB Descriptions of Plant Viruses No. 101.
Francki, R.I.B. (1985) **In** The Plant Viruses. Vol. 1. p. 1. ed. R.I.B. Francki. Plenum Press: New York.

broad bean necrosis virus. A *Furovirus*.
Inouye, T. and Nakasone, W. (1980) CMI/AAB Descriptions of Plant Viruses No. 223.
Brunt, A.A. and Shikata, E. (1986) **In** The Plant Viruses. Vol. 2. p. 305. ed. M.H.V. van Regenmortel and H. Fraenkel-Conrat. Plenum Press: New York.

broad bean severe chlorosis virus. An unclassified plant virus with rod-shaped particles 650-820 nm. long, 14 nm. wide. The virus is transmitted by aphids in the NON-PERSISTENT TRANSMISSION manner.
Thottappilly, G. *et al.* (1975) Phytopath. Zeitschrift **84**, 343.

broad bean stain virus. A *Comovirus*.
Gibbs, A.J. and Smith, H.G. (1970) CMI/AAB Descriptions of Plant Viruses No. 29.

Francki, R.I.B. *et al.* (1985) **In** Atlas of Plant Viruses. Vol. 2. p. 1. CRC Press: Boca Raton, Florida.

broad bean true mosaic virus. Synonym: ECHTES ACKERBOHNENMOSAIC VIRUS. A *Comovirus.* Gibbs, A.J. and Paul, H.L. (1970) CMI/AAB Descriptions of Plant Viruses No. 20.
Francki, R.I.B. *et al.* (1985) **In** Atlas of Plant Viruses. Vol. 2. p. 1. CRC Press: Boca Raton, Florida.

broad bean wilt virus. This virus has two serotypes; serotype 1 is the type member of the *Fabavirus* group.
Taylor, R.H. and Stubbs, L.L. (1972) CMI/AAB Descriptions of Plant Viruses No. 81.
Francki, R.I.B. *et al.* (1985) **In** Atlas of Plant Viruses. Vol. 2. p. 1. CRC Press: Boca Raton, Florida.

broad bean wilt virus group. *See* FABAVIRUS GROUP.

broad bean yellow band virus. A serotype of PEA EARLY BROWNING VIRUS.
Robinson, D.J. and Harrison, B.D. (1985) J. gen. Virol. **66**, 2003.

broad bean yellow ringspot virus. An unclassified plant virus with isometric particles 28 nm. in diameter.
Doi, Y. *Personal communication.*

broad bean yellow vein virus. A plant *Rhabdovirus.* Natsuaki, K.T. (1981) Ann. Phytopath. Soc. Japan **47**, 410.

broccoli necrotic yellows virus. A plant *Rhabdovirus,* subgroup 1; aphid-transmitted.
Campbell, R.N. and Lin, M.T. (1972) CMI/AAB Descriptions of Plant Viruses No. 85.
Francki, R.I.B. *et al.* (1985) **In** Atlas of Plant Viruses. Vol. 1. p. 73. CRC Press: Boca Raton, Florida.

brome mosaic virus. Type member of the *Bromovirus* group.
Lane, L.C. (1977) CMI/AAB Descriptions of Plant Viruses No. 180.
Francki, R.I.B. (1985) **In** The Plant Viruses. Vol. 1. p. 1. ed. R.I.B. Francki. Plenum Press: New York.

brome stem leaf mottle virus. *See* COCKSFOOT MILD MOSAIC VIRUS.

brome streak virus. A *Potyvirus.*
Francki, R.I.B. *et al.* (1985) **In** Atlas of Plant Viruses. Vol. 2. p. 183. CRC Press: Boca Raton, Florida.

bromelain. A mixture of proteolytic enzymes derived from pineapples; different enzymes are isolated from different parts of the plant. They have been used frequently to remove the haemagglutinin from INFLUENZA VIRUS.

bromo-deoxyuridine (5-bromo-2-deoxyuridine). A pyrimidine derivative which can become incorporated into DNA in place of thymidine. Of possible use in viral CHEMOTHERAPY.

Bromovirus group. (BROME MOSAIC VIRUS). Genus of MULTICOMPONENT plant viruses with small isometric particles, 26 nm. in diameter which sediment at 85S (pH5) and band in CsCl at

100nm

1.35 g/cc. The particles are stabilised by pH-dependent protein-protein interactions and protein-RNA links; they swell and become salt labile above pH7. Capsid structure is icosahedral (T=3), the structural subunit being coat protein (mw. 20 x 10^3). Genomic linear (+)-sense ssRNA comprises three species, RNA-1 (mw. 1.1 x 10^6; 3,234 nucleotides), RNA-2 (mw. 1.0 x 10^6; 2,865 nucleotides) and RNA-3 (mw. 0.7 x 10^6; 2,120 nucleotides). RNAs-1 and -2 are separately encapsidated; RNA-3 is encapsidated with the coat protein mRNA, RNA-4 (mw. 0.3 x 10^6; 876 nucleotides). The 5' termini of the RNAs have a CAP; the 3' termini have a tRNA-like structure which accepts tyrosine. Replication is in the cytoplasm and, for the genomic RNAs, is via distinct ds REPLICATIVE INTERMEDIATES; RNA-4 is derived from RNA-3. RNAs-1, -2 and -4 are the MONOCISTRONIC messengers for proteins of mw. 112, 107 and 20 (coat protein) x 10^3 respectively; RNA-3 is bicistronic encoding a protein of mw. 33 x 10^3 at the 5' end and having the coat protein cistron at the 3' end.

The natural host ranges are narrow. The particles are found in most cell types. Bromoviruses are readily transmissible mechanically. Some members are transmitted by beetles.
Matthews, R.E.F. (1982) Intervirology **17**, 173.
Lane, L.C. (1979) CMI/AAB Descriptions of

Plant Viruses No. 215. Francki, R.I.B. *et al.* (1985) **In** Atlas of Plant Viruses. Vol. 2, p. 69, CRC Press.: Boca Raton, Florida.
Francki, R.I.B. (1985) **In** The Plant Viruses. Vol. 1, p. 1, ed. R.I.B. Francki. Plenum Press: New York.

bruconha virus. Family *Bunyaviridae*, genus *Bunyavirus*.

Bryonia mottle virus. A possible *Potyvirus*. Francki, R.I.B. *et al.* (1985) **In** Atlas of Plant Viruses. Vol. 2. p. 183. CRC Press: Boca Raton, Florida.

BTI-EAA cells. Insect cell line from the saltmarsh caterpillar, *Estigmene acrea*.

budding. Method of release of enveloped virus particles from the cells in which they have grown. The viral nucleocapsid associates with an area of the cell membrane which closes around the virus as it leaves the cell. During budding the cellular proteins in that part of the membrane destined to become the virus coat are replaced by virus-coded proteins. In contrast the lipid in the viral envelope is host-cell derived.

buenaventura virus. Family *Bunyaviridae*, genus *Phlebovirus*.

buffalo poxvirus. Family *Poxviridae*, subfamily *Chordopoxvirinae*, genus *Orthopoxvirus*. Causes severe disease in India. Produces pocks on the chorioallantoic membrane.

buffer. A solution consisting of a mixture of an acid and a base which resists changes in pH and is therefore a suitable environment for ensuring the correct acidity or alkalinity. Many combinations of acids and bases are available to provide buffering at most required pHs.

bujaru virus. Family *Bunyaviridae*, genus *Phlebovirus*. Isolated from the rodent *Proechimys guyannensisoris* in Brazil.

bukalasa bat virus. Family *Flaviviridae*, genus *Flavivirus*. Isolated from bats in Africa.

bundle virion. Synonym for the MNPV subtype of subgroup A (NUCLEAR POLYHEDROSIS VIRUS) of the *BACULOVIRUS* genus.

Bung el Arab virus. Unclassified arthropod-borne virus. Isolated from a bird *Sylvia curruca* in Egypt.

Bunyamwera virus. Family *Bunyaviridae*, genus *Bunyavirus*. Isolated from mosquitoes in several African countries. Causes fever in man.

Bunyaviridae. A very large family of RNA viruses with spherical or oval particles, 90-100 nm. in diameter, with a lipid envelope and glycoprotein surface projections and three ribonucleocapsids composed of circular, helical strands 2.0-2.5 nm. in diameter. There are four genera,

 100nm

BUNYAVIRUS, PHLEBOVIRUS, NAIROVIRUS and *UUKUVIRUS*. The particles contain four proteins, two of which are external glycoproteins (G1, G2), a nucleoprotein (N) and a minor large protein (L). There are three species of (−)-sense ssRNA (L 3.5×10^6; M $1-2 \times 10^6$; S $0.4-0.8 \times 10^6$). The virus RNA is transcribed by the virion transcriptase into mRNA. At least one of the members possesses AMBISENSE EXPRESSION STRATEGY. Warm and cold-blooded vertebrates and arthropods are the natural hosts. Transmission is usually by a variety of vectors including mosquitoes, ticks, *Phlebotomus* spp. and other arthropods. Aerosol infection can also occur. The viruses can be grown in tissue culture.
Bishop, D.H.L. *et al.* (1980) Intervirology. **14**, 125.
Matthews, R.E.F. (1982) Intervirology **17**, 115.

Bunyavirus. (Bunyamwera, place in Uganda where type species was isolated.) A genus in the family *Bunyaviridae*, containing 100 or more viruses which have some antigenic cross-relationship. There are ten antigenic subgroups. Some may be transmitted transovarially.

buoyant density. The density which a virus or other macromolecule possesses when suspended in an aqueous solution of a heavy metal salt such as CsCl or a sugar such as sucrose. This is the density at which the macromolecule is in equilibrium and neither sinks nor floats. *See* CAESIUM CHLORIDE DENSITY GRADIENT CENTRIFUGATION.

burdock mosaic virus. An unclassified plant

virus with isometric particles; occurs in Japan. Doi, Y. *Personal communication.*

burdock mottle virus. An unclassified plant virus with rod-shaped particles; occurs in Japan. Doi, Y. *Personal communication.*

burdock stunt viroid. A VIROID.
Tien, P. and Cheu, W. (1987) The Viroids. p. 333. ed. T.O.Diener. Plenum Press: New York.

burdock yellows virus. A member of the *Closterovirus* subgroup 1.
Francki, R.I.B. *et al.* (1985) **In** Atlas of Plant Viruses. Vol. 2. p. 219. CRC Press: Boca Raton, Florida.

Burkitt's lymphoma virus. *See* EPSTEIN-BARR VIRUS.

burst size. The yield of infective virus particles obtained during a lytic one-step growth infection of a host cell. Generally used to describe PHAGE infections of prokaryotes. Yields of between 5-10 x 10^3 virus particles per bacterial cell can be obtained with some viruses (e.g. ssRNA phages).

Bushbush virus. Family *Bunyaviridae*, genus *Bunyavirus*. Isolated from mosquitoes in Trinidad and Brazil.

Bussuquara virus. Family *Flaviviridae*, genus *Flavivirus*. Isolated from man, sentinel monkeys *Alouatta*, sentinel mice and *Culex* sp. in Colombia, Panama and Brazil. Causes fever in man.

butterbur mosaic virus. A *Carlavirus*; occurs in Japan.
Doi, Y. *Personal communication.*

butterbur rhabdovirus. A plant *Rhabdovirus*; occurs in Japan.
Doi, Y. *Personal communication.*

Button willow virus. Family *Bunyaviridae*, genus *Bunyavirus*. Isolated from rabbits, hares and *Culicoides* sp. in California, New Mexico and Texas.

BV. Abbreviation sometimes used for 'budded virus', often referring to baculovirus particles which are not incorporated within occlusion bodies during infection and which acquire their envelopes by budding through the membrane of infected cells.

Bwamba virus. Family *Bunyaviridae*, genus *Bunyavirus*. Isolated from man and *Anopheles funestus* in several African countries. Causes fever in man.

C

C value. The amount of DNA in the haploid genome of a eukaryotic cell. Typical values are: bacteria 0.005 pg/cell; yeast 0.009 pg/cell; higher plants 10-50 pg/cell; vertebrates 1-10 pg/cell.

C3 spiroplasmavirus group. Proposed genus of phages isolated from *Spiroplasma*, which have short tails and resemble members of the *Podoviridae*. Proposed type member is phage SVC3.
Maniloff, J. *et al.* (1982) Intervirology **18**, 177.

C-type virus particles. Particles of a morphologically distinct group of RNA viruses often seen in association with cells in leukaemic tissues. They have a diameter of 90-110 nm. with a central core and a lipoprotein envelope covered with knob-like projections 8 nm. in diameter. The core appears to have CUBIC SYMMETRY and consists of an outer layer of ring-like subunits 6 nm. in diameter forming a hexagonal pattern and an inner membrane 3 nm. thick. The majority of RETROVIRUSES have c-type particles. *See* A-, B-, D-TYPE VIRUS PARTICLES.

Cabassou virus. Family *Togaviridae*, genus *Alphavirus*.

cacao necrosis virus. A *Nepovirus*.
Kenton, R.H. (1977) CMI/AAB Descriptions of Plant Viruses No. 173.
Francki, R.I.B. *et al.* (1985) **In** Atlas of Plant Viruses. Vol. 2. p. 23. CRC Press: Boca Raton, Florida.

cacao swollen shoot virus. Type member of the *Cacao Swollen Shoot Virus* group. Important virus of cacao in West Africa.
Brunt, A.A. (1970) CMI/AAB Descriptions of Plant Viruses No. 10.
Francki, R.I.B. *et al.* (1985) **In** Atlas of Plant Viruses. Vol. 2. p. 235. CRC Press: Boca Raton, Florida.

Cacao swollen shoot virus 'group'. (Named after type member, CACAO SWOLLEN SHOOT VIRUS). Unofficial group of plant viruses with BACILLIFORM particles 80-180 nm. long and 23-30 nm. in diameter; the particle dimensions vary between viruses. The particles are found in the cytoplasm. These viruses have narrow host ranges. Vectors

100nm

of members of this group include aphids (SEMIPERSISTENT TRANSMISSION), leafhoppers and mealybugs.
Francki, R.I.B. *et al.* (1985) **In** Atlas of Plant Viruses. Vol. 2. p. 237. CRC Press: Boca Raton, Florida.

Cacao virus. Family *Bunyaviridae*, genus *Phlebovirus*. Isolated from *Lutzomyia* sp. in Panama.

cacao yellow mosaic virus. A *Tymovirus*.
Brunt, A.A. (1970) CMI/AAB Descriptions of Plant Viruses No. 11.
Francki, R.I.B. *et al.* (1985) **In** Atlas of Plant Viruses. Vol. 1. p. 117. CRC Press: Boca Raton, Florida.
Hirth, L. and Girard, L. (1988) **In** The Plant Viruses. Vol. 3. p. 163. ed. R. Koenig. Plenum Press: New York.

Cache Valley virus. Family *Bunyaviridae*, genus *Bunyavirus*. Isolated in several states in the USA and Jamaica. Mosquito-borne. Antibodies found in many species including man but not known to cause disease.

cactus virus 2. A *Carlavirus*.
Francki, R.I.B. *et al.* (1985) **In** Atlas of Plant Viruses. Vol. 2. p. 173. CRC Press: Boca Raton, Florida.

cactus virus X. A *Potexvirus*.
Bercks, R. (1971) CMI/AAB Descriptions of Plant Viruses No. 58.

33

Francki, R.I.B. *et al.* (1985) **In** Atlas of Plant Viruses. Vol. 2. p. 159. CRC Press: Boca Raton, Florida.

caesium chloride density gradient centrifugation. Method for separating molecules or viruses according to their density. Sedimentation ceases when the molecules reach the position in the gradient which is the same as their own buoyant density. *See* ISOPYCNIC GRADIENT.

Caimito virus.   Family *Bunyaviridae*, genus *Phlebovirus*. Isolated from the fan fly *Lutzomyia ylephilator* in Panama.

calf diarrhoea virus. *See* CALF ROTAVIRUS.

calf rotavirus. Synonym: CALF DIARRHOEA VIRUS. Family *Reoviridae*, genus *Rotavirus*. Causes acute enteritis and diarrhoea. Can be grown in calf cell cultures. Related serologically to other rotaviruses.

Caliciviridae. (Latin 'calix' = cup, goblet.) A family of RNA viruses, roughly spherical in shape with a diameter of 35 nm., sedimenting at *c.*180S and banding in CsCl at 1.37 g/cc. They have no envelope or core but possess a unique

   100nm

morphology. Negatively stained preparations show dark cup-shaped depressions. Each capsid contains 180 copies of one protein, mw. 65-70 x 10^3 and one molecule of infectious (+)-sense ssRNA mw. *c.*2.8 x 10^6 with a VPg at the 5' end and a 3' polyadenylic acid sequence. Replication is complex; genome size ssRNA, two smaller ssRNAs (mw. 0.7 and 1.1 x 10^6) and a partially dsRNA are found in infected cells. The capsid protein is the major virus-induced protein. Virions mature in cytoplasm. Caliciviruses have been found in pigs, cats, sea lions and human infants. Transmission is probably mechanical. Matthews, R.E.F. (1982) Intervirology **17**, 133.

Calicivirus. Latin 'calix' = cup or goblet.) Only genus of the family *Caliciviridae*. Contains vesicular exanthema of swine virus, San Miguel sea lion virus and feline calicivirus, all of which occur as multiple serotypes. It is likely that Norwalk virus also belongs to this genus.

California encephalitis viruses.   Family *Bunyaviridae*, genus *Bunyavirus*. A group of viruses isolated from mosquitoes in California, Utah, New Mexico and Texas. Associated with a few cases of encephalitis in man.

California rabbit fibroma virus.   Family *Poxviridae*, subfamily *Chordopoxvirinae*, genus *Lepripoxvirus*. Causes fibromas in its natural host, *Sylvilagus* sp. In European rabbits it causes death without the typical signs of myxomatosis.

Callinectes 'baculovirus'.   Unclassified virus resembling a NON-OCCLUDED BACULOVIRUS, observed in the crab, *Callinectes sapidus*.
Johnson, P.T. (1978) Mar. Fish. Rev. **40**, 13.

Callinectes 'Reovirus'.   Unclassified reovirus-like agent observed in the crab, *Callinectes sapidus*.
Johnson, P.T. (1977) J. Invertebr. Pathol. **29**, 201.

Callinectes W2 virus. *See* W2 VIRUS.

Callistephus chinensis chlorosis virus.   A possible plant *Rhabdovirus*.
Francki, R.I.B. *et al.* (1985) **In** Atlas of Plant Viruses. Vol. 1. p. 73. CRC Press: Boca Raton, Florida.

calovo virus.   Family *Bunyaviridae*, genus *Bunyavirus*. Isolated from mosquitoes in Czechoslovakia, Yugoslavia and Austria. Causes fever in man.

camel pox virus.   Synonym: PHOTO-SHOOTUR VIRUS. Family *Poxviridae*, subfamily *Chordopoxvirinae*, genus *Orthopoxvirus*. Causes pustules around the lips and nose. Usually mild but in severe cases can cause abortion. Epidemics occur at regular intervals in the Middle East, North Africa, Pakistan and southern USSR. Can cause lesions in man. Grows in chorioallantoic membrane.

Campoletis sonorensis polydnavirus.   Type species of subgroup A POLYDNAVIRUS (*POLYDNAVIRIDAE*) isolated from the ichneumonid parasitoid *C. sonorensis*. Virus particles consist of uniform-length fusiform (cigar-shaped) nucleocapsids (330 x 85 nm.) surrounded by two concentric unit membranes, the inner derived by *de novo* synthesis within nuclei and the outer from the plasma membrane of infected parasitoid calyx epithelial cells. Particles have a buoyant density in CsCl of 1.20 g/cc. and contain about 25 polypeptides (mw. 15-80 x 10^3). The supercoiled circular dsDNA genome is multipartite, containing at least 25 components, not in equimolar amounts, (mw. 4-13.6 x 10^6; total mw. 135-170 x 10^6). Some DNA sequences are common to more than one molecule; one repeated 540bp section has been found in all segments of the genome. Whether the entire genome is encapsidated in one or several particles is not known. The virus replicates in the nuclei of calyx epithelial cells in female parasitoids (the calyx is a specialised region of the oviduct) and is secreted within the calyx lumen as part of the 'calyx fluid'. Viral DNA has also been detected in male parasitoid wasps, covalently linked to cellular DNA. Virus replication appears to cause no deleterious effects on the parasitoid, but the virus may directly or indirectly protect the parasitoid egg from encapsulation in the host of the parasitoid, *Heliothis virescens*. While virus replication does not appear to occur in the lepidopteran host of the parasitoid, the fact that virus-specific mRNA transcripts are expressed in parasitised *H. virescens* larvae within two hours of oviposition suggests that the virus plays some active role in this protection. Theilmann, D.A. and Summers, M.D. (1987) J. Virol. **61**, 2589.

canary avipoxvirus. Family *Poxviridae*, genus *Avipoxvirus*. Similar to fowlpox virus.

Canavalia maritima mosaic virus. A possible *Potyvirus*. Francki, R.I.B. *et al*. (1985) **In** Atlas of Plant Viruses. Vol. 2. p. 183. CRC Press: Boca Raton, Florida.

Candiru bunyavirus. Family *Bunyaviridae*, genus *Bunyavirus*, member of Phlebotomus fever group. Isolated from a man with fever in Brazil. Does not multiply in mosquitoes.

canid herpesvirus 1. Synonym: CANINE HERPESVIRUS, CANINE TRACHEO-BRONCHITIS VIRUS. Family *Herpesviridae*, possible member of subfamily *Alphaherpesvirinae*. Causes rhinitis and pneumonia in new-born puppies. Kennel cough in older animals. Grows in dog kidney cell cultures with cpe.

canine adeno-associated parvovirus. Family *Parvoviridae*, genus *Adeno-associated virus (AAV)* or *Dependovirus*.

canine adenovirus. *See* RUBARTH'S DISEASE VIRUS.

canine coronavirus. Synonym: GASTRO-ENTERITIS VIRUS OF DOGS. Family *Coronaviridae*, genus *Coronavirus*. Causes vomiting and diarrhoea. Related antigenically to porcine transmissible gastro-enteritis virus, which itself will infect dogs.

canine dermal papilloma virus. Synonym: CANINE PAPILLOMA VIRUS. Family *Papovaviridae*, genus *Papillomavirus*. Causes skin papillomas.

canine distemper virus. Family *Paramyxoviridae*, genus *Morbillivirus*. Infects dogs, foxes, wolves, raccoons and mink. Causes fever, nasal and ocular discharge, vomiting and diarrhoea in dogs. There is also CNS involvement, causing fits. The virus is closely related antigenically to measles and rinderpest. It can be grown in dog, ferret and monkey cell cultures and in eggs.

canine hepatitis virus. Synonym: KENNEL COUGH VIRUS, RUBARTH'S DISEASE VIRUS. Family *Adenoviridae*, genus *Mastadenovirus*. Causes fever, diarrhoea and vomiting and frequently death in dogs. In foxes, there is also acute encephalitis. Experimentally coyotes, wolves and raccoons can also be infected. The virus can be grown in dog, ferret, raccoon and pig cells.

canine herpesvirus. *See* CANID HERPESVIRUS.

canine papilloma virus. *See* CANINE DERMAL PAPILLOMA VIRUS.

canine parvovirus 1. Synonym: MINUTE VIRUS OF CANINES. Family *Parvoviridae*, genus *Parvovirus*. The virus was isolated from faeces but it is probably not pathogenic although antibodies are found in most dogs. Will grow in a canine cell line from the Walter Reed Institute.

canine parvovirus 2. Family *Parvoviridae*, genus *Parvovirus*. This virus emerged as a new infectious agent in 1978 when it caused a worldwide panzootic. It has since become enzootic in dogs throughout the world. The principal signs of disease are enteritis and myocarditis frequently followed by death.

canine tracheo-bronchitis virus. *See* CANID HERPESVIRUS.

canna yellow mottle virus. A member of the *Cacao Swollen Shoot Virus* group.

Francki, R.I.B. *et al.* (1985) **In** Atlas of Plant Viruses. Vol. 2. p. 235. CRC Press: Boca Raton, Florida.

cap. A sequence of methylated bases at the 5′ terminus of a eukaryotic mRNA molecule joined in the opposite orientation, i.e. 5′ to 5′ instead of 5′ to 3′. The sequence is $m^7G^{5'}ppp^5G$(or A)p...; in animal but not plant mRNAs the second G or A is often methylated. The cap interacts with various proteins involved with the initiation of translation.

Cape Wrath virus. Family *Reoviridae*, genus *Orbivirus*. Isolated from a female tick at Cape Wrath.

caper vein banding virus. A *Carlavirus*. Francki, R.I.B. *et al.* (1985) **In** Atlas of Plant Viruses. Vol. 2. p. 173. CRC Press: Boca Raton, Florida.

caper vein yellowing virus. A plant *Rhabdovirus*.
Franco, A.D. and Gallitelli, D. (1985) Phytop. Medit. **24**, 234.

Capillovirus group. (Latin 'capillus' = a hair). (Type member APPLE STEM GROOVING VIRUS). Genus of plant viruses comprising some of the viruses which were originally classified in the *CLOSTEROVIRUS*, subgroup 2. The particles are flexuous and filamentous, about 650 x 12 nm., and sediment at about 100S. They are composed

100nm

of a helix of a single polypeptide species (mw. *c.*27 x 10³), which gives an obvious cross-banding, and a single species of linear ssRNA (mw. 2.7 x 10⁶). The host ranges of members of this group are narrow. The viruses can be transmitted mechanically and are naturally transmitted through seed.

Capim virus. Family *Bunyaviridae*, genus *Bunyavirus*. Isolated from opossums and mosquitoes in S. America.

capping. Addition of the methylated base to the primary transcript in the nucleus while it is being SPLICED and POLYADENYLATED. *See* CAP.

caprine herpesvirus 1. *See* BOVID HERPESVIRUS 5.

caprinized strain of virus. Virus adapted to growth in goats.

Capripoxvirus. (Latin 'caper, capri' = goat.) A genus in the family *Poxviridae*, subfamily *Chordopoxvirinae*, consisting of viruses infecting ungulates. It contains GOATPOX, SHEEPPOX and LUMPY SKIN DISEASE VIRUSES. Causes papules, vesicles and pustules before forming scabs, usually on the udder, teats, scrotum and thighs. Widespread occurrence. Virus can be grown in lamb and goat kidney cells and in CAM, causing pocks.

capsid. Protein shell which surrounds the virus nucleic acid. The capsid has ICOSAHEDRAL, HELICAL or complex symmetry and may be enveloped (enclosed in a membrane) or non-enveloped. The capsid and nucleic acid form the NUCLEOCAPSID.

capsid polypeptide. Protein forming part of the capsid structure of a virus particle.

capsomere. Unit from which the virus capsid is built. The capsomeric unit is visible in the electron microscope and consists of groups of identical protein molecules. In capsids with icosahedral symmetry, the capsomeres at the twelve apices are termed PENTAMERS (because they have five neighbouring capsomeres). The other capsomeres, which have six neighbours, are called HEXAMERS.

capsule. The proteinaceous OCCLUSION BODY produced during GRANULOSIS virus infections (subgroup B, *BACULOVIRUS*), in which the virion is surrounded by a crystalline protein matrix (granulin).

Caraparu virus. Family *Bunyaviridae*, genus *Bunyavirus*. Transmitted by mosquitoes. Causes fever in man.

carbon dioxide-sensitivity virus. *See* SIGMAVIRUS.

carboxymethylcellulose. A cellulose derivative which is used for the separation of proteins by ion exchange chromatography.

Carcinus viruses. Unclassified viruses isolated from the crabs, *Carcinus maenas* and *C. mediterraneus,* include REOVIRUS- and RHABDOVIRUS-like particles as well as particles resembling NON-OCCLUDED BACULOVIRUSES.
Bergoin, M. *et al.* (1982) **In** Invertebrate Pathol-

ogy and Microbial Control. p. 523. Proc. IIIrd Internat. Colloq. Invertebr. Pathol.

Cardiovirus. (Greek 'kardia' = heart.) A genus in the family *Picornaviridae*. Consists of ENCEPHALOMYOCARDITIS VIRUS, MENGO VIRUS and MURINE ENCEPHALOMYELITIS VIRUS. Characterised by instability at pH6 in the presence of halide ions and of a long polycytidylic acid tract near the 5′ end of its RNA genome.

Carey Island virus. Family *Flaviviridae* genus *Flavivirus*. Isolated from bats in Malaysia.

Carlavirus group. (Sigla from CARNATION LATENT, the type virus). Genus of plant viruses with slightly flexuous rod-shaped particles, 600-700 nm. long and *c*.13 nm. wide which sediment at *c*.160S and band in CsCl at *c*.1.3 g/cc. The coat

100nm

protein subunits (mw. 32×10^3) are arranged in the particles in helical symmetry with pitch *c*.3.4 nm. Each particle contains one molecule of linear, (+)-sense ssRNA (mw. 2.7×10^6). Replication thought to be in the cytoplasm. Host ranges of individual viruses are rather narrow. Particles are found in most cell types. The viruses can be mechanically transmitted. Most members are transmitted by aphids in a NON-PERSISTENT TRANSMISSION manner.
Matthews, R.E.F. (1982) Intervirology **12**, 149.
Koenig. R. (1982) CMI/AAB Descriptions of Plant Viruses No. 259.
Francki, R.I.B. *et al.* (1985) **In** Atlas of Plant Viruses. Vol. 2. p. 173. CRC Press.: Boca Raton, Florida.

Carmovirus group. (Sigla of CARNATION MOTTLE VIRUS, the type member). Genus of plant viruses with small isometric particles, 30 nm. in diameter, which sediment at about 120S and band in CsCl at 1.35 g/cc. The capsid structure is thought to be icosahedral (T=3), the structural subunit being a coat protein (mw. 38×10^3). Each particle contains a single species of (+)-sense, linear ssRNA (mw. 1.4×10^6; 4003 nucleotides) which encodes five products. In order from the 5′ end these are: one of mw. 27×10^3, two potential READTHROUGH products of mw. 86 and 98×10^3, a putative product (mw. 7×10^3) and the coat protein; the latter two products are translated from

SUBGENOMIC mRNAS of 1.7 and 1.5kb respectively. Replication is in the cytoplasm and is via REPLICATIVE INTERMEDIATES. The host ranges of members of this group are wide among angiosperms. They are readily transmitted mechanically, natural transmission being associated with the soil.
Morris, T.J. and Carrington, J.C. (1988) **In** The Plant Viruses. Vol. 3, p. 73. ed. R. Koenig. Plenum Press: New York.

carnation bacilliform virus. A possible plant *Rhabdovirus*.
Matthews, R.E.F. (1982) Intervirology **12**, 113.

carnation cryptic virus. A member of the *Cryptoviru s* group, subgroup A.
Lisa, V. *et al.* (1986) CMI/AAB Descriptions of Plant Viruses No. 315.

carnation etched ring virus. A *Caulimovirus*.
Lawson, R.H. *et al.* (1977) CMI/AAB Descriptions of Plant Viruses No. 182.
Francki, R.I.B. *et al.* (1985) **In** Atlas of Plant Viruses. Vol. 1. p. 17. CRC Press: Boca Raton, Florida.

carnation Italian ringspot virus. A *Tombusvirus*.
Francki, R.I.B. *et al.* (1985) **In** Atlas of Plant Viruses. Vol. 1. p. 181. CRC Press: Boca Raton, Florida.
Martelli, G.P. *et al.* (1988) **In** The Plant Viruses. Vol 3, p. 13. ed. R. Koenig. Plenum Press: New York.

carnation latent virus. Type member of the *Carlavirus* group.
Wetter, C. (1971) CMI/AAB Descriptions of Plant Viruses No. 61.
Francki, R.I.B. *et al.* (1985) **In** Atlas of Plant Viruses. Vol. 2. p. 173. CRC Press: Boca Raton, Florida.

carnation mottle virus. Type member of the *Carmovirus* group.
Hollings, M. and Stone, O.M. (1970) CMI/AAB Descriptions of Plant Viruses No. 7.
Morris, T.J. and Carrington, J.C. (1988) **In** The Plant Viruses. Vol 3, p. 73. ed. R. Koenig. Plenum Press: New York.

carnation mottle virus group. Synonym: CARMOVIRUS GROUP.

carnation necrotic fleck virus. A member of the *Closterovirus* subgroup 1.
Inouye, T. (1974) CMI/AAB Descriptions of Plant Viruses No. 136.
Francki, R.I.B. *et al.* (1985) **In** Atlas of Plant Viruses. Vol. 2. p. 219. CRC Press: Boca Raton, Florida.

carnation ringspot virus. Type member of the *Dianthovirus* group.
Tremaine, J.H. and Dodds, J.A. (1985) AAB Descriptions of Plant Viruses No. 308.

carnation vein mottle virus. A *Potyvirus*.
Hollings, M. and Stone, O.M. (1971) CMI/AAB Descriptions of Plant Viruses No. 78.
Francki, R.I.B. *et al.* (1985) **In** Atlas of Plant Viruses. Vol. 2. p. 183. CRC Press: Boca Raton, Florida.

carnation yellow fleck virus. A member of the *Closterovirus* subgroup 1.
Francki, R.I.B. *et al.* (1985) **In** Atlas of Plant Viruses. Vol. 2. p. 219. CRC Press: Boca Raton, Florida.

carnation yellow stripe virus. A *Necrovirus*.
Gallitelli, D. *et al.* (1979) Phytopath. medit. **18**, 31.

carnivore pox virus. Family *Poxviridae,* not allocated to genus. A strain of cowpox virus differing in growth characteristics from the reference strains.

carrier culture. A cell culture which is persistently infected with a virus.

carrier state. The condition in which an animal is persistently infected with a virus, often without showing the signs of the disease associated with the virus. The virus can persist even in the presence of the specific neutralising antibody.

carrot latent virus. A plant *Rhabdovirus,* subgroup 2; aphid transmitted.
Francki, R.I.B. *et al.* (1985) **In** Atlas of Plant Viruses. Vol. 1. p. 73. CRC Press: Boca Raton, Florida.

carrot mosaic virus. A possible *Potyvirus*.
Francki, R.I.B. *et al.* (1985) **In** Atlas of Plant Viruses. Vol. 2. p. 183. CRC Press: Boca Raton, Florida.

carrot mottle virus. Unencapsidated RNA, dependent on CARROT REDLEAF VIRUS for aphid transmission.
Murant, A.F. (1974) CMI/AAB Descriptions of Plant Viruses No. 137.
Murant, A.F. *et al.* (1985) J. gen. Virol. **66**, 1575 and 2078.

carrot redleaf virus. A *Luteovirus*; considered to be a strain of BEET WESTERN YELLOWS VIRUS.
Waterhouse, P.M. and Murant, A.F. (1982) CMI/AAB Descriptions of Plant Viruses No. 249.
Francki, R.I.B. *et al.* (1985) **In** Atlas of Plant Viruses. Vol. 1. p. 137. CRC Press: Boca Raton, Florida.
Casper, R. (1988) **In** The Plant Viruses. Vol. 3. p. 235. ed. R. Koenig. Plenum Press: New York.

carrot temperate virus. A possible member of the *Cryptovirus group*, subgroup A.
Boccardo, G. *et al.* (1987) Adv. Virus Res. **32**, 171.

carrot thin leaf virus. A *Potyvirus*.
Howell, W.E. and Mink, G.I. (1980) CMI/AAB Descriptions of Plant Viruses No. 218.
Francki, R.I.B. *et al.* (1985) **In** Atlas of Plant Viruses. Vol. 2. p. 183. CRC Press: Boca Raton, Florida.

carrot yellow leaf virus. A member of the *Closterovirus* subgroup 1.
Francki, R.I.B. *et al.* (1985) **In** Atlas of Plant Viruses. Vol. 2. p. 219. CRC Press: Boca Raton, Florida.

cassava (African) mosaic virus. Synonym: CASSAVA LATENT VIRUS. Type member of the *Geminivirus*, subgroup B. The first geminivirus to be sequenced and shown to have a bipartite genome of 2779 and 2724 nucleotides.
Bock, K.R. and Harrison, B.D. (1985) AAB Descriptions of Plant Viruses No. 297.
Harrison, B.D. (1985) Ann. Rev. Phytopath. **23**, 55.

cassava common mosaic virus. A *Potexvirus*.
Costa, A.S. and Kitajima, E.W. (1972) CMI/AAB Descriptions of Plant Viruses No. 90.
Francki, R.I.B. *et al.* (1985) **In** Atlas of Plant Viruses. Vol. 2. p. 159. CRC Press: Boca Raton, Florida.

cassava green mottle virus. A *Nepovirus*.

100nm

Lennon, A.M. *et al.* (1987) Ann. Appl. Biol. **110**, 545.

cassava latent virus. *See* CASSAVA (AFRICAN) MOSAIC VIRUS.

cassava vein banding virus. A possible *Caulimovirus*.
Francki, R.I.B. *et al.* (1985) **In** Atlas of Plant Viruses. Vol. 1. p. 17. CRC Press: Boca Raton, Florida.

cassava virus X. A *Potexvirus*.
Harrison, B.D. (1986) Phytopath. **76**, 1075.

Cassia mild mosaic virus. A possible *Carlavirus*.
Francki, R.I.B. *et al.* (1985) **In** Atlas of Plant Viruses. Vol. 2. p. 173. CRC Press: Boca Raton, Florida.

Cassia yellow blotch virus. A *Bromovirus*.
Francki, R.I.B. (1985) **In** The Plant Viruses. Vol. 1. p. 1. ed. R.I.B. Francki. Plenum Press: New York.

cationic detergent. Detergent having positively-charged surface ions such as molecules containing a quaternary ammonium ion with a group of 12-24 carbon atoms attached to the nitrogen atom in the cation (e.g. cetyl-trimethyl ammonium bromide (CTAB)).

cattle plague virus. *See* RINDERPEST VIRUS.

Catu virus. Family *Bunyaviridae*, genus *Bunyavirus*. Isolated from man, monkeys, mice and mosquitoes in S. America.

cauliflower mosaic virus. Type member of the *Caulimovirus* group. The DNAs of three isolates have been sequenced (Cabb-S 8024, CM1841 8031 and D/H 8016 nucleotides). First plant virus shown to involve reverse transcription in its replication.
Shepherd, R.J. (1981) CMI/AAB Descriptions of Plant Viruses No. 243.
Francki, R.I.B. *et al.* (1985) **In** Atlas of Plant Viruses. Vol. 1. p. 17. CRC Press: Boca Raton, Florida.

Caulimovirus group. (Sigla from CAULIFLOWER MOSAIC, the type virus). Genus of the only plant viruses to contain dsDNA. The particles are found mainly in cytoplasmic proteinaceous INCLUSION BODIES, characteristic of members of this group, and are isometric, *c*.50 nm. in diameter, sediment at *c*.210S and band in CsCl at 1.37 g/cc. They are very stable and have no envelope, surface projections or core. The capsid comprises a single coat protein species (mw. 42 x 10^3) but degrades to several polypeptides with a major component (mw. 37 x 10^3) which is PHOSPHORYLATED and possibly GLYCOSYLATED. Each capsid contains a single molecule of relaxed circular dsDNA (mw. = 5 x 10^6; 8kbp) which has single-strand discontinuities at specific sites, one in the transcribed strand and one, two or three in the non-transcribed strand. The DNA of the type member codes for six proteins. Two major RNA species are transcribed in the nucleus from a mini-chromosome form of the virion DNA. One (1.9kb) is the mRNA for the major protein of the cytoplasmic inclusion bodies; the other (8.2kb) is full-length with a terminal redundancy of 180 nucleotides. Replication is by REVERSE TRANSCRIPTION and resembles in many respects that of RETROVIRIDAE; this stage occurs in the cytoplasmic inclusion bodies. The 8.2kb RNA is the template, is primed by a specific tRNA and is reverse-transcribed most probably by a virus-coded enzyme; second-strand priming is at purine-rich regions. The discontinuities in the virion DNA reflect the priming sites. The virus DNA does not specifically integrate into the host genome.

The host range for individual members is narrow. Particles of caulimoviruses are found in most cell types. Many members are transmissible mechanically. Most are transmitted by aphids in a SEMIPERSISTENT TRANSMISSION manner: aphid transmission requires a virus-coded transmission factor.
Matthews, R.E.F. (1982) Intervirology, **17**, 64.
Hull, R. (1984) CMI/AAB Descriptions of Plant Viruses No. 295.
Francki, R.I.B. *et al.* (1985) **In** Atlas of Plant Viruses. Vol. 1. p. 17. CRC Press: Boca Raton, Florida.

caviid herpesviruses. *See* GUINEA-PIG CYTOMEGALOVIRUS.

cDNA. Abbreviation for complementary DNA.

cebid herpesvirus. Species in the family Herpesviridae, isolated from marmosets and monkeys.

celery latent virus. Unclassified, rod-shaped particles, 860 nm. in length.
Luisoni, E. (1966) Atti Accad. Sci. Torino **100**, 541.

celery mosaic virus. A *Potyvirus*.
Shepherd, R.J. and Grogan, R.G. (1971) CMI/AAB Descriptions of Plant Viruses No. 50.
Francki, R.I.B. *et al.* (1985) **In** Atlas of Plant Viruses. Vol. 2. p. 183. CRC Press: Boca Raton, Florida.

celery yellow mosaic virus. A possible *Potyvirus*.
Francki, R.I.B. *et al.* (1985) **In** Atlas of Plant Viruses. Vol. 2. p. 183. CRC Press: Boca Raton, Florida.

celery yellow spot virus. A possible *Luteovirus*.
Francki, R.I.B. *et al.* (1985) **In** Atlas of Plant Viruses. Vol. 1. p. 137. CRC Press: Boca Raton, Florida.

cell culture A culture of cells in a liquid or soft gel medium *in vitro*. See ORGAN CULTURE, TISSUE CULTURE.

cell cycle. The cell cycle consists of metaphase (M phase) and interphase (S phase); DNA synthesis occurs during interphase. Between M and S phases is G_1 phase and between S and M is G_2 phase. Somatic cells are in G_0 phase. Most viruses can multiply in cells independently of cell division but there are exceptions, e.g. PARVOVIRUSES and BACULOVIRUSES require rapidly dividing cells.

cell fusion. The formation by fusion of cell membranes of multinucleate giant cells (syncytia). Virus-induced fusion from outside the cell can be caused by exposure to high multiplicity of large enveloped RNA viruses (e.g. SENDAI VIRUS) or some DNA viruses, even when they have been inactivated. Fusion from within the cell depends on the synthesis of virus macromolecules (but not virus particles themselves).

cell line. A culture of cells which can be subcultured indefinitely. It is derived from a primary culture at the time of first subculture. The term implies that such cultures comprise many lineages of cells present in the primary culture.

cell-mediated immunity. Immunity mediated by a variety of cells unrelated to the production of antibody.

cellular immunity. Immunity ascribed to various cellular functions other than those which produce antibody.

CELO virus. Acronym from CHICKEN EMBRYO LETHAL ORPHAN VIRUS. Family *Adenoviridae*, genus *Aviadenovirus*. Associated with several diseases in chickens, turkeys and pheasants. Replicates in chicken cell cultures. Oncogenic in newborn hamsters. Produces cpe in chicken cell cultures.

central dogma. The concept that genetic information is perpetuated as DNA and is transferred to protein via RNA. Thus it cannot be retrieved from the amino acid sequence of a protein. The discovery of reverse transcriptase by H. Temin and the synthesis of DNA from an RNA template is contrary to this concept of a unidirectional flow of information of DNA to RNA to protein.
Crick, F.H.C. (1970) Nature **227**, 561.

centrifuge. A machine for separating particles by centrifugal force. Usually referred to as low-speed centrifuges (capable of up to 20,000 rev/min.) and high-speed ultracentrifuges (capable of more than 50,000 rev/min.). See ANALYTICAL ULTRACENTRIFUGE.

Centrosema mosaic virus. A possible *Potexvirus*.
Francki, R.I.B. *et al.* (1985) **In** Atlas of Plant Viruses. Vol. 2. p. 159. CRC Press: Boca Raton, Florida.

Ceratitis picornavirus V. Unclassified small RNA virus from the Mediterranean fruit-fly, *C. capitata*. Virions are isometric, 24 nm. in diameter, with a buoyant density of 1.35 g/cc. in CsCl. They contain three structural polypeptides (mw. 27, 34 and 39 x 10^3) and an RNA genome. A possible INSECT PICORNAVIRUS.
Plus, N. and Cavalloro, R. (1983) **In** Fruit Flies of Economic Importance. p. 106. ed. R. Cavalloro. A.A. Balkema: Rotterdam.

Ceratitis reovirus I. Unclassified RNA virus (possible member of REOVIRIDAE) isolated from the Mediterranean fruit-fly, *C. capitata*. Virions are non-enveloped, isometric, 60 nm. in diameter with a double protein shell. Particles have a buoyant density in CsCl of 1.38 g/cc., contain RNA transcriptase activity and a genome of ten segments of dsRNA. The virus has many similarities to *Drosophila* F virus, but is serologically distinct and does not replicate in *Drosophila*. Although particle morphology resembles virions of the REOVIRUS genus, RNA segment and polypeptide mw. are distinct.

Plus, N. and Cavalloro, R. (1983) **In** Fruit Flies of Economic Importance. p. 106. ed. R. Cavalloro. A.A. Balkema: Rotterdam.

Cercopithecid herpesvirus 1. Synonym: HERPES B VIRUS, HERPES SIMIAE VIRUS. Family *Herpesviridae*, subfamily *Alphaherpesvirinae*, genus *suid herpesvirus 1 group*. A natural infection of Asiatic monkey; many Rhesus monkeys have antibodies. Infects man, causing vesicular lesions of the tongue and lips, acute encephalitis and, almost always, death. Young mice, chicks and guinea pigs can be infected experimentally.

Cercopithecid herpesvirus 2. Synonym: AFRICAN MONKEY CYTOMEGALOVIRUS, SA6 VIRUS. Family *Herpesviridae*, subfamily *Betaherpesvirinae*. Isolated from vervet monkey kidney and salivary gland cultures. Produces giant cells and eosinophilic inclusion bodies.

Cercopithecid herpesvirus 3. Synonym: SA8 VIRUS. Family *Herpesviridae*, subfamily *Alphaherpesvirinae*. Virus was isolated from vervet monkey kidneys and from a baboon. When injected into baby baboons i.c. it causes pneumonia. Serologically related to CERCOPITHECID HERPESVIRUS 1.

cereal chlorotic mottle virus. A plant *Rhabdovirus* subgroup 2; leafhopper transmitted. Greber, R.S. (1982) CMI/AAB Descriptions of Plant Viruses No. 251. Francki, R.I.B. *et al.* (1985) **In** Atlas of Plant Viruses. Vol. 1. p. 73. CRC Press: Boca Raton, Florida.

cereal striate virus. *See* BARLEY YELLOW STRIATE MOSAIC VIRUS.

cereal tillering disease virus. A *Fijivirus*. Francki, R.I.B. *et al.* (1985) **In** Atlas of Plant Viruses. Vol. 1. p. 47. CRC Press: Boca Raton, Florida.

CFT. Abbreviation for COMPLEMENT FIXATION TEST.

CFU. Abbreviation for COLONY FORMING UNITS.

Chaco virus. Family *Rhabdoviridae*; genus not established. Isolated from lizards in S. America. May be arthropod transmitted. Kills new-born mice and replicates in Vero cells.

Chagres virus. Family *Bunyaviridae*; genus *Phlebovirus*. Isolated from a man with fever in Panama and from mosquitoes. Causes cpe in primary Rhesus monkey, human amnion and mouse embryo cell cultures. Kills new-born mice when inoculated i.c.

chamois contagious ecthyma parapoxvirus. *See* CONTAGIOUS ECTHYMA VIRUS.

chamois papilloma virus. *See* CONTAGIOUS ECTHYMA VIRUS.

Chandipura virus. Family *Rhabdoviridae*, genus *Vesiculovirus*. Isolated from man with influenza-like illness in India and from sandflies and a hedgehog in Nigeria. Possible distant relationship with vesicular stomatitis virus. Grows well in BHK cells and kills young mice when inoculated i.c.

Changuinola virus. Family *Reoviridae*, genus *Orbivirus*. Isolated from *Phlebotomus* sp. and small rodents in Panama and from a man with fever.

Chaoborus 'baculovirus'. Unclassified virus, morphologically similar to a NON-OCCLUDED BACULOVIRUS, observed in the phantom midge *Chaoborus crystallinus* in Sweden. Particles observed in section contained a nucleocapsid (38-43 x 210-226 nm.) surrounded by a unit membrane. Infection was restricted to the mid gut epithelium. Similar particles have been observed in *C. astictopus*. Larsson, R. (1984) J. Invertebr. Pathol. **44**, 178.

Chara corallina virus. A possible member of the TOBAMOVIRUS GROUP. A virus of a green alga, *Chara corallina*. It has rigid rod-shaped particles, 530 nm. long, 18 nm. wide, which sediment at 230S. The coat protein subunits (mw. 17.5×10^3) are arranged in the particles in helical symmetry with pitch of 2.75 nm. Each particle contains one species of ssRNA (mw. 3.6×10^6). The virus is transmissible to *C. corallina* by injection, leading to chlorosis and death in ten days. Skotnicki, A. *et al.* (1976) Virology **75**, 457.

Charleville virus. Unclassified arthropod-borne

virus. Isolated from sandflies and a lizard in Charleville, Queensland, Australia.

chelating agent. (From Greek 'chelos' = crab's claw). Compounds which bind divalent cations, so inhibiting their activity. Can be used to inhibit biological interactions which require divalent cations, e.g. deoxyribonuclease, attachment of some viruses (e.g. foot-and-mouth disease virus) to cells. *See* ETHYLENEDIAMINETETRA-ACETIC ACID, ETHYLENEGLYCOLBIS(AMINOETHYLETHER)TETRA-ACETIC ACID.

Chelonid herpesvirus 1. Synonym: GREEN SEA-TURTLE HERPESVIRUS, GREY PATCH DISEASE OF TURTLES VIRUS. Family *Herpesviridae*, subfamily *Chordopoxvirinae*, not allocated to genus. Isolated from green sea-turtles kept in captivity in West Indies. Intranuclear inclusions and herpes-virus-like particles present in the lesions. Disease can be transmitted by a cell-free extract.

chemotherapeutic index. An index for assessing the potential of an antiviral compound. It is the ratio between the lowest effective antiviral concentration and the highest non-toxic concentration of that compound.

chemotherapy. The use of compounds which will inhibit the growth of infectious agents without unduly affecting host-cell metabolism e.g. ACYCLOVIR for herpesvirus infections, AZT for AIDS.

Chenuda virus. Family *Retroviridae*, subfamily *Oncovirinae*, genus *type C oncovirus* group, sub-genus avian type C oncovirus.

cherry leaf roll virus. A *Nepovirus*.
Jones, A.T. (1985) AAB Descriptions of Plant Viruses No. 306.
Francki, R.I.B. *et al.* (1985) **In** Atlas of Plant Viruses. Vol. 2. p. 23. CRC Press: Boca Raton, Florida.

cherry rasp leaf virus. A possible *Nepovirus*.
Stace-Smith, R. and Hansen, A.J. (1976) CMI/AAB Descriptions of Plant Viruses No. 159.
Francki, R.I.B. *et al.* (1985) **In** Atlas of Plant Viruses. Vol. 2. p. 23. CRC Press: Boca Raton, Florida.

cherry rugose mosaic virus. A strain of *Prunus Necrotic Ringspot Virus*.

Francki, R.I.B. (1985) **In** The Plant Viruses. Vol. 1. p. 1. ed. R.I.B. Francki. Plenum Press: New York.

chick embryo fibroblast (CEF) cells. A primary cell culture system prepared from chick embryos. Used for culture of many viruses e.g. RABIES VIRUS.

chickenpox virus. *See* VARICELLA ZOSTER VIRUS.

chickpea stunt virus. Synonym: BEAN LEAF ROLL VIRUS (PEA LEAF ROLL VIRUS).

chicory blotch virus. A possible *Carlavirus*.
Francki, R.I.B. *et al.* (1985) **In** Atlas of Plant Viruses. Vol. 2. p. 173. CRC Press: Boca Raton, Florida.

chicory virus X. A *Potexvirus*.
Gallitelli, D. and Franco, A.D. (1982) Phytopath. Z. **105**, 120.

chicory yellow mottle virus. A *Nepovirus*.
Quacquarelli, A. *et al.* (1974) CMI/AAB Descriptions of Plant Viruses No. 132.
Francki, R.I.B. *et al.* (1985) **In** Atlas of Plant Viruses. Vol. 2. p. 23. CRC Press: Boca Raton, Florida.

Chikungunya virus. Family *Togaviridae*, genus *Alphavirus*. Causes severe joint and back pains in man, followed several days later by a rash. Transmitted by *Aedes africanus* and *A. aegyptii*. Antibodies present in birds, monkeys and many mammals. Occurs in many parts of Africa and Asia. Kills suckling mice when inoculated i.c. and replicates in chick embryo fibro-blasts, Rhesus kidney and HeLa cells.

Chilibre virus. Family *Bunyaviridae*, genus *Phlebovirus*. Isolated from insects in Panama.

Chilo iridescent virus. Type species of the small iridescent insect virus group (*Iridovirus; Irido-viridae*), isolated from *Chilo suppressalis* (Lepidoptera). Originally referred to as insect irides-cent virus type 6. Virions have the characteristic morphological properties of the IRIDOVIRUS group (*c*.120 nm. in diameter) sediment at 3300S and contain about 26 polypeptides (mw. 10-230 x 10^3). The genome is a linear dsDNA (mw. 158 x 10^6; 209 kbp) which is circularly permuted and terminally redundant. The virus has a wide experimental host range within insects (including

some Lepidoptera, Coleoptera, Hymenoptera, Diptera and Orthoptera) and has been experimentally transmitted to some Crustacea.
Schnitzler, P. *et al.* (1987) Virology **160**, 66.

Chinese hamster ovary (CHO) cells. A primary cell culture system prepared from the ovaries of Chinese hamsters. Used for the culture of many viruses, e.g. the herpesviruses.

Chinese yam necrotic mosaic virus. A *Carlavirus*, occurs in Japan.
Doi, Y. *Personal communication.*

Chinook salmon virus. *See* SACRAMENTO RIVER CHINOOK SALMON VIRUS.

Chironomus entomopoxvirus. Type species of probable sub-genus C, insect poxviruses (ENTOMOPOXVIRINAE) from *C. luridus* (Diptera). Virions are brick-shaped, 330 x 230 x 110 nm., containing two LATERAL BODIES and a biconcave core. The genome is a single linear molecule of dsDNA, approximately 200 x 10⁶. Virions are occluded within proteinaceous occlusion bodies (2-8 μm) during infection. No spindle-shaped inclusion bodies are produced. Viruses with similar properties have been isolated from *C. attenuatus, Camptochironomus, Goeldichironomus* and *Aedes* spp. (Diptera).
Arif, B.M. (1984). Adv. Virus Res. **29**, 195.

chloramphenicol. An antibiotic (mw. 323.1) isolated from *Streptomyces venezuelae*. It acts as an analogue of phenylalanine and blocks the amino acid accepting site on phenylalanine tRNA.

chloramphenicol acetyl transferase (CAT) gene. The CAT gene product catalyses the acetylation of chloramphenicol, disrupting its antibiotic activity. The most commonly used type was obtained from the bacterial transposon, Tn9. The CAT gene is frequently used as a marker in genetic cloning experiments.

Chloriridovirus. (Greek 'chloros' = green.) Genus of the large iridescent virus group (IRIDO-VIRIDAE). Virions are isometric, approximately 180 nm. in diameter, reportedly containing about nine polypeptides (mw. 15-98 x 10³) and a linear dsDNA genome (mw. 240-288 x 10⁶). The type species is the regular (R) strain of MOSQUITO IRIDESCENT VIRUS (Iridescent virus type 3) from *Aedes taeniorhynchus*. Other insect iridescent

virus isolates included in the Chloriridovirus genus have been isolated from *Aedes* spp. (iridescent virus types 4, 5, 11, 12, 14 and 15), *Simulium* sp. (type 7), *Culicoides* sp. (type 8) and *Corethrella* sp. (type 13) (Diptera). A number of possible members have been observed in other Diptera (*see* IRIDESCENT VIRUSES). A virus with similar morphological properties has also been found in *Daphnia* sp. (Crustacea). A list of iridescent virus infections of insects is given in Appendix E.
Hall, D.W. (1985) **In** Viral Insecticides for Biological Control. p.163. ed. K. Maramorosch and K.E. Sherman. Academic Press: New York.

Chloris striate mosaic virus. A *Geminivirus*, subgroup A.
Francki, R.I.B. and Hatta, T. (1980) CMI/AAB Descriptions of Plant Viruses No. 221. Harrison, B.D. (1985) Ann. Rev. Phytopath. **23**, 55.

chlorotic. Yellow symptoms induced by viruses in plant leaves. Usually due to effects on the structure of the chloroplasts, resulting in reduced levels of chlorophyll.

Chondrilla juncea stunting virus. A possible plant *Rhabdovirus*, subgroup 2.
Francki, R.I.B. *et al.* (1985) **In** Atlas of Plant Viruses. Vol. 1. p. 73. CRC Press: Boca Raton, Florida.

Chordopoxvirinae. A subfamily in the family *Poxviridae*. It contains all the viruses infecting vertebrates and is subdivided into six genera: Avipox, Capripox, Leporipox, Orthopox, Parapox and Suipox viruses. The names are derived from Latin 'avis' = bird; Latin 'caper, capris' = goat; Latin 'lepus, lepori' = hare; Greek 'orthos' = straight; Greek 'para' = by the side of; Latin 'sus' = swine.

Choristoneura entomopoxvirus. Insect poxviruses isolated from the budworms *Choristoneura biennis, C. conflictana* and *C. diversana* (Lepidoptera), with characteristics of viruses of the Entomopoxvirus sub-genus B (type species *Amsacta* entomopoxvirus).
Arif, B.M. (1984) Adv. Virus Res. **29**, 195.

Choristoneura fumiferana nuclear polyhedrosis virus. Baculovirus (MNPV) isolated in Canada from the eastern spruce budworm, *C. fumiferana*; also infectious for other budworms including *C. occidentalis*. A number of closely-related genotypic variants have been isolated. Like *Autographa californica* NPV, the viral genome contains several regions of repeated sequences and the overall organisation of the two viral genomes shows similarities (colinearity) and extensive homology. The virus replicates *in vitro* in cell lines derived from the homologous host.

Arif, B.M. *et al.* (1985) Virus Res. **2**, 85.

chromatid. *See* CHROMOSOME.

chromatin. *See* CHROMOSOME.

chromatography. A method of separating and analysing chemical substances by preferential retention, either as a gas or liquid, on to a solid medium e.g. activated charcoal, silica gel, agarose beads with different pore sizes.

chromosome. A self-replicating nucleic acid molecule containing a number of genes. In bacteria the entire genome is contained within one ds circular DNA molecule. In eukaryotes chromosomes are linear dsDNA molecules and most organisms have a number of chromosomes which are characteristic of a particular species. When they are not undergoing mitosis or meiosis they normally interact with HISTONES to form chromatin. During the prophase stage of meiosis the chromatin forms thread-like structures termed chromatids.

chronic bee-paralysis virus. *See* BEE CHRONIC PARALYSIS VIRUS.

chronic bee-paralysis virus associate. *See* BEE CHRONIC PARALYSIS VIRUS ASSOCIATE.

chronic infection. An infection which lingers, often without symptoms, e.g. lymphocytic choriomeningitis in mice.

Chrysanthemum chlorotic mottle viroid. A VIROID.
Horst, R.K. (1987) **In** The Viroids. p. 291. ed. T.O. Diener. Plenum Press: New York.

Chrysanthemum frutescens rhabdovirus. A probable plant *Rhabdovirus.*
Francki, R.I.B. *et al.* (1985) **In** Atlas of Plant Viruses. Vol. 1. p. 73. CRC Press: Boca Raton, Florida.

Chrysanthemum   mild   mottle   virus.   A

Cucumovirus; occurs in Japan.
Doi, Y. *Personal communication.*

Chrysanthemum stunt viroid. A VIROID, 354 nucleotides, depending on isolate.

Lawson, R.H. (1987) **In** The Viroids. p. 247. ed. T.O. Diener. Plenum Press: New York.

Chrysanthemum vein chlorosis virus. A possible plant *Rhabdovirus.*
Matthews, R.E.F. (1982) Intervirology **17**, 113.

Chrysanthemum virus B. A *Carlavirus.*
Hollings, M. and Stone, O.M. (1972) CMI/AAB Descriptions of Plant Viruses No. 110.
Francki, R.I.B. *et al.* (1985) **In** Atlas of Plant Viruses. Vol. 2. p. 173. CRC Press: Boca Raton, Florida.

chymotrypsin. A protease which hydrolyses the peptide bond on the C-terminal side of valine, isoleucine, phenylalanine, tyrosine, methionine and some alanine residues. *See* TLCK.
Ambler, R.P. and Meadway, R.J. (1968) Biochem. J. **108**, 893.

circulative transmission. Synonym: PERSISTENT TRANSMISSION.

***cis-trans* test.** A genetic COMPLEMENTATION TEST first used in the genetic mapping of phage T4. is used to establish whether two mutations are in the same gene where the wild-type allele is dominant. If the mutations are in different genes there will be *trans*-complementation.
Benzer, S. (1955) Proc. Natl. Acad. Sci. USA **41**, 344.

cistron. The basic unit of genetic function. It usually refers to a gene or the coding region for a protein. The term derived from the *CIS-TRANS* TEST.
Benzer, S. (1959) Proc. Natl. Acad. Sci. USA **45**, 1607.

citrus crinkly leaf virus. An *Ilarvirus.*
Francki, R.I.B. (1985) **In** The Plant Viruses. Vol. 1. p. 1. ed. R.I.B. Francki. Plenum Press: New York.

citrus exocortis viroid. A VIROID, 371 nucleotides.
Semancik, J.S. (1980) CMI/AAB Descriptions of Plant Viruses No. 226.
Broadbent, P. and Garnsey, S.M. (1987) **In** The

Viroids. p. 235. ed. T.O. Diener. Plenum Press: New York.

citrus leaf rugose virus. An *Ilarvirus*.
Garnsey, S.M. and Gonsalves, D. (1976) CMI./ AAB Descriptions of Plant Viruses No. 164.
Francki, R.I.B. (1985) In The Plant Viruses. Vol. 2. p. 1. ed. R.I.B. Francki. Plenum Press: New York.

citrus leprosis virus. A possible plant *Rhabdovirus*; unenveloped particles.
Matthews, R.E.F. (1982) Intervirology **17**, 114.

citrus mosaic virus. An unclassified plant virus with isometric particles; occurs in Japan.
Doi, Y. *Personal communication.*

citrus tatter leaf virus. A possible *Closterovirus*.
Francki, R.I.B. *et al.* (1985) In Atlas of Plant Viruses. Vol. 2. p. 219. CRC Press: Boca Raton, Florida.

citrus tristeza virus. A member of the *Closterovirus* subgroup 1.
Price, W.C. (1970) CMI/AAB Descriptions of Plant Viruses No. 33.
Francki, R.I.B. *et al.* (1985) In Atlas of Plant Viruses. Vol. 2. p. 219. CRC Press: Boca Raton, Florida.

citrus variegation virus. An *Ilarvirus*.
Francki, R.I.B. (1985). In The Plant Viruses. Vol. 1. p. 1. ed. R.I.B. Francki. Plenum Press: New York.

clavelee virus. *See* SHEEP CAPRIPOX VIRUS.

cleavage. The cutting of nucleic acid or protein usually enzymatically, at specific sites. *See* RESTRICTION ENDONUCLEASE, PROTEASE.

Cleland's reagent. *See* DITHIOTHREITOL.

Clitoria mosaic virus. A *Potexvirus*.
Srivastava, B.N. *et al.* (1978) Ind. Phytopath. **31**, 248.

Clitoria yellow vein virus. A *Tymovirus*.
Bock, K.R. and Guthrie, E.J. (1977) CMI/AAB Descriptions of Plant Viruses No. 171.
Francki, R.I.B. *et al.* (1985) In Atlas of Plant Viruses. Vol. 1. p. 117. CRC Press: Boca Raton, Florida.

Hirth, L. and Girard, L. (1988) **In** The Plant Viruses. Vol. 3. p. 163. ed. R. Koenig. Plenum Press: New York.

Clo Mor virus. Family *Bunyaviridae*, genus *Nairovirus*.

clone. This term is used in a number of senses. As a noun it may mean a) a population of recombinant DNA molecules all carrying the same inserted sequence; b) a colony of micro-organisms containing a specific DNA fragment inserted into a vector; c) a population of cells or organisms of identical genotype. As a verb it means a) the use of *in vitro* recombination techniques to insert a particular DNA sequence into a vector; b) the selection of a unique virus isolate from individual · PLAQUES, POCKS or LESIONS or by limiting dilution.

Closterovirus group. (Greek 'kloster' = thread). Genus of plant viruses with very flexuous rod-shaped particles 600-2,000 nm. long, 12 nm. wide, which sediment at 96-130S and band in CsCl at 1.30-1.34 g/cc. The coat protein subunits

100nm

(mw. 23-27 x 10^3) are arranged in a helix with pitch of 3.4-3.7 nm. Each particle contains one molecule of linear, (+)-sense ssRNA (mw. 2.2-4.7 x 10^6). Replication is probably cytoplasmic. Individual viruses have a moderately wide host range. Virus particles are found mainly in the phloem cells and often form large crystalline arrays. These viruses are transmissible mechanically with difficulty. Some members are transmitted by aphids in a SEMIPERSISTENT TRANSMISSION manner. It has been suggested that closteroviruses be divided into two subgroups. Subgroup 1 (A) (type member BEET YELLOWS VIRUS) are typical closteroviruses with coat protein of mw. 23-25 x 10^3 and are mostly aphid-transmitted. Subgroup 2 (B) (type member APPLE CHLOROTIC LEAFSPOT VIRUS) viruses have a coat protein, mw. 27 x 10^3 and no known vectors; some members of this subgroup are now placed in the CAPILLOVIRUS group. The cytopathology of the two subgroups differs.
Matthews, R.E.F. (1982) Intervirology, **17**, 147.
Bar-Joseph, M. and Murant, A.F. (1982) CMI/

AAB Descriptions of Plant Viruses No. 260. Francki, R.I.B. *et al*. (1985) **In** Atlas of Plant Viruses. Vol. 2, p. 219. CRC Press.: Boca Raton, Florida.

cloudy wing particle. An unclassified small RNA virus from adult honey bees. Infection is characterised by a marked loss of transparency of the wings. Virions are spherical, 17 nm. in diameter and sediment at 49S. The particles have a buoyant density in CsCl of 1.38 g/cc. and contain a single polypeptide (mw. 19 x 10^3). The genome is ssRNA (mw. 0.45 x 10^6). The small size of the virus suggests that it could be a satellite virus but current evidence indicates that it replicates independently.
Bailey, L. *et al*. (1980). J. gen. Virol. **46**, 149.

clover (Croatian) virus. A possible *Potyvirus*. Francki, R.I.B. *et al*. (1985) **In** Atlas of Plant Viruses. Vol. 2. p. 183. CRC Press: Boca Raton, Florida.

clover enation virus. A possible plant *Rhabdovirus*, subgroup 2.
Francki, R.I.B. *et al*. (1985) **In** Atlas of Plant Viruses. Vol. 1. p. 73. CRC Press: Boca Raton, Florida.

clover mild mosaic virus. An unclassified plant virus with isometric particles 27-28 nm. in diameter which sediment at 121S. The virus is transmitted by aphids in the SEMIPERSISTENT TRANSMISSION manner.

clover yellow mosaic virus. A *Potexvirus*.
Bos, L. (1973) CMI/AAB Descriptions of Plant Viruses No. 111.
Francki, R.I.B. *et al*. (1985) **In** Atlas of Plant Viruses. Vol. 2. p. 159. CRC Press: Boca Raton, Florida.

clover yellow vein virus. A *Potyvirus*.
Hollings, M. and Stone, O.M. (1974) CMI/AAB Descriptions of Plant Viruses No. 131.
Francki, R.I.B. *et al*. (1985) **In** Atlas of Plant Viruses. Vol. 2. p. 183. CRC Press: Boca Raton, Florida.

clover yellows virus. A member of the *Closterovirus* subgroup 1.
Francki, R.I.B. *et al*. (1985) **In** Atlas of Plant Viruses. Vol. 2. p. 219. CRC Press: Boca Raton, Florida.

CMI/AAB. Abbreviation for Commonwealth Mycological Institute/Association for Applied Biologists who publish Plant Virus Descriptions.

CMP. Abbreviation for cytidine 5'-monophosphate.

CO_2-sensitivity virus. *See* SIGMAVIRUS.

Co Ar 1071 virus. Family *Bunyaviridae*, genus *Bunyavirus*.

Co Ar 3624 virus. Family *Bunyaviridae*, genus *Bunyavirus*.

coat (protein). The protective layer(s) which surround(s) the viral nucleic acid. In simple viruses, e.g. PICORNAVIRUSES, it is the capsid but in other viruses, e.g. ORTHOMYXOVIRUSES, it may consist of several layers of protein and lipids.

cocksfoot mild mosaic virus. Synonym: BROME STEM LEAF MOTTLE VIRUS. A possible *Sobemovirus*.
Huth, W. and Paul, H.L. (1972) CMI/AAB Descriptions of Plant Viruses No. 107.
Hull, R. (1988) **In** The Plant Viruses. Vol. 3. p. 113. ed. R. Koenig. Plenum Press: New York.

cocksfoot mottle virus. A possible *Sobemovirus*.
Catherall, P.L. (1970) CMI/AAB Descriptions of Plant Viruses No. 23.
Francki, R.I.B. *et al*. (1985) **In** Atlas of Plant Viruses. Vol. 1. p.153. CRC Press: Boca Raton, Florida.
Hull, R. (1988) **In** The Plant Viruses. Vol. 3. p. 113. ed. R. Koenig. Plenum Press: New York.

cocksfoot streak virus. A *Potyvirus*.
Catherall, P.L. (1971) CMI/AAB Descriptions of Plant Viruses No. 59.
Francki, R.I.B. *et al*. (1985) **In** Atlas of Plant Viruses. Vol. 2. p. 183. CRC Press: Boca Raton, Florida.

coconut cadang-cadang viroid. A VIROID with at least four molecular forms (246, 287, 492 and 574 nucleotides) which differ in size but not sequence complexity. Lethal to coconut palms. Found in the Philippines.
Randles, J.W. and Imperial, J.S. (1984) CMI/AAB Descriptions of Plant Viruses No. 287.
Randles, J.W. (1987) **In** The Viroids. p. 265. ed. T.O Diener. Plenum Press: New York.

coding capacity. The amount of protein that a

given DNA or RNA sequence can in theory encode. As a rough estimate a dsDNA of mw. 1 x 10⁶ can code for a protein of mw. 60-70 x 10³.

coding sequence. That portion of a nucleic acid that directly specifies the amino acid sequence of a protein product. Non-coding sequences include control regions such as PROMOTERS, and POLY-ADENYLATION SIGNALS.

codling moth granulosis virus. *See CYDIA PO-MONELLA* GRANULOSIS VIRUS.

codon. The set of three bases in a nucleic acid (listed under GENETIC CODE) which specifies an amino acid. Mitochondrial DNA uses some unusual codons. *See* ANTICODON, START CODON, STOP CODON.

cofactor. Additional (non-protein) components required by an enzyme for action.

coffee ringspot virus. A plant *Rhabdovirus*, subgroup 2; transmitted by mites.
Francki, R.I.B. *et al.* (1985) **In** Atlas of Plant Viruses. Vol. 1. p. 73. CRC Press: Boca Raton, Florida.

cohesive ends. Synonym: 'STICKY ENDS'.

coital exanthema virus. Synonym: EQUID HERPESVIRUS 3. Family *Herpesviridae*, probable member of subfamily *Alphaherpesvirinae*. Causes genital infection.

Col An 57389 virus. Family *Bunyaviridae*, genus *Bunyavirus*.

cold-adapted mutants. Mutants which can replicate at temperatures below the optimum for the wild type, e.g. INFLUENZA VIRUS can be adapted to replicate in eggs at 25˚C. This mutant has reduced replication efficiency at 37°C and hence may be suitable as a live vaccine.

cole latent virus. A possible *Carlavirus*. Francki, R.I.B. *et al.* (1985) **In** Atlas of Plant Viruses. Vol. 2. p. 173. CRC Press: Boca Raton, Florida.

colicin. A protein toxin produced by coliform bacteria, e.g. *E. coli*, which is toxic to other bacteria. Production of, and immunity to, a colicin is usually encoded by genes on a plasmid

termed a Col factor. These plasmids form the basis of a number of popular cloning vectors, e.g. pBR322.

coliphage. PHAGE isolated from *E. coli*.

coliphage lambda. *See* LAMBDA (λ) PHAGE.

coliphage T2. *See* T2 PHAGE.

coliphage T4. *See* T4 PHAGE.

coliphage T7. *See* T7 PHAGE.

collar. A morphological component in some TAILED PHAGES (e.g. T4 phage) attached to the connector (neck) that links the phage head to the tail. *See* T4 PHAGE.

Colocasia bacilliform virus. A member of the *Cacao Swollen Shoot Virus* group.
Francki, R.I.B. *et al.* (1985) Atlas of Plant Viruses. Vol. 2. p. 235. CRC Press: Boca Raton, Florida.

Colocasia bobone disease virus. A probable plant *Rhabdovirus*, subgroup 2; leafhopper transmitted.
Francki, R.I.B. *et al.* (1985) **In** Atlas of Plant Viruses. Vol. 1. p. 73. CRC Press: Boca Raton, Florida.

Colombian datura virus. A *Potyvirus*.
Francki, R.I.B. *et al.* (1985) **In** Atlas of Plant Viruses. Vol. 2. p. 183. CRC Press: Boca Raton, Florida.

colony. A cluster of cells derived from a single cell by division on a solid medium, e.g. agar.

colony forming units (CFU). The number of colonies formed per unit of volume or weight of a cell suspension

Colorado tick fever virus. Family *Reoviridae*, provisionally genus *Orbivirus*; unlike members of the *Orbivirus* genus, particles contain 12 segments of dsRNA. Causes fever, leucopenia, headache, limb pains and vomiting. Transmitted by bite of infected ticks. Virus also isolated from rodents, in which disease is inapparent. Occurs in parts of the USA. Hamsters can be infected i.p. Virus replicates in eggs and some tissue culture cells.

Columbia SK virus. A strain of ENCEPHALOMYO-CARDITIS VIRUS.

Columnea latent viroid. A VIROID.
Diener, T.O. (1987) **In** The Viroids. p. 297. ed. T.O. Diener. Plenum Press: New York.

Commelina mosaic virus. A *Potyvirus*. Francki, R.I.B. *et al.* (1985) **In** Atlas of Plant Viruses. Vol. 2. p. 183. CRC Press: Boca Raton, Florida.

Commelina virus X. A *Potexvirus*. Francki, R.I.B. *et al.* (1985) **In** Atlas of Plant Viruses. Vol. 2. p. 159. CRC Press: Boca Raton, Florida.

common cold virus. *See* HUMAN RHINOVIRUS.

Commonwealth Mycological Institute/Association of Applied Biologists Descriptions of Plant Viruses. A set of pamphlets, edited by B.D. Harrison and A.F. Murant, giving detailed descriptions of plant viruses and virus groups.

communicable disease. A disease which is easily transmitted horizontally between organisms, e.g. INFLUENZA.

Comovirus group. (Sigla from COWPEA MOSAIC, the type virus). Genus of MULTICOMPONENT plant viruses with small isometric particles, 28 nm. in diameter, which sediment at 118S (bottom (B) component), 98S (middle (M) component) and 58S (top (T) component which lacks nucleic acid); CsCl banding densities (g/cc.) are 1.44-1.46 (B), 1.41 (M) and 1.29 (T). The capsids have

 100nm

icosahedral symmetry (T=1), the structural subunit comprising one molecule each of polypeptides of mw. *c*.22 and 42 x 10^3. The genome is linear, (+)-sense ssRNA, B component containing RNA-1 (mw. 2.4 x 10^6; 5,889 nucleotides) and M component containing RNA-2 (mw. 1.4 x 10^6; 3,481 nucleotides); the 5′ terminus has a VPg (mw. 5 x 10^3), the 3′ terminus is polyadenylated. Replication is in the cytoplasm via distinct ds REPLICATIVE INTERMEDIATES; the polymerase is, at least in part, virus coded. Virus mRNAs are translated to give POLYPROTEINS (RNA-1, mw. 207 x 10^3; RNA-2, mw. 116 x 10^3) which are cleaved to give functional proteins, those from RNA-1 being involved primarily with replication and those from RNA-2 being primarily structural.

The host ranges of individual members are narrow. Particles are found in most cell types, infection being characterised by large cytoplasmic membranous inclusions. Comoviruses are readily transmissible mechanically. Some members are seed transmitted, others are transmitted by beetles.
Matthews, R.E.F. (1982) Intervirology **17**, 161.
Bruening, G. (1978) CMI/AAB Descriptions of Plant Viruses No. 199.
Francki, R.I.B. *et al.* (1985) **In** Atlas of Plant Viruses. Vol. 2, p. 1. CRC Press.: Boca Raton, Florida.

complement fixation test (CFT). A serological test often used for comparing different or related antigens, e.g. viruses. Complement is a multicomponent enzyme system present in an inactive state in serum. It becomes activated when a specific site of the Fc portion of certain antibody molecules becomes exposed as a result of antibody binding and is then able to bind to the Clq component. When an antibody is directed against a virus particle the complement will also be irreversibly bound to the virus-antibody complex; this removes complement from the antiserum. The amount of complement remaining is estimated by the reaction with antibody directed against membrane antigens of erythrocytes; the reaction leads to lysis of the erythrocytes and released haemoglobin can be estimated spectrophotometrically. From the amount of complement in the original serum the amount fixed by the virus-antibody complex can be estimated.

complementary base sequence. A nucleic acid sequence which is able to form a perfectly hydrogen-bonded duplex with the one to which it is complementary. Thus an mRNA molecule is complementary to the parts of the DNA strand from which it was transcribed, e.g. for the DNA base sequence ..CGATG.., ..GCTAC.. is the complementary sequence in the DNA strand and ..GCUAC.. in the complementary RNA strand.

complementary DNA (cDNA). An ssDNA molecule that is complementary in base sequence to the single strand from which it was transcribed. Often refers to DNA transcribed from RNA using REVERSE TRANSCRIPTASE. If this cDNA is made double-stranded and cloned it is described as a cDNA clone.

complementary RNA (cRNA). An ssRNA molecule that is complementary in base sequence to the single strand from which it was transcribed. Most ssRNA viruses (except RETROVIRUSES) use complementary RNA as intermediates in replication.

complementary strand. An ss nucleic acid molecule complementary in base sequence to the single strand from which it was transcribed.

complementation. The process by which one genome provides functions which another genome lacks. There are two types of complementation between viruses; a) intergenic, in which mutants defective in different genes assist one another and b) intragenic, in which mutants defective in the same gene produce a functional gene product.

complementation group. A group of viruses with mutations in the same codon and which cannot complement each other.

complementation test. A test to determine whether two mutants of a virus are defective in the same cistron. If there is no increase in virus yield from a mixed infection of the two mutants compared with the yield after infection with the individual mutants the two viruses are said to be in the same COMPLEMENTATION GROUP.

Con A. Abbreviation for CONCANAVALIN A.

concanavalin A. A lectin from the plant *Canavalia ensiformis*. It has affinity for terminal α-D-glucosyl residues and is used for purifying glycoproteins. It has been used in the purification of interferon.

concatemer. A long DNA molecule made up from repeated monomers of a single type joined to give a linear multimer with the monomers all in the same orientation. Phage λ replicates by a rolling circle mechanism to give concatemeric DNA which is cleaved at specific (COS) sites during packaging.

conditional lethal mutant. A mutant which will not replicate under conditions in which the wild type replicates, but will under other permissive conditions, e.g. different temperature.

congenital infection. Infection occurring before birth. There are two types of condition in congenital virus infections; a) affecting particular organs

depending upon the stage of foetal development during infection e.g. RUBELLA VIRUS or b) infection of every cell of the embryo and persistence of infection throughout adult life e.g. LYMPHOCYTIC CHORIOMENINGITIS VIRUS in mice.

conjugate. Used as a noun refers to the linking of two macromolecules or a molecule to a macromolecule, e.g. an enzyme to an antibody for use in ELISA.

conjugation. The transfer of a plasmid from one cell to another. The plasmid usually encodes most of the required functions.

conservative replication. A model of nucleic acid replication in which the parent double strand is preserved and both strands of the progeny molecule are newly synthesised. In REOVIRIDAE the progeny dsRNA strands are produced by a fully conservative mechanism. *See* SEMICONSERVATIVE REPLICATION.

contact infection. A disease transmitted by close mechanical contact between organisms e.g. common cold, herpes, AIDS.

contact inhibition. The inhibition of growth and/or movement when cells come into contact. Natural situation in many non-infected tissue culture systems. Infection with e.g. picornaviruses in mammalian cell line overcomes it.

contagion. A disease syndrome caused by an organism capable of transmission between individual hosts.

contagious ecthyma virus. Synonyms: CHAMOIS CONTAGIOUS ECTHYMA VIRUS, CHAMOIS PAPILLOMA VIRUS. Family *Poxviridae*, subfamily *Chordopoxvirinae*, genus *Parapoxvirus*. Causes lesions in sheep around mouth and lips, and sometimes on udder, thighs, anus and eyes. Occurs in many parts of world. Antigenically distinct from goat pox virus.

contagious pustular dermatitis of horses virus. *See* HORSE POX VIRUS.

continuous cell line. Cells with uniform morphology and capable of indefinite propagation *in vitro*. This condition may be induced by transformation of a primary cell culture, frequently of tumour tissue. Most do not show CONTACT INHIBITION.

continuous flow centrifugation. Centrifugation in a rotor which has a fluid seal which allows the continuous flow of a sample into and out of the rotor while it is rotating at high speed. Used for large-scale purification of viruses.

contour length. The length of a nucleic acid molecule as measured by electron microscopy. As the actual length measured depends on the strandedness of the nucleic acid (ss or ds) and on the method of preparation, internal size markers should be used. See KLEINSCHMIDT PROCEDURE.

conversion. A term usually applied to interactions between a temperate PHAGE and a prokaryote host where new properties, which have no obvious relation to the replication cycle of the phage, are conferred on the host by the genome of the PROPHAGE. These can include changes in colony morphology or pigmentation, modification of the antigenic properties of the host (e.g. *Salmonella* strains lysogenic for phage ε*15*) and effects on toxin production (e.g. *Corynebacterium diphtheriae* lysogenic for phage β).

core. The central part of a virion enclosed by a capsid and comprising protein and the viral nucleic acid genome.

Coronaviridae. (Latin 'corona' = crown.) A family of RNA viruses comprising a single genus (CORONAVIRUS) with pleomorphic particles, 75-160 nm. in diameter. The particles have a lipid

100nm

envelope with club-shaped surface projections 12-24 nm. in length. The ribonucleoprotein is seen as a helical structure 11-13 nm. diameter or as strands 9 nm. in diameter. There are four to six proteins: the surface projections are glycosylated. The RNA comprises one molecule of (+)-sense ssRNA, mw. 5-6 x 10^6 with a poly A tail at the 3´ end. During replication ssRNA species are produced. These comprise the virion RNA and shorter forms in which the sequence of each RNA is contained within every larger RNA species (nested set). Several virus-specific non-structural proteins are found in infected cells. The host range includes man, mouse, pig, dog, cow and rats. Transmission is probably mechanical.
Matthews, R.E.F. (1982) Intervirology **17**, 102.

Coronavirus. The single genus of the family *Coronaviridae*. Contains AVIAN INFECTIOUS BRONCHITIS, HUMAN CORONA, MURINE HEPATITIS, PORCINE HAEMAGGLUTINATING ENCEPHALITIS and TRANSMISSIBLE PORCINE GASTRO-ENTERITIS VIRUSES.

Corriparta virus. Family *Reoviridae*, genus *Orbivirus*. Isolated from culicines and birds in Northern Australia.

Corticoviridae. (Latin 'cortex, corticis' = bark, crust). A family of viruses of prokaryotes including large (60 nm. in diameter) isometric DNA-containing PHAGES with an outer protein capsid and an internal lipoprotein layer. Few virus iso-

100nm

lates have been found with these properties. The type species is PM2 PHAGE (isolated from a bacterium of the genus, *Alteromonas*). Phage 06N 58P (from *Vibrio*) is a possible member of this virus group. Contains a single genus, CORTICOVIRUS.

Corticovirus. See CORTICOVIRIDAE.

cos site. The sticky ends of certain phage DNA molecules. It usually refers to λ which has 12 base sticky ends formed during packaging of CONCATAMERIC λ DNA. The λ-'ter' gene product cuts the ds cos sequence asymmetrically to give the sticky ends.

cosmid. A plasmid vector which contains the COS SITE of phage λ and one or more selectable markers; the cos site enables *in vitro* packaging to take place in phage. Cosmids are capable of being used for cloning large (up to 35kbp) DNA fragments at high efficiency.

Cotesia marginiventris virus. Unclassified long, filamentous virus associated with the reproductive tract of the braconid parasitoid *Cotesia* (= *Apanteles*) *marginiventris*. Virions are 50-98 nm. in width and about 865 nm. long, straight or slightly flexuous. The virus is morphologically distinct from polydnaviruses and, unlike polydnaviruses, it replicates in the tissues of the lepidopteran hosts of the parasitoid. It is also morphologically distinct from the other parasitoid virus (*see Cotesia melanoscela* virus) known to infect host larvae.
Styer, E.L., *et al.* (1987) J. Invertebr. Pathol. **50**, 302.

Cotesia melanoscela viruses. Viruses isolated from the braconid wasp, *C. melanoscela*, a parasitoid of the gypsy moth. They include a typical POLYDNAVIRUS and an unclassified virus (designated CmV2). The CmV2 virion consists of a large quasicylindrical nucleocapsid surrounded by two unit membranes. The genome is probably a single dsDNA molecule of approximately 125kbp rather than the polydisperse genome characteristic of polydnaviruses. Virus replication occurs in the ovarian calyx and in some other tissues of both male and female parasitoids. CmV2 also replicates in tissues of the caterpillar hosts of the parasitoid and in lepidopteran cell cultures.
Stoltz, D.B. *et al.* (1988) Virology **162**, 311.

Cotia virus. Family *Poxviridae*, not assigned to genus. Isolated from man, sentinel mice and mosquitoes in S. America. Replicates in human embryo lung cells and in cell lines.

cotton anthocyanosis virus. A possible *Luteovirus*.
Francki, R.I.B. *et al.* (1985) **In** Atlas of Plant Viruses. Vol. 1. p. 137. CRC Press: Boca Raton, Florida.

cotton leaf crumple virus. A probable *Geminivirus*, subgroup B.
Harrison, B.D. (1985) Ann. Rev. Phytopath. **23**, 55.

councilman bodies. Collections of necrotic hyaline cells in the liver of individuals infected with YELLOW FEVER VIRUS. They stain in a characteristic manner with eosin.

counts per minute (cpm). A measure of the radioactivity of a sample. It is the basic output from a radioactivity counter and does not allow for quenching of counts by components in the sample. *See* DISINTEGRATIONS PER MINUTE.

covalently closed circular DNA (cccDNA). A form of dsDNA in which both strands are circular i.e. do not have free ends. Also known as form I DNA. *See* SUPERCOILED DNA.

cow parsnip mosaic virus. A probable plant *Rhabdovirus*, subgroup 2.
Francki, R.I.B. *et al.* (1985) **In** Atlas of Plant Viruses. Vol. 1. p. 73. CRC Press: Boca Raton, Florida.

Cowbone Ridge virus. Family *Flaviviridae*, genus *Flavivirus*. Isolated from a cotton rat in Florida.

cowpea aphid-borne mosaic virus. Synonym: ADZUKI BEAN MOSAIC VIRUS. A *Potyvirus*.
Bock, K.R. and Conti, M. (1974) CMI/AAB Descriptions of Plant Viruses No. 134.
Francki, R.I.B. *et al.* (1985) **In** Atlas of Plant Viruses. Vol. 2. p. 183. CRC Press: Boca Raton, Florida.

cowpea chlorotic mottle virus. A *Bromovirus*.
Bancroft, J.B. (1971) CMI/AAB Descriptions of Plant Viruses No. 49.
Francki, R.I.B. (1985) **In** The Plant Viruses. Vol. 1. p. 1. ed. R.I.B. Francki. Plenum Press: New York.

cowpea green vein-banding virus. A *Potyvirus*.
Lin, M.T. *et al.* (1979) Fitopat. Bras. **4**, 120.

cowpea mild mottle virus. A *Carlavirus*.
Brunt, A.A. and Kenton, R.H. (1974) CMI/AAB Descriptions of Plant Viruses No. 140.
Francki, R.I.B. *et al.* (1985) **In** Atlas of Plant Viruses. Vol. 2. p. 173. CRC Press: Boca Raton, Florida.

cowpea mosaic virus. Type member of the *Comovirus* group.
van Kammen, A. and de Jager, C.P. (1978) CMI/AAB Descriptions of Plant Viruses No. 197.
Francki, R.I.B. *et al.* (1985) **In** Atlas of Plant Viruses. Vol. 2. p. 1. CRC Press: Boca Raton, Florida.

cowpea mottle virus. A possible member of the *Carmovirus* group.
Bozarth, R.F. and Shoyinka, S.A. (1979) CMI/AAB Descriptions of Plant Viruses No. 212.
Francki, R.I.B. *et al.* (1985) **In** Atlas of Plant Viruses. Vol. 2. p. 235. CRC Press: Boca Raton, Florida.
Morris, T.J. and Carrington, J.C. (1988) **In** The Plant Viruses. Vol 3. p. 73. ed. R. Koenig. Plenum Press: New York.

cowpea ringspot virus. A possible *Cucumovirus*.
Francki, R.I.B. (1985) **In** The Plant Viruses. Vol. 1. p. 1. ed. R.I.B. Francki. Plenum Press: New York.

cowpea severe mosaic virus. A *Comovirus*.

de Jager, C.P. (1979) CMI/AAB Descriptions of Plant Viruses No. 209.
Francki, R.I.B. *et al.* (1985) **In** Atlas of Plant Viruses. Vol. 2. p. 1. CRC Press: Boca Raton, Florida.

cowpox virus. Family *Poxviridae*, subfamily *Chordopoxvirinae*, genus *Orthopoxvirus*. Causes papules and then vesicles, followed by crusting on teats and udders of cows. Infects hands of milk attendants. Forms basis of Jenner's original smallpox vaccine. Many species including guinea pigs, mice, monkeys and rabbits can be infected. Can be grown in chorioallantoic membrane and in many cell lines.

Coxsackie viruses. Family *Picornaviridae*, genus *Enterovirus*. Cause herpangina, aseptic meningitis, pleurodynia, myalgia, orchitis and myocarditis. Divided into groups A and B on basis of disease signs in suckling mice. Those in group A cause flaccid paralysis whereas those in group B cause a spastic paralysis. Both groups grow in tissue culture cells. Coxsackie B5 virus is closely related antigenically to swine vesicular disease virus. The group first assumed importance because it was mistakenly considered to be poliovirus.

Cp 169 cells. Insect cell line derived from the codling moth, *Cydia pomonella*, that is non-permissive for the replication of *Autographa californica* nuclear polyhedrosis virus.

cpe. Abbreviation for CYTOPATHIC EFFECT.

cpm. Abbreviation for COUNTS PER MINUTE.

CPV. Common abbreviation for cytoplasmic polyhedrosis virus, but also used on occasion for BEE CHRONIC PARALYSIS VIRUS.

Crawley virus. Family *Reoviridae*, genus *Reovirus*. Isolated from birds.

cricket paralysis virus: CrPV. A member of the PICORNAVIRIDAE not yet assigned to a specific genus; first isolated from the Australian field cricket (*Teleogryllus oceanicus*; Orthoptera). Virions resemble vertebrate PICORNAVIRUSES in many properties being isometric, 27 nm. in diameter, sedimenting at 167S and having a buoyant density in CsCl of 1.34 g/cc. However, only three major structural polypeptides have been observed (mw. 35, 34, 30 x 10³) and the presumed viral replicase is larger than replicases from vertebrate picornaviruses. The (+)-sense ssRNA genome has a mw. of 2.5-2.8 x 10⁶. CrPV is serologically related to DROSOPHILA C VIRUS but no sequence homology with this virus has been detected. IgM antibodies to CrPV have been found in several mammalian species and the virus appears to cross-react serologically with ENCEPHALOMYOCARDITIS VIRUS. However, direct evidence obtained so far indicates that CrPV only infects invertebrates. Identical or closely-related viruses have been isolated from Lepidoptera, Orthoptera and Hymenoptera.
Moore, N.F. (1985) J. gen. Virol. **66**, 647.

Crimean-Congo haemorrhagic fever virus. Family *Bunyaviridae*, genus *Nairovirus*. Isolated from febrile patients in Belgian Congo. Occurs widely in Africa, Crimea, causing high mortality. Symptoms include purpuric rash and severe haemorrhaging in lungs, gastro-intestinal tract and kidneys, accompanied by shock and renal insufficiency.

crimson clover latent virus. A *Nepovirus*. Francki, R.I.B. *et al.* (1985) **In** Atlas of Plant Viruses. Vol. 2. p. 23. CRC Press: Boca Raton, Florida.

Crinum virus. A possible *Potyvirus*. Francki, R.I.B. *et al.* (1985) **In** Atlas of Plant Viruses. Vol. 2. p. 183. CRC Press: Boca Raton, Florida.

cRNA. Abbreviation for COMPLEMENTARY RNA

cross-hybridisation. Hybridisation between complementary nucleic acids from different sources. Percentage cross-hybridisation can provide a measure of the relatedness of two nucleic acids.

cross-protection. The protection conferred on a host by infection with one strain of a virus which prevents infection by a closely-related strain. Usually applied to plant or animal viruses (*see* SUPERINFECTION EXCLUSION). Mild strains of a virus have been used to protect against infection with severe strains e.g. TOMATO MOSAIC VIRUS. The phenomenon is also used to assess relatedness of virus strains.

cross-reactivation. Synonym: MARKER RESCUE.

crossed immunoelectrophoresis. A variation of

ROCKET IMMUNOELECTROPHORESIS in which a mixture of antigens is first separated by electrophoresis in an agarose gel and then electrophoresed at right angles into a gel containing antibodies. It is used to resolve complex antibody-antigen systems and to quantify the antigens present by analysing the area enclosed within the precipitin arcs formed during electrophoresis in the second dimension.
Vestergaard, B.F. *et al.* (1977) J. Virol. **24**, 82.

cryptogram. A cipher used to record certain basic properties of a virus. It consists of four pairs of symbols; 1st. pair: type of nucleic acid and strandedness; 2nd. pair: mw. of nucleic acid (x 10⁶) and percent nucleic acid in infective particle; 3rd pair: outline of particle and shape of nucleocapsid; 4th. pair: kind of host infected and kind of vector, e.g. INFLUENZA VIRUS A R/1:2/1:S/E:V/O; TOBACCO MOSAIC VIRUS R/1:2/5:E/E:S/O where R = RNA, 1 = ss; 2 = mw. 2 x 10⁶, 1 and 5 are the % nucleic acid; S = spherical, E = elongate; V = vertebrate, S = seed plant, O = no known vector. Gibbs, A.J. *et al.* (1966) Nature **209**, 450.

cryptotope. *See* EPITOPE.
Jerne, N.K. (1960) Ann. Rev. Microbiol. **14**, 341.

Cryptovirus group. ('cryptic' = lack of symptoms). Genus of plant viruses with isometric particles, 30 or 38 nm. in diameter, which sediment at *c.*120S. The structural subunit is a coat

 100nm

protein mw. 53-63 x 10³. The particles contain dsRNA, usually two or three species with mw. ranging from 1.6-0.6 x 10⁶. There are two subgroups, A (type member WHITE CLOVER CRYPTIC VIRUS 1) which has smooth particles 30 nm. in diameter and B (type member WHITE CLOVER CRYPTIC VIRUS 2) which has particles 38 nm. in diameter with prominent subunits. Cryptic viruses cause no apparent symptoms in plants. They are not transmitted mechanically or by grafting but are transmitted very efficiently through seed. Cryptic viruses are also known as temperate viruses.
Francki, R.I.B. *et al.* (1985) **In** Atlas of Plant Viruses. Vol. 2, p. 244. CRC Press.: Boca Raton, Florida.
Boccardo, G. *et al.* (1987) Adv. Virus Res. **31**, 171.

crystalline array virus. Unclassified small

RNA virus isolated from the grasshopper *Melanoplus bivittatus* (Orthoptera). Virions are isometric, 13 nm. in diameter (from thin section studies) and sediment at 42S. Particles contain 18-22% RNA and occur in dense arrays in the cytoplasm of infected cells. Although the small size of the virus suggests that it could be a satellite virus, it does not occur in association with a larger helper virus.
Jutila, J.W. *et al.* (1970) J. Invertebr. Pathol. **15**, 225.

cubic phages. Viruses, isolated from prokaryotes, whose capsids have cubic (icosahedral) symmetry. These include members of the MICROVIRIDAE (e.g. øX174), CORTICOVIRIDAE (e.g. PM2), TECTIVIRIDAE (e.g. PRD1), LEVIVIRIDAE (e.g. MS2), CYSTOVIRIDAE (e.g. ø6) and the unclassified phage SiI (*see* PHAGE).
Ackermann, H.W (1978) **In** Handbook of Microbiology. Vol II, p. 639 ed. A.I. Laskin and H.A. Lechevalier. CRC Press: Boca Raton, Florida.

cubic symmetry. One of the major groups of regular (Platonic) polyhedra. There are three types of cubic symmetry: a) tetrahedral symmetry which has four 3-fold axes and three 2-fold axes, b) octahedral symmetry which has three 4-fold axes, four 3-fold axes and six 2-fold axes and c) ICOSAHEDRAL SYMMETRY which has twelve 5-fold axes, ten 3-fold axes and fifteen 2-fold axes. Many isometric viruses have icosahedral symmetry.

cucumber fruit streak virus. Synonym: CUCUMBER LEAF SPOT VIRUS.

cucumber green mottle mosaic virus. A *Tobamovirus.*
Hollings, M. *et al.* (1975) CMI/AAB Descriptions of Plant Viruses No. 154.
Okada, Y. (1986) **In** The Plant Viruses. Vol. 2. p. 267. ed. M.H.V. van Regenmortel and H. Fraenkel-Conrat. Plenum Press: New York.

cucumber leaf spot virus. A possible member of the *Tombusvirus* group.
Weber, I. (1986) AAB Descriptions of Plant Viruses No. 319.
Morris, T.J. and Carrington, J.C. (1988) **In** The Plant Viruses. Vol. 3. p. 73. ed. R. Koenig. Plenum Press: New York.

cucumber mosaic virus (CMV). Type member of the *Cucumovirus* group. The virus has a very wide host range with more than 470 species from 67 families being reported as natural hosts. The symptoms are most commonly mosaics but in some hosts deformation of leaves or necrosis can occur. CMV is of moderate economic importance but in some crops can cause severe damage. Its occurrence is widespread and it is found in both temperate and tropical countries. The virus is very variable and numerous 'strains' have been described. The designation of strain is confused because of the effects that a virus-dependent satellite RNA (CARNA-5) can have on symptoms. Strains Q, S, Y, D, W, Imperial and Price's No. 6 are widely recognised. CARNA-5 is a linear ssRNA of 335-336 nucleotides.

Francki, R.I.B. *et al*. (1979) CMI/AAB Descriptions of Plant Viruses No. 213.

Francki, R.I.B. (1985) **In** The Plant Viruses. Vol. 1. p. 1. ed. R.I.B. Francki. Plenum Press: New York.

cucumber necrosis virus. A possible member of the *Tombusvirus* group.

Dias, H.F. and McKeen, C.D. (1972) CMI/AAB Descriptions of Plant Viruses No. 82.

Rochon, D'A. and Tremaine, J.H. (1988) J. gen. Virol. **69**, 395.

Morris, T.J. and Carrington, J.C. (1988) **In** The Plant Viruses. Vol. 3, p. 73. ed. R. Koenig. Plenum Press: New York.

cucumber pale fruit viroid. A VIROID.

Diener, T.O. (1987) **In** The Viroids. p. 261. ed. T.O. Diener. Plenum Press: New York.

cucumber soil-borne virus. A possible member of the *Tombusvirus* group.

Francki, R.I.B. *et al*. (1985) **In** Atlas of Plant Viruses. Vol. 2. p. 235. CRC Press: Boca Raton, Florida.

cucumber yellows virus. A possible *Closterovirus*.

Francki, R.I.B. *et al*. (1985) **In** Atlas of Plant Viruses. Vol. 2. p. 219. CRC Press: Boca Raton, Florida.

Cucumovirus group. (Sigla from CUCUMBER MOSAIC, the type virus). Genus of MULTICOMPONENT plant viruses with small isometric particles, 29 nm. in diameter, which sediment at 99S and band in CsCl at 1.37 g/cc when fixed. Virus

particles are stabilised primarily by protein-RNA interactions; they are sensitive to ribonuclease. Capsid structure is icosahedral (T=3), the structural subunit being a single protein species (mw. 24 x 10^3). Genomic linear (+)-sense ssRNA comprises three species, RNA-1 (mw. 1.27 x 10^6; 3,389 nucleotides), RNA-2 (mw. 1.13 x 10^6; 3,035 nucleotides) and RNA 3 (mw. 0.82 x 10^6; 2,216 nucleotides). RNAs-1 and -2 are separately encapsidated; RNA-3 is encapsidated with the coat protein mRNA, RNA-4 (mw. 0.35 x 10^6; 1,049 nucleotides). The 5′ termini of the RNAs have a CAP; the 3′ termini have a tRNA-like structure which accepts tyrosine. Replication is in the cytoplasm and, for the genomic RNAs, is via distinct ds REPLICATIVE INTERMEDIATES; RNA-4 is derived from RNA-3. RNAs-1, -2 and -4 are MONOCISTRONIC messengers for proteins of mw. 105, 120 and 24 (coat protein) x 10^3 respectively; RNA-3 is bicistronic encoding a protein of mw. 34 x 10^3 at the 5′ end and having the coat protein cistron at the 3′ end.

The host range varies from wide for the type member to narrow for the other members. Particles are found in most cell types. Cucumoviruses are readily mechanically transmitted. They are seed transmitted in several hosts and are transmitted by aphids in the NON-PERSISTENT TRANSMISSION manner.

Matthews, R.E.F. (1982) Intervirology **17**, 171.

Francki, R.I.B. *et al*. (1985) **In** Atlas of Plant Viruses. Vol. 2. p. 53. CRC Press.: Boca Raton, Florida.

Francki, R.I.B. (1985) **In** The Plant Viruses. Vol. 1. p. 1. ed. R.I.B. Francki. Plenum Press: New York.

culture collection. A repository of cultures of characterised viruses, bacteria, cells and other organisms. Used for reference and comparison with new isolates. The AMERICAN TYPE CULTURE COLLECTION is one of the main repositories for viruses.

culture medium. Solution, usually containing various inorganic salts, sugars, amino acids, antibiotics and (for animal viruses) sera, used for culturing cells.

curing. Loss of a LYSOGENIC phage from a bacterial culture thus converting the culture to a non-lysogenic state.

cyanogen bromide. A chemical which reacts with methionine, converting it to homoserine

100nm

lactone, thus splitting the peptide chain on the C-terminal side of that amino acid. It is used to obtain peptides with C-terminal methionine residues from proteins.

Steeres, E. *et al.* (1965) J. Biol. Chem. **240**, 2478.

Cyanomyovirus. (From 'cyano', Greek 'kyanos' = blue and 'myo', Greek 'myos' = muscle). Proposed genus of the *MYOVIRIDAE* including phages with long contractile tails isolated from blue-green algae (cyanobacteria). The proposed type species is cyanophage AS-1 isolated from unicellular cyanobacteria (*Anacystis* and *Synechococcus*). Virions of AS-1 are of the A1 morphotype (*see* PHAGE), with a 90 nm. diameter isometric head and an extended tail of 244 x 23 nm. contracting to 93 nm. The base plate is 40 nm. wide and bears short tail pins. Particles sediment at 754S and have a buoyant density in CsCl of 1.49 g/cc. Thirty structural proteins have been detected. The genome is linear ds DNA (mw. 57 x 10^6). The virus replicates in the nucleoplasm of the host. Other members of the genus are cyanophages N1, A-1(L) and A-2.

Safferman, R.S. *et al.* (1983) Intervirology **19**, 61.

cyanophages. Viruses isolated from blue-green algae (cyanobacteria), morphologically similar to many bacteriophages. All isolates to date resemble TAILED PHAGES and have been provisionally grouped into three proposed genera: CYANO-MYOVIRUS (*MYOVIRIDAE*), CYANOSTYLOVIRUS (*Styloviridae*, now superseded by *SIPHOVIRIDAE*) and CYANOPODOVIRUS (*PODOVIRIDAE*).

Safferman, R.S. *et al.* (1983) Intervirology **19**, 61.

Cyanopodovirus. ('cyano', Greek 'kyanos' = blue and 'podo', Greek 'podos' = foot). Proposed genus of the *PODOVIRIDAE* including phages with short tails isolated from blue-green algae (cyanobacteria). The proposed type species is cyanophage LPP-1 isolated from filamentous cyanobacteria (*Lyngbya, Plectonema* and *Phormidium*). Virions of LPP-1 are of the C1 morphotype (*see* PHAGE), with a 59 nm. diameter isometric head and a tail 15-20 nm. long, 15 nm. wide. Particles sediment at 548S and have a buoyant density in CsCl of 1.48 g/cc. Ten structural proteins have been observed (major head proteins mw. 39 and 13 x 10^3; major tail protein mw. 80 x 10^3). The genome is linear dsDNA (mw. 27 x 10^6). The virus replicates in the peripheral region of the host cell, displacing the photosynthetic lamellae. Other members of the genus are cyanophages LPP-2, SM-1, A-4(L) and AC-1.

Safferman, R.S. *et al.* (1983) Intervirology **19**, 61.

cyanostylovirus. ('cyano', Greek 'kyanos' = blue and 'stylo', Greek 'stylos' = pillar). Proposed genus of the *STYLOVIRIDAE* (now superseded by *SIPHOVIRIDAE*), including phages with long, noncontractile tails isolated from blue-green algae (cyanobacteria). The proposed type species is cyanophage S-1 isolated from unicellular cyanobacteria (*Synechococcus*). Virions of S-1 are of the B1 morphotype (*see* PHAGE), with a 50 nm. diameter head and a rigid noncontractile tail 140 nm. long. Particles sediment at 353S and have a buoyant density in CsCl of 1.50 g/cc. Thirteen structural proteins have been observed (major proteins mw. 39, 11 and 10 x 10^3). The genome is linear dsDNA (mw. 23-26 x 10^6). Other members of the genus are cyanophages S-2L and SM-2 .

Safferman, R.S. *et al.* (1983) Intervirology **19**, 61.

cybrid. The product of the fusion of a cell with a cytoplast (a cell from which the nucleus has been deleted). The resulting cell may contain cytoplasmic components, e.g. A-type virus particles, not under the control of the cell genome.

cycas necrotic stunt virus. A *Nepovirus*, occurs in Japan.

Kusunoki, M. *et al.* (1986) Ann. Phytopath. Soc. Japan **52**, 302.

cyclic AMP. A compound derived from ATP by the action of the enzyme adenyl cyclase. It is an important regulatory molecule in higher eukaryotes, being a mediator of hormone action while in prokaryotes it is involved in the catabolite repression of gene expression.

cycloheximide. 3-[2-(3,5-dimethyl-2-oxocyclohexyl)-2-hydroxyethyl]glutarimide. An antibiotic isolated from certain strains of *Streptomyces griseus*. It is a reversible inhibitor of protein synthesis in eukaryotic cells but not in prokaryotic systems.

Cydia pomonella granulosis virus. Baculovirus (Subgroup B) isolated in Mexico from the codling

moth, *C. pomonella*. The virus is highly infectious for neonate larvae with an LD_{50} of *c*. one virus particle and an LT_{50} of approximately four days at 25˚C. The virus is infectious for a small number of closely-related insect species within the family Tortricidae. It has received considerable attention as a selective biological control agent of codling moth and several trial commercial products have been developed (e.g. SAN 404 and DECYDE). A product is currently marketed on a small scale in Switzerland. Isolates of the virus from the UK and USSR are genetically very closely related, though not identical, to the original Mexican isolate. Codling moth GV is one of the few GVs where a productive infection has been recorded *in vitro* in cell cultures of the homologous host.
Crook, N.E. *et al*. (1985) J. gen. Virol. **66**, 2423.

Cylindocapsa geminella 'virus'.   Virus-like particles found in a green alga. The particles are isometric, 200-230 nm. in diameter with a complex multilayered capsid comprising at least ten protein species (mw. $10\text{-}160 \times 10^3$) surrounding dsDNA (mw. $175\text{-}190 \times 10^6$). Infectivity has not been shown.
Stanker, L.H. *et al*. (1981) Virology **114**, 357.

Cymbidium mosaic virus.   A *Potexvirus*.
Francki, R.I.B. (1970) CMI/AAB Descriptions of Plant Viruses No. 27.
Francki, R.I.B. *et al*. (1985) **In** Atlas of Plant Viruses. Vol. 2. p. 159. CRC Press: Boca Raton, Florida.

Cymbidium ringspot virus.   A *Tombusvirus*.
Hollings, M. and Stone, O.M. (1977) CMI/AAB Descriptions of Plant Viruses No. 178.
Francki, R.I.B. *et al*. (1985) **In** Atlas of Plant Viruses. Vol. 1. p. 181. CRC Press: Boca Raton, Florida.
Martelli, G.P. *et al*. (1988) **In** The Plant Viruses. Vol. 3. p. 13. ed. R. Koenig. Plenum Press: New York.

Cynara virus. A plant *Rhabdovirus*, subgroup 1.
Francki, R.I.B. *et al*. (1985) **In** Atlas of Plant Viruses. Vol. 1. p. 73. CRC Press: Boca Raton, Florida.

Cynodon mosaic virus.   A possible *Carlavirus*.
Francki, R.I.B. *et al*. (1985) **In** Atlas of Plant Viruses. Vol. 2. p. 173. CRC Press: Boca Raton, Florida.

cynosorus mottle virus.   An unclassified plant virus with isometric particles.
Hull, R. (1988) The Plant Viruses. Vol. 3. p. 113. ed. R. Koenig. Plenum Press: New York.

Cypovirus.   Unofficial name for CYTOPLASMIC POLYHEDROSIS VIRUS GROUP.

Cystoviridae. ('cysto', Greek 'kystis' = bladder, sack). A family of phages in which virions are enveloped and contain a segmented dsRNA. genome. ø6 PHAGE is the only member.

100nm

Cystovirus. *See* ø6 PHAGE.

cytidine. The nucleoside of cytosine and ribose. *See* NUCLEIC ACID.

cytidine 5'-triphosphate. The triphosphate of the nucleoside cytidine. *See* NUCLEIC ACID.

cytochrome *c*. A basic protein with mw. ranging from $12.1\text{-}12.6 \times 10^3$ depending upon source. Used in the electron microscopy of nucleic acids as it binds to them and renders them detectable in the electron microscope. *See* KLEINSCHMIDT PROCEDURE.

cytocidal. Causes cell death.

cytomegalovirus.   *See* BETAHERPESVIRINAE. A subfamily in the *Herpesviridae*. Isolated principally from the salivary glands and kidneys of many species, including man, mouse and guinea pig.

cytopathic effect (cpe). Changes in the microscopic appearance of cultured cells often seen following virus infection. Can consist of changes in cell morphology, e.g. rounding up, cell fusion or the production of intracellular structures, e.g. INCLUSION BODIES.

cytoplasm. The protoplasm of an animal or plant cell external to the nucleus and other organelles.

cytoplasmic polyhedrosis virus group. Genus of the *REOVIRIDAE* containing viruses isolated only from arthropods. The viruses differ from other genera within the *Reoviridae* by having virions only 50-65 nm. in diameter, with a single shell which carries twelve spikes (*c*.20 nm. long) located at the icosahedral vertices. During replication many virions are occluded singly or, more usually, in large numbers within proteinaceous occlusion bodies (polyhedra). Unlike most other reoviruses, the RNA transcriptase and associated RNA-capping enzymes in the virion do not require prior treatment with proteolytic enzymes for their activation. Virions generally contain five polypeptides (mw. 31-151 x 10^3) and the OCCLUSION BODY matrix protein (POLYHEDRIN) varies in mw. from 25-37 x 10^3 depending on the virus isolate. Twelve virus 'types' have been defined (from 68 isolates examined) by their distinctive ELECTROPHEROTYPE, antigenic properties and RNA homology studies. The type species is *BOMBYX MORI* CYTOPLASMIC POLYHEDROSIS VIRUS (type 1). Other CPV 'types' are: type 2 (from *Inachis io*); type 3 (from *Spodoptera exempta*); type 4 (from *Actias selene*); type 5 (from *Trichoplusia ni*); type 6 (from *Biston betularia*);

type 7 (from *Noctua* (=*Triphaena*) *pronuba*); type 8 (from *Abraxas grossulariata*); type 9 (from *Agrotis segetum*); type 10 (from *Aporophyla lutulenta*); type 11 (from *Spodoptera exigua*); type 12 (from *Spodoptera exempta*). CPV isolates have been recorded in more than 200 arthropod species, predominantly from Lepidoptera, but also Diptera, Hymenoptera and the crustacean *Simocephalus expinosus*. Virus replication is confined to the cytoplasm of cells of the gut epithelium. A list of insect hosts in which CPVs have been recorded is given in Appendix B.

Payne, C.C. and Mertens, P.P.C. (1983) **In** The *Reoviridae*, p. 425, ed. W.K. Joklik. Plenum Press: New York.

cytosine. A constituent pyrimidine base of DNA and RNA. *See* NUCLEIC ACID.

cytoskeleton. The protein fibres which make up the structural framework of a cell.

D

D'Aguilar virus. Family *Reoviridae*, genus *Orbivirus*. Isolated from *Culicoides brevitarsis* in Australia. Antibodies are present in cattle and sheep.

D-loop. Abbreviation for displacement or displaced loop. It is a structure formed in supercoiled DNA which has been incubated, under the appropriate conditions, with a short ssDNA fragment which is homologous for part of the supercoiled molecule. The displaced strand of the supercoiled molecule, the D-loop, is then available for manipulation e.g. site-specific mutagenesis. D-loop formation is an initial step in the replication of many circular dsDNA molecules. See R-LOOP.

D-type virus particles. A morphologically defined group of enveloped RNA virus particles found both intra- and extra-cellularly. The intracellular form is ring-shaped, 60-90 nm. diameter and generally more abundant near the plasma membrane. The extracellular form is 100-120 nm. diameter, has an electron-dense nucleoid eccentrically located and has shorter surface spikes than those of the B-TYPE PARTICLES. See A-, B-, C-TYPE VIRUS PARTICLES.

Dacus oleae viruses. *See* OLIVE FRUIT FLY VIRUSES.

daffodil mosaic virus. A possible *Potyvirus*. Francki, R.I.B. *et al.* (1985) **In** Atlas of Plant Viruses. Vol. 2. p. 183. CRC Press: Boca Raton, Florida.

dahlia mosaic virus. A *Caulimovirus*. Brunt, A.A. (1971) CMI/AAB Descriptions of Plant Viruses No. 51. Francki, R.I.B. *et al.* (1985) **In** Atlas of Plant Viruses. Vol. 1. p. 17. CRC Press: Boca Raton, Florida.

Dakar bat virus. Family *Flaviviridae*, genus *Flavivirus*. Isolated from insectivorous bats in many parts of Africa.

dalton. A term which is incorrectly used as a unit for the molecular weight of a compound. Molecular weight is the ratio of the weight of a molecule of the compound compared with an atom of hydrogen. Named after Joseph Dalton, a nineteenth century chemist. *See* MOLECULAR WEIGHT.

dandelion carlavirus. A *Carlavirus*. Dijkstra, J. *et al.* (1985) Neth. J. Plant Path. **91**, 77.

dandelion latent virus. A *Carlavirus*. Francki, R.I.B. *et al.* (1985) **In** Atlas of Plant Viruses. Vol. 2. p. 173. CRC Press: Boca Raton, Florida.

dandelion yellow mosaic virus. A possible member of the *Parsnip Yellow Fleck Virus* group. Bos, L. *et al.* (1983) Neth. J. Plant Path. **89**, 207. Murant, A.F. (1988) **In** The Plant Viruses. Vol. 3. p. 273. ed. R. Koenig. Plenum Press: New York.

Dane particle. The original name for the complete infectious human HEPATITIS B VIRUS particle. Dane, D. *et al.* (1970) Lancet **i**, 695.

Danish plum line pattern virus. An *Ilarvirus*. Francki, R.I.B. (1985) **In** The Plant Viruses. Vol. 1. p. 1. ed. R.I.B. Francki. Plenum Press: New York.

Daphne virus S. A *Carlavirus*; occurs in Japan. Doi, Y. *Personal communication*.

Daphne virus X. A possible *Potexvirus*. Forster, R.L. and Milne, K.S. (1978) CMI/AAB Descriptions of Plant Viruses No. 195. Francki, R.I.B. *et al.* (1985) **In** Atlas of Plant Viruses. Vol. 2. p. 159. CRC Press: Boca Raton, Florida.

Daphne virus Y. A possible *Potyvirus*. Francki, R.I.B. *et al.* (1985) **In** Atlas of Plant

Viruses. Vol. 2. p. 183. CRC Press: Boca Raton, Florida.

'Dark Cheeks'. Lethal disease of the locusts *Locusta migratoria* and *Schistocerca gregaria* induced following their exposure to the nuclear polyhedrosis virus (NPV) of *Spodoptera littoralis*. This surprising result provides the first evidence for the transmission of a baculovirus (NPV) from one insect order to another (from Lepidoptera to Orthoptera).
Bensimen. A. *et al.* (1987) J. Invertebr. Pathol. **50**, 254.

dark field microscopy. A method of microscope illumination in which the illuminating beam is a hollow cone of light formed by an opaque stop at the centre of the condenser large enough to prevent direct light from entering the objective. The specimen is seen with light scattered or diffracted by it.

Darna trima virus. Member of *TETRAVIRIDAE*.

dasheen mosaic virus. A *Potyvirus*.
Zettler, F.W. *et al.* (1978) CMI/AAB Descriptions of Plant Viruses No. 191.
Francki, R.I.B. *et al.* (1985) **In** Atlas of Plant Viruses. Vol. 2. p. 183. CRC Press: Boca Raton, Florida.

Dasychira pudibunda virus. Possible member of *TETRAVIRIDAE*.

Datura 437 virus. A possible *Potyvirus*.
Francki, R.I.B. *et al.* (1985) **In** Atlas of Plant Viruses. Vol. 2. p. 183. CRC Press: Boca Raton, Florida.

Datura mosaic virus. A possible *Potyvirus*.
Francki, R.I.B. *et al.* (1985) **In** Atlas of Plant Viruses. Vol. 2. p. 183. CRC Press: Boca Raton, Florida.

Datura shoestring virus. A *Potyvirus*.
Francki, R.I.B. *et al.* (1985) **In** Atlas of Plant Viruses. Vol. 2. p. 183. CRC Press: Boca Raton, Florida.

DBM paper. Abbreviation for DIAZOBENZYLOXYMETHYL PAPER.

Decyde. Trials product of the Granulosis Virus of the codling moth, *Cydia pomonella*, produced in 1986 by MicroGeneSys, West Haven, Connecticut, USA.

defective interfering (DI) particles. Virus particles which lack part of the genome nucleic acid of the standard virus and which interfere with the replication of the standard virus. The nucleic acids in DI particles depend upon the standard virus genome for their replication. DI particles are produced by growing virus at high multiplicity of infection. *See* VON MAGNUS PHENOMENON.

defective phages. Phage virus particles or subviral particles which are unable to multiply in host cells. Certain types of particulate BACTERIOCINS have been named 'killer particles'.

defective virus. Virus which lacks part of its genome or some function and thus is unable to replicate fully. *See* DEFECTIVE INTERFERING PARTICLES, VON MAGNUS PHENOMENON.

degeneracy of the code. Most amino acids are coded for by several codons which differ in the identity of the third base. Thus changes in nucleic acid sequence do not necessarily result in changes in amino acid sequence. *See* GENETIC CODE, WOBBLE HYPOTHESIS.

deletion mutant. A mutant generated by the loss of one or more nucleotides from a coding sequence.

denaturation. The destruction of secondary and tertiary structure of a protein, nucleic acid or virus by physical or chemical means. For nucleic acids the hydrogen bonds maintaining the double-stranded nature can be denatured by, e.g. heat, formamide, low ionic strength (*see* MELTING TEMPERATURE). For proteins the structure can be denatured by a detergent, e.g. SODIUM DODECYL SULPHATE.

Dendrobium leaf streak virus. A possible plant *Rhabdovirus*; unenveloped particles.
Francki, R.I.B. *et al.* (1985) **In** Atlas of Plant Viruses. Vol. 1. p. 73. CRC Press: Boca Raton, Florida.

Dendrobium mosaic virus. A possible *Potyvirus*.
Francki, R.I.B. *et al.* (1985) **In** Atlas of Plant Viruses. Vol. 2. p. 183. CRC Press: Boca Raton, Florida.

Dendrolimus spectabilis cytoplasmic poly-

hedrosis virus. The only cytoplasmic polyhedrosis virus to have been produced on a commercial scale for biological pest control. The virus was registered in 1974 in Japan as a microbial insecticide for the control of pine caterpillar. Katagiri, K. (1981) In Microbial Control of Pests and Plant Diseases 1970-1980. p. 433. ed. H.D. Burges. Academic Press: New York.

Dengue virus. Synonym: BREAKBONE FEVER VIRUS. Family *Flaviviridae*, genus *Flavivirus*. Causes an acute febrile illness with severe headache, back and limb pains and rash. Infection of a person immune to one serotype with virus from a second of the four serotypes often causes haemorrhagic fever. Occurs in Australia and many Pacific areas and in Caribbean. Transmitted by *Aedes aegyptii*. Virus grows in suckling mouse brain and in several cell types. No good experimental model to study efficacy of vaccines. Currently under intensive study, sponsored mainly by World Health Organisation.

Denhardt's solution. A solution comprising 0.02% Ficoll, 0.02% polyvinyl pyrrolidone and 0.02% bovine serum albumin used in the preincubation of nitrocellulose and nylon filters. This treatment prevents non-specific binding of a probe thus reducing the background in NORTHERN and SOUTHERN BLOTS. Denhardt, D.T. (1966) Biochem. Biophys. Res. Comm. **23**, 641.

density (buoyant density). The density at which macromolecules will band in an ISOPYCNIC GRADIENT. As well as being related to the PARTIAL SPECIFIC VOLUME of the macromolecule, the value is affected by factors such as hydration of the macromolecule, the effect of the material making the gradient on the hydration and binding of that material to the macromolecule. Viruses with lipid coats have lower buoyant density than those without. The buoyant density of DNA is dependent upon the G + C content according to the formula: $\rho = 0.0988(G + C) + 1.6541$
Where ρ is the density in g/cc
$(G + C)$ is the fraction of the nucleotides which are guanine and cytidine
Typical banding densities (g/cc) of viruses, proteins and nucleic acids are:

	CsCl	Cs_2SO_4
Protein	1.28	1.24
Virus (5% RNA)	1.32	1.27
Virus (20% DNA)	1.35	-
Virus (20% RNA)	1.36	1.30
Virus (40% RNA)	1.50	1.38
DNA (50% G + C)	1.70	1.42
RNA	1.90	1.63

n.b. Banding density of some viruses (e.g. foot-and-mouth disease virus) may be higher than expected due to particle porosity.

density gradient. A gradient of a solute in a solvent used to support macromolecules during their fractionation. Usually applied to the separation of macromolecular species by centrifugation or electrophoresis. See RATE ZONAL CENTRIFUGATION, ISOPYCNIC CENTRIFUGATION.

densonucleosis virus. See DENSOVIRUS.

Densovirus. (Latin 'densus' = thick, compact). Genus of the PARVOVIRIDAE which includes some of the smallest known insect-pathogenic viruses. Virions are isometric, about 20 nm. in diameter, and sediment at about 120S. The particles contain four polypeptides (mw. 92×10^3, 67×10^3, 58×10^3 and 49×10^3) and a linear ssDNA genome, mw. 1.7-2.0×10^6 with inverted terminal repetitions, resembling the genome organisation of the DEPENDOVIRUS genus. Complementary (+ and −) strands are packaged in separate virions in about equal frequency. The viruses replicate without a HELPER VIRUS in the nuclei of midgut cells (some *Densovirus* isolates) or most tissues of larvae, nymphs and adults where they produce nuclear hypertrophy and characteristic dense, Feulgen-positive nuclear inclusions. The type species is *GALLERIA MELLONELLA* DENSOVIRUS. Other isolates include densoviruses from Lepidoptera, Diptera and Orthoptera. Some common antigens have been detected as well as genetic homology between *Galleria* and *Junonia* densoviruses. A densovirus from *Bombyx mori* shares a region of homology (300 nucleotides) with other parvoviruses. A list of insect hosts in which densoviruses have been recorded is given in Appendix C. Berns, K.I. (1984) The Parvoviruses. Plenum Press: New York. Bando, H. *et al.* (1987) J. Virol. **61**, 553.

deoxycholate. Abbreviation for sodium deoxycholate ($C_{24}H_{39}O_4Na$) (mw. 414.6) an anionic detergent.

deoxyribonuclease (DNase). An enzyme which degrades DNA. Often refers to DNase1, an endonuclease which digests ss and dsDNA to give oligonucleotides containing a 5′-phosphate.

62   deoxyribonucleic acid (DNA)

Dependent upon the divalent cation, Mg^{2+}, for activity. It is usually prepared from bovine pancreas.

deoxyribonucleic acid (DNA). A linear chain of deoxyribonucleotides. It is ds in some DNA viruses and ss in others. In dsDNA the two strands are anti-parallel and arranged in a double helix. Purine bases of one strand form HYDROGEN BONDS with pyrimidine bases of the other. This maintains a constant width to the helix of 2.0 nm. In A-DNA the base pairs are slanted with respect to the helix axis and are pulled away from the centre. Thus A-DNA is like a ribbon wrapped around a central cylinder. In the B-form the dsDNA adopts a right-handed helical conformation with each chain making a complete turn every 10 bases (3.4 nm.). In Z-DNA the double helix winds in a left-hand manner. When purines and pyrimidines alternate the bases give a zig-zag formation, hence the name Z-DNA. It is ds in some viruses and ss in others. See NUCLEIC ACID.

Dependovirus. See ADENO-ASSOCIATED VIRUS.

Dera Ghazi Khan virus. Family *Bunyaviridae*, genus *Nairovirus*. Isolated from a tick in Pakistan.

Desmodium mosaic virus. A possible *Potyvirus*.
Francki, R.I.B. *et al.* (1985) **In** Atlas of Plant Viruses. Vol. 2. p. 183. CRC Press: Boca Raton, Florida.

Desmodium yellow mottle virus. A *Tymovirus*.
Scott, H.A. (1976) CMI/AAB Descriptions of Plant Viruses No. 168.
Francki, R.I.B. *et al.* (1985) **In** Atlas of Plant Viruses. Vol. 1. p. 117. CRC Press: Boca Raton, Florida.
Hirth, L. and Girard, L. (1988) **In** The Plant Viruses. Vol. 3. p. 163. ed. R. Koenig. Plenum Press: New York.

determinant. Synonym: ANTIGENIC SITE or EPITOPE.

Dhori virus. Family *Bunyaviridae*, genus *Bunyavirus*. Isolated from ticks in Egypt, India and USSR. Antibodies have been found in camels.

DI particle. Abbreviation for DEFECTIVE INTERFERING PARTICLE.

Diabrotica 'baculovirus'. Unclassified virus, morphologically similar to a NON-OCCLUDED BACULOVIRUS observed in haemocytes of the cucumber beetle, *Diabrotica undecimpunctata*. Rod-shaped nucleocapsids (230 x 52 nm.) are enveloped singly by a trilaminar unit membrane. A distinct filamentous virus-like particle (up to 2000 nm. x 25 nm.) is also observed in mid gut cells of the same insect species.
Kim, K.S. and Kitajima, E.W. (1984) J. Invertebr. Pathol. **43**, 234.

dialysis. A process of selective diffusion through a membrane. It is usually used to separate low molecular weight solutes which diffuse through the membrane from the colloidal and high molecular weight solutes which do not.

Dianthovirus group. (*DIANTHUS*, the generic name of carnation). (Type member CARNATION RINGSPOT VIRUS). Genus of MULTICOMPONENT plant viruses with isometric particles, 31-34 nm. in diameter, which sediment at 135S and band in CsCl at 1.37 g/cc. The capsids are made of a single

100nm

protein species (mw. 40×10^3) and contain two (+)-sense ssRNA species (mw. 1.5 and 0.5×10^6). The larger RNA codes for the coat protein. Dianthoviruses have wide host ranges. Virus particles are found in most cell types. They are easily transmitted mechanically. They are naturally transmitted through soil.
Matthews, R.E.F. (1982) Intervirology **17**, 160.
Francki, R.I.B. *et al.* (1985) **In** Atlas of Plant Viruses. Vol. 2. p. 47. CRC Press.: Boca Raton, Florida.

diazobenzyloxymethyl (DBM) paper. An activated paper used to bind nucleic acid covalently for hybridisation procedures. It is particularly useful for binding RNA in NORTHERN BLOTTING experiments.
Alwine, J.C. *et al.* (1979) Meths. Enzymol. **68**, 220.

dideoxy method. See SANGER METHOD.

dideoxyribose. A sugar constituent of nucleotides which resembles deoxyribose except that it lacks a hydroxyl group at position 3 (for structure *see* NUCLEIC ACID). Thus nucleotides based on it can be incorporated into a polynucleotide chain but the lack of the hydroxyl group prevents addition of further nucleotides. *See* SANGER METHOD.

diethylaminoethylcellulose (DEAE-cellulose). An ion exchange medium used in chromatography. It has a pK of about 9.5. It is used in the chromatography of proteins and nucleic acids.

differential host. A host for separating one virus from a mixture of viruses; it is susceptible to one virus but not to the other(s).

diffusion coefficient. A measure of the rate at which a solute moves along a gradient from a higher concentration to a lower one. The rate is proportional to the concentration gradient and the constant of proportionality is the diffusion coefficient, designated D and given in cm^2sec^{-1}. It can be measured in an analytical ultracentrifuge and the observed value is usually corrected to water at 20°C. The coefficient is used in the SVEDBERG EQUATION and, since it is related directly to the STOKES RADIUS of the macromolecule, it can be used for calculating the hydrated radius (r_h):

$$r_h = \frac{kT}{6\eta D}$$

where k = Boltzmann's constant (1.38×10^6 ergK^{-1}

T = 293˙k
η = viscosity (0.01 poise)

Digitaria streak virus. A *Geminivirus*, subgroup A; related to maize streak virus.
Donson, J. *et al.* (1987) Virology **161**, 160.

Digitaria striate virus. A probable plant *Rhabdovirus*, subgroup 1; leafhopper transmitted. Francki, R.I.B. *et al.* (1985) **In** Atlas of Plant Viruses. Vol. 1. p. 73. CRC Press: Boca Raton, Florida.

dilution end point. 1. A property of a virus in which a series of dilutions is made of material from an infected plant or animal, usually with phosphate buffer, and the infectivity of each dilution measured. The DEP is the greatest dilution which retains infectivity. DEPs are usually in the range of 10^{-1} to 10^{-7}. 2. The greatest dilution of an antibody which gives a measurable reaction with an antigen in a serological test.

dimethyl POPOP. 1,4-di-2-(4-methyl-5-phenyloxazolyl)benzene. A secondary solute used in SCINTILLATION FLUIDS to shift wavelength.

dimethyl sulphoxide (DMSO). A solvent with several uses. It is used for dissolving both organic and inorganic chemicals, as a solvent for substances being applied to cells in tissue culture and in the preparation of cells for storage in liquid nitrogen. It can also be used to strand-separate ds nucleic acids.

Dioscorea green banding virus. A possible *Potyvirus*.
Francki, R.I.B. *et al.* (1985) **In** Atlas of Plant Viruses. Vol. 2. p. 183. CRC Press: Boca Raton, Florida.

Dioscorea latent virus. A possible *Potexvirus*.
Francki, R.I.B. *et al.* (1985) **In** Atlas of Plant Viruses. Vol. 2. p. 159. CRC Press: Boca Raton, Florida.

Dioscorea trifida virus. A possible *Potyvirus*.
Francki, R.I.B. *et al.* (1985) **In** Atlas of Plant Viruses. Vol. 2. p. 183. CRC Press: Boca Raton, Florida.

diphenylamine. A chemical used for the colorimetric determination of DNA; RNA does not give a colour reaction. *See* ORCINOL.
Burton, K. (1956) Biochem. J. **62**, 315.

Diplocarpon rosae virus (DrV). A probable member of the *Partitivirus* group.

disintegrations per minute (dpm). A direct measurement of radioactivity. It is the actual number of radioactive disintegrations in the sample and is estimated from the COUNTS PER MINUTE measured in a radioactivity counter and the quenching of counts in the sample. 1μCi gives 2.2×10^6 dpm; 1 Becquerel is one disintegration per second.

distemper virus. *See* CANINE DISTEMPER VIRUS.

disulphide bond. A bond which forms spontaneously when two sulphydryl (-SH) groups of cysteine side chains of a protein are close together and are oxidised by some reagent. They give additional stability to the structure of proteins.

dithiothreitol (DTT). (CHOHCH$_2$SH)$_2$. Also

known as Cleland's reagent; a reducing reagent. Used for maintaining sulphydryl groups in a reduced state and, hence, for disrupting DISULPHIDE BONDS. As it has little tendency to be oxidised by air it is a particularly useful reducing agent.

Cleland, W.W. (1964) Biochemistry **3**, 480.

DMSO. Abbreviation for DIMETHYL SULPHOXIDE.

DNA. Abbreviation for DEOXYRIBONUCLEIC ACID.

DNA-binding protein. A protein which binds specifically to DNA. This property is often determined by specific amino acid sequences. DNA-binding proteins have many functions, e.g. maintaining DNA in an ss form for transcription or replication. Several viruses, e.g. ADENOVIRUS encode DNA binding proteins.

DNA-dependent DNA polymerases (I, II and III, alpha, beta, gamma). Enzymes which synthesise DNA from a DNA template. In eukaryotes there are three major groups of DNA polymerases. DNA polymerase α is the enzyme mainly involved in the replication of chromosomal DNA, DNA polymerase β is mainly involved in DNA repair, and DNA polymerase γ is considered to be the enzyme involved in the replication of mitochondrial DNA. In prokaryotes there are also three groups of DNA polymerases: DNA polymerase I which is involved in chromosomal replication and has both 3′-5′ (proof reading) and 5′-3′ (excision repair in nick translation) exonuclease activities: DNA polymerase II which is involved in repair of UV damage to DNA: DNA polymerase III which is also involved in chromosomal replication. See KLENOW FRAGMENT.

DNA-dependent RNA polymerases. An enzyme which transcribes RNA from a DNA template. The enzyme initiates and terminates at specific sites (promoter region and termination site). In eukaryotes there are three forms of DNA-dependent RNA polymerase, I mediating the synthesis of ribosomal RNA, II mRNA and III tRNA.

DNA endonuclease. See ENDONUCLEASE, RESTRICTION ENDONUCLEASE.

DNA exonuclease. See EXONUCLEASE, DNA-DEPENDENT DNA POLYMERASE.

DNA gyrase. An enzyme which introduces negative superhelical twists into relaxed closed circular DNA; also known as topoisomerase 2. Several DNA phages (e.g. T4) require the use of this enzyme during replication.

DNA ligase. An enzyme which catalyses phosphodiester bond formation between the 5′-phosphate of one oligonucleotide and the 3′-hydroxyl of another. It is involved in DNA synthesis by the SEMICONSERVATIVE method and in linking DNA in genetic manipulation. The ligase produced by phage T4 is widely used in genetic manipulation.

DNA primase. See PRIMASE.

DNA sequencing. See MAXAM AND GILBERT METHOD, SANGER METHOD.

DNase. Abbreviation for DEOXYRIBONUCLEASE.

DNV. Abbreviation sometimes used for DENSOVIRUS.

dock mottling mosaic virus. A possible *Potyvirus*.

Francki, R.I.B. *et al.* (1985) **In** Atlas of Plant Viruses. Vol. 2. p. 183. CRC Press: Boca Raton, Florida.

dolichos enation mosaic virus. A serotype of TOBACCO MOSAIC VIRUS.

double-stranded (ds) nucleic acid. Nucleic acid with two antiparallel strands hydrogen-bonded together. Usually both strands are either DNA or RNA but RNA:DNA duplexes can occur. The genome of viruses of several groups are either dsDNA or dsRNA.

Douglas fir tussock moth nuclear polyhedrosis virus. See ORGYIA PSEUDOTSUGATA NUCLEAR POLYHEDROSIS VIRUS.

Douglas virus. Family *Bunyaviridae*, genus *Bunyavirus*.

Drosophila A virus. An unclassified small isometric virus (possible INSECT PICORNAVIRUS) isolated from *Drosophila melanogaster* (Diptera, Insecta). Virions are isometric, 30 nm. in diameter, with a buoyant density in CsCl of 1.37 g/cc. Nucleic acid type is probably RNA. Particles contain three structural proteins (mw. 72.4 x 10^3, 41.2 x 10^3, 31.6 x 10^3) and are serologically

distinct from DROSOPHILA C VIRUS and DROSOPHILA P VIRUS.
Plus, N. *et al.* (1976). Intervirology **7**, 346.

Drosophila C virus (DCV). A member of the *PICORNAVIRIDAE* not yet assigned to a specific genus; first isolated from the fruit-fly, *D. melanogaster* (Diptera; Insecta). Virions resemble vertebrate PICORNAVIRUSES in many properties, being isometric, 27 nm. in diameter, sedimenting at 153S with a buoyant density in CsCl of 1.34 g/cc. and containing four major structural proteins (mw. 31×10^3, 30×10^3, 28×10^3 and 8×10^3). The virus contains a (+)-sense RNA genome of one piece of ssRNA (mw. 2.8×10^6), polyadenylated at the 3´ end with a VPg at the 5´ end. Several geographical isolates have been obtained which are distinct in their pathogenicity in *D. melanogaster* and by cDNA-hybridisation. The virus has similar physicochemical properties to CRICKET PARALYSIS VIRUS. The virus will infect honey bees (Hymenoptera) and Mediterranean fruit fly (*Ceratitis capitata*; Diptera) when experimentally injected.
Moore, N.F. *et al.* (1985) J. gen. Virol. **66**, 647.

Drosophila F virus. Unclassified RNA virus (possible member of *REOVIRIDAE* most closely resembling members of the *REOVIRUS* genus) isolated from the fruit fly, *D. melanogaster*. Virions are non-enveloped, 60-69 nm. in diameter with a double protein shell, buoyant density in CsCl of 1.38 g/cc. and about eight structural proteins. The dsRNA. genome consists of ten segments. Morphologically similar viruses isolated from established lines of *D. melanogaster* cells appear similar if not identical to *Drosophila* F virus. CERATITIS REOVIRUS I has some similarities but is serologically distinct.
Plus, N. *et al.* (1981) Annls. Virol. **132**, 261.

Drosophila line 1 cells. Insect cell line derived from *Drosophila melanogaster*, susceptible to infection by certain nodaviruses.

Drosophila P virus. An unclassified small RNA virus (possible INSECT PICORNAVIRUS) isolated from *D. melanogaster*. Virions are isometric, 25-30 nm. in diameter and a buoyant density in CsCl of 1.36 g/cc.; they contain three structural proteins (mw. 48×10^3, 29.4×10^3 and 26×10^3) and an RNA genome. It is serologically distinct from DROSOPHILA A VIRUS and DROSOPHILA C VIRUS.
Plus, N. *et al.* (1976) Intervirology **7**, 346.

Drosophila retrovirus. Possible retrovirus-like particles observed in association with *Drosophila melanogaster* cell cultures.

Drosophila RS virus. Unclassified small RNA virus isolated from *D. melanogaster*. Virions have unusual ellipsoidal morphology, similar to that of BEE CHRONIC PARALYSIS VIRUS although the particles are regular in size, have a buoyant density of only 1.26 g/cc. in CsCl and contain two structural polypeptides (mw. 19.5×10^3 and 45×10^3).
Longworth, J.F. (1978) Adv. Virus Res. **23**, 103.

Drosophila S virus. An unclassified virus (possibly a *REOVIRUS*) isolated from *D. simulans* populations with abnormal bristles. At least nine dsRNA segments have been observed in infected flies, distinct from the patterns of *Drosophila* F virus and *Ceratitis* reovirus 1.
Lopez-Ferber, M. *et al.* (1987) Abstr. VII Internat. Congr. Virol., 206.

Drosophila X virus. Member of *Birnavirus* genus, isolated from *D. melanogaster*.

ds. Abbreviation for double-stranded.

DSIR-HA-1179 cells. Insect cell line derived from the black beetle, *Heteronychus arator*, which supports replication of the non-occluded baculovirus from *Oryctes rhinoceros*.

duck enteritis virus. *See* ANSERID HERPESVIRUS 1.

duck hepatitis virus. Family *Picornaviridae*, genus *Enterovirus*. Causes economically important disease of ducklings with haemorrhagic necrosis of the liver and high mortality. Virus grows in hens' eggs, killing the embryo, and in duck cell cultures without causing cpe.

duck plague virus. *See* ANSERID HERPESVIRUS 1.

Dugbe virus. Family *Bunyaviridae*, genus *Nairovirus*. Isolated from cattle and ticks in several parts of Africa and from humans with fever.

dulcamara mottle virus. A *Tymovirus*.
Francki, R.I.B. *et al.* (1985) **In** Atlas of Plant Viruses. Vol. 1. p. 117. CRC Press: Boca Raton, Florida.
Hirth, L. and Girard, L. (1988) **In** The Plant Viruses. Vol. 3. p. 163. ed. R. Koenig. Plenum Press: New York.

duplex. Synonym for double-stranded.

Dutch plum line pattern virus. *See* APPLE MO-
SAIC VIRUS.

Duvenhage virus. Family *Rhabdoviridae*, genus
Lyssavirus. First isolated from the brain of a man
dying from a rabies-like illness following a bite
from a bat. Resembles but is not identical to rabies
virus. Mice can be infected experimentally.
Grows in BHK cells.

Meredith, C.D. *et al.* (1971) S. Afr. med. J. **45**,
767.

E

Eagle's medium (basal medium). Simple defined medium; used in growing many animal cell lines, e.g. HELA CELLS.
Eagle, H. (1955) J. exp. Med. **102**, 595.

early genes. Viral genes which are expressed early in the replication cycle, e.g. the T ANTIGEN genes of SV40. Early gene products are usually involved in the replication of the viral nucleic acid. *See* LATE GENES.

Eastern equine encephalomyelitis virus. Family *Togaviridae*, genus *Alphavirus*. Infects man and horses, causing fever and involvement of the CNS. Mortality is high. Transmission is by mosquito. Maintained in the wild in birds, small rodents, reptiles and amphibia, without showing any signs of disease but outbreaks in pheasants have caused high mortality. Found in many parts of the USA and in the Caribbean. Many species, including mice, guinea pigs, goats, chickens and snakes can be infected experimentally. Can be grown in eggs and a variety of cell cultures.

Ebola virus. Family *Filoviridae*. Causes a severe, often fatal, haemorrhagic fever. In addition, it causes headache, limb and back pains, diarrhoea, vomiting and extensive bleeding from which recovery is slow. Extensive outbreak occurred, with high mortality in Sudan and Zaire, in 1976. Rodents may be the reservoir of the virus. Can be isolated by inoculation of guinea pigs or by infection of Vero cells.

ECBO viruses (Enteric Cytopathic Bovine Orphan). Synonym: BOVINE ENTEROVIRUSES. Picornaviruses isolated in cultures from bovine tissues or excreta. There are many serotypes. None is associated with disease.

ECHO virus. Family *Picornaviridae*, genus *Enterovirus*. Originally Enteric Cytopathic Human Orphan because they were regarded as being viruses in search of a disease. However, they can cause a variety of symptoms including aseptic meningitis, encephalitis, exanthema respiratory disease, gastro-intestinal disturbance, pericarditis and myocarditis.

echtes ackerbohnenmosaic virus. *See* BROAD BEAN TRUE MOSAIC VIRUS.

eclipse period (phase). The time between the penetration of a virus into a cell, when it appears to lose its infectivity, and the appearance of newly synthesised infectious virus.

Ectromelia virus. Family *Poxviridae*, subfamily *Chordopoxvirinae*, genus *Orthopoxvirus*. A latent infection of mice, present in many mouse units. Injection into mice i.p. causes hepatitis. Injection into the skin of rabbits and guinea pigs causes skin lesions which can be prevented by immunisation with vaccinia virus. Can be grown in the chorioallantoic membrane of eggs, producing pocks and in many cell cultures.

ED$_{50}$. Abbreviation for MEDIAN EFFECTIVE DOSE.

Edgehill virus. Family *Flaviviridae*, genus *Flavivirus*.

Edman degradation. A method for sequencing peptides. The N-terminal amino acid is removed by cleavage of the peptide bond with trifluoroacetic acid. If the N-terminal residue has been labelled (e.g. with dansyl chloride) it can be identified by its chromatographic properties. Otherwise the amino acid composition of the remaining peptide is compared with that of the original peptide and the terminal amino acid determined by deduction.
Edman, P. (1956) Acta Chem. Scand. **10**, 761.
Gray, W.R. (1972) Meth. Enzymol. **25**, 121.

EDTA. Abbreviation for ETHYLENEDIAMINE-TETRA-ACETIC ACID.

efficiency of plating (eop). The relative efficiency of infection in susceptible cells. It is determined by dividing the plaque count by the total number of virus particles in the inoculum.
Ellis, E.L. and Delbruck, M. (1939) J. gen. Physiol. **22**, 365.

Eg An 1825-61. Family *Bunyaviridae*, genus *Uukuvirus*.

eggplant mild mottle virus. A possible *Carlavirus*.
Francki, R.I.B. *et al.* (1985) **In** Atlas of Plant Viruses. Vol. 2. p. 173. CRC Press: Boca Raton, Florida.

eggplant mosaic virus. A *Tymovirus*.
Gibbs, A.J. and Harrison, B.D. (1973) CMI/AAB Descriptions of Plant Viruses No. 124.
Francki, R.I.B. *et al.* (1985) **In** Atlas of Plant Viruses. Vol. 1. p. 117. CRC Press: Boca Raton, Florida.
Hirth, L. and Girard, L. (1988) **In** The Plant Viruses. Vol. 3. p. 163. ed. R. Koenig. Plenum Press: New York.

eggplant mottled crinkle virus. A *Tombusvirus*.
Francki, R.I.B. *et al.* (1985) **In** Atlas of Plant Viruses. Vol. 1. p. 181. CRC Press: Boca Raton, Florida.
Martelli, G.P. *et al.* (1988) **In** The Plant Viruses. Vol. 3. p. 13. ed. R. Koenig. Plenum Press: New York.

eggplant mottled dwarf virus. A plant *Rhabdovirus*, subgroup 2.
Martelli, G.P. and Russo, M. (1973) CMI/AAB Descriptions of Plant Viruses No. 115.
Francki, R.I.B. *et al.* (1985) **In** Atlas of Plant Viruses. Vol. 1. p. 73. CRC Press: Boca Raton, Florida.

EGTA. Abbreviation for ETHYLENEGLYCOLBIS-(AMINOETHYLETHER)TETRA-ACETIC ACID.

Egtved virus. Synonym: VIRAL HAEMORRHAGIC SEPTICAEMIA VIRUS. Family *Rhabdoviridae*, not assigned to genus. Causes a severe and often fatal haemorrhagic septicaemia in European salmonids. Named after Danish village where virus was first isolated. Rainbow trout are the main natural host but other trout can be infected experimentally. The fish lose appetite and swim abnormally, with swollen abdomen and haemorrhages. The virus can be grown in cultures of fish cells.

Egypt bee virus. Unclassified small RNA virus (possible INSECT PICORNAVIRUS) isolated from adult honey bees in Egypt. Virions are isometric, 30 nm. in diameter, sediment at 165S and have a buoyant density in CsCl of 1.37 g/cc. The particles contain RNA and three structural proteins (mw. 41, 30 and 25 x 10^3).
Bailey, L. *et al.* (1979) J. gen. Virol. **43**, 641.

EHD virus. Acronym for EPIZOOTIC HAEMORRHAGIC DISEASE OF DEER VIRUS.

EID$_{50}$/HA ratio. A means for measuring the proportion of defective particles present in a virus preparation, e.g. influenza virus. It is the ratio between the infective titre measured in eggs (ID$_{50}$) and the haemagglutinating (HA) titre of the virus preparation.

Elcar. Commercial product of *Heliothis* spp. nuclear polyhedrosis virus used for the control of the lepidopteran pests *H. virescens* and *H. zea*. The virus was produced by Sandoz Inc., but production was discontinued in 1982.

elderberry carlavirus. A *Carlavirus*.
Dijkstra, J. and van Lent, J.W.M. (1983) CMI/AAB Descriptions of Plant Viruses No. 263.
Francki, R.I.B. *et al.* (1985) **In** Atlas of Plant Viruses. Vol. 2. p. 173. CRC Press: Boca Raton, Florida.

elderberry latent virus. A possible member of the CARMOVIRUS group.
Jones, A.T. (1974) CMI/AAB Descriptions of Plant Viruses No. 127.
Francki, R.I.B. *et al.* (1985) **In** Atlas of Plant Viruses. Vol. 2. p. 235. CRC Press: Boca Raton, Florida.
Morris, T.J. and Carrington, J.C. (1988) **In** The Plant Viruses. Vol. 3. p. 73. ed. R. Koenig. Plenum Press: New York.

elderberry ring mosaic virus. A *Carlavirus*, occurring in Japan.
Doi, Y. *Personal communication.*

elderberry vein clearing virus. A plant *Rhabdovirus*, occurs in Japan.
Doi, Y. *Personal communication.*

electroendosmosis. The movement of fluids through a gel or other solid matrix during electrophoresis. Under certain conditions it will significantly affect the migration of macromolecules.

electrofocusing. A method for separating protein molecules by gel or density gradient electrophoresis in which a pH gradient has been established. Each protein species moves to a pH approximating to its ISOELECTRIC POINT.

electron microscope. An instrument for providing magnified images of substances. It uses an electron beam which is focused by electromagnets and provides a resolving power more than 1000 times that of a light microscope. *See* SCANNING ELECTRON MICROSCOPE, TRANSMISSION ELECTRON MICROSCOPE.

electron spin resonance. Magnetic resonance arising from the magnetic moment of unpaired electrons in a paramagnetic substance. Used in determining the structure of organic compounds.

electropherogram. A picture (photograph or autoradiograph) which shows the distribution of proteins or nucleic acids which have been separated by gel electrophoresis.

electropherotype. A characteristic profile of proteins or nucleic acids separated by electrophoresis and often used to distinguish different virus strains. For example, members of the REOVIRIDAE can readily be distinguished from one another by electrophoretic separation on agarose or polyacrylamide gels of their dsRNA genome segments.

electrophoresis. The separation of charged macromolecules, either in free solution or in a liquid in the pores of a matrix, by the application of an electrical potential difference. The movement of the macromolecules is due to their charge but the separation may also be due to their STOKES' RADIUS. *See* GEL ELECTROPHORESES.

electrophoretic mobility. The relative rate of movement of a macromolecule upon electrophoresis. Usually applied to gel or paper electrophoresis where the mobility is inversely related to the STOKES' RADIUS of the macromolecule.

electroporation. A method by which nucleic acids or virus particles can be introduced into protoplasts or cells by creating transient pores in the plasma membrane using an electric pulse.
Zimmermann, U. (1982) Biochem. Biophys. Acta **694**, 227.

elementary bodies. Small round extracellular

virus-associated structures visualised by light microscopy of stained scrape preparations from skin lesions of SMALLPOX, VACCINIA or VARICELLA ZOSTER.

elephant poxvirus. Family *Poxviridae*, not assigned to genus.

ELISA. A serological method in which the antigen is immobilised on a solid matrix and is detected by an antibody to which an enzyme has been linked; the enzyme is detected by the production of colour on reaction with a substrate. There are many variations of this method. In the most commonly used version, antibodies fixed to a microtitre plate are used to trap the antigen. The trapped antigen is, in turn, detected by homologous antibodies which themselves may be detected by an enzyme linked to anti-gamma globulin antibodies raised in a different animal or to PROTEIN A. The enzymes most commonly used are horseradish peroxidase which gives a brown colour with ortho-dianisidine and alkaline phosphatase which gives a yellow colour with p-nitrophenyl phosphate.
Clark, M.F. and Bar-Joseph, M. (1984) **In** Methods in Virology, Vol. 7, p.51, ed. K. Maramorosch and H. Koprowski. Academic Press: New York.

elm mottle virus. An *Ilarvirus*.
Jones, A.T. (1974) CMI/AAB Descriptions of Plant Viruses No. 139.
Francki, R.I.B. (1985) **In** The Plant Viruses. Vol. 1. p. 1. ed. R.I.B. Francki. Plenum Press: New York.

elongation factor. Protein, forming part of the ribosomal binding complex, which promotes elongation of a polypeptide chain.

elution of virus. Release of virus particles either from the cell surface, e.g. ORTHOMYXOVIRUSES from erythrocytes by the action of NEURAMINIDASE or from a solid support such as an ion exchange column. Used in the purification of viruses.

EMC virus. *See* ENCEPHALOMYOCARDITIS VIRUS.

empty particles. Virus particles which do not contain any nucleic acid, e.g. the top component of TURNIP YELLOW MOSAIC VIRUS. Can be identified by negative staining and electron microscopy (the stain penetrates the particle where the nucleic acid is missing) and by having a lower BUOYANT DENSITY than complete virus particles (i.e. similar to that of protein).

enation. A symptom caused by certain plant viruses in which there are small outgrowths on the plant. In PEA ENATION MOSAIC VIRUS the outgrowths are on the underside of the veins on leaves.

encephalomyocarditis virus. Family *Picornaviridae*, genus *Cardiovirus*. A natural infection of rodents in which it does not cause disease. The virus has been isolated from many species, including man. Antibodies are found in many species. During the last few years, several outbreaks of fatal disease have occurred in pigs. Several species can be infected experimentally, including mice, hamsters and guinea pigs. The virus grows to high titre in many cell cultures. Unusually for a picornavirus, it agglutinates RBC at very high dilution.

endemic. A disease which persists in a given locality.

endive virus. A probable plant *Rhabdovirus*. Matthews, R.E.F. (1982) Intervirology **17**, 113.

endocytosis. An active process in which extracellular materials are introduced into the cytoplasm of cells by PINOCYTOSIS.

endogenous Drosophila line virus. A possible NODAVIRUS (*NODAVIRIDAE*) detected in a subline of *Drosophila melanogaster* line 1 cells, related to but distinguishable from BLACK BEETLE VIRUS. Friesen, P. *et al.* (1980) J. Virol. **35**, 741.

endogenous virus. A virus causing a persistent infection and/or whose genome integrates with cellular DNA and thus is transmitted vertically e.g. RETROVIRUSES.

endonuclease. An enzyme which cleaves a polynucleotide chain internally.

endoplasmic reticulum. A membrane system in the cytoplasm of cells that functions in synthesis and intracellular distribution of proteins. *See* ROUGH MEMBRANE, SMOOTH MEMBRANE.

engulfment. Synonym: PINOCYTOSIS.

enhancement. An increased yield of one virus in joint infections with another virus.

Enseada virus. Family *Bunyaviridae*, genus *Bunyavirus*.

Entebbe bat virus. Family *Flaviviridae*, genus *Flavivirus*. Isolated from salivary glands of bats in Uganda.

enteritis of mink virus. *See* FELINE PANLEUCOPENIA VIRUS.

Enterovirus. (Greek 'enteron' = intestine). Genus of family PICORNAVIRIDAE, type species human poliovirus 1. Virions are stable at acid pH and have a buoyant density in CsCl of 1.33-1.34 g/cc. Viruses infect vertebrates including man (e.g. polioviruses 1-3), Coxsackie A viruses 1-24, Coxsackie B viruses 1-6, ECHOviruses 1-9, 11-27, 29-34, enteroviruses 68-72), monkeys (simian enteroviruses 1-18), pigs, (porcine enteroviruses 1-8) and cows (bovine enteroviruses 1-7). Hepatitis A virus is a probable member. Viruses primarily infect gastro-intestinal tract but also multiply in nerve, muscle, etc.

entomopathogenic virus. A virus which causes disease in insects.

Entomopoxvirinae. (Greek 'entomon' = insect). A subfamily in the family *Poxviridae*, containing poxviruses which replicate only in insects. There is a single genus, ENTOMOPOXVIRUS.

Entomopoxvirus. Genus of pox-like viruses (family POXVIRIDAE) isolated from insects. Virions are brick-shaped or ovoid (170-250 nm. x 300-400 nm.) with an external coat containing lipid and tubular or globular protein subunits, enclosing one or two LATERAL BODIES and a core containing the genome. Virions are often occluded within large proteinaceous OCCLUSION BODIES (SPHEROIDS) up to 24 μm in diameter. The protein lattice of the spheroids (SPHEROIDIN) is composed of a single polypeptide species (mw. $102\text{-}110 \times 10^3$). Virions may contain at least 40 structural polypeptides (mw. $12\text{-}250 \times 10^3$) and at least four enzyme activities: nucleotide phosphohydrolase, acidic and neutral deoxyribonucleases and DNA-dependent RNA polymerase. The genome is a linear dsDNA (mw. $123\text{-}136 \times 10^6$) with a low G+C content (16-26%). Virus replication occurs exclusively in the cytoplasm, mainly in fat body tissue, following ingestion of virus. Three probable sub-genera have been defined, based on morphological differences between virions: A: type species, MELOLONTHA ENTOMOPOXVIRUS; B: type species: AMSACTA ENTOMOPOXVIRUS; C: type species: CHIRONOMUS ENTOMOPOXVIRUS.

Viruses in sub-genus A have been isolated from Coleoptera; B from Lepidoptera and Orthoptera, C from Diptera. DNAs from members of sub-genus B are smaller (123-136 x 10^6) than DNAs from the other sub-genera (c.20 x 10^6). No serological relationships nor DNA homology have been observed between entomopoxviruses and other members of the *Poxviridae*. A list of insect hosts in which entomopoxviruses have been recorded is given in Appendix D.
Arif, B.M. (1984) Adv. Virus Res. **29**, 195.

envelope. A membrane layer surrounding a virus particle, e.g. ORTHOMYXO- OR PARAMYXO-VIRUS. It usually contains virus-coded proteins.

enzootic. A disease persisting within animals in a defined geographic region.

enzyme conjugate. Usually an antibody preparation to which an enzyme is linked covalently. The enzyme is chosen to produce a colour reaction with an appropriate substrate; enzymes commonly used include alkaline phosphatase and horseradish peroxidase. Used in ENZYME-LINKED IMMUNOSORBENT ASSAY and WESTERN BLOTTING.

enzyme elevating virus. *See* RILEY VIRUS.

enzyme-linked immunosorbent assay. *See* ELISA.

epidemic. A sudden increase in the incidence of a disease, affecting a large number of animals or plants and spreading over a large area.

epidemic jaundice virus. *See* HEPATITIS A VIRUS.

epidemiology. The study of factors involved in the spread of diseases.

Epilachna varivestis reovirus. Unclassified reovirus-like agent observed in tissues of the Mexican bean beetle, *Epilachna varivestis* (Coleoptera: Coccinellidae).
Kitajima, E.W. *et al.* (1985) J. Invertebr. Pathol. **46**, 83.

episome. An extrachromosomal element which replicates in a cell independent of the chromosome. Thus most viruses are episomal as are plasmids.

epitope. The structure of an ANTIGEN which elicits the formation of specific antibodies, e.g. a grouping of amino acid sequences on a protein, or between adjacent protein subunits; also termed ANTIGENIC SITE. Various types of epitope have been distinguished: cryptotope which is buried inside the antigen and becomes reactive only after denaturation or dissociation of the antigen; neotope which arises from changes in folding of the polypeptide chain; metatope which is formed from adjacent protein molecules during assembly of structures from subunits.
van Regenmortel, M.H.V. (1966) Adv. Virus Res. **12**, 207.

epizootic. A disease affecting many susceptible animals in one region simultaneously; widely diffuse and rapidly spreading.

epizootic haemorrhagic disease of deer virus. Synonym: EHD VIRUS. (Family *Reoviridae*, genus *Orbivirus*). Causes haemorrhages, coma and death in Virginian white-tailed deer. Occurs in USA and Canada. Probably transmitted by *Culicoides*.

Epstein-Barr virus. Synonym: BURKITT'S LYMPHOMA VIRUS, GLANDULAR FEVER VIRUS. Family *Herpesviridae*, subfamily *Gammaherpesvirinae*, not assigned to genus. Isolated from Burkitt's lymphoma of African children. A widely occurring human infection, mainly of children. The disease is a serious problem in some parts of China. Primary infection of young adults causes infectious mononucleosis. The virus can be grown in human cell cultures.

EPV. Abbreviation for ENTOMOPOXVIRUS.

equid herpesvirus. Synonym: EQUINE RHINOPNEUMONITIS VIRUS, EQUINE ABORTION VIRUS. Family *Herpesviridae*, subfamily *Alphaherpesvirinae*, genus *suid alpha herpesvirus*. Causes acute respiratory disease in young horses. Can cause abortion. Transmitted by aerosol route. Horses infected experimentally may develop genital vesicular exanthema. Experimentally infected guinea pigs abort. Can be grown in the CAM or in the yolk sac and amnion and in several different cell cultures.

equid herpesvirus 3. *See* COITAL EXANTHEMA VIRUS.

equilibrium density gradient centrifugation. Synonym: ISOPYCNIC GRADIENT.

equine abortion virus. *See* EQUID HERPESVIRUS.

equine adenovirus. Family *Adenoviridae*, genus *Mastadenovirus*. Isolated from horses with respiratory disease. Often causes death in young animals from pneumonia.

equine arteritis virus. Synonym: EQUINE INFECTIOUS ARTERITIS VIRUS, PINK EYE VIRUS. Family *Flaviviridae*, genus *Flavivirus*. Causes fever, conjunctivitis, rhinitis, oedema, enteritis and colitis. In pregnant mares, the foetus may become infected and abortion occur. Disease can be fatal. Virus grows in horse kidney cell cultures.

equine infectious anaemia virus. Family *Retroviridae*, subfamily *Lentivirinae*. Causes acute or chronic infection in horses. Acute infection typified by fever, anaemia, nasal discharge and oedema. Disease is usually fatal. Grows in cell cultures of embryonic equine tissue and in leucocyte cultures.

equine infectious arteritis virus. *See* EQUINE ARTERITIS VIRUS.

equine influenza virus. Family *Orthomyxoviridae*, genus *Influenza virus A*. Causes respiratory infection of horses which is important in the horse-racing industry. A vaccine is available.

equine papilloma virus. Family *Papovaviridae*, genus *Papillomavirus*. Causes papillomas on nose and lips. Can be transmitted experimentally.

equine rhino-pneumonitis virus. *See* EQUID HERPESVIRUS.

equine rhinovirus. Family *Picornaviridae*, genus *Rhinovirus*. Causes rhinitis and pharyngitis. Spreads rapidly. Several other species can be infected.

Erysimum latent virus. A *Tymovirus*.
Shukla, D.D. and Gough, G.H. (1980) CMI/AAB Descriptions of Plant Viruses No. 222.
Francki, R.I.B. *et al.* (1985) **In** Atlas of Plant Viruses. Vol. 1. p. 117. CRC Press: Boca Raton, Florida.
Hirth, L. and Girard, L. (1988) **In** The Plant Viruses. Vol. 3. p. 163. ed. R. Koenig. Plenum Press: New York.

ET$_{50}$. Abbreviation for MEDIAN EFFECTIVE TIME.

ethidium bromide. A chemical which intercalates between the base pairs of nucleic acid, either in ds nucleic acid or due to the secondary structure of ss nucleic acid. It has two main uses: a) in the detection of nucleic acids in gels or gradients as it fluoresces brightly in UV light; its maximum excitation is at 300 nm. wavelength, maximum emission at 450 and 590 nm.; b) in separating covalently closed circular from linear and open circular DNA in CsCl ISOPYCNIC GRADIENTS. The covalently closed DNA binds less ethidium bromide than does the relaxed DNA and, as its density in the gradient is reduced less, it bands below the open-circular and linear DNA.

ethylenediaminetetra-acetic acid (EDTA). Mw. of disodium salt = 372.2. A chelating agent with high affinity for Mg^{2+} ions. Used in solutions to remove divalent cations, inhibit DNase and to act as a bacteriostat.

ethyleneglycolbis(aminoethylether)tetra-acetic acid (EGTA). Mw. 380.4. A chelating agent with, at its optimum pH of 6.5-7.0, high affinity for Ca^{2+} (log K$_{Ca}$ = 11.0) and less affinity for Mg^{2+} (log K$_{Mg}$ = 5.2). It is thus used for removing Ca^{2+} in the presence of Mg^{2+}. Used in the preparation of rabbit reticulocyte lysate for *in vitro* translation in inhibiting the micrococcal nuclease used to remove endogenous mRNAs.
Marhol, M. and Cheng, K.L. (1970) Anal. Chem. **42**, 652.

Eubenangee virus. Family *Reoviridae*, genus *Orbivirus*. Isolated from mosquitoes in Northern Australia. Antibodies found in wallabies and kangaroos.

eucharis mosaic virus. A possible *Potyvirus*.

Francki, R.I.B. *et al.* (1985) **In** Atlas of Plant Viruses. Vol. 2. p. 183. CRC Press: Boca Raton, Florida.

eukaryote. An organism which has the following cellular features: a) a genome divided into a number of chromosomes which are contained within the nucleus; b) the cytoplasm contains other membrane-bound organelles as well as the nucleus; c) gene expression does not involve the grouping of genes into operons, and d) three DNA-dependent RNA polymerase species with different specificities. *See* PROKARYOTE.

eukaryotic initiation factors. Proteins which are essential for the translation of mRNAs. Their functions include the recognition of the 5′-terminal CAP structure and binding of the 40S ribosome subunit to the mRNA. There are at least nine eukaryotic initiation factors.
Pain, V.M. (1986) Biochem. J. **235**, 625.

Euonymus fasciation virus. A possible plant *Rhabdovirus*.
Francki, R.I.B. *et al.* (1985) **In** Atlas of Plant Viruses. Vol. 1. p. 73. CRC Press: Boca Raton, Florida.

Euonymus mosaic virus. A *Rhabdovirus*, occurs in Japan.
Doi, Y. *Personal communication.*

Eupatorium yellow vein virus. A probable *Geminivirus*, subgroup B.
Harrison, B.D. (1985) Ann. Rev. Phytopath. **23**, 55.

Euphorbia mosaic virus. A *Geminivirus*, subgroup B.
Harrison, B.D. (1985) Ann. Rev. Phytopath. **23**, 55.

Euphorbia ringspot virus. A possible *Potyvirus*.
Francki, R.I.B. *et al.* (1985) **In** Atlas of Plant Viruses. Vol. 2. p. 183. CRC Press: Boca Raton, Florida.

European wheat striate mosaic virus. A possible member of the *Tenuivirus* group.
Hoppe, W. (1987) Proc. Czechoslovak. Plant Virologists **8**, 371.

Everglades virus. Family *Togaviridae*, genus *Alphavirus*. Isolated from the cotton mouse, rat and mosquitoes in the Everglades National Park. Can cause infection of man.

exon. In many eukaryotic genes there are regions of intervening non-coding sequences (introns) which separate regions of coding sequence (exons). The initial transcript from such genes is processed by SPLICING to remove the introns and to produce the mRNA. Many DNA-containing viruses, e.g. SV40, ADENOVIRUS, have exons.

exonuclease. An enzyme which requires a free end in order to digest a RNA or DNA molecule. There are both 5′- or 3′-exonucleases.

exponential growth phase. The phase of viral or bacterial replication at which the number of particles or cells increases at a constant rate.

extinction coefficient. Synonym: SPECIFIC ABSORBANCE.

Eyach virus. Family *Reoviridae*, genus *Orbivirus*. Isolated from ticks in Germany. Antigenically related to Colorado tick fever.

F

F factor. *See* F PLASMID.

F plasmids. Conjugative plasmids (mw. 63 x 10^6) which determine F+ and F– mating type in *Escherichia coli* and other enterobacteria (e.g. *Shigella, Proteus, Salmonella*) and mediate transfer of chromosomal genes between these bacteria. Code for sex pili (F pili); hollow tubes which protrude up to 20 μm from the cell. Cells containing the plasmid are denoted F+ (male). Members of the LEVIVIRUS group (e.g. MS2 PHAGE) and some INOVIRUSES (e.g. fd PHAGE) are specific for enterobacteria as they adsorb exclusively to the F PILUS.
Hardy, K. (1986) Bacterial Plasmids. 2nd. ed. Thomas Nelson: Walton-upon- Thames.
Fiers, W. (1977) **In** Comprehensive Virology. Vol. 13. p. 69. ed. H. Fraenkel-Conrat and R.R. Wagner. Plenum Press: New York.

Fab fragment. The antigen-binding fragment of IMMUNOGLOBULIN.

Fabavirus group. (Latin 'faba' = bean, after type member, BROAD BEAN WILT VIRUS GROUP, serotype 1). Genus containing viruses which resemble COMOVIRUSES in the structure and composition of the virus particles but differ in cytopathology (crystals and tubular arrays of virus particles) and in being aphid transmitted. The particles are isometric, about 30 nm. in diameter, and sediment as three components, 113-126S (bottom (B) component), 93-100S (middle (M) component) and 56S (top (T) component which lacks nucleic acid).

   100nm

The capsids comprise two polypeptide species of mw. *c*.27 and 43 x 10^3. The genome is two species of linear (+)-sense ssRNA, B component containing RNA 1 (mw. 2.1 x 10^6) and M component containing RNA 2 (mw. 1.5 x 10^6). The host ranges are wide. Viruses of this group are easily transmissible mechanically and are transmitted by aphids in the NON-PERSISTENT TRANSMISSION manner.
Lisa, V. and Boccardo, G. (in press). **In** The Plant Viruses: Viruses with Bipartite RNA Genomes and Isometric Particles. ed. B. Harrison and A. Murant. Plenum Press: New York.

Facey's Paddock virus. Family *Bunyaviridae*, genus *Bunyavirus*.

Farallon virus. Family *Bunyaviridae*, genus *Nairovirus*.

Fc fragment. The crystallisable constant fragment of IMMUNOGLOBULIN.

fd phage. Proposed type member of the *INOVIRUS* genus. Virions are flexuous rods, about 870 nm. x 6 nm. Particles contain a few copies of a minor protein (mw. 56 x 10^3) present at one tip of the particle, which are involved in adsorption of the virus to the F-pilus of the host. The major protein has a mw. of about 5.2 x 10^3. The genome is circular ssDNA (mw. 1.9 x 10^6; 6389 nucleotides) with at least eight genes. During infection, replicative form (RF) DNA is produced and transcribed by host enzymes. Single-stranded progeny DNA is produced by displacement from RF DNA. The virus is specific for bacteria containing the F PLASMID. Virus infection does not lyse the host. Close relatives include f1 and M13 phages.
Ray, D.S. (1977) **In** Comprehensive Virology. Vol. 7. p. 105. ed. H. Fraenkel-Conrat and R.R. Wagner. Plenum Press: New York.

felid herpesvirus 1. Family *Herpesviridae*, subfamily *Alphaherpesvirinae*, genus not allotted. Causes nasal discharge, lacrimation and fever in kittens. Virus replicates in the mucous membranes of the nose, larynx and trachea and in the conjunctiva and genital tract. Causes lesions in cell cultures prepared from cat kidney, lung and testes.

feline calicivirus. Synonym: FELINE RHINO-
TRACHEITIS VIRUS. Family *Caliciviridae*, genus
Calicivirus. Causes rhinitis, conjunctivitis, ulceration and pneumonia. Can be fatal. Replicates
in feline kidney cells. Occurs as several serotypes.

feline coronavirus. Synonym: FELINE INFECTIOUS
PERITONITIS VIRUS. Family *Coronaviridae*, genus
Coronavirus. Causes loss of appetite, wasting
and abdominal distension due to peritonitis, pleu-
risy and necrotic inflammatory lesions. The disease is fatal.

feline infectious peritonitis virus. *See* FELINE
CORONAVIRUS.

feline leukaemia oncovirus. Family *Retroviridae*, subfamily *Oncovirinae*, genus *Type C
oncovirus*, sub-genus *Mammalian type C oncovirus*. A common infection of cats causing leukaemia and/or sarcomas. Depresses immune
system leading to a variety of infections. Virus
replicates in cells of feline, human, canine and pig
origin. Vaccination prevents infection.

feline panleucopenia virus. Synonyms: ATAXIA
OF CATS VIRUS, FELINE PARVOVIRUS. Family *Parvoviridae*, genus *Parvovirus*. Infects all *Felidae*.
Causes a severe febrile illness with vomiting and
diarrhoea. Infected animals may excrete virus for
prolonged periods. After an initial leucocytosis,
there is a progressive fall in circulating lymphocytes and polymorphs, leading to lethargy and
anorexia. Virus can be grown in kitten kidney
cells, particularly if they are rapidly dividing.

feline rhino-tracheitis virus. *See* FELINE
CALICIVIRUS.

feline syncytial virus. A species in the family
Retroviridae, subfamily *Spumavirinae*. Isolated
from normal cats. Not known to cause disease.
Grows in feline embryo cell cultures.

fern virus. A possible *Potyvirus*.
Francki, R.I.B. *et al.* (1985) **In** Atlas of Plant
Viruses. Vol. 2. p. 183. CRC Press: Boca Raton,
Florida.

Festuca leaf streak virus. A plant *Rhabdovirus*,
subgroup 1.
Francki, R.I.B. *et al.* (1985) **In** Atlas of Plant
Viruses. Vol. 1. p. 73. CRC Press: Boca Raton,
Florida.

Festuca mottle virus. Synonym: COCKSFOOT
MILD MOSAIC VIRUS.
Hull, R. (1988) **In** The Plant Viruses. Vol. 3. p.
113. ed. R. Koenig. Plenum Press: New York.

Festuca necrosis virus. A member of the
Closterovirus subgroup 1.
Francki, R.I.B. *et al.* (1985) **In** Atlas of Plant
Viruses. Vol. 2. p. 219. CRC Press: Boca Raton,
Florida.

Feulgen stain. A histochemical stain made up
from basic fuchsin and sulphurous acid. Stains
CHROMATIN containing THYMIDINE. Also known as
Schiff's reagent.

few-polyhedra variant. Plaque variant of NU-
CLEAR POLYHEDROSIS VIRUS which produces few or
no polyhedra in nuclei of infected insect cell
cultures. Arises as a spontaneous mutant of MANY
POLYHEDRA VARIANT ('wild-type'), apparently by
a transposon-like insertion into the viral genome
of host cell DNA sequences.
Fraser, M.J. *et al.* (1985) Virology **145**, 356.

fibroblast. The stellate connective tissue cell
type found in fibrous tissue. Important type of cell
for use in cell culture, e.g. HELA CELLS.

fig virus S. A *Carlavirus*, occurs in Japan.
Doi, Y. *Personal communication.*

figwort mosaic virus. A *Caulimovirus*.
Francki, R.I.B. *et al.* (1985) **In** Atlas of Plant
Viruses. Vol. 1. p. 17. CRC Press: Boca Raton,
Florida.

Fiji disease virus. *See* SUGARCANE FIJI DISEASE
VIRUS.

Fijivirus group. Synonym: PLANT REOVIRUS
SUBGROUP 2. (SUGARCANE FIJI DISEASE VIRUS, the
type member). One of the two genera of plant
viruses in the family *REOVIRIDAE*. The particles are
isometric, 65-71 nm. in diameter with 12 external
knobs 8-16 nm. long, 11 nm. in diameter situated
one on each 5° axis. This structure breaks down

100nm

readily to release cores 54 nm. in diameter, which have 12 spikes 8 nm. long, 14-19 nm. in diameter. The particles contain ten species of dsRNA (mw. 1.0-2.9 x 10⁶). Replication is in cytoplasmic VIRO-PLASMS consisting of a matrix containing filaments. Fijiviruses multiply both in higher plants (confined to Graminae) and in insects (plant hoppers). Particles are found in most cell types. Natural transmission is by Delphacid plant hoppers in which the relationship is in the PERSISTENT TRANSMISSION manner and propagative; the virus can be transmitted transovarially.
Matthews, R.E.F. (1982) Intervirology **17**, 85.

filamentous phages. *See* INOVIRUS.

filaree red leaf virus. A possible *Luteovirus*. Francki, R.I.B. *et al.* (1985) **In** Atlas of Plant Viruses. Vol. 1. p. 137. CRC Press: Boca Raton, Florida.

Filoviridae. (Latin 'filo' = thread, filament.) A family consisting of the two viruses Marburg and Ebola, which cause haemorrhagic fever in man. Except for their extreme length, the viruses have a morphology similar to that of members of the family *Rhabdoviridae*. The length is highly variable and can be as great as 14,000 nm. but is usually about 800-1000 nm.; the diameter is 80 nm. The particles are enveloped with spikes *c*.7

Casper, R. (1988) **In** The Plant Viruses. Vol. 3. p. 235. ed. R. Koenig. Plenum Press: New York.

filterable. The ability of a solute in a solvent to pass through a filter. The early recognition of the small size of virus particles was due to their filterability through diatomite or glazed porcelain filters with pore sizes too small to allow the passage of bacteria. D. Ivanowski (1892) and M.W. Beijerink (1898) showed that the particles of TOBACCO MOSAIC VIRUS would pass through a Chamberland filter-candle.
Beijerink, M.W. (1898) Verhand. Kon. Akad. Weten. Amsterdam **6**, 3.
Ivanowski, D. (1892) St. Petersb. Acad. Imp. Sci. Bull. **37**, 67.

fin isolate virus. Family *Reoviridae*, genus *Orbivirus*.

finch paramyxovirus. Family *Paramyxoviridae*, genus *Paramyxovirus*.

finger millet mosaic virus. A plant *Rhabdovirus*, subgroup 2; leafhopper transmitted. Francki, R.I.B. *et al.* (1985) **In** Atlas of Plant Viruses. Vol. 1. p. 73. CRC Press: Boca Raton, Florida.

fingerprinting. A procedure for characterising

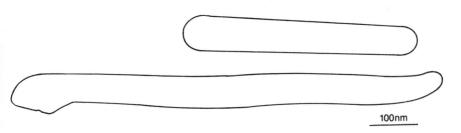

100nm

nm. in length and 10 nm. apart. The virus RNA has a mw. of *c*.4.2 x 10⁶, has (–)-sense and is therefore non-infectious. There are at least five virus proteins with mw. of *c*. 190, 125, 104, 40 and 26 x 10³. Both viruses are highly virulent for man and several species of monkey.
Kiley, M.P. *et al.* (1982) Intervirology **18**, 24.

Filaree red leaf virus. A possible *Luteovirus*. Francki, R.I.B. *et al.* (1985) **In** Atlas of Plant Viruses. Vol. 1. p. 137. CRC Press: Boca Raton, Florida.

DNA, RNA or proteins by electrophoretic or chromatographic analysis of specific fragments, e.g. uniformly- or terminally-labelled RNA is digested using various ribonucleases (often T1 or pancreatic) and the products separated by electrophoresis in two dimensions. The oligoribonucleotides are then detected by autoradiography.

flacherie. A term applied to a disease condition of the silkworm, *Bombyx mori*, caused by a complex of pathogens and physiological deficiencies.

The most common viral pathogens involved are CYTOPLASMIC POLYHEDROSIS VIRUS, INFECTIOUS FLACHERIE VIRUS and INA-FLACHERIE VIRUS.

flacherie 1 virus. Synonym: INFECTIOUS FLACHERIE VIRUS.

Flanders virus. Family *Rhabdoviridae*, not allotted to genus. Isolated from mosquitoes and from an ovenbird in USA. Kills new-born mice when inoculated i.c.

Flaviviridae. (Latin 'flavus' = yellow.) Formerly a genus in the family *Togaviridae*. A family of enveloped RNA viruses with spherical particles 40-50 nm. in diameter, sedimenting at *c*.200S and with surface projections. There are

 100nm

three proteins, the glycosylated surface projection (mw. 53-63 x 10^3), the nucleocapsid protein (mw. 13.5 x 10^3) and a small protein (mw. *c*.8 x 10^3). The RNA is ss (+)-sense (mw. 4.0-4.6 x 10^6). Replication is in the cytoplasm. Most members multiply in arthropods as well as in vertebrates. Westaway, E.G. *et al.* (1985) Intervirology. **24**, 183.

Flavivirus. (Latin 'flavus' = 'yellow'.) The only genus of the family *Flaviviridae*. Contains many members, the most important being those causing dengue haemorrhagic fever, Japanese encephalitis, louping ill, Murray Valley encephalitis, St. Louis encephalitis, tick-borne encephalitis and yellow fever.

flexal virus Be An 293022. Family *Arenaviridae*, genus *Arenavirus*.

flock house virus. A possible NODAVIRUS (NODAVIRIDAE) isolated from the grass grub, *Costelytra zealandica* (Coleoptera) in New Zealand. Scotti, P.D. *et al.* (1983) Arch. Virol. **75**, 181.

fluorescein isothiocyanate (FITC). ($C_{12}H_{11}NO_5S$) (mw. 389.4). A fluorescent compound used for labelling proteins or nucleic acids. It is excited by light of wavelength in the range 450-490 nm. and emits in the range 520-560 nm.

fluorescence microscope. A compound light microscope which is arranged to admit radiation of specific wavelengths (e.g. UV) to the specimen which then fluoresces.

fluorescent antibody. An antibody which is labelled with a fluorescent dye, e.g. FLUORESCEIN ISOTHIOCYANATE. It can then be used in conjunction with a fluorescence microscope to detect viral antigen in cells. *See* ENZYME CONJUGATE.

5-fluorodeoxyuridine (Fudr). A pyrimidine analogue which is a reversible inhibitor of DNA synthesis. Phosphorylation by THYMIDINE KINASE converts it to an analogue of thymidylic acid. Used in the treatment of CYTOMEGALOVIRUS.

fluorography. Photography of an image produced on a fluorescent screen. Now more widely used to describe a technique in which ^{3}H-labelled molecules can be detected in chromatograms and polyacrylamide gels. The scintillator PPO is introduced into the chromatogram or gel which is then exposed to photographic film. Laskey, R.A. and Mills, A.D. (1975) Eur. J. Biochem. **56**, 335.

5-fluorouracil. *See* BASE ANALOGUE.

FMDV. *See* FOOT-AND-MOUTH DISEASE VIRUS.

focus-forming assay. An assay for non-cytolytic transforming viruses, e.g. ROUS SARCOMA VIRUS based on the transformation changing the morphology of the cells in tissue culture. The sites of growth of these modified cells show up as foci. A similar test can be used for the assay of DEFECTIVE INTERFERING PARTICLES of a cytolytic virus (e.g. LYMPHOCYTIC CHORIOMENINGITIS VIRUS) in which the DI particles prevent the virus destruction of the cells by the helper virus.

focus-forming units (f.f.u.) Units of quantification for the FOCUS-FORMING ASSAY.

foetal calf serum. A frequent constituent of culture media used for growing animal cells or tissue cultures. *See* SERUM-FREE MEDIUM.

foot-and-mouth disease virus. Synonyms: APHTHOVIRUS, FMDV, VIRUS AFTOSA, LE VIRUS DE LA FIEVRE APHTEUSE, MAUL- UND KLAUENSEUCHEVIRUS. Family *Picornaviridae*, genus *Aphthovirus*. Only species of genus. First virus shown to cause disease of animals (1897). Produces blisters on tongue and feet of cattle, pigs, sheep and goats. Highly contagious and debilitating disease. Very important economically because of productivity losses and restrictions it causes on trading. This has led to extensive vaccination programmes where disease is endemic. Control in areas with sporadic outbreaks is by slaughter. Occurs worldwide except in Australasia, Britain, Japan and North America. Antigenic variation is an important consideration in vaccination (seven serotypes, O, A, C, SAT1, SAT2, SAT3 and Asia 1).

Formvar. Trade name for polyvinyl formal used for making films for grids for electron microscopy.

formycin. 7-amino-3-(β-D-ribofuranosyl)-pyrazolo[4,3-d]pyrimidine. An analogue of adenosine which acts as a nucleoside antibiotic.

formycin B. 3-(β-D-ribofuranosyl)pyrazolo[4,3-d]-6(H)-7-pyrimidine. An analogue of inosine isolated from *Streptomyces lavendulae* and *Nocardia interforma*. Acts as a nucleoside antibiotic.

Fort Morgan virus. Family *Togaviridae*, genus *Alphavirus*.

fowl adenovirus. Family *Adenoviridae*, genus *Aviadenovirus*.

fowl diphtheria virus. See FOWLPOX VIRUS.

fowlpox virus. Synonym: FOWL DIPHTHERIA VIRUS. Family *Poxviridae*, subfamily *Chordopoxvirinae*, genus *Avipoxvirus*. Causes lesions which are followed by scabbing. Sometimes there are eye lesions. Transmitted by contact or by mosquitoes.

foxtail mosaic virus. A *Potexvirus*. Short, M.N. (1983) CMI/AAB Descriptions of Plant Viruses No. 264. Francki, R.I.B. *et al.* (1985) In Atlas of Plant Viruses. Vol. 2. p. 159. CRC Press: Boca Raton, Florida.

FP variant. Abbreviation for FEW POLYHEDRA VARIANT.

frame shift. 1. A mutation caused by insertion or deletion of one or more nucleotides whose effect is to change the reading frame of a codon giving a changed amino acid sequence starting at the mutated codon. 2. The change from the reading frame of one cistron to that of an overlapping cistron, e.g. the retroviruses. GAG-POL GENES of some retroviruses

frangipani mosaic virus. A *Tobamovirus*. Varma, A. and Gibbs, A.J. (1978) CMI/AAB Descriptions of Plant Viruses No. 196. Brunt, A.A. (1986) In The Plant Viruses. Vol. 2. p. 283. eds. M.H.V. van Regenmortel and H. Fraenkel-Conrat. Plenum Press: New York.

Fraser Point virus. Family *Bunyaviridae*, genus *Nairovirus*.

Freesia mosaic virus. A possible *Potyvirus*. Francki, R.I.B. *et al.* (1985) In Atlas of Plant Viruses. Vol. 2. p. 183. CRC Press: Boca Raton, Florida.

Freesia streak virus. A possible *Potyvirus*. Francki, R.I.B. *et al.* (1985) In Atlas of Plant Viruses. Vol. 2. p. 183. CRC Press: Boca Raton, Florida.

freeze-drying. See LYOPHILISATION.

freeze fracture. A method for preparing samples for electron microscopy. They are frozen rapidly at very low temperature and then the brittle material is fractured. The exposed surfaces may be etched to reveal further details.

Freund's adjuvant. A mixture of mineral oil and emulsifier (and, in the complete adjuvant, killed mycobacteria) with which an antigen is emulsified before intramuscular or subcutaneous injection. The antigen is released slowly into the blood stream, often leading to the production of a higher antibody titre.

Friend leukaemia virus. Family *Retroviridae*, subfamily *Oncovirinae*, genus *Type C oncovirus*. Originally obtained from spleen of Swiss mouse which had been injected at birth with cell-free extract of Ehrlich ascites tumour cells. Adult mice die a few weeks after injection.

Frijoles virus. Family *Bunyaviridae*, genus *Phlebovirus*. Isolated from insects in Panama.

frog virus 4. See RANID HERPESVIRUS 2.

Fuchsia latent virus. A possible *Carlavirus*. Francki, R.I.B. *et al.* (1985) In Atlas of Plant

Viruses. Vol. 2. p. 173. CRC Press: Boca Raton, Florida.

Furovirus group. (Sigla from fungus transmitted rod-shaped). (Type member SOIL-BORNE WHEAT MOSAIC VIRUS). Genus of MULTICOMPONENT plant viruses with rod-shaped particles. The type member has two nucleoprotein components of length 110-160 nm. and 300 nm., width 20 nm.

100nm

The particles are composed of a single species of coat protein (mw. 19.7×10^3) encapsidating two genomic linear (+)-sense ssRNA species (mw. 0.86-1.23 and 2.28×10^6). Host ranges are narrow. Furoviruses are mechanically transmissible. They are naturally transmitted by fungi (Plasmodiophorales). Many viruses in this group were originally classified as tobamoviruses but are distinguished from them by their multicomponent nature, coat protein mw. and fungal transmission.

Shirako, Y. and Brakke, M.K. (1984) J. gen. Virol. **65**, 119.

Brunt, A.A. (1986) In The Plant Viruses. Vol. 2. p. 305. ed. M.H.V. van Regenmortel and H. Fraenkel-Conrat. Plenum Press: New York.

fusion (of cells). Phenomenon caused by some enveloped viruses (e.g. SENDAI VIRUS). The ability of inactivated Sendai virus to fuse cells has been used in the production of hybrid cells.

G

G+C content. The total guanine (G) + cytosine content (C) of a nucleic acid but usually refers to dsDNA. Because the triple hydrogen bond between G and C (*see* figure under BASE PAIR) is more stable than the double hydrogen bond between A and T, the G+C content is a good measure of physical properties such as melting temperature; it also affects banding density in isopycnic gradients. G+C contents of dsDNA viruses vary from about 25% (ENTOMOPOX VIRUS) to 75% (e.g. HERPES VIRUS). *See* DENSITY, MELTING TEMPERATURE.

Gaeumannomyces graminis virus 019/6-A (GgV-019/6A). Type member of the *PARTITIVIRUS* group.
Buck, K.W. (1986) In Fungal Viruses. p. 221. ed. K.W. Buck. CRC Press: Boca Raton, Florida.

Gaeumannomyces graminis virus 87-1-H (GgV-87-1-H). A probable member of the *Totivirus* group.
Buck, K.W. (1986) In Fungal Viruses. p. 221. ed. K.W. Buck. CRC Press: Boca Raton, Florida.

Gaeumannomyces graminis virus T1-A (GgV-T1-A). A member of the *PARITITVIRUS* group.
Buck, K.W. (1986) In Fungal Viruses. p. 221. ed. K.W. Buck. CRC Press: Boca Raton, Florida.

gag-pol gene. The POLYPROTEIN expression of the adjacent GAG and POL GENES of RETROVIRUSES.

gag gene. Abbreviation of group-specific antigen gene. It is the 5´ gene on RETROVIRUS genomes and is translated to give the precursor protein for the internal proteins of those viruses.

gal virus. *See* GALLUS ADENO-LIKE VIRUS.

galinsoga mosaic virus. A member of the *Carmovirus* group.
Behncken, G.M. *et al.* (1982) CMI/AAB Descriptions of Plant Viruses No. 252.

Morris, T.J. and Carrington, J.C. (1988) In The Plant Viruses. Vol. 3. p.73. ed. R. Koenig. Plenum Press: New York.

Galleria mellonella densovirus. Type species of the DENSOVIRUS genus (*PARVOVIRIDAE*), isolated from the wax moth *G. mellonella* (Lepidoptera). Virus particles are approximately 23 nm. in diameter, although two distinct size classes have been reported, 24 and 21 nm. in diameter. Virions contain four structural proteins (mw. *c*.49, 58.5, 67 and 92 x 10^3), which in total size exceed the coding capacity of the genome. However, all proteins share extensive sequence homologies. The genome is single-stranded DNA (mw. 1.9-2.2 x 10^6), with inverted terminal repetitions of 60-380 base pairs. Particles package complementary (+) or (–)-sense strands. The virus host range is restricted to *G. mellonella* where it multiplies in almost all the tissues (except the mid gut) causing nuclear hypertrophy and death.
Kawase, S. (1985) In Viral Insecticides for Biological Control. p. 197. ed. K. Maramorosch and K.E. Sherman. Academic Press: New York.

Galleria mellonella nuclear polyhedrosis virus. NPV (BACULOVIRUS subgroup A) isolated from the waxmoth, *G. mellonella.* Closely related to the prototype baculovirus, *Autographa californica* MNPV (AcMNPV); it is regarded as a genotypic variant of this virus. In mixed infections with AcMNPV, the virus forms stable recombinants.
Croizier, G. and Quiot, J.M. (1981) Ann. Virol. **132**, 3.

gallid herpesvirus 1. *See* MAREK'S DISEASE VIRUS.

gallid herpesvirus 2. Synonym: TURKEY HERPESVIRUS. Family *Herpesviridae*, subfamily *Gammaherpesvirinae*. Isolated from turkeys. Grows in avian cell cultures. Protects fowls against Marek's disease and is used widely as a highly successful live vaccine.

gallus adeno-like virus. Synonym: GAL VIRUS. Family *Adenoviridae*, genus *Aviadenovirus*. Isolated from chicken cell cultures. Causes death of chick embryos experimentally but is not associated with natural disease.

Gamboa virus. Family *Bunyaviridae*, genus *Bunyavirus*. Isolated from mosquitoes in Panama.

gamma globulin. Any of the serum proteins with antibody activity.

Gammaherpesvirinae. A subfamily in the family *Herpesviridae*. Members cause lymphoproliferative disease.

Gan Gan virus. Family *Bunyaviridae*, genus *Bunyavirus*.

Ganjam virus. Family *Bunyaviridae*, genus *Nairovirus*. Isolated from ticks in India. Known to cause fever in man.

garland chrysanthemum temperate virus. A possible member of the *Cryptovirus* group, subgroup A.
Boccardo, G. *et al.* (1987) Adv. Virus Res. **32**, 171.

garlic mosaic virus. A possible *Carlavirus*. Francki, R.I.B. *et al.* (1985) **In** Atlas of Plant Viruses. Vol. 2. p. 173. CRC Press: Boca Raton, Florida.

garlic yellow streak virus. A possible *Potyvirus*. Francki, R.I.B. *et al.* (1985) **In** Atlas of Plant Viruses. Vol. 2. p. 183. CRC Press: Boca Raton, Florida.

gasping disease virus. *See* AVIAN INFECTIOUS BRONCHITIS VIRUS.

gastro-enteritis virus of dogs. *See* CANINE CORONAVIRUS.

gastro-enteritis virus of man. Family *Parvoviridae*, genus *Parvovirus*. One of many viruses which cause gastro-enteritis in man.

gattine. A disease syndrome of the silkworm, *Bombyx mori*, which may in part be caused by an unknown virus.

gel diffusion. *See* IMMUNODIFFUSION.

gel electrophoresis. ELECTROPHORESIS of macromolecules in a matrix of POLYACRYLAMIDE, AGAROSE or similar gel. The gel is chosen to have a uniform and determinable pore size which separates the macromolecules.

gel filtration. A type of column chromatography which separates molecules on the basis of size. The higher molecular weight molecules pass through the column first, the smaller molecules entering pores in the gel making up the column and thus being retarded.

Geminivirus group. (Latin 'gemini' = twins, from the characteristic double particles). Genus of the only group of plant viruses to contain circular ssDNA and which is divided into three subgroups (A, B and C). Members of subgroups A and C have genomes of only a single DNA species (2,687-2,749 and 2,993 nucleotides respectively) and are transmitted by leafhoppers, those of subgroup A (type member MAIZE STREAK VIRUS) infecting only monocotyledons, whereas those of subgroup C (type member BEET CURLY TOP VIRUS) infect dicotyledons. Members of subgroup B (type member CASSAVA (AFRICAN) MOSAIC VIRUS) have genomes of two DNA species (2,588-2,779 and 2,508-2,724 nucleotides), are transmitted by whitefly and infect only dicotyledons. The particles of members of each subgroup are geminate, 18 x 20 nm., consisting of two incomplete icosahedra with T=1 symmetry. They comprise

Subgroup A   Subgroup B

Subgroup C

22 capsomeres which are made up from subunits of mw. $28\text{-}34 \times 10^3$. Replication is thought to occur in the nucleus where particles accumulate in large aggregates.

Most individual geminiviruses have narrow host ranges. Particles are found mainly in the phloem and occasionally in other cell types. Some can be mechanically transmitted. Both leafhopper transmission of members of subgroups A and C and whitefly transmission of members of subgroup B are in the PERSISTENT TRANSMISSION manner.
Matthews, R.E.F. (1982) Intervirology **17**, 76.

Francki, R.I.B. *et al.* (1985) **In** Atlas of Plant Viruses. Vol. 1. p. 33. CRC Press: Boca Raton, Florida.

Harrison, B.D. (1985) Ann. Rev. Phytopath. **23**, 55.

gene. The unit of hereditary function. The nucleic acid which has the information for the expression of a functional protein or RNA molecule.

gene cloning. The cloning of the nucleic acid sequence of a gene. It is often done in an expression vector so that the gene product can be obtained from the cells in which the clone is propagated.

gene expression. The transcription of mRNA from the DNA sequence of a gene and the subse-

quent translation of that mRNA to give the protein gene product. Less strictly it can mean the transcription step alone.

genetic code. The arrangements of three nucleotides (codon) each of which specifies a single amino acid. The code is non-overlapping and so a single nucleotide change in a gene may only change one amino acid. The code is degenerate as 64 codons specify 20 amino acids and thus many amino acids are determined by more than one triplet. *See* START CODON, STOP CODON.

genetic complementation. *See* COMPLEMENTATION.

genetic engineering. Synonym: GENETIC MANIPULATION. The use of *in vitro* techniques to produce

genetic code

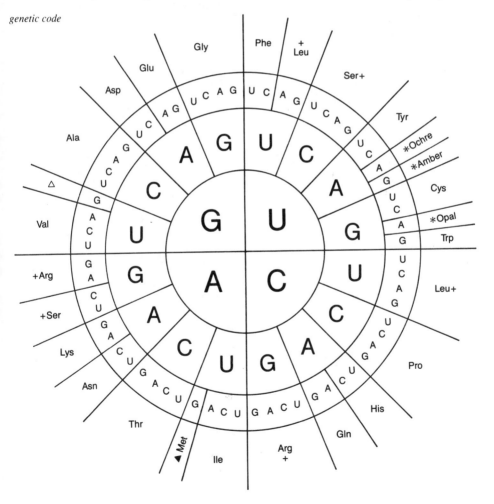

DNA molecules containing novel combinations of genes or other sequences.

genetic manipulation. *See* GENETIC ENGINEERING.

genetic map. A graphic representation of the linear arrangement of genes on a chromosome or genome. In large chromosomes the positions of the genes are determined by percentages of recombinations in linkage experiments. In smaller viral genomes they are determined by HETERODUPLEX ANALYSIS of mRNAs or by sequencing.

genetic marker. A mutation in a gene which allows its phenotypic identification.

genetic reassortment. *See* REASSORTMENT.

genetic transmission. *See* VERTICAL TRANSMISSION.

genome. The genetic information in a virus or cell. For a virus it is either DNA or RNA, but never both. DNA and RNA viral genomes may be either double- or single-stranded, circular or linear, unsegmented or segmented, (+)- or (−)-sense (if single-stranded) or AMBISENSE; ds circular RNA genomes have not yet been found.

genomic masking. *See* PHENOTYPIC MIXING.

genotype. The genetic constitution of an organism.

Gerbera symptomless virus. A possible plant *Rhabdovirus*.
Francki, R.I.B. *et al.* (1985) **In** Atlas of Plant Viruses. Vol. 1. p. 73. CRC Press: Boca Raton, Florida.

German measles virus. *See* RUBELLA VIRUS.

Germiston virus. Family *Bunyaviridae*, genus *Bunyavirus*. Isolated from man, rodents and mosquitoes in several countries in Africa. Causes fever in man. Natural hosts are sheep, cattle and goats.

Getah virus. Family *Togaviridae*, genus *Alphavirus*. Isolated from mosquitoes in several Pacific countries. Antibodies found in several species.

ghost. A term applied to TAILED PHAGE particles in which the contents of the head have been lost.

This can be induced by osmotic rupture, when phages are rapidly diluted from solutions of high ionic strength with water. Ghost particles are used in studies of virus adsorption and of effects on the host other than those attributable to virus replication.
Mathews, C.K. (1977) **In** Comprehensive Virology. Vol. 7. p. 179. ed. H. Fraenkel-Conrat and R.R. Wagner. Plenum Press: New York.

Gibbon ape leukaemia virus. Family *Retroviridae*, subfamily *Oncovirinae*, genus type C oncovirus, sub-genus mammalian type C oncovirus. Causes leukaemia in gibbons. Related antigenically to the simian sarcoma viruses.

glandular fever virus. *See* EPSTEIN-BARR VIRUS.

gloriosa fleck virus. A plant *Rhabdovirus*; occurs in Japan.
Araki, M. *et al.* (1985) Ann. Phytopath. Soc. Japan **51**, 632.

gloriosa stripe mosaic virus. A possible *Potyvirus*.
Francki, R.I.B. *et al.* (1985) **In** Atlas of Plant Viruses. Vol. 2. p. 183. CRC Press: Boca Raton, Florida.

Glycine mosaic virus. A *Comovirus*.
Francki, R.I.B. *et al.* (1985) **In** Atlas of Plant Viruses. Vol. 2. p. 1. CRC Press: Boca Raton, Florida.

Glycine mottle virus. A member of the *Carmovirus* group.
Behncken, G.M. and Dale, J.L. (1984) Intervirology. **21**, 154.
Morris, T.J. and Carrington, J.C. (1988) **In** The Plant Viruses. Vol. 3. p.73. ed. R. Koenig. Plenum Press: New York.

glycoprotein. A protein containing at least one carbohydrate group covalently attached to an amino acid.

glycosylated. Adjective to describe a protein containing at least one carbohydrate group.

glyoxal. (OHC-CHO). A chemical used to maintain nucleic acids in a denatured state during electrophoresis.
McMaster, G.K. and Carmichael, G.C. (1977) Proc. Natl. Acad. Sci. USA **74**, 4835.

GMP. Abbreviation for guanosine 5´-phosphate.

gnotobiotic. Adjective describing germ-free condition especially that in which experimental animals are inoculated with a known microorganism.

goat capripoxvirus. Synonym: GOAT POXVIRUS. Family *Poxviridae*, subfamily *Chordopoxvirinae*, genus *Capripoxvirus*. Causes epidermal lesions which proceed through the papule, vesicle and pustule stages to scab formation. Lesions found on many parts of body. Transmissible to sheep. Grows on chorioallantoic membrane, producing pocks and in cultures of sheep and goat kidney cells.

goat poxvirus. *See* GOAT CAPRIPOXVIRUS.

Goldberg-Hogness box. *See* TATA BOX.

Gomphrena rhabdovirus. A plant *RHABDOVIRUS*, subgroup 2.
Francki, R.I.B. *et al.* (1985) **In** Atlas of Plant Viruses. Vol. 1. p. 73. CRC Press: Boca Raton, Florida.

Gonometa virus. A member of the *PICORNAVIRIDAE*, not yet assigned to a specific genus, isolated from *Gonometa podocarpi* (Lepidoptera) from Uganda. Virions resemble vertebrate picornaviruses in many properties, being isometric, 32 nm. in diameter, having a buoyant density in CsCl of 1.35 g/cc., containing four structural proteins (mw. 36.5, 32, 29 and 12 x 10^3) and an ssRNA genome. Particles sediment at 180S. As with CRICKET PARALYSIS VIRUS, IgM antibodies to *Gonometa* virus have been detected in sera from several mammalian species. The explanation for this phenomenon is not known.
Longworth, J.F. (1978) Adv. Virus Res. **23**, 103.

goose hepatitis virus. Synonym: GOOSE PARVOVIRUS. Family *Parvoviridae*, genus *Parvovirus*. Causes haemorrhagic disease. Experimentally, injection of young goslings causes haemorrhagic liver disease and pericarditis, leading to death within ten days. Occurs widely in North America and Europe. Replicates in allantoic cavity of goose and Muscovy duck eggs.

goose parvovirus. *See* GOOSE HEPATITIS VIRUS.

Gordil virus. Family *Bunyaviridae*, genus *Phlebovirus*. Isolated from grass mouse and ger-bils in Central Africa.

Grace's medium. A general-purpose medium for the cultivation of insect cells.
Grace, T.D.C. (1962) Nature (Lond.) **195**, 788.

Grand Arbaud virus. Family *Bunyaviridae*, genus *Uukuvirus*. Isolated from ticks in France.

granule. Synonym for CAPSULE, the OCCLUSION BODY produced during infections by GRANULOSIS viruses.

granulin. The protein surrounding the virus particle of GRANULOSIS VIRUSES which constitutes the major part of the occlusion body (CAPSULE) produced during virus replication. It is a virus-coded polypeptide soluble at alkaline pH, with a mw. of 25-30 x 10^3. Closely-related in structure and function to POLYHEDRIN.
Rohrmann, G.F. (1986) J. gen. Virol. **67**, 1499.

granulosis virus (GV). Subgroup B of the *Baculovirus* genus. Virions usually contain one rod-shaped nucleocapsid (30-60 x 260-360 nm.) surrounded by an envelope. In general, each virion is individually occluded during the replication cycle within a proteinaceous occlusion body (*c*.300 x 500 nm.). Like the occlusion bodies of the closely-related NUCLEAR POLYHEDROSIS VIRUS (NPV) group, GV occlusion bodies help to preserve virus viability for many years outside the insect host. Most of the biochemical properties of the virions are shared with members of *Baculovirus* subgroup A (*see* BACULOVIRUS and NUCLEAR POLYHEDROSIS VIRUS) including a large circular supercoiled dsDNA genome, about 25 virion polypeptides (mw. 10-160 x 10^3) and an occlusion body matrix protein (granulin) of approximate mw. 29 x 10^3. GVs have only been isolated from Lepidoptera where they predominantly infect fat body tissue. Biochemical events in GV replication have been little studied (*see* NUCLEAR POLYHEDROSIS VIRUS) mainly because few GVs have been propagated successfully *in vitro*. Unlike NPVs, GV infection causes early disruption of the nuclear envelope. Some nucleocapsids gain their envelope by budding through the plasma membrane; others acquire the membrane by *de novo* synthesis and are then surrounded by granulin, to generate occlusion bodies. Virus infection is generally lethal, producing >10^{11} occlusion bodies from larger larvae. GVs have been isolated from >200 species of Lepidoptera. Each isolate often has very high host specificity, infecting only one insect species or a group of closely-related species. There is no formal nomenclature for each virus isolate although, conventionally, viruses have been named after the host from which they were first isolated. The type species is TRICHOPLUSIA NI GV. Other notable examples include GVs from the codling moth (see CYDIA POMONELLA GV), potato moth (see PHTHORIMAEA OPERCULELLA GV), cabbage white butterfly (see PIERIS SPP. GV) and Indian meal moth (see PLODIA INTERPUNCTELLA GV), many of which have high pathogenicity for their hosts and have been tested for use as biological insecticides. Virus names are often abbreviated in the literature e.g. CpGV for C. pomonella GV. See Appendix A for a list of insect hosts in which GV infections have been recorded.

Tweeten, K.A. et al. (1981) Microbiol. Revs. 45, 379.

Granados, R. R. and Federici, B.A. (1986). The Biology of Baculoviruses. Vols. I and II. CRC Press: Boca Raton, Florida.

grapevine Ajinashika virus. A possible Luteovirus; occurs in Japan.
Doi, Y. Personal communication.

grapevine Algerian latent virus. A Tombusvirus.
Martelli, G.P. et al. (1988) In The Plant Viruses. Vol. 3. p. 13. ed. R. Koenig. Plenum Press: New York.

grapevine Bulgarian latent virus. A Nepovirus.
Martelli, G.P. et al. (1978) CMI/AAB Descriptions of Plant Viruses No. 186.
Francki, R.I.B. et al. (1985) In Atlas of Plant Viruses. Vol. 2. p. 23. CRC Press: Boca Raton, Florida.

grapevine chrome mosaic virus. A Nepovirus.
Martelli, G.P. and Quaquerelli, A. (1972) CMI/AAB Descriptions of Plant Viruses No. 103.
Francki, R.I.B. et al. (1985) In Atlas of Plant Viruses. Vol. 2. p. 23. CRC Press: Boca Raton, Florida.

grapevine fanleaf virus. A Nepovirus.
Hewitt, W.B. (1970) CMI/AAB Descriptions of Plant Viruses No. 28.
Francki, R.I.B. et al. (1985) In Atlas of Plant Viruses. Vol. 2. p. 23. CRC Press: Boca Raton, Florida.

grapevine leafroll virus. A possible Clostero-

virus.
Francki, R.I.B. et al. (1985) In Atlas of Plant Viruses. Vol. 2. p. 219. CRC Press: Boca Raton, Florida.

grapevine stem-pitting associated virus. See GRAPEVINE VIRUS A.

grapevine virus A. Synonym: GRAPEVINE STEM-PITTING ASSOCIATED VIRUS. A member of the Closterovirus, subgroup 1.
Francki, R.I.B. et al. (1985) In Atlas of Plant Viruses. Vol. 2. p. 219. CRC Press: Boca Raton, Florida.

grass carp virus. Family Rhabdoviridae, not assigned to genus. Isolated from grass carp which die in eight to nine days after infection, with major haemorrhages.

grasserievirus. See BOMBYX MORI NUCLEAR POLYHEDROSIS VIRUS.

Great Island virus. Family Reoviridae, genus Orbivirus. Isolated from ticks and sea birds on Great Island, Newfoundland.

green islands. Non-chlorotic regions in a leaf showing mosaic symptoms. At least in TOBACCO MOSAIC VIRUS infections, they contain much less virus than do the chlorotic regions.

green sea-turtle herpesvirus. See CHELONID HERPESVIRUS 1.

Grey Lodge virus. Family Rhabdoviridae, unassigned to genus.

grey patch disease of turtles virus. See CHELONID HERPESVIRUS 1.

groundnut crinkle virus. A possible Carlavirus.
Francki, R.I.B. et al. (1985) In Atlas of Plant Viruses. Vol. 2. p. 183. CRC Press: Boca Raton, Florida.

groundnut eyespot virus. A possible Potyvirus.
Francki, R.I.B. et al. (1985) In Atlas of Plant Viruses. Vol. 2. p. 183. CRC Press: Boca Raton, Florida.

groundnut mild mottle virus. A Potyvirus.
Zeyong, X. et al. (1983) Plant Dis. 67, 1029.

groundnut rosette assistor virus. A *Luteovirus.*
Francki, R.I.B. *et al.* (1985) **In** Atlas of Plant
Viruses. Vol. 1. p. 137. CRC Press: Boca Raton,
Florida.
Casper, R. (1988) **In** The Plant Viruses. Vol. 3. p.
235. ed. R. Koenig. Plenum Press: New York.

groundnut rosette virus. Unencapsidated RNA,
dependent on GROUNDNUT ROSETTE ASSISTOR VIRUS
for aphid transmission. The two viruses together
cause groundnut rosette disease, an important
disease in central Africa.
Reddy, D.V.R. *et al.* (1985) Ann. appl. Biol. **107**,
65.

group-specific antigen. An antigen specific to a
group of viruses. *See* TYPE-SPECIFIC ANTIGEN.

Gryllus 'baculovirus'. Unclassified virus resembling a NON-OCCLUDED BACULOVIRUS infecting
nuclei of fat body cells in the field cricket *Gryllus
campestris.* Also infectious for *G. bimaculatus*
and *Teleogryllus* spp.
Huger, A.M. (1985) J. Invertebr. Pathol. **45**, 108.

GTP. Abbreviation for guanosine 5´-
triphosphate.

Guajara virus. Family *Bunyaviridae*, genus
Bunyavirus. Isolated from sentinel mice in S.
America.

Guama virus. Family *Bunyaviridae*, genus
Bunyavirus. Isolated from mosquitoes, rodents,
bats, marsupials and man. Found in Brazil,
Trinidad, Surinam, French Guiana and Panama.
Causes fever in man.

guanidine; guanidine hydrochloride. A chemical that selectively inhibits the replication of the
RNA of some members of the *Picornaviridae*,
e.g. poliovirus, foot-and-mouth disease virus, by
affecting the RNA polymerase.

guanine. A constituent purine base of DNA and
RNA. *See* NUCLEIC ACID.

guanosine. A nucleoside of guanine and ribose.
See NUCLEIC ACID.

guanylation. Synonym: CAPPING.

guanylyl transferase. An enzyme found in certain virions (e.g. REOVIRUS, VACCINIA) which
catalyses the addition of guanosine 5´-mono-

phosphate from guanosine 5´-triphosphate to the
5´ terminus of nascent RNA molecules thus forming the 5´-terminal CAP structure of viral mRNAs.
See METHYL TRANSFERASE.

guar symptomless virus. A possible *Potyvirus.*
Francki, R.I.B. *et al.* (1985) **In** Atlas of Plant
Viruses. Vol. 2. p. 183. CRC Press: Boca Raton,
Florida.

Guaratuba virus. Family *Bunyaviridae*, genus
Bunyavirus. Isolated from mice, hamsters, birds
and mosquitoes in Brazil.

guarnieri bodies. Inclusion bodies in VACCINIA
VIRUS infected cells.

Guaroa virus. Family *Bunyaviridae*, genus
Bunyavirus. Isolated from mosquitoes in Colombia, Brazil and Panama. Causes fever in man.

guinea grass mosaic virus. A *Potyvirus.*
Thouvenal, J-C. *et al.* (1978) CMI/AAB Descrip-
tions of Plant Viruses No. 190.
Francki, R.I.B. *et al.* (1985) **In** Atlas of Plant
Viruses. Vol. 2. p. 183. CRC Press: Boca Raton,
Florida.

guinea pig cytomegalovirus. Synonym: CAVIID
HERPESVIRUS 1. Family *Herpesviridae*, subfamily
Betaherpesvirinae, genus *murine cytomegalovirus.* Silent infection of guinea pigs but inclusion
bodies are found in cells of the salivary gland. Can
be passed experimentally, giving fatal meningitis
(i.c.) or pneumonia (intratracheally). Grows in
primary guinea pig fibroblast cultures, producing
enlarged cells with nuclear inclusions.

guinea pig oncovirus. Family *Retroviridae*,
subfamily *Oncovirinae*, genus *Type C Oncovirus*,
sub-genus *Mammalian Type C Oncovirus.* An
endogenous virus which can be induced in cell
cultures from guinea pigs with 5-bromode-
oxyuridine.

Gumboro disease virus. *See* INFECTIOUS BURSAL
DISEASE VIRUS.

GV. Common abbreviation for GRANULOSIS
VIRUS.

Gynura latent virus. A possible *Carlavirus.*
Francki, R.I.B. *et al.* (1985) **In** Atlas of Plant
Viruses. Vol. 2. p. 173. CRC Press: Boca Raton,
Florida.

Gypchek.   Preparation of LYMANTRIA DISPAR NUCLEAR POLYHEDROSIS VIRUS produced and registered for use by the US Forest Service for the control of gypsy moth.

gypsy moth nuclear polyhedrosis virus.   *See* LYMANTRIA DISPAR NUCLEAR POLYHEDROSIS VIRUS.

Gyrinus 'baculovirus'.   Unclassified virus resembling a NON-OCCLUDED BACULOVIRUS infecting gut cells of the whirligig beetle, *Gyrinus natator*. Gouranton, J. (1972) J. Ultrastruct. Res. **39**, 281.

H

H1 virus. Family *Parvoviridae*, genus *Parvovirus*. Isolated from human tumour HEp1.

H3 virus. Family *Parvoviridae*, genus *Parvovirus*. Isolated from a human tumour which had been transplanted in rats.

Haden virus. *See* BOVINE PARVOVIRUS.

haemadsorption. The attachment of red blood cells to the surface of virus-infected cells due to virus-encoded products, usually glycoproteins, being incorporated into the infected cell membrane. Some viruses which bud from the cell surface, e.g. ORTHO- AND PARA-MYXOVIRUSES give this property to infected cells. Adsorption of erythrocytes can be used to identify infected cultures.

haemadsorption inhibition. A serological test in which haemadsorption is inhibited by the interaction of the antibodies with the surface of the haemadsorbing virus.

haemagglutinating encephalomyelitis virus of pigs. Synonym: PORCINE HAEMAGGLUTINATING ENCEPHALITIS VIRUS. Family *Coronaviridae*, genus *Coronavirus*. Causes encephalomyelitis in pigs with high mortality in young animals. Virus grows in primary pig kidney cell cultures giving multi-nucleate giant cells.

haemagglutination. The clumping of red blood cells. A large number of membrane-bound animal viruses haemagglutinate a wide variety of red blood cells, each virus favouring certain cells from certain animals. Used as a quick, quantitative assay for certain viruses, e.g. INFLUENZA VIRUS

haemagglutination inhibition test (HI test). Used for the detection of antibodies to haemagglutinating viruses. The antibodies react with the viral haemagglutinin thus preventing haemagglutination.

haemagglutinin. A glycoprotein which is present as spikes on certain membrane-bound viruses (e.g. MYXO- and PARAMYXO-VIRUSES) and binds to neuraminic acid-containing receptors on cells. In INFLUENZA VIRUS it is serotype-specific; its three-dimensional structure has been determined. When binding to red blood cells causes HAEM-AGGLUTINATION.
Wilson, I.A. *et al.* (1981) Nature, **289**, 366.

haemorrhagic encephalopathy of rats virus. Family *Parvoviridae*, genus *Parvovirus*. Isolated by injection of new-born rats with brain and spinal cord extracts from Lewis rats treated with cyclophosphamide. Causes haemorrhages and necrosis in the spinal cord. Injection i.c. into new-born hamsters causes fatal infection.

haemorrhagic enteritis of turkey virus. Family *Adenoviridae*, genus *Aviadenovirus*. Causes depression with bloody droppings. Intestine is filled with blood. Often fatal disease.

haemorrhagic septicaemia of trout virus. Synonym: EGTVED VIRUS. Family *Rhabdoviridae*, not assigned to genus. Causes haemorrhagic septicaemia in many species of trout. Can be grown in trout ovarian cells and fat head minnow cells.

hairless black syndrome virus. A disease of the honeybee, *Apis mellifera*, serologically-indistinguishable from BEE CHRONIC PARALYSIS VIRUS, but apparently causing distinct histopathology and inducing hairlessness and a black appearance in adult bees.
Rinderer, T.E. and Green, T.J. (1976) J. Invertebr. Pathol. **27**, 403.

hare fibroma virus. Family *Poxviridae*, subfamily *Chordopoxvirinae*, genus *Lepripoxvirus*. Causes fibromas in hares in Northern Italy and Southern France. Transmissible to rabbits. Related serologically to myxoma virus.

Hart Park virus. Family *Rhabdoviridae*, not assigned to genus. Isolated from mosquitoes in California. Multiplies in new-born mice inoculated i.c.

hart's-tongue fern virus. The first virus to be identified in ferns. It has been found in *Phyllitis scolopendrium* (hart's-tongue fern) and in various other ferns in England. It did not infect the angiosperms tested. The particles are rod-shaped, of two lengths, 320 nm. and 135 nm., and are 22 nm. wide. The virus is mechanically transmitted with difficulty and is transmitted in the soil, possibly by nematodes. The properties so far determined suggest that it might belong to the *TOBRAVIRUS* group.
Hull, R. (1968) Virology **35**, 333.

HAT medium. A cell culture medium containing hypoxanthine, aminopterin and thymidine. As aminopterin inhibits *de novo* synthesis of purines and pyrimidines HAT medium is used to select cells which have THYMIDINE KINASE and hypoxanthine-guanine phosphoribosyl transferase activities. Used in monoclonal antibody production to select hybridomas from unfused myeloma cells.

Hazara virus. Family *Bunyaviridae*, genus *Nairovirus*. Isolated from a tick in Pakistan.

head. The structural component of TAILED PHAGES which contains the DNA genome. The head may be isometric (e.g. λ phage) or elongated, apparently by the addition of extra rows of capsomeres (e.g. T4 PHAGE).

headful packaging. Mechanism for DNA packaging which occurs during replication in T4 and other phages whose DNA is in the circularly permuted form. Empty phage head structures are filled with DNA from a concatemeric precursor. After completion of one 'headful', the remaining DNA is cut and the filling of a second head begins. The amount of DNA packaged in each head includes a complete set of viral genes. This commonly produces a collection of genomes which are circularly permuted and terminally redundant.
Ritchie, D.A. (1983) **In** Topley and Wilson's Principles of Bacteriology, Virology and Immunity. Vol. 1. p. 177. ed. G. Wilson, A. Miles and M.T. Parker. Edward Arnold: London.

HeLa cells. Epithelial cells derived from a cervical adenocarcinoma from HEnrietta LAx, but the pseudonyms HElen LAne and HElen LArson were used to protect her identity when she was alive. Susceptible to many viruses, e.g. POLIO-VIRUS 1 and ADENOVIRUS TYPE 3.
Gey, G.O. *et al.* (1952) Cancer Res. **12**, 264.

Helenium virus S. A *Carlavirus*.
Koenig, R. and Lesemann, D-E. (1983) CMI/AAB Descriptions of Plant Viruses No. 265.
Francki, R.I.B. *et al.* (1985) **In** Atlas of Plant Viruses. Vol. 2. p. 173. CRC Press: Boca Raton, Florida.

Helenium virus Y. A possible *Potyvirus*.
Francki, R.I.B. *et al.* (1985) **In** Atlas of Plant Viruses. Vol. 2. p. 183. CRC Press: Boca Raton, Florida.

helical symmetry. A form of capsid structure found in many RNA viruses in which the protein subunits which interact with the nucleic acid form a helix; the adjacent subunits along the helix and between turns of the helix interact to give greater stability. For given helix characteristics (pitch etc.) the length of the particle is determined by the size of nucleic acid encapsidated. All rod-shaped plant viruses, e.g. TOBACCO MOSAIC VIRUS, POTATO VIRUS Y, have coat protein subunits arranged in helical symmetry; the nucleocapsid proteins of ORTHO- AND PARA-MYXOVIRUSES and of RHABDO-VIRUSES are arranged helically around the genomic RNA. *See* CUBIC SYMMETRY, ICOSA-HEDRAL SYMMETRY.

Heliothis zea nuclear polyhedrosis virus. Baculovirus (Subgroup A) isolated from the cotton bollworm, *H. zea,* in the USA. The virus is of the SNPV type and is highly infectious for *H. virescens, H. punctigera* and *H. armigera* as well as the homologous host. The virus has been extensively evaluated as a selective biological control agent for *Heliothis* spp. on cotton, tobacco, sorghum and vegetable crops. The virus was first registered for use in the USA in 1975 and was marketed as ELCAR by Sandoz Inc. until 1982, when the product was discontinued. A range of closely-related genotypic variants of the virus has been isolated from *H. zea* and *H. armigera*. Although less easy to propagate in cell culture than the prototype *BACULOVIRUS, Autographa californica* NPV, successful infection *in vitro* has been achieved in several cell lines derived from *H. zea*.
Ignoffo, C.M. and Couch, T. (1981) **In** Microbial Control of Pests and Plant Diseases 1970-1980. p. 330. ed. H.D. Burges. Academic Press: London.

Helminthosporium maydis virus. Type member of the *Helminthosporium maydis virus group*.
Buck, K.W. (1986) **In** Fungal Virology. p. 1. ed. K.W. Buck. CRC Press: Boca Raton, Florida.

Helminthosporium maydis virus group. (Named after the type member). A possible genus of a fungal virus with isometric particles, 48 nm. in diameter, which sediment at 283S. The capsid consists of a single coat protein species (mw. 121 x 10^3) and each contains a single molecule of dsRNA (mw. 5.7 x 10^6).

Helminthosporium victoriae 145S virus. A possible member of the *Penicillium chrysogenum virus* group.
Ghabrial, S.A. (1986) **In** Fungal Virology. p. 163. ed. K.W. Buck. CRC Press: Boca Raton, Florida.

Helminthosporium victoriae 190S virus. A possible member of the *Totivirus* group.
Ghabrial, S.A. (1986) **In** Fungal Virology, p. 163. ed. K.W. Buck. CRC Press: Boca Raton, Florida.

Helminthosporium victoriae virus A (HvV-A). Synonym: HELMINTHOSPORIUM VICTORIAE 190S VIRUS.

Helminthosporium victoriae virus B (HvV-B). Synonym: HELMINTHOSPORIUM VICTORIAE 145S VIRUS.

helper virus. A virus which provides the factor(s) needed by a defective virus for replication. For certain plant virus complexes the term refers to the virus which confers insect transmissibility to the complex.

henbane mosaic virus. A *Potyvirus*.
Govier, D.A. and Plumb, R.T. (1972) CMI/AAB Descriptions of Plant Viruses No. 95.
Francki, R.I.B. *et al.* (1985) **In** Atlas of Plant Viruses. Vol. 2. p. 183. CRC Press: Boca Raton, Florida.

Hepadnaviridae. A proposed family consisting of DNA viruses which infect man, woodchucks, ground squirrels and Pekin ducks. The virus infecting man is composed of a double shell particle (the Dane particle), 42 nm. in diameter, but small spherical particles, 22nm. in diameter (the surface antigen), are also present in the plasma of

 100nm

carriers. The virus particle consists of a 27 nm. icosahedral nucleocapsid (the core particle) containing one major polypeptide species surrounded by a detergent-sensitive envelope. The envelope protein is similar to, if not identical with, the 22 nm. particle which occurs naturally in the sera of infected patients. The surface particles contain at least seven polypeptides. The viral genome is circular, has a mw. of 1.6 x 10^6 and is partially ds and partially ss. It can integrate into chromosome DNA. The virus causes acute and chronic hepatitis, hepatocellular carcinoma, immune complex disease, polyarteritis and aplastic anaemia.
Gust, I.A. *et al.* (1986) Intervirology 25, 14.

hepatitis A virus. Synonym: EPIDEMIC JAUNDICE VIRUS, INFECTIOUS HEPATITIS VIRUS. Family *Picornaviridae*, genus *Enterovirus*. Causes 'short incubation' hepatitis. Usually caused by water or food-borne infection. Chimpanzees and marmosets are susceptible to experimental infection. Can be grown in cell culture.
Gust, I.D. *et al.* (1983) Intervirology 20, 1.

hepatitis B virus. Synonym: SERUM HEPATITIS VIRUS. Family *Hepadnaviridae*. Causes 'long incubation' hepatitis. Infection usually results from inoculation of serum but virus can be transmitted by sexual contact. Chronic carriers are common. Non-human primates can be infected. First virus disease for which a genetically engineered vaccine became available.
Gust, I.D. *et al.* (1986) Intervirology 25, 14.

hepatitis non A non B virus. Many cases of hepatitis in man are not associated with either hepatitis A or hepatitis B viruses. Recent work has shown that a calicivirus is implicated in the water-transmitted disease and a togavirus in the post-transfusion disease.
Bradley, D. *et al.* (1986) J. gen. Virol. **69**, 731.

hepatopancreatic parvo-like virus. Possible member of *Parvoviridae* observed in shrimps (*Penaeus* spp.; Crustacea).
Lightner, D.V. and Redman, R.M. (1985) J. Invertebr. Pathol. **45**, 47.

HEPES. N-2-hydroxyethylpiperazine-N'-2-ethanesulphonic acid. (mw. 238.3). A biological buffer, pK_a 7.55, with a pH range of 6.6-8.6.
Good, N. *et al.* (1966) Biochem. **5**, 467.

Heracleum latent virus. A possible member of

the *Closterovirus* subgroup 2.
Bem, F. and Murant, A.F. (1980) CMI/AAB Descriptions of Plant Viruses No. 228.
Francki, R.I.B. *et al.* (1985) **In** Atlas of Plant Viruses. Vol. 2. p. 219. CRC Press: Boca Raton, Florida.

Herpes B virus. *See* CERCOPITHECID HERPESVIRUS 1.

Herpes simiae virus. *See* CERCOPITHECID HERPESVIRUS 1.

Herpes simplex virus. Synonym: HUMAN (ALPHA) HERPESVIRUS 1. Family *Herpesviridae*, subfamily *Alphaherpesvirinae*, genus *Human Herpes Virus group*. Causes 'cold sores', particularly in young children. The virus can also pass along nerves and become latent in ganglia from which it is reactivated by stimuli such as colds and sunlight. Can be treated successfully with Acyclovir. *See* INDUCTION.

Herpesviridae. (Greek 'herpes, herpetos' = creeping, crawling creature, from nature of lesions.) A family of enveloped DNA viruses comprising three subfamilies, ALPHA, BETA and GAMMA HERPESVIRINAE and five genera, HUMAN HERPESVIRUS, SUID HERPESVIRUS, HUMAN CYTOMEGALOVIRUS group, LYMPHOPROLIFERATION VIRUS group and one unnamed genus. The particles are roughly spherical, 120-200 nm. in diameter and with a buoyant density of 1.20-1.29 g/cc in CsCl. There are more than 20 structural

100nm

proteins with mw. 12-220 x 10^3. Each particle contains one molecule of linear dsDNA, mw. 80-150 x 10^6. Replication of the DNA is nuclear but the mRNA from the transcripts is translated in the cytoplasm. Each virus has its own host range. Herpes viruses occur in both man and cold-blooded vertebrates and in invertebrates. Some viruses induce neoplasia. Transmission is usually by contact but it can occur by other routes.
Matthews, R.E.F. (1982) Intervirology **17**, 47.

Herpesvirus 3. *See* CHICKENPOX VIRUS.

Hershey circles. Circles of λ phage DNA pro-

duced by the annealing of the complementary, 12-nucleotide single-stranded ends of the DNA (after A.D. Hershey).

heteroduplex analysis. The study of the structures produced by the hybridisation of two ssDNA molecules derived from different sources. If they have complementary or near complementary sequences double-stranded molecules will be formed. If there are regions of non-complementarity single-stranded regions will remain. The study is usually by electron microscopy using the KLEINSCHMIDT PROCEDURE.

heterokaryon. A hybrid cell formed by the fusion of two cells from different species. *See* HOMOKARYON.

hexamer. A group of six protein subunits on the triangular faces of capsids with ICOSAHEDRAL SYMMETRY.

hexon. The arrangement of protein subunits on the triangular faces of ADENOVIRUS particles.

Hibiscus chlorotic ringspot virus. A member of the *Carmovirus* group.
Waterworth, H.E. (1980) CMI/AAB Descriptions of Plant Viruses No. 227.
Morris, T.J. and Carrington, J.C. (1988) **In** The Plant Viruses. Vol. 3. p. 73. ed. R. Koenig. Plenum Press: New York.

Hibiscus latent ringspot virus. A *Nepovirus*.
Brunt, A.A. *et al.* (1981) CMI/AAB Descriptions of Plant Viruses No. 233.
Francki, R.I.B. *et al.* (1985) **In** Atlas of Plant Viruses. Vol. 2. p. 23. CRC Press: Boca Raton, Florida.

Hibiscus yellow mosaic virus. A *Tobamovirus*. Occurs in Japan.
Doi, Y. *Personal communication.*

high pressure (performance) liquid chromatography (HPLC). A method for separating peptides, oligonucleotides, etc., with high resolution.

high voltage electrophoresis. Electrophoresis at potential differences of more than 1,000 volts. Used in nucleic acid sequencing and in paper electrophoresis of nucleotides.

Hinze virus. *See* LEPORID HERPESVIRUS.

Hippeastrum latent virus. Synonym: NERINE LATENT VIRUS.

Hippeastrum mosaic virus. A *Potyvirus*.
Brunt, A.A. (1973) CMI/AAB Descriptions of Plant Viruses No. 117.
Francki, R.I.B. *et al.* (1985) **In** Atlas of Plant Viruses. Vol. 2. p. 183. CRC Press: Boca Raton, Florida.

Hirt supernatant. A method for separating viral DNA from cellular DNA by lysing cells with sodium dodecyl sulphate in the presence of 1M NaCl. The cellular DNA precipitates leaving viral DNA in the supernatant.
Hirt, B. (1967) J. mol. Biol. **26**, 365.

histones. Basic proteins found in cell nuclei in close association with DNA forming CHROMATIN. They contain tyrosine but little or no tryptophan. There are five main classes differing in their relative content of lysine and arginine. Cells infected with CARDIOVIRUSES have altered histone composition. DNA viruses which use the eukaryotic host transcription system, e.g. cauliflower mosaic virus, are thought to interact with host histones.

histopathology. The branch of pathology dealing with tissue changes associated with disease.

HIV. *See* HUMAN IMMUNODEFICIENCY VIRUS.

HOB-mutant. Mutant of the 1A clone of *AUTOGRAPHA CALIFORNICA* NUCLEAR POLYHEDROSIS VIRUS which kills the insect host, *Trichoplusia ni*, more rapidly than the wild-type strain.
Wood, H.A. *et al.* (1981) J. invertebr. Pathol. **38**, 236.

hog cholera virus. Synonym: SWINE FEVER. Family *Togaviridae*, genus *Pestivirus*. A highly contagious disease of pigs causing fever, vomiting, diarrhoea, haemorrhages and frequently death. Occurs in most parts of the world. Calves, goats and deer can be infected experimentally. Can be grown in pig cells. Serologically related to bovine viral diarrhoea virus.

Holcus lanatus yellowing virus. A possible plant *Rhabdovirus*.
Francki, R.I.B. *et al.* (1985) **In** Atlas of Plant Viruses. Vol. 1. p. 73. CRC Press: Boca Raton, Florida.

Holcus streak virus. A possible *Potyvirus*.
Francki, R.I.B. *et al.* (1985) **In** Atlas of Plant Viruses. Vol. 2. p. 183. CRC Press: Boca Raton, Florida.

Holcus transitory mottle virus. Synonym: COCKSFOOT MILD MOSAIC VIRUS.

homokaryon. Hybrid cell formed by the fusion of two cells of the same species. *See* HETEROKARYON.

homologous antiserum. A serum containing antibodies raised against a specific antigen and which will react with that antigen.

homology. The degree of relatedness between the nucleotide sequences of two nucleic acid molecules or the amino acid sequences of two protein molecules. Hybridisation experiments can produce useful information but, for critical analyses, sequence data are needed.

honeysuckle latent virus. A *Carlavirus*.
Brunt, A.A. and van der Meer, F.A. (1984) CMI/AAB Descriptions of Plant Viruses No. 289.
Francki, R.I.B. *et al.* (1985) **In** Atlas of Plant Viruses. Vol. 2. p. 173. CRC Press: Boca Raton, Florida.

honeysuckle yellow vein mosaic virus. A *Geminivirus*, transmitted by whitefly.
Harrison, B.D. (1985) Ann. Rev. Phytopath. **23**, 55.

hop American latent virus. A *Carlavirus*.
Barbara, D.J. and Adams, A.N. (1983) CMI/AAB Descriptions of Plant Viruses No. 262.
Francki, R.I.B. *et al.* (1985) **In** Atlas of Plant Viruses. Vol. 2. p. 173. CRC Press: Boca Raton, Florida.

hop latent virus. A *Carlavirus*.
Barbara, D.J. and Adams, A.N. (1983) CMI/AAB Descriptions of Plant Viruses No. 261.
Francki, R.I.B. *et al.* (1985) **In** Atlas of Plant Viruses. Vol. 2. p. 173. CRC Press: Boca Raton, Florida.

hop mosaic virus. A *Carlavirus*.
Barbara, D.J. and Adams, A.N. (1981) CMI/AAB Descriptions of Plant Viruses No. 241.
Francki, R.I.B. *et al.* (1985) **In** Atlas of Plant Viruses. Vol. 2. p. 173. CRC Press: Boca Raton, Florida.

hop stunt viroid. A VIROID, 297 nucleotides.
Shikata, E. (1987) In The Viroids. p. 279. ed. T.O.
Diener. Plenum Press: New York.

hop trefoil cryptic virus 1. A member of the
Cryptovirus group, subgroup A.
Boccardo, G. *et al.* (1987) Adv. Virus Res. **32**,
171.

hop trefoil cryptic virus 2. A member of the
Cryptovirus group, subgroup B.
Boccardo, G. *et al.* (1987) Adv. Virus Res. **32**,
171.

hop virus A. A strain of *Apple Mosaic Virus.*
Francki, R.I.B. (1985) In The Plant Viruses. Vol.
1. p. 1. ed. R.I.B. Francki. Plenum Press: New
York.

hop virus B. A strain of *Prunus Necrotic Ringspot Virus.*
Francki, R.I.B. (1985) In The Plant Viruses. Vol.
1. p. 1. ed. R.I.B. Francki. Plenum Press: New
York.

Hordeivirus group. (Latin 'hordeum' = barley).
(Type member BARLEY STRIPE MOSAIC VIRUS).
Genus of MULTICOMPONENT plant viruses with
rigid rod-shaped particles, 100-150 nm. long and
20 nm. in diameter, which sediment at 175-200S.

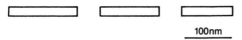

100nm

The coat protein subunits (mw. 21 x 10^3) are
arranged in the particles in helical symmetry with
pitch *c*.2.5 nm. There are two to four nucleoprotein components each containing a unique species
of RNA. The infective genome comprises three
species of linear (+)-sense ssRNA, RNA-1 (mw.
1.43-1.5 x 10^6), RNA-2 (mw. 1.24-1.35 x 10^6)
RNA-3 (mw. 1.10-1.24 x 10^6); in some strains the
sizes of RNA-2 and -3 are very similar which led
to the belief that the type virus was bipartite.
SUBGENOMIC RNAs may also be encapsidated.
The 5′ termini of the RNAs have a CAP; the 3′
termini have tRNA activity accepting tyrosine
and there is a short internal poly(A) sequence near
the 3′ terminus. RNA-1 is a MONOCISTRONIC messenger for a protein mw. 120 x 10^3, RNA-2
encodes the coat protein at its 5′ end and probably
another protein, RNA-3 is bicistronic, the 5′
protein having mw. 75 x 10^3, the 3′ protein (mw.
19 x 10^3) being translated from a SUBGENOMIC RNA

(RNA-4, mw. 0.28 x 10^6). Host ranges of mem-
bers are narrow and mainly restricted to
Graminae. Particles are found in most cell types.
Hordeiviruses are mechanically transmissible
and are transmitted through seed.
Matthews, R.E.F. (1982) Intervirology **17**, 178.
Francki, R.I.B. *et al.* (1985) In Atlas of Plant
Viruses. Vol. 2. p. 133. CRC Press: Boca Raton,
Florida.
Carroll, T.W. (1986) In The Plant Viruses. Vol. 2.
p. 373. ed. M.H.V. van Regenmortel and H.
Fraenkel-Conrat. Plenum Press: New York.
Atabekov, J.G. and Dolja, V.V. (1986) In The
Plant Viruses. Vol. 2. p. 397. ed. M.H.V. van Re-
genmortel and H. Fraenkel-Conrat. Plenum
Press: New York.

horizontal transmission. Transmission of a
virus, or other pathogen, between animals at any
age after birth. *See* TRANSMISSION, VERTICAL
TRANSMISSION.

horse pox virus. Synonym: CONTAGIOUS PUSTU-
LAR DERMATITIS OF HORSES VIRUS. Family *Poxviridae*, subfamily *Chordopoxvirinae*, genus
Orthopoxvirus. Causes lesions on lips, buccal
mucosa and nose with fever and drooling of
saliva. Reported to cause lesions on fingers of
people working with horses.

horsegram yellow mosaic virus. A *Geminivirus*, transmitted by whitefly.
Muniyapa, V. *et al.* (1987) J. Phytopath. **119**, 81.

horseradish peroxidase. An enzyme derived
from the plant, horseradish, *Armoracia rusti-
cana.* Used in ELISA tests to give the colour reaction with its substrate, e.g. brown with 3′, 3′-dian-
isidine.

host. An organism or cell culture in which a given
virus can replicate.

host range. A listing of species of hosts which are
susceptible to a given virus (or other pathogen). A
critical host range listing should include species
which are resistant to the virus. Can be used in
identifying and characterising viruses.

hot spot. A region usually within a gene where
mutations occur at an unusually high frequency.

housefly virus. Unclassified REOVIRUS-like
agent, containing 10 segments of double-
stranded RNA (total mw. 17-18 x 10^6) isolated

from the housefly, *Musca domestica,* in Australia. The electrophoretic profile of genome segments and serological properties of the virus suggest that it is distinct from other members of the REOVIRIDAE.
Moussa, A.Y. *et al.* (1982) Aust. J. biol. Sci **35**, 669.

HPB-SL-26 cells. Insect cell line from the cotton leafworm, *Spodoptera littoralis,* susceptible to infection by *S. LITTORALIS* NUCLEAR POLYHEDROSIS VIRUS.

HPS-1 virus. An unclassified RNA virus isolated from cultured cells of *Drosophila melanogaster.* Virions are unenveloped, isometric (36 nm. in diameter), contain two proteins (mw. of major protein 120×10^3) and a single segment of dsRNA (*c*.6 kbp).
Scott, M.P. *et al.* (1980) Cell **22**, 929.

HTLV III. *See* HIV, HUMAN T-CELL LYMPHOTROPIC VIRUS TYPE III.

Huacho virus. Family *Reoviridae,* genus *Orbivirus.* Isolated from ticks in Peru.

Hughes group viruses. Family *Bunyaviridae,* genus *Nairovirus.* Isolated from ticks and sea birds.

human alphaherpesvirus 2. Synonyms: HERPES FEBRILIS, HERPES SIMPLEX TYPE 2. Family *Herpesviridae,* subfamily *Alphaherpesvirinae,* genus *Herpesvirus Group.* Similar to herpesvirus 1 except that it is usually transmitted sexually. The virus causes genital lesions and it may cause carcinoma of the cervix. It is very closely related antigenically to herpesvirus 1.

Human cytomegalovirus group. A genus in the subfamily *Betaherpesvirinae.* Contains viruses which infect the mouse, rat, pig and guinea pig.

human embryo lung cells. Non-transformed diploid cells having a finite life span. Used for vaccine production for e.g. RABIES VIRUS.
Hayflick, L. and Moorhead, P.S. (1961) Exp. Cell Res. **25**, 585.

Human herpesvirus 1 group. A genus in the subfamily *Alphaherpesvirinae.* Contains the viruses causing human herpes, types 1 and 2 and bovine mammillitis.

human herpesvirus 3. *See* VARICELLA ZOSTER VIRUS.

human herpesvirus 4. *See* EPSTEIN-BARR VIRUS.

human immunodeficiency virus (HIV). Synonyms: HTLV III, LAV. Family *Retroviridae,* subfamily *Lentivirinae.* Causative agent of AIDS. The name was proposed and accepted by ICTV in 1986 to minimise the confusion caused by the two names HTLV III and LAV. It can be grown in lymphocyte cultures without causing cpe.

human rhinovirus. *See* COMMON COLD FEVER VIRUS.

human T-cell lymphotropic virus type I. Family *Retroviridae,* subfamily *Lentivirinae.* Causes adult T-cell leukaemia which is common in Japan.

human T-cell lymphotropic virus type III. *See* HIV. The name given by workers at the National Cancer Institute, USA, to isolates of the virus causing AIDS.
Gallo, R.C. *et al.* (1984) Science **224**, 500.

humoral immunity. Immunity conferred by antibodies in extracellular fluids including the serum and lymph.

hundskrankheit virus. *See* SANDFLY FEVER VIRUS.

hyacinth mosaïc virus. A possible *Potyvirus.* Francki, R.I.B. *et al.* (1985) **In** Atlas of Plant Viruses. Vol. 2. p. 183. CRC Press: Boca Raton, Florida.

hybrid arrested translation. A method for identifying the proteins encoded by a cloned DNA sequence. The mRNA preparation is hybridised with the cloned DNA and only mRNA species homologous to the DNA will anneal to it. Comparison of *in vitro* translation products of annealed with unannealed mRNAs will identify the proteins, the production of which is inhibited by hybrid formation. *See* HYBRID RELEASED TRANSLATION.
Paterson, B.M. (1977) Proc. Natl. Acad. Sci. USA. **74**, 4370.

hybrid released (selected) translation. A method used to identify proteins encoded by a

cloned DNA. A preparation of mRNA is hybridized to the cloned DNA immobilised on a solid matrix such as nitrocellulose. The mRNA homologous to the DNA is retained on the filter and can then be removed by melting the RNA:DNA duplex. The purified RNA is then translated *in vitro* and the protein product(s) identified, often by gel electrophoresis.
Goldberg, M.L. *et al*. (1979) Meths. Enzymol. **68**, 206.

hybridisation. 1. In molecular biology it refers to the formation of stable duplexes between complementary sequences by way of Watson-Crick base-pairing. *See* ANNEALING. 2. In genetics and breeding it means the formation of a novel diploid organism by normal sexual processes or by protoplast fusion.

hybridoma. A hybrid cell line produced from the fusion of a normal lymphocyte with a myeloma cell. After selection and cloning a hybridoma cell line will produce a MONOCLONAL ANTIBODY.

Hydra viridis/Chlorella virus. A virus from a *Chlorella*-like green alga which has a symbiotic relationship with the digestive cells of *Hydra viridis*. The virus particles are isometric, 185 nm. in diameter, contain dsDNA (mw. 136 x 10^6) and up to 19 proteins (mw. 10.3-82 x 10^3). The algal cells lyse immediately they are isolated from the *Hydra* and it is thought that the virus may be important in the symbiotic relationship.
van Etten, J.L. *et al*. (1981) Virology **113**, 704.

Hydrangea mosaic virus. An *Ilarvirus*.
Francki, R.I.B. (1985) In The Plant Viruses. Vol. 1. p. 1. ed. R.I.B. Francki. Plenum Press: New York.

Hydrangea ringspot virus. A *Potexvirus*.
Koenig. R. (1973) CMI/AAB Descriptions of Plant Viruses No. 114.
Francki, R.I.B. *et al*. (1985) In Atlas of Plant Viruses. Vol. 2. p. 159. CRC Press: Boca Raton, Florida.

hydration. The incorporation of water into a complex molecule. Proteins, nucleic acids and virus particles are hydrated to varying extents and this affects their hydrodynamic properties. The amount of hydration is affected by the composition of the solvent in which the molecules are suspended. The water structure surrounding virus particles is disrupted by chaotropic ions such as

Cl⁻; this can lead to disruption of the virus particles.

hydrogen bonding. A non-covalent bond formed between the hydrogen atom in an -O-H or -N-H group and an oxygen or nitrogen atom (*see* figure under BASE PAIR). These bonds are relatively weak but are crucial in maintaining the secondary structure of nucleic acids and proteins. Hydrogen bonds maintain the double-helical structure of DNA.

hydrophobia virus. *See* RABIES VIRUS.

hydroxyapatite. A form of calcium phosphate often used to bind nucleic acids. The binding depends upon the nature of the nucleic acid and on its secondary structure. Thus, under certain conditions it will bind supercoiled DNA but not relaxed DNA and under other conditions it will bind ds and not ss DNA.
Bernardi, G. (1971) Meths. Enzymol. **21**, 95.

hyperchromicity. The increase in the absorbance of light of 260 nm. wavelength by nucleic acid at its melting temperature. It is an indication of the amount of base-pairing in the nucleic acid.

hyperimmune serum. Serum from an animal which has received two or more injections of a foreign antigen for the purpose of producing a reagent for use in serology.

hypersensitive. The state of being abnormally sensitive. In virology it refers to an extreme reaction to a virus, e.g. the formation of local lesions or the necrotic response of a leaf to a plant virus.

hypertrophy. Increase in cell size causing an increase in the size of an organ or tissue, e.g. mumps virus infection of the lymph nodes.

Hypochoeris mosaic virus. A possible member of the *Furovirus* group.
Brunt, A.A. and Stace-Smith, R. (1983) CMI/AAB Descriptions of Plant Viruses No. 273.
Brunt, A.A. and Shikata, E. (1986) In The Plant Viruses. Vol. 2. p. 305. ed. M.H.V. van Regenmortel and H. Fraenkel-Conrat. Plenum Press: New York.

hypochromicity. Reduction in absorbance of light at 260 nm. wavelength by complementary strands of nucleic acid which are hybridising or nucleic acid which is increasing its secondary structure.

Hyposoter exiguae virus. Formerly classified as the type species of proposed subgroup D 1, BACULOVIRUS genus. Now included as a member of subgroup A, POLYDNAVIRUS genus. Virions contain a polydisperse supercoiled dsDNA genome. The virus was isolated from the parasitoid *Hyposoter exiguae* (Hymenoptera; Ichneumonidae) (*see* POLYDNAVIRUS).
Krell, P.J. and Stoltz, D.B. (1980) Virology **101**, 408.

HYPR virus. Family *Flaviviridae*, genus *Flavivirus*. Isolated from a boy with encephalitis in Czechoslovakia. Causes frequent infections in Hungary, Poland, Yugoslavia, Austria, Bulgaria, Sweden and Finland.

Hz-1 'Baculovirus'. Unclassified virus, closely resembling the NON-OCCLUDED BACULOVIRUS group in morphological and biochemical properties. First discovered in a persistently infected *Heliothis zea* cell culture (IMC-Hz-1) where it replicates in the nucleus. Virions are loosely-enveloped, rod-shaped nucleocapsids with two predominant size classes of nucleocapsid (367-400 x 40 nm. and 775-815 x 40 nm.). Particles contain at least 28 structural proteins (mw. 14-153 x 10^3) and a large circular dsDNA genome (mw. 154 x 10^6; 235 kbp), the largest reported for any baculovirus. The virus elicits both productive (lytic) and persistent infections in several lepidopteran cell cultures but has no known natural host organism. Defective particles (containing genome deletions of up to 91 kbp) are involved in the establishment of persistent infections. Unlike many baculoviruses, virions are released from cells by lysis of nuclear and cytoplasmic membranes rather than by budding. Some limited DNA homology has been reported with AUTOGRAPHA CALIFORNICA NUCLEAR POLYHEDROSIS VIRUS.
Burand, J.P. and Wood, H.A. (1986) J. gen. Virol. **67**, 167.

I

I plasmids. Plasmids coding for I pilus. Adsorption of phages If1 and If2 (INOVIRUS) is specific to strains of enterobacteria (e.g. *E. coli*, *Shigella* and *Salmonella*) carrying the I pilus.
Hardy, K. (1986) Bacterial Plasmids. 2nd. ed. Thomas Nelson: Walton-upon-Thames.

Ibaraki virus. Family *Reoviridae*, genus *Orbivirus*. Isolated from cattle in Japan. Probably arthropod-borne. Causes severe disease in cattle. Grows in bovine cell cultures and in yolk sac of embryonated eggs.

Icoaraci virus. Family *Bunyaviridae*, genus *Phlebovirus*. Isolated from rodents and sentinel mice in Brazil.

icosadeltahedron. A deltahedron with icosahedral symmetry. A deltahedron is a polyhedron which has 20 x T equilateral triangles on its surface, T being the triangulation number. In small icosadeltahedra, e.g. POLIOVIRUS, the faces are flat whereas in larger icosadeltahedra, e.g. ADENOVIRUS, the faces are not flat. *See* TRIANGULATION NUMBER.

icosahedral cytoplasmic deoxyriboviruses. Vernacular name for *IRIDOVIRIDAE*.

icosahedral symmetry. One of the types of CUBIC SYMMETRY, found in isometric virus CAPSIDS and generating the ability to make a range of virus capsids with different numbers of structural subunits (CAPSOMERES); this is in contrast to HELICAL SYMMETRY in which the structural subunit is the protein monomer. The simplest form (e.g. phage ø) has 60 identical subunits arranged to give 20 triangular faces and 12 pentameric vertices and has twofold, threefold and fivefold rotational symmetry. For larger icosahedra the number of subunits in the triangular faces is increased in a regular manner. *See* ICOSADELTAHEDRON, QUASI-EQUIVALENCE THEORY, TRIANGULATION NUMBER.

icosahedron. *See* ICOSAHEDRAL SYMMETRY.

ICTV. Abbreviation for the INTERNATIONAL COMMITTEE ON TAXONOMY OF VIRUSES.

ID$_{50}$. Abbreviation for MEDIAN INFECTIVE DOSE.

IgA. *See* IMMUNOGLOBULIN.

IgG. *See* IMMUNOGLOBULIN.

IgM. *See* IMMUNOGLOBULIN.

Ilarvirus group. (Sigla from isometric labile ringspot). (Type member TOBACCO STREAK VIRUS). Genus of MULTICOMPONENT plant viruses with quasi-isometric particles, 26-35 nm. in diameter, those of different components differing in size.

The components sediment at rates ranging from 117S to 78S and, when fixed, band in CsCl at 1.36 g/cc. The particles are stabilised primarily by protein-RNA interactions, the components having similar protein:RNA ratios. The capsids are composed of a single polypeptide species mw. 19-30 x 10³. Genomic linear (+)-sense ssRNA comprises three species RNA-1 (mw. 1.1-1.3 x 10⁶), RNA-2 (mw. 1.18-0.89 x 10⁶) and RNA-3 (mw. 0.91-0.7 x 10³); RNAs-1 and -2 are separately encapsidated; RNA-3 is encapsidated with the SUBGENOMIC RNA which is the messenger for the coat protein (RNA-4, mw. 0.3 x 10⁶). RNAs -1, -2 and -3 plus either coat protein or RNA-4 are required for infectivity. RNAs-1, -2, and -4 are MONOCISTRONIC messengers for proteins of mw. 120, 100 and coat protein respectively; RNA-3 is bicistronic encoding a protein of mw. 34 x 10³ at the 5′ end and having the coat protein cistron at the 3′ end. The host ranges of ilarviruses are wide. Particles are found in most cell types. They are readily mechanically transmitted. Some are

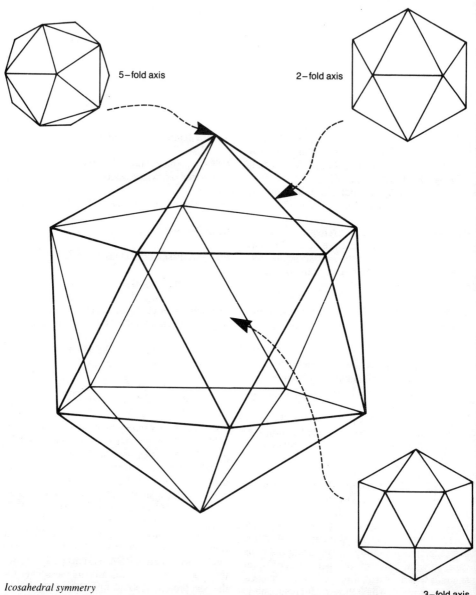

5–fold axis

2–fold axis

Icosahedral symmetry

3–fold axis

transmitted by seeds and by pollen.
Matthews, R.E.F. (1982) Intervirology **17**, 175.
Fulton, R.W. (1983) CMI/AAB Descriptions of
Plant Viruses No. 275.
Francki, R.I.B. *et al*. (1985) **In** Atlas of Plant
Viruses. Vol. 2. p. 81. CRC Press Inc.: Boca
Raton, Florida.
Francki, R.I.B. (1985) **In** The Plant Viruses. Vol.
2. p. 1. ed. R.I.B. Francki. Plenum Press: New
York.

Ilesha virus.   Family *Bunyaviridae*, genus
Bunyavirus. Isolated from mosquitoes in several
parts of Africa.

Ilheus virus.   Family *Flaviviridae*, genus

Flavivirus. Isolated from mosquitoes in central and South America. Can cause encephalitis in man. Antibodies found in man, horses and birds. Mice injected i.c. develop encephalitis. Virus grows in chorioallantoic membrane and in chick embryo and hamster kidney cells.

IMC-Hz-1 cells. Insect cell line derived from *Heliothis zea* and persistently infected by the non-occluded Baculovirus, Hz-1.

IMC-Hz-1-NOV. *See* HZ-1 BACULOVIRUS.

immobilised DNA or RNA. Nucleic acid which has been linked to nitrocellulose or activated paper. *See* NORTHERN BLOTTING, SOUTHERN BLOTTING, DIAZOBENZYLOXYMETHYL PAPER.

immortalisation. It has two meanings: a) the continued growth of cells in culture beyond the time that it would have normally been expected to cease, e.g. infection of human cells with EPSTEIN-BARR VIRUS; b) the cloning of cDNA in a bacterial plasmid or the production of monoclonal antibodies in a hybridoma; this refers to the ability to obtain faithful copies at will.

immune response. The response of the immune system to the injection of foreign proteins or carbohydrates. It can be divided conveniently into humoral (antibody) and T-cell responses. In most virus diseases, the production of specific antibody provides protection against reinfection but there are instances where antibody alone is insufficient.

immunisation. Rendering an organism immune to a specific disease. Usually performed by injecting preparations into the organism which will induce antibodies against the causal agent of the disease.

immunity. The condition of a living organism whereby it resists and overcomes an infection or disease.

immunocytochemistry. The identification of the sites of ANTIGENS in cells by the use of ANTIBODIES to which a reporter molecule e.g. ferritin, gold or a fluorescent dye, is attached.

immunodiffusion. A serological procedure in which ANTIGENS and ANTIBODIES in solution are permitted to diffuse towards each other through a gel matrix. The interaction between the antigen and antibody is manifested by a 'PRECIPITIN' line produced by the precipitation of the antigen-antibody complex. There are two major types of test – single diffusion in which one reactant diffuses into a gel containing the other reactant and double diffusion in which both reactants diffuse into a gel initially free of them. *See* RADIAL IMMUNODIFFUSION, OUCHTERLONY GEL DIFFUSION TEST.

immuno-electron microscopy. The use of ANTIBODIES to detect ANTIGENS, usually virus particles, in the electron microscope. There are two basic, but not exclusive, forms of the technique: a) immunosorbent electron microscopy, in which virus particles are entrapped on an antibody-coated electron microscope grid and b) 'decoration', in which the antibody adhering to the virus particle is visualised in the electron microscope.
Derrick, K.S. (1973) Virology **56**, 652.
Milne, R.G. and Luisoni, E. (1977) **In** Methods in Virology, Vol. **6**, 265, ed. K. Maramorosch and H. Koprowski. Academic Press: New York.

immunelectro-osmophoresis. A serological test based on the migration of ANTIGENS to the anode and ANTIBODIES to the cathode by electrophoresis. The antigen and antibody are placed in wells in an agarose gel and electrophoresed towards each other until they interact and form a precipitate. It is a very rapid method of antigen detection.
Ragetli, H.W. and Weintraub, M. (1964) Science **144**, 1023.

immunoelectrophoresis. A technique in which an ANTIGEN mixture is separated into its components by electrophoresis in a supporting medium (e.g. agar gel). ANTISERUM is placed in a trough parallel to the path of electrophoretic migration and immunoprecipitation lines allowed to develop. It is a powerful analytical method for resolving complex mixtures of antigens.

immunofluorescence. A technique in which fluorescent dyes (e.g. fluorescein) are linked to ANTIBODIES that are used for detecting ANTIGEN in cells and thin sections by fluorescence microscopy.

immunogen. A substance which induces the production of specific ANTIBODIES in a suitable animal. It is usually protein and/or carbohydrate.

immunoglobulin. A set of proteins produced in the immune response of animals. Several classes of immunoglobulins have the same basic structure of two identical light (L) polypeptide chains and two heavy (H) chains linked together by non-covalent forces and disulphide bonds (*see* figure). There are five classes distinguished on the basis of five different types of H chain, IgG, IgA, IgM, IgD and IgE; the heavy chains are termed γ, α,μ, ξ and ε respectively. The concentration of IgD and IgE in animal serum is very low. IgG is the most common, comprising about 75% of all immunoglobulins; it has a mw. of 150×10^3 and will fix COMPLEMENT. IgM has a mw. of 900×10^3 and also fixes complement. IgA has a mw. of 160×10^3 and does not fix complement.

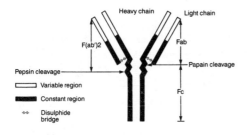

immunogold labelling. The linking of colloidal gold molecules to ANTIBODIES. The gold molecules can then be detected by electron microscopy of samples containing ANTIGEN which have been treated with the labelled antibody.
Beesley, J.E. and Betts, M.P. (1984) Proc. Roy. Microsc. Soc. **19**, 36.

immunological drift. *See* ANTIGENIC DRIFT.

immunoperoxidase. The linking of HORSERAD-ISH PEROXIDASE to antibodies. The antibodies are then used in tests such as ELISA or WESTERN BLOT-TING and in cytological studies: the presence of the peroxidase is detected by reaction with a substrate which gives a colour.

immunoprecipitation. The precipitation of anti-gen-antibody complexes that forms the basis of several serological tests. When this occurs in solution the test is called a PRECIPITIN TEST; when the reactants diffuse towards each other in a gel it is known as IMMUNODIFFUSION. *See* PRECIPITATION REACTION.

immunosorbent electron microscopy. *See* IMMUNOELECTRON MICROSCOPY.

in vitro. Relating to biological reactions outside the living cell or organism.

in vivo. Relating to a biological reaction taking place within a living cell or organism.

ina-flacherie virus. Probable DENSOVIRUS (*PARVOVIRIDAE*) isolated from the silkworm, *Bombyx mori*. Particles are isometric, 22 nm. in diameter containing four structural proteins (mw. 77, 70, 57 and 50×10^3) which share some common sequences and a DNA genome which is double-stranded when extracted (ds form mw. 3.4×10^6).
Bando, H. *et al.* (1984) Arch. Virol. **80**, 209.

Inachis io Cytoplasmic Polyhedrosis Virus. Cytoplasmic polyhedrosis virus (CPV) isolated from the peacock butterfly, *Inachis io* (Nymph-alidae, Lepidoptera), in the UK. The virus is the type member of 'type 2' CPVs. It is unrelated to *Bombyx mori* (type 1) CPV on the basis of RNA electropherotype, serological properties and nucleic acid homology. Viruses of similar elec-tropherotype have been observed in other Lepi-doptera (*see* APPENDIX B).
Payne, C.C. and Mertens, P.P.C. (1983) **In** The *Reoviridae*. p. 425. ed. W.K. Joklik. Plenum Press: New York.

inactivation. The loss of the ability of a virus to initiate infection. It is used in the production of vaccines where viruses are inactivated by treat-ment with certain chemicals or by physical condi-tions such as heat or irradiation. Viruses are also inactivated by specific antibodies, a process termed neutralisation. This is used in a test for the identification of a virus or measuring the titre of an antibody. *See* KILLED VACCINE.

inapparent infection. An infection which does not give obvious symptoms.

inclusion body. Any matrix or array of virus particles, abnormal proteinaceous bodies, or ar-eas of abnormal staining within a virus-infected cell. Most of such areas are sites of virus synthesis (e.g. poxvirus B type inclusions), late degenera-tive changes in cells, crystalline aggregates of virions or accumulations of virus-coded products and are readily detected by light microscopy. They have limited use in the diagnosis of some vertebrate viruses (*see* NEGRI BODY) and plant viruses (*see* PINWHEEL INCLUSION). The term is also used for the virus-containing POLYHEDRA, CAP-

SULES AND SPHEROIDS produced during infections of insects with NUCLEAR POLYHEDROSIS, CYTOPLASMIC POLYHEDROSIS, GRANULOSIS and ENTOMOPOX VIRUSES (though occlusion body is the preferred term). *See* A-TYPE INCLUSION BODY, B-TYPE INCLUSION BODY.

inclusion body protein. A term sometimes used for the main constituent of an inclusion body, e.g. POLYHEDRIN.

incomplete virus. *See* DEFECTIVE VIRUS.

Indian peanut clump virus. A possible *Furovirus.*
Brunt, A.A. and Shikata, E. (1986) **In** The Plant Viruses. Vol. 2. p. 305. ed. M.H.V. van Regenmortel and H. Fraenkel-Conrat. Plenum Press: New York.

indicator cell. A cell which reacts in a characteristic manner to infection with a specific virus.

indicator plant. A plant species which gives characteristic symptoms to a specific virus. Used in virus diagnosis.

Indonesian soybean dwarf virus. A *Luteovirus.*
Iwaki, M. *et al.* (1980) Plant Disease **64**, 1027.

induction. The activation of a latent virus infection. The activation can occur spontaneously, (e.g. HERPES SIMPLEX VIRUS latent in dorsal root ganglia and activated to give 'cold sores'), by changing growing conditions of infected cells or by a variety of exogenous stimuli, e.g. certain chemicals (activation of EPSTEIN-BARR VIRUS by halogenated derivatives of uridine), UV light (induces the production of mature phage from a bacterial cell line containing prophage). *See* LYSOGENY.

infectious anaemia of horses virus. Synonym: SWAMP FEVER VIRUS. *See* INFECTIOUS EQUINE ANAEMIA VIRUS.

infectious bovine rhinotracheitis virus. Synonym: INFECTIOUS PUSTULAR VULVO-VAGINITIS VIRUS. Family *Herpesviridae*. Infects cattle worldwide. Causes acute disease of respiratory tract with high mortality but can cause mild infection. Young goats infected experimentally develop fever and rabbits develop encephalitis with paralysis. Virus can be grown in bovine embryo cell cultures and in pig, sheep, goat and horse kidney cell cultures.

infectious bursal disease virus. Synonym: AVIAN NEPHROSIS VIRUS, GUMBORO DISEASE VIRUS. Family *Birnaviridae*, genus *Birnavirus*. Causes a lymphoproliferative disease of chickens, involving the bursa of Fabricius, thus impairing development of the immune system. Virus can be grown in cell cultures and in eggs.

infectious dropsy of carp virus. *See* RHABDOVIRUS CARPIO, SPRING VIRAEMIA OF CARP VIRUS.

infectious equine anaemia virus. Synonym: SWAMP FEVER VIRUS. Family *Retroviridae*, subfamily *Lentivirinae*. Causes acute infection in horses, resulting in fever, anaemia, nasal discharge and oedema. Remissions occur but disease is usually fatal. Virus can be grown in equine embryonic tissue cultures.

infectious equine arteritis virus. Family *Togaviridae*, genus *Rubivirus*. Causes fever, conjunctivitis, rhinitis, oedema, enteritis and colitis. A cause of abortion. Can be grown in horse kidney and hamster kidney cells.

infectious flacherie virus. Unclassified small RNA virus (possible INSECT PICORNAVIRUS) isolated from the silkworm, *Bombyx mori*. Virions are isometric, 26-27 nm. in diameter, sediment at 183S and have a buoyant density in CsCl of 1.375 g/cc. The particles contain four major structural proteins (35.2, 33, 31.2 and 11.6 x 10^3) and a (+)-sense uncapped ssRNA genome (mw. 2.4 x 10^6), with a poly A tract.
Hashimoto, Y. *et al.* (1984) Microbiologica **7**, 91.
Moore, N.F. *et al.* (1985) J. gen. Virol. **66**, 647.

infectious haematopoietic necrosis virus. Family *Rhabdoviridae*, not allotted to genus. Causes necrosis of haematopoietic tissue of spleen and kidney of young trout and salmon. Highly contagious. Virus grows in fat head minnow cells.

infectious hepatitis. An old name for HEPATITIS A VIRUS and the disease associated with it.

infectious hepatitis virus. *See* HEPATITIS A VIRUS.

infectious mononucleosis virus. *See* EPSTEIN-BARR VIRUS.

infectious pancreatic necrosis virus. Family *Birnaviridae*, genus *Birnavirus*. Causes acute, lethal disease of young trout by infecting pancreatic tissue. Highly contagious. Replicates in various fish tissue culture cells.

infectious particle. A virus particle which contains the complete viral genome and thus can infect a susceptible cell.

infectious pustular vulvo-vaginitis virus. *See* INFECTIOUS BOVINE RHINOTRACHEITIS VIRUS.

infectious unit. The smallest number of virus particles that, in theory, can cause infection. For most viruses, those that contain the full viral genome, it is one particle but for multipartite viruses it can be two or three particles. *See* ONE-HIT KINETICS, MULTI-HIT KINETICS.

infectivity. The ability of a virus to replicate within the cells of its host.

influenza virus. Family *Orthomyxoviridae*. There are three genera, A, B and C. Causes influenza and pneumonia in man, often in epidemic proportions. Different strains infect birds,

myxoviridae.

Ingwavuma virus. Family*Bunyaviridae*, genus *Bunyavirus*. Isolated from pigs in Thailand and birds in several countries in Africa; mosquito-borne.

Inini virus.   Family *Bunyaviridae*, genus *Bunyavirus*.

initiation. The start of synthesis of a polypeptide or nucleic acid chain. *See* START CODON, EUKARYOTIC INITIATION FACTORS.

initiation codon. *See* START CODON.

Inkoo virus.   Family *Bunyaviridae*, genus *Bunyaviruś*. Isolated from mosquitoes in Finland. Antibodies found in man and several other host species. Causes fever in man.

Inoviridae. (Greek 'ino', 'inos' = muscle.) A family of rod-shaped phages containing two genera; *INOVIRUS* (filamentous phages) and *PLECTROVIRUS* (Mycoplasma virus type 1 phages). Virions are unenveloped rods, 84-1950 nm. in length (depending on virus strain) and include phage morphotypes F1 and F2 (*see* PHAGE). The

MVL51 phage

fd phage

100nm

horses and pigs. Good evidence that pig strains infect man and vice versa. Some strains infect mice and ferrets. Grows well in eggs. A highly variable virus antigenically, making control by vaccination difficult. Recombination occurs frequently between species.

influenza virus A. A species in the genus *Influenza virus*, family *Orthomyxoviridae*.

influenza virus A-porcine. Synonym: SWINE INFLUENZA VIRUS.

influenza virus B. A species in the genus *Influenza virus*, family *Orthomyxoviridae*.

Influenza virus C. A genus in the family *Ortho-*

viral genome is circular ssDNA. Cells infected with these viruses continue to function more or less normally. Virus particles are extruded through cell membranes and the host is not lysed. Ray, D.S. (1977) **In** Comprehensive Virology. Vol. 7, p. 105 ed. H. Fraenkel-Conrat and R.R. Wagner. Plenum Press: New York.
Maniloff, J. *et al.* (1982) Intervirology **18**, 177.

Inovirus. Genus of the *INOVIRIDAE* which contains flexible filamentous rod-shaped viruses isolated from bacteria. Particles are unenveloped rods 760-1950 nm. long x 6 nm. wide (PHAGE morphotype F1). Virions sediment at 41-45 S and have a buoyant density in CsCl of approximately 1.29 g/cc. Particles contain two proteins and a circular ssDNA genome (mw. 1.9-2.7×10^6).

Most members adsorb to sex pili and are therefore specific for male bacteria (Enterobacteria, *Pseudomonas, Vibrio, Xanthomonas*). The proposed type member is fd PHAGE.

Ray, D.S. (1977) In Comprehensive Virology. Vol. 7. p. 105 ed. H. Fraenkel-Conrat and R.R. Wagner. Plenum Press: New York.

Ackermann, H.W. and Du Bow, M.S. (1987) Viruses of Prokaryotes. CRC Press: Boca Raton, Florida.

Insect Parvovirus group. Vernacular name for *DENSOVIRUS*.

insect picornaviruses. Small RNA viruses of invertebrates, the majority of which are unclassified but are commonly grouped together because of their general resemblance to PICORNAVIRUSES in size, buoyant density, mw. of RNA and protein content (3-4 polypeptides). Only three insect picornaviruses are formally assigned to the *Picornaviridae*; CRICKET PARALYSIS VIRUS, *DROSOPHILA C VIRUS* and *GONOMETA* VIRUS. Other possible members with at least three structural proteins include *DROSOPHILA A VIRUS, DROSOPHILA P VIRUS*, INFECTIOUS FLACHERIE VIRUS, *RHOPALOSIPHUM PADI* VIRUS, APHID LETHAL PARALYSIS VIRUS, KAWINO VIRUS, *LYMANTRIA NINAYI* VIRUS, BEE SLOW PARALYSIS VIRUS, SACBROOD VIRUS, THAI SACBROOD VIRUS, EGYPT BEE VIRUS, KASHMIR BEE VIRUS (including isolates from South Australia, New South Wales and Queensland), BERKELEY BEE PICORNAVIRUS, CERATITIS PICORNAVIRUS V, PSEUDOPLUSIA PICORNAVIRUS. Other viruses which contain RNA and are of similar size but differ in other properties include ARKANSAS BEE VIRUS, TERMITE PARALYSIS VIRUS, BEE ACUTE PARALYSIS VIRUS, BEE VIRUS X and BLACK QUEEN CELL VIRUS.

Moore, N.F. *et al.* (1985) J. gen. Virol. **66**, 647.

insert. Term used for the foreign DNA cloned into a bacterial plasmid or other gene vector.

integrase. Virus-induced enzyme which mediates the integration of the viral DNA into the host genome. Found, for example, in infections with TEMPERATE PHAGES and RETROVIRUSES.

integration. The process of insertion of viral DNA into the host genome. It usually involves a virus-coded enzyme, the INTEGRASE. The viral DNA then comes under the host nucleic acid replication mechanism.

intercalation. The insertion of planar molecules between the adjacent base pairs of dsDNA or dsRNA or ss nucleic acid with secondary structure. Intercalating agents will inhibit the replication and transcription of DNA. They will also reduce the buoyant density of DNA and RNA in isopycnic gradients. *See* ETHIDIUM BROMIDE.

interference. Prevention of the replication of one virus by another. It may be due, for example, to the blocking of virus attachment sites or the presence of defective interfering particles which block the replication of the competent ones. *See* CROSS PROTECTION.

interferon(s). Animal cell proteins formed in response to virus infection and which cause other cells to resist infection. They are also induced by other agents including dsRNA and synthetic ribonucleotide poly(I)-poly(C). There are three classes of human interferons; α produced mainly by leucocytes; β by fibroblasts in response to virus infection or dsRNA and γ by lymphocytes in response to mitogens.

intermolecular recombination. Recombination due to the reassortment of species of nucleic acid between viruses whose genomes are segmented, e.g. segments of RNA of REOVIRUSES, INFLUENZA. Also termed GENETIC REASSORTMENT. *See* PHENOTYPIC MIXING.

International Committee on Taxonomy of Viruses (ICTV). A committee of the Virology Division of the International Union of Microbiological Societies which controls the nomenclature and classification of viruses. It has seven subcommittees which deal with specific areas, e.g. viruses of vertebrates, plant viruses, viruses of invertebrates.

intramolecular recombination. Recombination leading to the exchange of sequences between molecules of nucleic acid.

intraperitoneal. A route often used for the injection of e.g. antigens and viruses into an animal (via the peritoneum).

intravenous. A route sometimes used for the injection of e.g. antigens into the venous system of an animal. Usually the animal responds by relatively rapid production of antibodies but the antibody titre often remains high for only a short time as the antigen is rapidly diluted in the bloodstream.

intron. A region of DNA which is transcribed into RNA initially but is then removed by SPLICING, in the formation of functional RNA. There are four types of introns, those in nuclear mRNA, those in nuclear tRNA, group II introns in the mRNA of mitochondria and chloroplasts and group I introns in various RNAs including mitochondrial messenger and tRNAs, chloroplast RNA genes and some nuclear RNA genes. *See* EXON.

ion exchange chromatography. A chromatographic procedure in which the stationary phase consists of an ion exchange resin which may be acidic or basic. Used for the purification of charged molecules, e.g. the separation of immunoglobulins on DEAE cellulose.

IPL-41 medium. Cell culture medium developed for the growth of insect cells in culture (e.g. the IPLB-SF-21 cell line) and for the large-scale production of nuclear polyhedrosis viruses. Weiss, S.A. & Vaughn, J.L. (1986) In The Biology of Baculoviruses, Vol II, p. 63. ed. R.R. Granados and B.A. Federici. CRC Press: Boca Raton, Florida.

IPLB-21 cells. *See* *i*IPLB-SF-21 CELLS.

IPLB-HZ-1075 cells. Insect cell line derived from *Heliothis zea*, susceptible to infection by *H. zea* nuclear polyhedrosis virus.

IPLB-LD-652Y cells. Insect cell line derived from the gypsy moth, *Lymantria dispar*, permissive for the replication of *L. dispar* nuclear polyhedrosis virus (NPV) and semi-permissive for *Autographa californica* NPV.

IPLB-SF-21 cells. Insect ovarian cell line from the fall armyworm, *Spodoptera frugiperda*, susceptible to infection by *AUTOGRAPHA CALIFORNICA NUCLEAR POLYHEDROSIS VIRUS* (NPV) and some other baculoviruses including *Trichoplusia ni* MNPV and *S. frugiperda* MNPV.

iridescent viruses. (*Iridoviridae*). Large unenveloped, isometric dsDNA-containing viruses, 120-180 nm. in diameter, isolated from insects. Infected insects generally appear iridescent. An interim classification system assigned a type number to each new virus isolate. Thirty-two 'types' were listed before the system was discontinued. At least 23 other isolates have also been recorded. Types 1, 2, 6, 9, 10 and 16-29 have been assigned to the small iridescent insect virus group (*see* IRIDOVIRUS) and types 3-5, 7, 8 and 11-15 to the large iridescent insect virus group (*see* CHLORIRIDOVIRUS). Iridescent viruses have also been observed in non-insect invertebrate hosts including *Armadillidium vulgare* (virus type 31), *Porcellio dilatatus* (type 32), *P. scaber* and *Simocephalus expinosus* (type 20) (Crustacea); *Nereis diversicolor* (type 27) (Annelida) and *Octopus vulgaris* (Gastropoda). A list of insect hosts in which iridescent virus infections have been recorded is given in Appendix E.
Hall, D.W. (1985) In Viral Insecticides for Biological Control. p. 163. ed. K. Maramorosch and K.E. Sherman. New York: Academic Press.

Iridoviridae = icosahedral cytoplasmic deoxyriboviruses. (Greek 'iris, iridos' = iridescent; from the appearance of accumulations of virus.) A family of four genera infecting vertebrate and invertebrate hosts. Two genera (IRIDOVIRUS and CHLORIRIDOVIRUS) contain viruses isolated from invertebrates; certain members of the IRIDOVIRUS genus experimentally infect a wide range of insects and crustacea. Viruses in the genera RANAVIRUS and LYMPHOCYSTIVIRUS have been isolated from amphibia and fish, respectively. Virus particles are icosahedral (125-300 nm. in diameter) and contain a spherical nucleoprotein core surrounded by a lipid membrane and a layer of morphological protein subunits. Virions sedi-

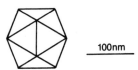

100nm

ment between 1300-4450 S, and have a buoyant density of 1.16-1.35 g/cc. (depending on genus). Particles contain many proteins and several virion-associated enzymes including a protein kinase. The genome is a single molecule of linear dsDNA (mw. $100\text{-}250 \times 10^6$) which has been shown in some strains to be terminally redundant and circularly permuted. Virus morphogenesis occurs in the cytoplasm though functions in the host cell nucleus are required for DNA transcription and replication. Virus entry rapidly inhibits host macromolecular synthesis. The AFRICAN SWINE FEVER group is no longer classified as a genus within the *Iridoviridae*.

Goorha, R. and Granoff, A. (1979) In Comprehensive Virology. Vol. 14. p. 347 ed. H. Fraenkel-Conrat and R.R. Wagner. Plenum Press: New York.

Delius, H. *et al.* (1984) J. Virol. **49**, 609.

Iridovirus. (Greek 'iris, iridos' = goddess whose sign was the rainbow, hence iridescent.) Genus of the small iridescent insect virus group (*Iridoviridae*). Virions are icosahedral (*c*.120 nm. in diameter), composed of a dense central core surrounded by a unit membrane which is in turn enclosed within a protein capsid. Some virus isolates carry small fibril-like attachments to the capsid subunits. Particles sediment at 2200-3300S and have a buoyant density of 1.33 g/cc. Virus particles contain about 26-28 polypeptides (mw. 17.5-300 x 10^3) and a number of virus-associated enzymes including DNA-dependent RNA polymerase, nucleotide phosphohydrolase and protein kinase. The genome is a linear dsDNA (mw. *c.* 130 x 10^6) which in some strains has been shown to be circularly permuted and terminally redundant. Virus infection probably occurs by ingestion, followed by replication in the cytoplasm of susceptible tissues (e.g. fat body cells). The type species is CHILO IRIDESCENT VIRUS (iridescent virus type 6). Particle size has been used as the main criterion for distinguishing *Iridovirus* members from those included in the *Chloriridovirus* genus. A number of isolates assigned to the Iridovirus genus are serologically related to each other but not to the Chloriridovirus type species. Apart from the type species other insect iridescent virus isolates included in the Iridovirus genus have been isolated from *Tipula* sp. (types 1 and 25), *Sericesthis* sp. (type 2), *Wiseana* sp. (type 9), *Witlesia* sp. (type 10), *Costelytra* sp. (type 16), *Pterostichus* sp. (type 17), *Opogonia* sp. (type 18), *Odontria* sp. (type 19), *Heliothis* sp. (type 21), *Simulium* sp. (type 22), *Heteronychus* sp. (type 23), *Apis* sp. (type 24), an Ephemaropteran (type 26), *Lethocerus* sp. (type 28), *Tenebrio* sp. (type 29). Morphologically-similar viruses were isolated from the marine worm *Nereis* sp. (type 27) and the crustacean *Simocephalus* sp. (type 20). A list of iridoviruses infecting insects is given in Appendix E.

Hall, D.W. (1985) In Viral Insecticides for Biological Control. p.163. ed. K. Maramorosch and K. E. Sherman. Academic Press: New York.

Iris fulva mosaic virus. A possible *Potyvirus*.

Barnett, O.W. (1986) CMI/AAB Descriptions of Plant Viruses No. 310.

Francki, R.I.B. *et al.* (1985) In Atlas of Plant Viruses. Vol. 2. p. 183. CRC Press: Boca Raton, Florida.

Iris germanica leaf stripe virus. A possible plant *Rhabdovirus*, subgroup 1.

Francki, R.I.B. *et al.* (1985) In Atlas of Plant Viruses. Vol. 1. p. 73. CRC Press: Boca Raton, Florida.

Iris mild mosaic virus. A *Potyvirus*.

Brunt, A.A. (1986) CMI/AAB Descriptions of Plant Viruses No. 324.

Francki, R.I.B. *et al.* (1985) In Atlas of Plant Viruses. Vol. 2. p. 183. CRC Press: Boca Raton, Florida.

Iris severe mosaic virus. A *Potyvirus*.

Francki, R.I.B. *et al.* (1985) In Atlas of Plant Viruses. Vol. 2. p. 183. CRC Press: Boca Raton, Florida.

Irituia virus. Family *Reoviridae*, genus *Orbivirus*. Isolated from rodents in Brazil.

Isachne mosaic virus. A possible *Potyvirus*.

Francki, R.I.B. *et al.* (1985) In Atlas of Plant Viruses. Vol. 2. p. 183. CRC Press: Boca Raton, Florida.

Isfahan virus. Family *Rhabdoviridae*, genus *Vesiculovirus*. Isolated from the sand fly in Iran. Neutralising antibody found in man and other host species in several areas of that country. Resembles the New Jersey serotype of vesicular stomatitis virus more closely than the other vesiculoviruses.

isiolovirus. *See* SHEEP CAPRIPOX VIRUS.

isoelectric focusing (electrofocusing). A separation technique in which mixtures of proteins and/or viruses are resolved into their components by subjecting them to an electric field in a supporting gel or stabilised solution in which a pH gradient is established. The proteins or viruses migrate to the positions in the gel which have a pH equivalent to their ISOELECTRIC POINTS.

isoelectric point. The pH value of a solution in which a given macromolecule (usually a protein or virus) does not move in an electric field. At this pH the net surface charge is zero.

isolate. A sample (e.g. of virus) from a defined source.

isometric particle. A virus particle with identical linear dimensions which thus appears spherical. *See* ICOSAHEDRAL SYMMETRY.

isometric phages. *See* CUBIC PHAGES.

isopycnic gradient. A separation technique in which a sample containing macromolecules is centrifuged through a gradient of increasing density. The macromolecules migrate until they reach the region where their density is equal to that in the gradient. Gradients may be formed either by layering and diffusion or by the effects of centrifugal force and diffusion on small heavy metal ions, e.g. Cs⁺. *See* RATE ZONAL GRADIENT.

Israel turkey meningoencephalitis virus. Family *Flaviviridae*, genus *Flavivirus*. Isolated from turkeys in Israel.

Itaituba virus. Family *Bunyaviridae*, genus *Phlebovirus*.

ivy vein clearing virus. A possible plant *Rhabdovirus*, subgroup 1.
Francki, R.I.B. *et al.* (1985) In Atlas of Plant Viruses. Vol. 1. p. 73. CRC Press: Boca Raton, Florida.

IZD-MB(0507) cells. Insect cell line from the cabbage moth, *Mamestra brassicae*, susceptible to infection by M. BRASSICAE NUCLEAR POLYHEDROSIS VIRUS.

J

Jacareacanga virus. Family *Reoviridae*, genus *Orbivirus*.

Jamestown Canyon virus. Family *Bunyaviridae*, genus *Bunyavirus*. Isolated from mosquitoes, horse flies and deer flies in Wisconsin and Colorado, USA and in Canada. White-tailed deer may be host.

Japanant virus. Family *Reoviridae*, genus *Orbivirus*. Isolated from mosquitoes and bats in New Guinea.

Japanese B encephalitis virus. Synonym: JAPANESE B VIRUS. Family *Flaviviridae*, genus *Flavivirus*. Infects man, birds, pigs and horses. Probably transmitted by mosquito. Causes fever in man and a small percentage have encephalitis. Horses also have encephalitis. Occurs in SE Asia and is regarded as a serious disease in man in China and Japan. Can be grown in eggs, various mammalian cell cultures and in mosquito cell cultures. Experimentally, i.c. inoculation of mice, monkeys and hamsters causes encephalitis.

Japanese B virus. *See* JAPANESE B ENCEPHALITIS VIRUS.

Japanese haemagglutinating virus. *See* SENDAI VIRUS.

Jatropha mosaic virus. A probable *Geminivirus*, subgroup B.
Harrison, B.D. (1985) Ann. Rev. Phytopath. **23**, 55.

JC virus. Family *Papovaviridae*, genus *Polyomavirus*. Isolated from brain of patient with progressive multifocal leucoencephalopathy. Antibodies found in man, suggesting it is a common infection. Can be grown in primary human foetal glial cell cultures. Experimentally, virus is highly oncogenic in new-born hamsters.

Jerry Slough virus. Family *Bunyaviridae*, genus *Bunyavirus*. Isolated from mosquitoes in California.

JHM virus. *See* MOUSE HEPATITIS VIRUS.

Johnston Atoll virus. Unclassified arbovirus. Isolated from ticks on Johnston Atoll and in Australia and New Zealand.

Joinjakaka virus. Family *Rhabdoviridae*, unassigned to genus. Isolated from culicines in New Guinea.

jonquil mild mosaic virus. *See* NARCISSUS LATE SEASON YELLOWS VIRUS.

Juan Diaz virus. Family *Bunyaviridae*, genus *Bunyavirus*.

Jugra virus. Family *Flaviviridae*, genus *Flavivirus*. Isolated from bats and mosquitoes in Malaysia.

Junin virus. Synonym: ARGENTINIAN HAEMORRHAGIC FEVER VIRUS. Family *Arenaviridae*, genus *Arenavirus*. Causes haemorrhagic disease of agricultural workers in Argentina, with fever, leucopenia and exanthema. Mortality rate up to 15%. Isolated from wild rodents and mites. Transmission to man is probably from urine and faeces of rodents. Disease in rodents is chronic and persistent. Guinea pigs and primates can be infected experimentally, showing haemorrhagic disease similar to that in man. Virus can be grown in wide range of cells, usually giving chronically infected cultures.

Jurona virus. Family *Bunyaviridae*, genus *Bunyavirus*. Isolated from mosquitoes in Brazil.

Jutiapa virus. Family *Flaviviridae*, genus *Flavivirus*. Isolated from cotton rats in Guatemala.

K

K virus. Family *Papovaviridae*, genus *Polyomavirus*. Causes silent infection of wild mice but when injected into very young mice, causes fatal pneumonia. Can be grown in mouse lung cell cultures.

Kadam virus. Family *Flaviviridae*, genus *Flavivirus*. Isolated from ticks in Uganda.

Kaeng Khoi virus. Family *Bunyaviridae*, genus *Bunyavirus*. Isolated from bats and rats in Thailand.

Kaffir pox virus. *See* VARIOLA MINOR VIRUS.

Kaikalur virus. Family *Bunyaviridae*, genus *Bunyavirus*. Isolated from mosquitoes at Kaikalur in India.

Kairi virus. Family *Bunyaviridae*, genus *Bunyavirus*. Isolated from mosquitoes in Trinidad, Brazil and Colombia.

Kaisodi virus. Family *Bunyaviridae*, genus *Bunyavirus*. Isolated from ticks and a ground thrush in Mysore, India.

Kakalur virus. Family *Bunyaviridae*, genus *Bunyavirus*. Isolated from mosquitoes in Maikolur, India. Kills mice injected i.c.

Kao Shuan virus. Family *Bunyaviridae*, genus *Nairovirus*. Isolated from ticks in Taiwan.

Karimabad virus. Family *Bunyaviridae*, genus *Phlebovirus*. Isolated from Phlebotomus in Iran and Pakistan.

Karshi virus. Family *Flaviviridae*, genus *Flavivirus*. Isolated from ticks in the Karshi desert in the USSR. Causes paralysis in young mice.

Kasba virus. Family *Reoviridae*, genus *Orbivirus*. Isolated from ticks in India.

Kashmir bee virus. Unclassified small RNA virus (possible INSECT PICORNAVIRUS) isolated from the European honey bee, *Apis mellifera*, but probably originating in *A. cerana* from Kashmir. Virions are isometric, 30 nm. in diameter, sediment at 172S and have a buoyant density in CsCl of 1.37 g/cc. Particles contain three structural proteins (mw. 41.1, 37.3, 24 and 5 x 10^3) and an RNA genome; serologically related to, but distinct from SOUTH AUSTRALIA KASHMIR BEE VIRUS, NEW SOUTH WALES KASHMIR BEE VIRUS and QUEENSLAND KASHMIR BEE VIRUS.

Bailey, L. and Woods, R.D. (1977) J. gen. Virol. **37**, 175.

Bailey, L. *et al.* (1979) J. gen. Virol. **43**, 641.

Kata virus. *See* PESTE DES PETITS RUMINANTS VIRUS.

Kawino virus. Unclassified small RNA virus (a possible INSECT PICORNAVIRUS) isolated from the mosquito, *Mansonia uniformis*, near the village of Kawino, Kenya. Virus particles are isometric, 28 nm. in diameter, sediment at 165S and have a buoyant density in CsCl of 1.33 g/cc. The virus contains four structural polypeptides (mw. 33, 30, 27 and 7 x 10^3) and an ssRNA genome (mw. 2.6 x 10^6). Unlike certain other insect picornaviruses, the viral RNA appears to lack a poly A tract.

Moore, N.F. *et al.* (1985) J. gen. Virol. **66**, 647.

kb. Abbreviation for KILOBASE.

kbp. Abbreviation for KILOBASE PAIRS.

Kedong virus. *See* SHEEP CAPRIPOX VIRUS.

Kedougou virus. Family *Flaviviridae*, genus *Flavivirus*. Isolated from mosquitoes. Antibodies found in man but no disease reported.

kelp fly virus. An unclassified small RNA-containing virus isolated from the kelp fly, *Chaetocoelopa sydneyensis* (Diptera) in New South

Wales, Australia. Virus particles are isometric, 29 nm. in diameter and are readily distinguished from other small RNA insect viruses by the presence of surface projections (*c*.8 nm. in length) located on the fivefold axes. The virions sediment at 158S and have a buoyant density in CsCl of 1.425 g/cc. at pH 7.0. The virus has two major structural polypeptides (mw. 73 and 29.4 x 10^3) and an ssRNA genome (mw. *c*.3.5 x 10^6). The virus appears to cause no symptoms in kelp flies; it also replicates in larvae of the wax moth, *Galleria mellonella* (Lepidoptera).
Scotti, P.D. *et al.* (1976) J. gen. Virol. **30**, 1.

Kemerovo virus. Family *Reoviridae*, genus *Orbivirus*. Isolated from ticks and also from two people with fever in Siberia and from a bird in Egypt. Antibodies found in several mammalian species in Siberia.

Kenai virus. Family *Reoviridae*, genus *Orbivirus*.

Kennedya virus Y. A possible *Potyvirus.* Francki, R.I.B. *et al.* (1985) **In** Atlas of Plant Viruses. Vol. 2. p. 183. CRC Press: Boca Raton, Florida.

Kennedya yellow mosaic virus. A *Tymovirus.* Gibbs, A.J. (1978) CMI/AAB Descriptions of Plant Viruses No. 193.
Francki, R.I.B. *et al.* (1985) **In** Atlas of Plant Viruses. Vol. 1. p. 117. CRC Press: Boca Raton, Florida.
Hirth, L. and Girard, L. (1988) **In** The Plant Viruses. Vol. 3. p. 163. ed. R. Koenig. Plenum Press: New York.

kennel cough virus. *See* CANINE HEPATITIS VIRUS.

Kern Canyon virus. Family *Rhabdoviridae*, unassigned to genus. Isolated from a bat in California.

Ketapang virus. Family *Bunyaviridae*, genus *Bunyavirus*. Isolated from mosquitoes in Malaya. Antibodies found in man but not associated with disease.

Keuraliba virus. Family *Rhabdoviridae*, unassigned to genus. Isolated from gerbils and rodents in Senegal.

Keystone virus. Family *Bunyaviridae*, genus *Bunyavirus*. Isolated from mosquitoes in Tampa

Bay area of Florida and in several other southern states. May be maintained in cotton-tail rabbits, grey squirrels and cotton rats.

Khasan virus. Family *Bunyaviridae*, genus *Bunyavirus*. Isolated from ticks in the USSR Pathogenic when injected i.c. into young mice. Grows in chick, duck and green monkey cell cultures.

Kilham rat virus. Synonym: LATENT RAT VIRUS. Causes a latent infection of rats. Experimentally it is fatal for new-born hamsters given i.p. In older hamsters it stunts growth with abnormal development of bones. Infection of pregnant hamsters and rats causes congenital abnormalities. Grows in rat cell cultures.

killed vaccine (virus). Vaccine comprising virus which has been inactivated, usually by treatment with a chemical such as an imine or formalin, e.g. foot-and-mouth disease vaccine.

killer particles. Particulate BACTERIOCINS which resemble bacteriophage particles.

kilobase (pairs) (kb, kbp). One thousand nucleotides in a polynucleotide chain. Used as a measure of the size of a nucleic acid molecule. Kb refers to ss nucleic acid, kbp to ds nucleic acid.

kinase. Any enzyme that catalyses phosphorylation reactions. *See* POLYNUCLEOTIDE KINASE, PROTEIN KINASE.

Kirk virus. Family *Papovaviridae*, genus *Parvovirus*. Isolated from Detroit 6 cells which had been inoculated with plasma from an individual who had received infectious hepatitis serum.

Klamath virus. Family *Rhabdoviridae*, unassigned to genus. Isolated from a mouse in Klamath County, Oregon. Day-old mice die when inoculated i.c. Grows in BHK 21 cells.

Kleinschmidt procedure. A technique in which small amounts of nucleic acid are coated with a basic protein, e.g. cytochrome *c*, and spread on a denatured protein monolayer at an air-water interface. After being picked up on an electron microscope grid, the nucleic acid molecules are shadowed with a heavy metal. When viewed in the electron microscope, ss and ds nucleic acid molecules can be distinguished, contour lengths measured and features such as R-LOOPS and D-

LOOPS examined.
Kleinschmidt, A.K. and Zahn, R.K. (1959) Zeit. Naturforsch. **14b**, 77.
Kleinschmidt, A.K. (1968) Meths. Enzymol. **12B**, 361.

Klenow fragment. The larger of the two fragments of *E. coli* DNA POLYMERASE I formed after limited proteolytic cleavage. It retains the 5'-3' polymerase and the 3'-5' exonuclease activities and is used extensively in the SANGER METHOD for sequencing DNA and in infilling STICKY ENDS and gaps in DNA.

Koch's postulates. A set of criteria proposed by Robert Koch in 1882 and used for assessing whether a given micro-organism is the causal agent of a disease. They are: 1) the organism must be associated with all cases of the disease, in a logical pathological relationship to the disease and its symptoms; 2) the organism must be obtained in pure culture; 3) an injection of the organism into a host must reproduce the disease in that host; 4) the organism must be recovered from that host.

For viruses the criteria have been modified by Rivers, T.M. (1937) J. Bact. **33**, 1, to: 1) isolation of virus from the diseased host; 2) cultivation of the virus in experimental hosts or cells; 3) filterability of the pathogen; 4) production of a comparable disease in the original or related host species; 5) re-isolation of the virus; 6) in the case of an animal virus, the detection of a specific immune response.

Kokobera virus. Family *Flaviviridae*, genus *Flavivirus*. Isolated from mosquitoes in Queensland and New Guinea.

Koongol virus. Family *Bunyaviridae*, genus *Bunyavirus*. Isolated from mosquitoes in Queensland and New Guinea. Antibodies commonly found in cattle.

Korean haemorrhagic fever virus. Family *Bunyaviridae*. Causes severe haemorrhagic fever in man, with damage to kidneys. First seen in Korea but also occurs in the USSR, Eastern Europe and Scandinavia. Can be passaged in *Aedes agrarius*.

kotonkan virus. Family *Rhabdoviridae*, unas-signed to genus. Isolated from mosquitoes in Nigeria. Antibodies found in man and domestic animals. May be cause of fever. Related antigenically to Obodhiang and Mokola viruses.

Koutango virus. Family *Flaviviridae*, genus *Flavivirus*. Isolated from rodents in Senegal and Central African Republic.

Kowanyama virus. Family *Bunyaviridae*, genus *Bunyavirus*. Isolated from mosquitoes in Northern Queensland. Antibodies found in man, domestic fowl, horses and kangaroos.

Kumlinge virus. Family *Flaviviridae*, genus *Flavivirus*. Isolated from ticks, squirrels, voles, hares, thrushes in Finland. Causes fever with encephalitis in man.

Kunitachi virus. Family *Paramyxoviridae*, genus *Paramyxovirus*. Isolated from a budgerigar in Japan. Causes fatal disease on injection into budgerigars. Grows in eggs and chick embryo cell cultures.

Kunjin virus. Family *Flaviviridae*, genus *Flavivirus*. Isolated from mosquitoes in Australia, Borneo and Sarawak. Can cause fever in man.

Kununurra virus. Family *Rhabdoviridae*, genus unassigned. Isolated from mosquitoes at Kununurra, Western Australia.

Kuru virus. Belongs to the SPONGIFORM ENCEPHALOPATHY GROUP OF AGENTS. Not a conventional virus. Causes subacute degeneration of the brain in man. Occurrence restricted to Papua, New Guinea. Resembles scrapie and Creutzfeld Jakob disease in its pathology.

Kyasanur Forest disease virus. Family *Flaviviridae*, genus *Flavivirus*. Isolated from ticks in Mysore, India. Causes fever, headache, limb pains, prostration, conjunctivitis, diarrhoea, vomiting and intestinal haemorrhaging. Antibodies present in several wild mammalian species. Experimentally, mice develop encephalitis when injected with virus. Rhesus and bonnet monkeys develop viraemia without showing disease.

Kyzylagach virus. Family *Togaviridae*, genus *Alphavirus*.

L

L1 Acholeplasmavirus group. Alternative name for Mycoplasma virus type 1 phages (*PLECTROVIRUS*). Type member (L1 strain L51) is MVL51 PHAGE.
Maniloff, J. *et al.* (1982) Intervirology **18**, 177.

L1 Mycoplasma phage strain L51. Synonym: MVL51 PHAGE.

L2 Acholeplasmavirus group. Alternative name for Mycoplasma virus type 2 phages (*PLASMAVIRUS*). Type member (L2 strain L2) is L2 PHAGE.
Maniloff, J. *et al.* (1982) Intervirology **18**, 177.

L2 Mycoplasma phage strain L2. Synonym: L2 PHAGE.

L2 phage. PLASMAVIRUS type species isolated from *Acholeplasma laidlawii*. Virions are enveloped, pleomorphic and quasi-spherical, about 74-132 nm. in diameter. Three morphological forms of the virus have been recognised, L2-I (74 nm. in diameter), L2-II (88 nm.), L2-III (132 nm. in diameter). No clearly defined nucleocapsid is present and the virion envelope is derived by budding from infected cell membranes. Virus infectivity is sensitive to detergents and organic solvents. All particles contain at least eight proteins (mw. 19-70 x 10^3). The genome is circular supercoiled dsDNA (mw. 7.8 x 10^6; 11.6 kbp). Particles L2-I and L2-II each contain one genome copy, while L2-III contains two to three copies. The productive cycle of L2 does not cause lysis, and is followed by the establishment of LYSOGENY.
Maniloff, J. *et al.* (1982) Intervirology **18**, 177.
Poddar, S.K. *et al.* (1985) Intervirology **23**, 208.

L3 Acholeplasmavirus group. Proposed genus of phages isolated from *Acholeplasma*, which have short tails and resemble members of the *Podoviridae*. Proposed type member is phage L3.
Maniloff, J. *et al.* (1982) Intervirology **18**, 177.

L3 phage. Proposed type member of L3 Acholeplasmavirus group (*Podoviridae*). Virions have an isometric head (60 nm. in diameter) with a short (20 nm.) tail and tail fibres. Particles have a buoyant density of 1.48 g/cc. in CsCl. The genome is linear dsDNA (mw. *c.*26 x 10^6). Isolated from *Acholeplasma laidlawii*.
Maniloff, J. *et al.* (1982) Intervirology **18**, 177.

L172 phage. Unclassified virus isolated from *Acholeplasma laidlawii*, originally thought to be closely-related to L2 phage (the type member of the PLASMAVIRUS genus). However, the L172 genome is a circular ssDNA (14kb; cf. dsDNA of L2 phage; 11.6kbp) and the two viruses have no detectable DNA homology. Virions are globular, 60-80 nm. in diameter, frequently with a protuberance. At least seven structural proteins are present (mw. 15-71 x 10^3).
Dybvig, K. *et al.* (1985) J. Virol. **53**, 384.

La Crosse virus. Family *Bunyaviridae*, genus *Bunyavirus*. First isolated from a woman with a fatal meningoencephalitis in Wisconsin. Many other cases reported in Wisconsin and Minnesota. Can be isolated from mosquitoes. Antibodies are found in small forest rodents.

labelling. A term usually referring to the attachment of reporter groups to a PROBE nucleic acid or to a protein or other macromolecule. The reporter group can be radioactive or non-radioactive (*see* BIOTIN) and can be used to trace the macromolecule.

laboratory strains. Isolates (e.g. viruses) which have been propagated in the laboratory for some time. They may differ from the original WILD TYPE from which they were derived.

Laburnum yellow vein virus. A possible plant *Rhabdovirus*, subgroup 2.
Francki, R.I.B. *et al.* (1985) **In** Atlas of Plant

Viruses. Vol. 1. p. 73. CRC Press: Boca Raton, Florida.

lactic dehydrogenase virus. *See* RILEY VIRUS.

Laelia red leafspot virus. A possible plant *Rhabdovirus*.
Francki, R.I.B. *et al.* (1985) **In** Atlas of Plant Viruses. Vol. 1. p. 73. CRC Press: Boca Raton, Florida.

lag phase. The correct use of the term is for bacterial growth where it is the period of physiological activity and diminished cell division following the addition of an inoculum of a bacterium to a new culture medium. For viruses it is sometimes used to refer to the LATENT PERIOD.

Lagos bat virus. Family *Rhabdoviridae*, genus *Lyssavirus*. Isolated from a fruit bat in Nigeria. Pathogenic for adult mice, dogs and Rhesus monkeys when inoculated. Related antigenically to rabies virus.

lambda (λ) phage. Type species of the only officially-recognised genus of the SIPHOVIRIDAE; PHAGES with long non-contractile tails (phage morphotype B1). Virions have isometric heads, approximately 60 nm. in diameter with a tail of approximately 160 nm. Particles sediment at 388S, have a buoyant density in CsCl of 1.49 g/cc. and a mw. of about 60×10^6. Nine structural proteins (mw. 17×10^3-130×10^3) have been detected. The genome is linear dsDNA with a mw. of 31×10^6 and 12-nucleotide long, single-stranded sticky ends which allow the genome to circularise. The length of the circular DNA is 48502 base pairs. The complete nucleotide sequence is known, and genetic, restriction and functional maps have been prepared. It is a temperate phage. Particles adsorb to the wall of host cells and inject their DNA, which circularises and replicates. By mid-latent period the DNA either replicates by producing concatemers and subsequently virus particles (lytic cycle) or integrates into the host chromosome between the *gal* and *bio* genetic loci (lysogenic cycle). The balance between lytic or lysogenic infection is controlled by host and viral factors. Establishment of the lysogenic state is influenced by a phage-coded insertion-promoting protein and a phage-coded repression of mRNA synthesis and is affected by the number of infecting phages and by temperature and culture conditions. Insertion of the viral genome occurs at precise sites on the host chromosome, resulting in a unique circular permutation of the phage DNA. Excision of the integrated DNA occurs when repression fails and can be induced by agents which inhibit host chromosome replication and induce DNA repair functions (e.g. UV light). Excision is generally a precise mechanism, restoring the intact phage DNA but abnormal excision can lead to the production of TRANSDUCING PHAGE where phage genes are replaced by bacterial DNA. In lytic infections, viral DNA replication occurs in two modes. There is an early mode in which circular ds λ DNA is used as a template for the production of circular daughter molecules which can be transcribed but not packaged into virions. In the latter mode concatemeric DNA produced late in infection by a rolling circle mechanism is packaged into virions. The concatemers are cut at specific *cos* sites which mark the cohesive ends of mature genomes. The production of mature virus leads to cell lysis. λ phage is used as a cloning vector for foreign DNA inserts of up to 8×10^6 mw., as considerable regions of the central part of the genome can be deleted without seriously affecting the ability of the virus to reproduce. The *cos* sites can be used in the construction of hybrid λ and plasmid DNAs (cosmids) which can be packaged into λ particles after the insertion of much larger (18-27×10^6) mw. sections of foreign DNA. In the latter case the molecule replicates as a plasmid after 'infection' of the host. The phage was originally isolated from *E. coli* but will infect other enterobacteria.
Weisberg, R.A. *et al.* (1977) **In** Comprehensive Virology. Vol. 8. p. 197. ed. H. Fraenkel-Conrat and R.R. Wagner. Plenum Press: New York.
Daniels, D.L. *et al.* (1984) **In** Genetic Maps 1984. Vol. 1. ed. S.J. O'Brien. Cold Spring Harbor Laboratory: New York.
Ritchie, D.A. (1983) **In** Topley and Wilson's Principles of Bacteriology, Virology and Immunity. Vol. 1. p. 177. ed. G. Wilson, A. Miles and M.T. Parker. Edward Arnold: London.

Lamium mild mosaic virus. A member of the *Fabavirus* group.
Francki, R.I.B. *et al.* (1985) **In** Atlas of Plant Viruses. Vol. 2. p. 1. CRC Press: Boca Raton, Florida.

Langat virus. Family *Flaviviridae*, genus *Flavivirus*. Isolated from ticks in Malaya. Antibodies found in several species including man and rats. Can be grown in chick embryo fibroblasts.

Langur virus. Family *Retroviridae*, subfamily *Oncovirinae*, proposed genus *Type D Oncovirus*. Isolated from a langur. Related antigenically to Mason-Pfizer monkey virus.

Lanjan virus. Family *Bunyaviridae*, genus *Bunyavirus*. Isolated from ticks in Malaya.

lapine parvovirus. Synonym: RABBIT PARVO-VIRUS. Family *Parvoviridae*, genus *Parvovirus*. Isolated from rabbit faeces but does not cause disease. Grows in rabbit kidney cell cultures.

lapinised virus. Virus adapted to growth in rabbits.

large iridescent insect virus group. Vernacular name for *CHLORIRIDOVIRUS* genus.

Lassa fever virus. Family *Arenaviridae*, genus *Arenavirus*. Causes a severe and often fatal disease which involves headache, fever, severe limb and back pains, diarrhoea, vomiting and severe prostration. Occurs mainly in West Africa. Mild cases of the disease also occur. Natural host is the field rat *Mastomys natalensis*. Monkeys and guinea pigs can be infected experimentally. Virus growth in Vero cells can be detected by fluorescent antibody.

late genes. Genes in a viral nucleic acid which are expressed late in the virus replication cycle. They are frequently the genes coding for capsid proteins. *See* EARLY GENES.

latency. The stage of an infectious disease, other than the incubation period, in which no symptoms are expressed in the host.

latent infection. An infection which does not normally produce symptoms, e.g. ADENOVIRUS in adenoidal tissue, VARICELLA ZOSTER VIRUS in neuronal cells of ganglia. Latency can be broken by INDUCTION.

latent period (phase). The time in the virus infection cycle between the apparent disappearance of the infecting virus and the appearance of newly synthesised virus.

latent rat virus. *See* KILHAM RAT VIRUS.

latent virus. *See* OCCULT VIRUS.

lateral body. Part of the structure of vertebrate POXVIRUS particles. In vertebrate poxviruses the core of the particles appears biconcave with two lateral bodies in the concavities; in some insect poxviruses the core is kidney-shaped with one lateral body. The function of lateral bodies is not known.

latex agglutination. A serological test in which the ANTIBODY or ANTIGEN is adsorbed on to polystyrene latex particles which are then incubated with the other reactant. Positive reactions show as aggregates of latex particles that can be detected readily by the naked eye.

Latino virus. Family *Arenaviridae*, genus *Arenavirus*. Isolated from a rodent in Bolivia and Brazil. Related to the American Haemorrhagic Fever Viruses.

Launea arborescens stunt virus. A possible plant *Rhabdovirus*.
Matthews, R.E.F. (1982) Intervirology **17**, 114.

LAV. *See* LYMPHADENOPATHY-ASSOCIATED VIRUS, HUMAN IMMUNODEFICIENCY VIRUS.

LC$_{50}$. Abbreviation for MEDIAN LETHAL CONCENTRATION.

LCM virus. Acronym from LYMPHOCYTIC CHORIOMENINGITIS VIRUS.

LD$_{50}$. Abbreviation for MEDIAN LETHAL DOSE.

leader sequence. a) Synonym: SIGNAL PEPTIDE; b) The 5′ non-coding part of an mRNA.

leafhopper A virus. Virus with particles resembling those of Fijiviruses (*Reoviridae*) isolated from and multiplying in the leafhopper, *Cicadulina (=Cicadula) bimaculata*. At one time the virus was believed to infect maize (*Zea mays*) plants inducing maize wallaby ear disease, but recent studies have shown that the virus does not multiply in the plants.
Franki, R.I.B. and Boccardo, G. (1983) In The *Reoviridae*. p. 505. ed. W.K. Joklik. Plenum Press: New York.

leaky mutant. Mutant which either has some residual WILD-TYPE activity under 'non-permissive' conditions or reverts readily to the wild type.

Lebombo virus. Family *Reoviridae*, genus *Orbivirus*. Isolated from mosquitoes in Africa.

Lecontvirus. Preparation of nuclear polyhedrosis virus produced and provisionally registered for use in Canada by the Forest Pest Management Institute for the control of redheaded pine sawfly, *Neodiprion lecontei*.

lectins. Plant proteins which bind to certain sugar residues in glycoproteins. *See* CONCANAVALIN A.

Lednice virus. Family *Bunyaviridae*, genus *Bunyavirus*. Isolated from ticks in the town of Lednice, Moravia. Grows in goose and chick embryo cell cultures. Kills new-born mice.

leek yellow stripe virus. A *Potyvirus*.
Bos, L. (1981) CMI/AAB Descriptions of Plant Viruses No. 240.
Francki, R.I.B. *et al.* (1985) **In** Atlas of Plant Viruses. Vol. 2. p. 183. CRC Press: Boca Raton, Florida.

leek yellows virus. A *Luteovirus*; occurs in Japan.
Doi, Y. *Personal communication.*

legume yellows virus. A *Luteovirus*; considered to be a strain of BEAN LEAFROLL VIRUS.
Francki, R.I.B. *et al.* (1985) **In** Atlas of Plant Viruses. Vol. 1. p. 137. CRC Press: Boca Raton, Florida.
Casper, R. (1988) **In** The Plant Viruses. Vol. 3. p. 235. ed. R. Koenig. Plenum Press: New York.

lemon-scented thyme leaf chlorosis virus. A possible plant *Rhabdovirus*, subgroup 2.
Francki, R.I.B. *et al.* (1985) **In** Atlas of Plant Viruses. Vol. 1. p. 73. CRC Press: Boca Raton, Florida.

Lenny virus. Family *Poxviridae*, subfamily *Chordopoxvirinae*, genus *Orthopoxvirus*. Isolated from a patient with smallpox in Nigeria. Produces pocks on chorioallantoic membrane and replicates in rabbit skin. Grows in Vero cells. Similar to vaccinia and buffalo pox viruses.

Lentivirinae. (Latin 'lentus' = slow.) Subfamily of the family *Retroviridae*. The most important members are those causing AIDS and Maedi/Visna.

lentogenic strain. Mild or avirulent strain of a virus.

leporid herpesvirus. Synonym: HINZE VIRUS.

Family *Herpesviridae*, subfamily *Gammaherpesvirinae*, unnamed genus. Indigenous in cotton-tail rabbits. Causes a lymphoproliferative disease in young animals.

Leporipoxvirus. (Latin 'lepus, leporis' = hare.) A genus in the subfamily *Chordopoxvirinae* consisting of those viruses which infect leporids and squirrels: hare fibroma, rabbit (Shope) fibroma and squirrel fibroma viruses. Type species is myxoma virus.

lettuce big vein virus. An unclassified plant virus with rod-shaped particles 320-360 nm. long and 18 nm. wide. The virus is transmitted by the Phycomycete fungus, *Olpidium* sp. *See* TOBACCO STUNT VIRUS.
Kuwata, S. *et al.* (1983) Ann. Phytopath. Soc. Japan **49**, 246.

lettuce infectious yellows virus. A possible *Closterovirus* transmitted by whiteflies.
Duffus, J.E. *et al.* (1986) Phytopath. **76**, 97.

lettuce mosaic virus. A *Potyvirus*. Important disease of lettuce as it is seed-transmitted.
Tomlinson, J.A. (1970) CMI/AAB Descriptions of Plant Viruses No. 9.
Francki, R.I.B. *et al.* (1985) **In** Atlas of Plant Viruses. Vol. 2. p. 183. CRC Press: Boca Raton, Florida.

lettuce necrotic yellows virus. Type member of the plant *Rhabdovirus* group, subgroup 1; transmitted by aphids.
Francki, R.I.B. and Randles, J.W. (1970) CMI/AAB Descriptions of Plant Viruses No. 26.
Francki, R.I.B. *et al.* (1985) **In** Atlas of Plant Viruses. Vol. 1. p. 73. CRC Press: Boca Raton, Florida.

lettuce speckles mottle virus. Unencapsidated RNA, dependent on BEET WESTERN YELLOWS VIRUS for aphid transmission.
Falk, B.W. *et al.* (1979) Virology **96**, 239.

Leviviridae. (Latin 'levi, levis' = 'light'). A family of phages containing a single genus of isometric viruses with a (+)-sense ssRNA genome (*see* LEVIVIRUS).

◯   ____100nm____

Levivirus. ssRNA phages with isometric unenveloped particles (morphotype E1; *see* PHAGE) about 23 nm. in diameter, which sediment at 78-82S and have a buoyant density in CsCl of 1.46-1.47 g/cc. Virions contain 180 copies of the major capsid protein (mw. 12-14 x 10^3) and one copy of 'A' protein (mw. 35-44 x 10^3) which is required for maturation and infectivity. The genome is a linear (+)-sense ssRNA (mw. $c.1.2$ x 10^6) which, unlike some RNA viruses from eukaryotic hosts, is neither polyadenylated at the 3' end nor carries a 5' methylated 'cap'. The RNA has extensive secondary structure which is important in the regulation of gene expression. Three genes have been defined, coding for the major coat protein, A protein and RNA replicase. Members of serogroup I also code for a lysis protein. Viral progeny RNA and mRNA synthesis proceeds via a replicative intermediate (RI). Infection proceeds after specific adsorption to the F pilus of male gram negative bacteria and results in lysis of the host with the release of thousands of virions. The type species is MS2 PHAGE and the genus is the only one presently classified within the *Leviviridae*. Five serogroups have been established which correlate with other properties of the viruses. Other members include f2 and Qβ (*see* Appendix F).

Fiers, W. (1979) In Comprehensive Virology. Vol. 13. p. 69. ed. H. Fraenkel-Conrat and R.R. Wagner. Plenum Press: New York.
Ritchie, D.A. (1983) In Topley & Wilson's Principles of Bacteriology, Virology and Immunity. Vol. 1. p. 177. ed. G. Wilson, A. Miles and M.T. Parker. Edward Arnold: London.
Ackermann, H.W. and Du Bow, M.S. (1987) Viruses of Prokaryotes. CRC Press: Boca Raton, Florida.

ligase. *See* DNA LIGASE, RNA LIGASE.

ligation. The process of joining two linear nucleic acid molecules via a phosphodiester bond by e.g. using ligase enzymes.

light scattering. The process by which energy which has been removed from a beam of light energy and is re-emitted without appreciable change in wavelength. In a suspension of virus particles it can affect the absorption spectrum of UV light.

lilac chlorotic leafspot virus. A member of the *Closterovirus* subgroup 1.
Brunt, A.A. (1979) CMI/AAB Descriptions of Plant Viruses No. 202.
Francki, R.I.B. *et al.* (1985) In Atlas of Plant Viruses. Vol. 2. p. 219. CRC Press: Boca Raton, Florida.

lilac mottle virus. A *Carlavirus*.
Francki, R.I.B. *et al.* (1985) In Atlas of Plant Viruses. Vol. 2. p. 173. CRC Press: Boca Raton, Florida.

lilac ring mottle virus. An *Ilarvirus*.
van der Meer, F.A. and Huttinga, H. (1979) CMI/AAB Descriptions of Plant Viruses No. 201.
Francki, R.I.B. (1985) In The Plant Viruses. Vol. 1. p. 1. ed. R.I.B. Francki. Plenum Press: New York.

lilac ringspot virus. A *Carlavirus*, occurs in Japan.
Doi Y. *Personal communication.*

lily symptomless virus. A *Carlavirus*.
Allen, T.C. (1972) CMI/AAB Descriptions of Plant Viruses No. 96.
Francki, R.I.B. *et al.* (1985) In Atlas of Plant Viruses. Vol. 2. p. 173. CRC Press: Boca Raton, Florida.

lily virus X. A *Potexvirus*.
Francki, R.I.B. *et al.* (1985) In Atlas of Plant Viruses. Vol. 2. p. 159. CRC Press: Boca Raton, Florida.

linear DNA. One of several topological forms of DNA. The two strands of linear dsDNA may be free, bound to a specific protein (e.g. ADENOVIRUS) or closed by a hairpin loop (e.g. in some situations with the formation of cDNA by reverse transcription).

Lineweaver-Burk plots. A method for the accurate determination of V_{max} (the maximum rate of a reaction) and K_m (the Michaelis-Menten constant) of an enzyme for a particular reaction.

lipoprotein. A conjugated protein containing a lipid or group of lipids.

liposome. Lipid vesicle used to introduce biological molecules (virus particles or nucleic acid) into cells. Liposomes may be uni- or multi-lamellar and of differing net surface charge depending on the method of production and their composition.
Poste, G. and Papahadjopoulos, D. (1976) Methods in Cell Biology. Vol. 14. p. 23. ed. D.M. Prescott. Academic Press: New York.

Lipovnik virus. Family *Reoviridae*, genus *Orbivirus*. Isolated from ticks in Czechoslovakia. Antibodies found in many people.

lisianthus necrosis virus. A *Necrovirus*. Iwaki, M. *et al.* (1987) Phytopath. **77**, 867.

live attenuated virus. Vaccine comprising virus with low virulence, e.g. the Sabin poliovirus vaccine strains, yellow fever vaccine 17D.

local infection. The infection of a few cells of a host which is usually prevented by the host response from spreading systemically.

local lesion. Symptom sometimes found in a plant leaf inoculated with a virus (or bacterium) caused by the death of, or changes in, cells around the original entry point of the pathogen. Local lesions may be necrotic or chlorotic. In necrotic lesions the cells around the initially infected cell die, frequently limiting the virus and forming a brown or black spot. Local lesion production depends upon the virus-host combination and may be diagnostic for certain viruses. It is also the main method used for assessing the infectivity of plant virus preparations. *See* RINGSPOT, HYPERSENSITIVE.

Lokern virus. Family *Bunyaviridae*, genus *Bunyavirus*. Isolated from the hare, rabbit and mosquito in California.

Lolium enation virus. Synonym: OAT STERILE DWARF VIRUS.

Lolium rhabdovirus. A possible plant *Rhabdovirus*.
Francki, R.I.B. *et al.* (1985) **In** Atlas of Plant Viruses. Vol. 1. p. 73. CRC Press: Boca Raton, Florida.

lollipops a) Abnormal particles produced by T-even phages in the presence of the amino acid analogue, canavine; b) stem-loop structures formed by nucleic acid molecules which have inverted terminal repeat sequences. *See* POLYHEADS.
Uhlenhopp, E.L. *et al.* (1974) J. mol. Biol. **89**, 689.

Lone Star virus. Family *Bunyaviridae*, genus *Bunyavirus*. Isolated from ticks in Kentucky. Antibodies found in raccoons.

longevity *in vitro*. A property sometimes assessed for plant viruses in which the infectivity of extracted sap, stored at room temperature for various times, is measured. Different viruses survive this treatment for different lengths of time, e.g. APPLE (TULARE) MOSAIC VIRUS about one hour, TOBACCO MOSAIC VIRUS more than one year.

Lonicera latent virus. Synonym: HONEYSUCKLE LATENT VIRUS.

Lotus streak virus. A plant *Rhabdovirus*, occurs in Japan.
Yamashita, S. *et al.* (1985) Ann. Phytopath Soc. Japan **51**, 627.

louping ill virus. Family *Flaviviridae*, genus *Flavivirus*. Causes disease in sheep and occasionally cattle in several parts of Britain. Fever is followed by incoordination, paralysis and often death. Man in close contact can become infected; symptoms include meningitis and encephalitis. Several wild mammals are infected subclinically.

LOVAL. Abbreviation sometimes used for 'larval occluded virus alkali-liberated', referring to those baculovirus particles which have been incorporated within occlusion bodies (OBs) during infection of insect larvae and subsequently released by dissolving the OBs in alkali.

LT$_{50}$. Abbreviation for MEDIAN LETHAL TIME.

Lu III virus. Family *Parvoviridae*, genus *Parvovirus*. Isolated in laboratory from cultures of human lung cells. Can be grown in HeLa cells. Causes intestinal haemorrhages in new-born hamsters and abortion in pregnant hamsters.

lucerne Australian latent virus. A *Nepovirus*.
Jones, A.T. and Forster, R.L.S. (1980) CMI/AAB Descriptions of Plant Viruses No. 225.
Francki, R.I.B. *et al.* (1985) **In** Atlas of Plant Viruses. Vol. 2. p. 23. CRC Press: Boca Raton, Florida.

lucerne enation virus. A plant *Rhabdovirus*, subgroup 2; transmitted by aphids.
Francki, R.I.B. *et al.* (1985) **In** Atlas of Plant Viruses. Vol. 1. p. 73. CRC Press: Boca Raton, Florida.

lucerne transient streak virus. A member of the *Sobemovirus* group.
Forster, R.L.S. and Jones, A.T. (1980) CMI/AAB

Descriptions of Plant Viruses No. 224. Francki, R.I.B. *et al.* (1985) **In** Atlas of Plant Viruses. Vol. 1. p. 153. CRC Press: Boca Raton, Florida.
Hull, R. (1988) **In** The Plant Viruses. Vol. 3. p. 113. ed. R. Koenig. Plenum Press: New York.

Lucilia cuprina viruses. Unclassified virus-like particles observed in the sheep blowfly, *Lucilia cuprina.* Three types of particle have been observed; VLP1, structurally similar to BEE CHRONIC PARALYSIS VIRUS and *DROSOPHILA* RS virus; VLP2, structurally similar to HOUSEFLY VIRUS; VLP3, isometric particles *c.*40 nm. in diameter.
Binnington, K.C. *et al.* (1987) J. Invertebr. Pathol. **49**, 175.

Lucke virus. *See* RANID HERPESVIRUS 1.

Lukuni virus. Family *Bunyaviridae,* genus *Bunyavirus.* Isolated from mosquitoes in Trinidad and Brazil.

Lumbo virus. Family *Bunyaviridae,* genus *Bunyavirus.* Isolated from mosquitoes in California.

lumpy skin disease virus. Synonym: NEETHLING VIRUS. Family *Poxviridae,* subfamily *Chordopoxvirinae,* genus *Capripoxvirus.* Causes fever, skin nodules and lesions in viscera of cattle in Africa. Grows in eggs and calf and lamb embryo cell cultures.

lupin yellow vein virus. A possible plant *Rhabdovirus.*
Francki, R.I.B. *et al.* (1985) **In** Atlas of Plant Viruses. Vol. 1. p. 73. CRC Press: Boca Raton, Florida.

Luteovirus group. (Latin 'luteus' = yellow, from a common symptom). (Type member BARLEY YELLOW DWARF VIRUS). Genus of plant viruses with isometric particles, 25-30 nm. in diameter which sediment at 115-127S. The capsids are

 100nm

made up of subunits of a single polypeptide species (mw. 24 x 10³) and contain a single species of linear (+)-sense ssRNA (mw. 2.0 x 10⁶). Host range varies with member, some infecting a wide range of either monocotyledons or dicotyledons, others having a restricted host range. Virus particles are mainly confined to the phloem. Luteoviruses are usually not mechanically transmissible. They are transmitted by aphids in the PERSISTENT TRANSMISSION manner.
Matthews, R.E.F. (1982) Intervirology **17**, 140.
Francki, R.I.B. *et al.* (1985) **In** Atlas of Plant Viruses. Vol. 1. p. 137. CRC Press: Boca Raton, Florida.
Casper, R. (1988) **In** The Plant Viruses. Vol. 3, p. 235. ed. R. Koenig. Plenum Press: New York.

Lychnis ringspot virus. A *Hordeivirus.*
Francki, R.I.B. *et al.* (1985) **In** Atlas of Plant Viruses. Vol. 2. p.133. CRC Press: Boca Raton, Florida.

Lymantria dispar nodavirus. *See* NODAVIRUS.

Lymantria dispar nuclear polyhedrosis virus. Baculovirus (Subgroup A) isolated from the gypsy moth, *L. dispar,* in the USA, Europe and USSR. The virus is of the MNPV type and a range of related genotypic variants have been isolated from different geographical regions. The virus was registered for use in 1978 by the US Forest Service as a selective biological control agent for gypsy moth under the name Gypchek. It is related to an NPV isolated from the nun moth, *L. monacha,* and will replicate *in vitro* in cell lines derived from *L. dispar.*
Lewis, F.B. (1981) **In** Microbial Control of Pests and Plant Diseases 1979-1980. p. 363. ed. H.D. Burges. Academic Press: London.

lymphadenopathy-associated virus (LAV). The name given by French workers at the Pasteur Institute to the first reported isolate of the virus now known to cause AIDS. This virus was recovered from a person with lymphadenopathy (enlarged lymph nodes) who was also in a group at high risk for AIDS.
Barré-Sinoussi, F. *et al.* (1983) Science, **220**, 868.

Lymphocryptovirus. A genus in the family *Herpesviridae,* subfamily *Gammaherpesvirinae.* HUMAN HERPESVIRUS 4 (EPSTEIN-BARR VIRUS) is the type species and cercopithecine herpesvirus 12 and pongine herpesvirus 1 are other members.

lymphocystis virus. Family *Iridoviridae, Lymphocystivirus* group. Chronic disease affecting several species of fish, producing tumours.

lymphocyte. A type of white blood cell which is involved in the immune response. Antibodies are produced by B lymphocytes whereas T lymphocytes give cell-mediated immunity.

lymphocytic choriomeningitis virus. Family *Arenaviridae*, genus *Arenavirus*. Causes a natural inapparent infection of mice but the virus has also been isolated from several species including man. In man the infection may be inapparent or influenza-like or sometimes may cause meningitis. In guinea pigs there is generalised disease which is often fatal. Grows in a variety of cell cultures and on the chorioallantoic membrane of eggs.

lymphoproliferative disease of turkeys virus. Family *Retroviridae*, subfamily *Oncovirinae*, genus *Type C Oncovirus* group. Occurs naturally in turkeys, causing a considerable level of mortality. Can be transmitted by injection of serum from infected birds.

lymphotropic virus. Family *Papovaviridae*, genus *Polyoma virus*.

lyophilisation. Synonym: FREEZE-DRYING. Rapid freezing of a material at low temperature followed by rapid dehydration by sublimation in a high vacuum. A method used to preserve biological specimens or to concentrate macromolecules with little or no loss of activity.

lysate. The product of LYSIS.

lysis. The rupturing of a host cell, generally observed at the end of a lytic cycle of virus infection. Lysis is common in bacterial infections by virulent phages and during the lytic cycle of replication by temperate phages, and is induced by lysozyme-like enzymes which degrade the cell wall. Cell lysis by virus forms the basis of most plaque assays for measuring virus infectivity. When a very large number of particles of certain phages (e.g. T-EVEN PHAGES) are adsorbed by a bacterium, the cell lyses almost immediately without virus multiplication (known as 'lysis from without').
Ritchie, D.A. (1983) **In** Topley and Wilson's Principles of Bacteriology, Virology and Immunity. Vol. 1. p. 177. ed. G. Wilson, A. Miles and M.T. Parker. Edward Arnold: London.

lysogen. A bacterial or other prokaryotic cell carrying phage in a state of LYSOGENY. Lysogens are immune to superinfection by the same virus (e.g. K-12 strain of *E. coli* which is lysogenic for

λ phage cannot be infected by λ).

lysogenic. Term descriptive of a bacterial strain having a PROPHAGE integrated into its genome. Activation of the prophage, either spontaneously or due to certain stimuli, causes the phage to replicate and destroy the bacterial cell.

lysogeny. The state where a DNA-containing phage maintains a long-term non-lytic relationship with its host in a non-replicating form (*see* PROPHAGE). This can be achieved by the integration of the viral genome into the host chromosome (e.g. λ phage) and the transmission of the integrated DNA (prophage) to progeny cells during division. Alternatively, lysogeny can occur by plasmid formation (e.g. P1 phage) and the viral DNA is then perpetuated as an extrachromosomal element. The cell carrying the non-replicating viral DNA is called a lysogen. In λ phage, insertion of the viral DNA occurs between specific regions of homology in the circular viral DNA and host DNA. P2 phage can integrate into several sites in the chromosome while Mu phage can integrate anywhere in the bacterial genome or other DNA present in the infected cell (e.g. plasmid). Suppression of λ viral replication is ensured by the synthesis of a virus-specific repressor of mRNA synthesis. The proportion of cells entering lysogeny increases with the multiplicity of infection and the concentration of magnesium and cyclic AMP. Lysogenic cultures of bacteria may spontaneously lyse at low frequency (e.g. 1 in 10^6). The frequency of lysis (arising from excision of the viral DNA and subsequent replication) can be increased by 'inducing' agents (e.g. UV light) which inhibit DNA replication. Excision of viral DNA is usually precise but errors can lead to the production of defective transducing phage containing bacterial DNA sequences.
Weisberg, R.A. *et al.* (1977) **In** Comprehensive Virology. Vol. 8. p. 197. ed. H. Fraenkel-Conrat and R.R. Wagner. Plenum Press: New York.
Ritchie, D.A. (1983) **In** Topley and Wilson's Principles of Bacteriology, Virology and Immunity. Vol. 1. p. 177. ed. G. Wilson, A. Miles and M.T. Parker. Edward Arnold: London.

lysozyme. An enzyme which degrades the rigid cell wall material (peptidoglycan) of many bacteria. Used in the preparation of plasmids from bacteria which, after lysozyme treatment, are lysed with detergents.

Lyssavirus. (Greek 'lyssa' = rage, rabies.) A

genus in the family *Rhabdoviridae*. Contains rabies virus and the antigenically related Duvenhage, Lagos Bat and Mokola viruses, and the more distantly related kotonkan and Obodhiang viruses.

lyssavirus. *See* RABIES VIRUS.

lytic cycle. A term usually applied to the productive replicating cycle in virus infections (particularly phage) which ends in the production of progeny virus particles and their release from the host, most usually by lysis of the host cell wall. For VIRULENT PHAGES this is the only course of infection. TEMPERATE PHAGE can also enter into a non-productive interaction with the host cell (*see* LYSOGENY).

lytic phages. *See* VIRULENT PHAGES.

M

M13 phage. Filamentous phage containing a circular ssDNA genome (*see* INOVIRUS), which infects male *E. coli* cells (bearing the F pilus) without lysing them. The virus has been much used as a cloning vector for foreign DNA to produce cloned fragments of DNA in a single-stranded form that can then be sequenced by the SANGER METHOD. The foreign DNA is inserted into the purified replicative form (RF) of the virus which can be isolated in the same way as a plasmid. Derivatives of M13 RF DNA have been constructed (mp2-mp19; 'mp' from Max-Planck, the Institute where they were developed) containing part of the *E. coli lac* (= β-galactosidase) operon and suitable restriction enzyme sites. Successful insertion of foreign DNA into the viral derivatives can be detected visually as the insertion interrupts the β-galactosidase gene and causes recombinant phage plaques to be white rather than blue when suitable substrates (X-GAL) are incorporated into the medium.
Oliver, S.G. and Ward, J.M. (1985) A Dictionary of Genetic Engineering. Cambridge: Cambridge University Press.

Machlovirus group. (Sigla from maize chlorotic after the type virus MAIZE CHLOROTIC DWARF VIRUS). Genus of plant viruses with isometric particles, 30 nm. in diameter, which sediment at 180S and band in CsCl at 1.51 g/cc. The capsids

 ———— 100nm

are composed of subunits of two polypeptide species (mw. 18 and 30 x 10^3). Each particle contains one molecule of ssRNA (mw. 3.2 x 10^6). Host range is narrow. Particles are found mainly in phloem cells. Machloviruses are not transmitted mechanically but by leafhoppers in the SEMI-PERSISTENT TRANSMISSION manner.
Matthews, R.E.F. (1982) Intervirology **17**, 137.
Francki, R.I.B. *et al.* (1985) **In** Atlas of Plant Viruses. Vol. 1. p. 111. CRC Press: Boca Raton, Florida.
Gingery, R.E. (1988) **In** The Plant Viruses. Vol. 3. p. 259. ed. R. Koenig. Plenum Press: New York.

Machupo virus. Synonym: BOLIVIAN HAEMORRHAGIC FEVER VIRUS. Family *Arenaviridae*, genus *Arenavirus*. Causes severe haemorrhaging and frequently death. Sporadic outbreaks occur. Natural host is the rodent *Calomys callosus*.

Maclura mosaic virus. A possible *Potyvirus*.
Koenig. R. *et al.* (1981) CMI/AAB Descriptions of Plant Viruses No. 239.
Francki, R.I.B. *et al.* (1985) **In** Atlas of Plant Viruses. Vol. 2. p. 183. CRC Press: Boca Raton, Florida.

macromolecule. A molecule having a molecular weight in the range of a few thousands to many millions.

Macropipus paralysis virus. Unclassified REOVIRUS-like agent ('P-virus') isolated from the swimming crab, *Macropipus depurator*. Virions sediment at 430S and contain ten segments of dsRNA. Often associated with a bunya-like virus ('S-virus'), which has a genome of three pieces of ssRNA.
Bergoin, M. *et al.* (1982) **In** Invertebrate Pathology and Microbial Control. p. 523. Proc. IIIrd Internat. Colloq. Invertebr. Pathol.

mad itch virus. *See* AUJESZKY'S DISEASE VIRUS.

Madrid virus. Family *Bunyaviridae*, genus *Bunyavirus*. Isolated from a man with fever and from sentinel mice, rats and mosquitoes in Panama. Antibodies found in a few humans.

Maedi virus. Family *Retroviridae*, subfamily *Lentivirinae*. Causes chronic pulmonary disease of sheep. Similar to VISNAVIRUS. Often referred to as Maedi-Visna group.

Maguari virus. Family *Bunyaviridae*, genus *Bunyavirus*. Isolated from a horse in Guyana and from mosquitoes and sentinel mice in Brazil. Antibodies found in cattle, sheep and birds.

mahogany hammock virus. Family *Bunyaviridae*, genus *Bunyavirus*. Isolated from mosquitoes and a cotton rat in Florida.

main drain virus. Family *Bunyaviridae*, genus *Bunyavirus*. Isolated from hares and mosquitoes in California. Antibodies found in cattle and sheep.

maize chlorotic dwarf virus. Type member of the *Machlovirus* group.
Gingery, R.E. *et al.* (1978) CMI/AAB Descriptions of Plant Viruses No. 194.
Gingery, R.E. (1988) **In** The Plant Viruses. Vol. 3. p. 259. ed. R. Koenig. Plenum Press: New York.

maize chlorotic dwarf virus group. *See* MACHLOVIRUS GROUP.

maize chlorotic mottle virus. A possible *Sobemovirus*.
Gordon, D.T. *et al.* (1984) CMI/AAB Descriptions of Plant Viruses No. 284.
Hull, R. (1988) **In** The Plant Viruses. Vol. 3. p. 113. ed. R. Koenig. Plenum Press: New York.

maize dwarf mosaic virus. *See* SUGARCANE MOSAIC VIRUS.

maize mosaic virus. A probable plant *Rhabdovirus*, subgroup 2; transmitted by leafhopper.
Herold, F. (1972) CMI/AAB Descriptions of Plant Viruses No. 94.
Francki, R.I.B. *et al.* (1985) **In** Atlas of Plant Viruses. Vol. 1. p. 73. CRC Press: Boca Raton, Florida.

maize rayado Colombiano virus. A strain of MAIZE RAYADO FINO VIRUS.

maize rayado fino virus. Synonym: BRAZILIAN CORN STREAK VIRUS. Type member of the *Marafivirus* group.
Gamez, R. (1980) CMI/AAB Descriptions of Plant Viruses No. 220.
Gamez, R. and Leon, P. (1988) **In** The Plant Viruses. Vol. 3. p. 213. ed. R. Koenig. Plenum

Press: New York.

maize rayado fino virus group. Synonym: MARAFIVIRUS GROUP.

maize rough dwarf virus. A *Phytoreovirus*.
Lovisolo, O. (1971) CMI/AAB Descriptions of Plant Viruses No. 72.

maize sterile stunt virus. A possible plant *Rhabdovirus*, subgroup 1; transmitted by leafhopper.
Francki, R.I.B. *et al.* (1985) **In** Atlas of Plant Viruses. Vol. 1. p. 73. CRC Press: Boca Raton, Florida.

maize streak virus. Type member of the *Geminivirus* group, subgroup A. The genome comprises one species of ssDNA of 2687 nucleotides. The virus causes an important disease of maize in Africa.
Bock, K.R. (1974) CMI/AAB Descriptions of Plant Viruses No. 133.
Harrison, B.D. (1985) Ann. Rev. Phytopath. **23**, 55.

maize stripe virus. A member of the *Tenuivirus* group.
Gingery, R.E. (1985) AAB Descriptions of Plant Viruses No. 300.

maize white line mosaic virus. An unclassified plant virus with isometric particles, 35 nm. in diameter which sediment at 117S and band in CsCl at 1.35 g/cc. The capsids are composed of coat protein subunits of mw. $32\text{-}35 \times 10^3$ and contain a single species of ssRNA (mw. 1.25×10^6).
de Zoeten, G.A. and Reddick, B.B. (1984) CMI/AAB Descriptions of Plant Viruses No. 283.

Makonde virus. *See* UGANDA S VIRUS.

Malaya disease. A disease of the coconut rhinoceros beetle (*Oryctes rhinoceros*) caused by a NON-OCCLUDED BACULOVIRUS (*see* ORYCTES RHINOCEROS VIRUS).

malignant catarrhal fever virus. Synonym: BOVID HERPESVIRUS 3. Family Herpesviridae, subfamily Alphaherpesvirinae. Causes a fatal disease of cattle with fever, acute inflammation of nasal and oral membranes and involvement of pharynx and lungs. Disease can be transmitted experimentally to cattle and rabbits. Can be grown in bovine thyroid and adrenal cell cultures.

Malva sylvestris rhabdovirus. A possible plant *Rhabdovirus*.
Matthews, R.E.F. (1982) Intervirology **17**, 114.

Malva vein clearing virus. A possible *Potyvirus*.
Francki, R.I.B. *et al.* (1985) **In** Atlas of Plant Viruses. Vol. 2. p. 183. CRC Press: Boca Raton, Florida.

Malva veinal necrosis virus. A possible *Potexvirus*.
Francki, R.I.B. *et al.* (1985) **In** Atlas of Plant Viruses. Vol. 2. p. 159. CRC Press: Boca Raton, Florida.

Malva yellows virus. A *Luteovirus*; considered to be a strain of BEET WESTERN YELLOWS VIRUS.
Francki, R.I.B. *et al.* (1985) **In** Atlas of Plant Viruses. Vol. 1. p. 137. CRC Press: Boca Raton, Florida.
Casper, R. (1988) **In** The Plant Viruses. Vol. 3. p. 235. ed. R. Koenig. Plenum Press: New York.

Mamestra brassicae nuclear polyhedrosis virus. A BACULOVIRUS (Subgroup A) isolated from the cabbage moth, *M. brassicae*. The virus is of the MNPV type and a range of related genotypic variants has been isolated from different geographical locations within Europe. The most extensively studied genotype is closely related (90% homology) to an MNPV isolated from the American bollworm, *Heliothis armigera*. The virus has a relatively broad host range, infecting at least 20 species of Noctuidae (Lepidoptera). The virus is produced on a commercial basis (*see* 'MAMESTRIN') in France as a selective biological control agent for *M. brassicae*, and used in Africa for the control of the cotton pests *H. armigera* and *Diparopsis watersi*. The virus replicates in some cell lines derived from *M. brassicae*.
Allaway, G.P. and Payne, C.C. (1984) Ann. appl. Biol. **105**, 29.

Mamestrin. Commercial product of the NUCLEAR POLYHEDROSIS VIRUS of the cabbage moth, *Mamestra brassicae*, from Calliope, Béziers, France. The preparation is used for the control of several noctuid pests including *M. brassicae* and the cotton pests, *Heliothis armigera* and *Diparopsis watersi*.

Mammalian type C Oncoviruses. Family *Retroviridae*, genus *Oncovirinae*. A subgenus in the genus *Type C Oncovirus*.

Manawa virus. Family *Bunyaviridae*, genus *Bunyavirus*. Isolated from ticks in West Pakistan.

Manawatu virus. *See* NODAVIRUS.

many polyhedra variant. Plaque variant of nuclear polyhedrosis virus which produces many polyhedra in infected nuclei in insect cell cultures. *See* FEW POLYHEDRA VARIANT.

Manzanilla virus. Family *Bunyaviridae*, genus *Bunyavirus*. Isolated from a monkey in Trinidad.

map. As a verb: to determine the position of genes or restriction endonuclease sites on a DNA. Gene mapping is done by mating experiments to determine the frequency of recombination between them. The further genes are apart the greater the frequency of recombination. Restriction endonuclease mapping is by sequential cutting of the DNA and measurement of the size of the resultant fragments. As a noun: a diagram showing the relative positions of genes or restriction endonuclease sites on a nucleic acid.

Mapputta virus. Family *Bunyaviridae*, genus *Bunyavirus*. Isolated from mosquitoes in Queensland, Australia. Antibodies found in man, cattle, horses, pigs, kangaroos and rats.

Maprik virus. Family *Bunyaviridae*, genus *Bunyavirus*. Isolated from mosquitoes in New Guinea.

Maracuja mosaic virus. A *Tobamovirus*.
Brunt, A.A. (1986) **In** The Plant Viruses. Vol. 2. p. 283. eds. M.H.V. van Regenmortel and H. Fraenkel-Conrat. Plenum Press: New York.

Marafivirus group. (Sigla from maize rayado fino, the type member). Genus of plant viruses with isometric particles, 31 nm. in diameter, which sediment as two components (120 and 54S) and form two bands in CsCl (1.46 and 1.28 g/cc). The capsids are made up of two polypeptide

    100nm

species (mw. 29 and 22 x 10^3) in the ratio of 1:3. Particles contain one species of linear (+)-sense ssRNA (mw. 2.0-2.4 x 10^6). Host ranges are narrow and confined to Graminae. Virus particles

are found mainly in chlorenchyma and phloem parenchyma cells. Members are not transmitted mechanically. They are transmitted by leafhoppers in the PERSISTENT TRANSMISSION manner; they replicate in the leafhopper vector.
Gomez, R. and Leon, P. (1988) In The Plant Viruses. Vol. 3. p. 212. ed. R. Koenig. Plenum Press: New York.

Marburg virus. Member of the family *Filoviridae*. Causes a severe and often fatal disease of man. There is a sudden onset of fever, head and limb pains, diarrhoea, vomiting and confused mental state. Cardiac and renal failure with haemorrhaging develops. First reports could all be traced to contact with tissue from a batch of African green monkeys trapped in Uganda. Some secondary cases occurred, probably transmitted by blood, from infected patients. The virus causes fatal infection in monkeys and guinea pigs. It can be grown in a variety of tissue culture cells. The virus particles have some resemblance to rhabdoviruses but have now been allotted to a new family, *Filoviridae*.

Marco virus. Family *Rhabdoviridae*, not allotted to genus. Isolated from a lizard in Brazil. Pathogenic when injected into new-born mice. Replicates in mosquitoes infected experimentally. Grows well in Vero cells.

Marek's disease. Synonym: GALLID HERPESVIRUS 1. Family *Herpesviridae*, subfamily *Gammaherpesvirinae*. A natural infection of several species of bird causing progressive paralysis. Lymphoid tumours are produced. Can be grown in avian cultures in which the virus is cell-aassociated. An economically important disease of fowl which has led to the development of a highly successful attenuated vaccine.

marigold mottle virus. A possible *Potyvirus*. Francki, R.I.B. *et al.* (1985) In Atlas of Plant Viruses. Vol. 2. p. 183. CRC Press: Boca Raton, Florida.

Marituba virus. Family *Bunyaviridae*, genus *Bunyavirus*. Isolated from sentinel monkeys, mice and mosquitoes in Brazil. Causes fever in man. Antibodies found in several species of rodent living in forests.

marker rescue. The production of infective progeny virus from co-infections between one virus which lacks a specific gene function and another related or unrelated virus.

marsupial papilloma virus. *See* QUAKKA POXVIRUS.

Mason-Pfizer monkey virus. Family *Retroviridae*, subfamily *Oncovirinae*, type D retrovirus group. Isolated from a mammary carcinoma in a Rhesus monkey and also from placental tissue of normal animals. It has not been transmitted experimentally in monkeys. Transforms Rhesus foreskin cell cultures.

mast cell. A connective-tissue cell with numerous large basophilic metachromatic granules in the cytoplasm.

Mastadenovirus. (Greek 'mastos' = breast.) A genus in the family *Adenoviridae* consisting of those viruses isolated from mammals. The species take their name from the host species. They do not cross-react with the aviadenoviruses in serological tests.

matrix protein. A term used for several different types of protein. In INCLUSION BODIES it is the major protein making up the structure of the inclusion. In ORTHO- AND PARA-MYXOVIRUSES and RHABDOVIRUSES it refers to the protein between the viral membrane and the nucleocapsid.

Matruh virus. Family *Bunyaviridae*, genus *Bunyavirus*. Isolated from birds in Egypt and Italy.

Maus-Elberfeld virus. Family *Picornaviridae*, genus *Cardiovirus*. A virus which is closely related serologically to ENCEPHALOMYOCARDITIS VIRUS of mice.

Maxam and Gilbert method. A method of sequencing DNA using chemical base-specific modification and cleavage. The DNA is end-labelled (*see* POLYNUCLEOTIDE KINASE) and the various reactions are set up for cleavages at specific nucleotides. The cleaved labelled fragments are then separated on a polyacrylamide gel which gives a sequence ladder. *See* SANGER METHOD. Maxam, A. and Gilbert, W. (1980) Meth. Enzymol. **65**(1), 497.

Mayaro virus. Family *Togaviridae*, genus *Alphavirus*. Causes fever and severe headaches.

measles virus. Synonyms: ROUGEOLE VIRUS,

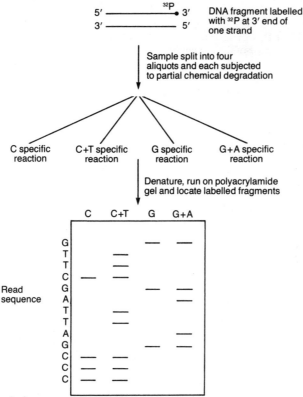

Maxam and Gilbert method

RUBEOLA VIRUS, MORBILLI VIRUS. Family *Paramyxoviridae*, genus *Morbillivirus*. Causes an acute febrile illness in children. Associated with cough, coryza and conjunctivitis, spots on the buccal mucosa and rash on the head and neck, spreading to the remainder of the body. Encephalitis also occurs in some cases. Can cause death in under-nourished children. Sub-acute sclerosing panencephalitis, a progressive degenerative disease of the central nervous system, is regarded as due to chronic infection with measles. Monkeys are susceptible and develop a disease similar to that seen in man. Virus can be adapted to grow in mice, ferrets and hamsters. Can be grown in human or monkey kidney cell cultures and in several other cell lines and eggs. An effective attenuated vaccine is in widespread use. The virus is closely related antigenically to those causing canine distemper, rinderpest and peste des petits ruminants.

mechanical inoculation. Inoculation of plants by rubbing sap or other viral extract on to the leaves. Often an abrasive such as celite (diatomaceous earth) or carborundum is included in the inoculum. The virus enters through small wounds made in the leaf cuticle and the walls of epidermal cells.

median effective dose (ED$_{50}$). The amount of, e.g. a drug, required to produce a response in 50% of the subjects to whom it is given.

median effective time (ET$_{50}$). The time taken, e.g. for a drug, to produce a response in 50% of the subjects to whom it has been given.

median infective dose (ID$_{50}$). The dose, e.g. of a virus, which, on average, will infect 50% of the individuals to whom it is administered. These may be animals, tissue cultures or eggs; with eggs the term EID$_{50}$ (egg infective dose) is often used.

median lethal concentration (LC$_{50}$). The concentration of an agent (e.g. virus) which is required to kill 50% of the subjects to whom it is administered.

median lethal dose (LD$_{50}$). The dose of a substance which is fatal to 50% of the test animals.

median lethal time (LT$_{50}$). The period of time required for 50% of a group of organisms to die following a specific dose of an injurious agent, e.g. virus, drug or radiation.

median survival time (ST$_{50}$). The period of time at which half the subjects have died following the administration of an injurous agent (e.g. virus, drug or radiation).

median tissue culture infective dose (TCID$_{50}$). The dose (e.g. of a virus) which, on average, will infect 50% of susceptible tissue culture cells.

Melandrium yellow fleck virus. A possible *Bromovirus*.
Hollings, M. and Horvath, J. (1981) CMI/AAB Descriptions of Plant Viruses No. 236.
Francki, R.I.B. (1985) In The Plant Viruses. Vol. 1. p. 1. ed. R.I.B. Francki. Plenum Press: New York.

Melao virus. Family *Bunyaviridae*, genus *Bunyavirus*. Isolated from mosquitoes in Trinidad and Brazil.

Melilotus latent virus. A plant *Rhabdovirus*, subgroup 2.
Francki, R.I.B. *et al.* (1985) In Atlas of Plant Viruses. Vol. 1. p. 73. CRC Press: Boca Raton, Florida.

Melilotus mosaic virus. A possible *Potyvirus*.
Francki, R.I.B. *et al.* (1985) In Atlas of Plant Viruses. Vol. 2. p. 183. CRC Press: Boca Raton, Florida.

Melolontha entomopoxvirus. Type species of probable subgenus A insect poxviruses (*Entomopoxviridae*) from the cockchafer, *M. melolontha* (Coleoptera). Virions are oval, 450 x 250 nm., containing a unilaterally concave core and one lateral body. Particle surface has globular subunits 22 nm. in diameter. The genome is a single linear molecule of dsDNA, mw. approximately 200 x 10^6. During replication, virions are occluded in large occlusion bodies (SPHEROIDS) 10-24 μm in diameter. Spindle-shaped inclusions devoid of virions are also produced. Viruses with similar properties have been isolated from *Othnonius, Demodena, Geotrupes, Dermolepida, Aphodius, Anomala, Phyllopertha* spp. and *Figulus* sp. (Coleoptera).
Arif, B.M. (1984) Adv. virus Res. **29**, 195.

melon necrotic spot virus. A member of the *Carmovirus* group.
Hibi, T. and Furuki, I. (1985) AAB Descriptions of Plant Viruses No. 302.
Morris, T.J. and Carrington, J.C. (1988) In The Plant Viruses. Vol. 3. p.73. ed. R. Koenig. Plenum Press: New York.

melon rugose mosaic virus. A *Tymovirus*.
Jones, P. *et al.* (1986) Ann. Appl. Biol. **108**, 303.

melon variegation virus. A possible plant *Rhabdovirus*, subgroup 1.
Francki, R.I.B. *et al.* (1985) In Atlas of Plant Viruses. Vol. 1. p. 73. CRC Press: Boca Raton, Florida.
Hollings, M. and Horvath, J. (1981) CMI/AAB Descriptions of Plant Viruses No. 236.
Francki, R.I.B. (1985) In The Plant Viruses. Vol. 1. p. 1. ed. R.I.B. Francki. Plenum Press: New York.

melting temperature. The temperature at which a DNA or RNA molecule denatures into separate single strands without secondary structure. It is usually applied to ds nucleic acid and is characteristic for each nucleic acid species and is dependent on the guanosine + cytosine (G+C) content of the nucleic acid as well as the salt concentration of the solution. It is higher for RNA than for DNA. For dsDNA these parameters are related by the following formulae:
$$Tm = 69.3 + 0.41(G+C)\%$$
$$Tm = \frac{(G+C)\%}{2.44} + 81.5 + 16.6\log M$$
Where Tm = melting temperature °C
M = ionic strength of solution
See HYPERCHROMICITY.

membrane. 1) A lipid bilayer which separates the internal contents of a cell or organelle from its surroundings. Proteins are distributed at the surfaces or completely traverse the bilayer. Membrane-bound viruses (e.g. RHABDOVIRUSES, MYXOVIRUSES) modify and acquire cell membranes as they mature. 2) Term sometimes used for nitrocellulose or nylon filters used in blotting.

Mengo virus. Family *Picornaviridae*, genus *Cardiovirus*. A virus which is closely related serologically to ENCEPHALOMYOCARDITIS VIRUS of mice.

Mermet virus. Family *Bunyaviridae*, genus *Bunyavirus*. Isolated from birds in Illinois and Texas.

Merodon equestris virus. An unclassified DNA-containing virus isolated from the large narcissus bulb fly, *Merodon equestris* (Syrphidae, Diptera), in France. The elongated rod-shaped particles are composed of a nucleocapsid (650 nm. long) surrounded by an envelope. Infection is characterised by salivary gland enlargement and gonadal atrophy (*see* TSETSE FLY VIRUS). Amargier, A. *et al.* (1979) C.R. hebd. Acad. Sci. D **289**, 481.

MES. 2-(N-morpholino)-ethane sulphonic acid (mw. 213.3). A biological buffer, pK_a 6.15, with a pH range 5.2-7.2. Good, N. *et al.* (1966) Biochem. **5**, 467.

Mesoleius 'baculovirus'. A putative NON-OCCLUDED BACULOVIRUS isolated from females of the ichneumonid parasitoid *Mesoleius tenthredinis*. Virus particles consist of a nucleocapsid (65 x 235 nm.) surrounded by one (and sometimes two) unit membrane(s). The DNA genome is circular (mw. *c.* 100 x 10⁶) and unipartite, as opposed to the multipartite genomes of many viruses isolated from parasitoid Hymenoptera (*see* POLYDNAVIRUS). Stoltz, D.B. (1981) Canad. J. Microbiol. **27**, 116.

messenger RNA (mRNA). An RNA molecule which is translated by ribosomes to give a protein. By definition it is positive-stranded.

methyl transferase. Enzyme activity found in REOVIRUS particles which is involved in 'CAPPING' of viral mRNAs. Thought to be the product of gene 12 in Reovirus.

methylated bovine serum albumin. Material used as a matrix in the separation of protein molecules by chromatography.

methylation. The addition of methyl groups to nucleic acids. Methylation, by methylase enzymes, of specific nucleotides within the target site of a restriction endonuclease (termed modification) can protect the DNA against cleavage by that enzyme and is the means by which bacteria protect their own DNA against the restriction endonucleases they encode. *See* S-ADENOSYL-L-METHIONINE.

methylene blue. A photoreactive dye which has been used as a vital stain for cells and for the detection of nucleic acids in gel electrophoresis.

MEV. Abbreviation sometimes used to describe the MNPV subtype of NUCLEAR POLYHEDROSIS VIRUSES where the majority of enveloped virions contain more than one nucleocapsid.

mibuna temperate virus. A possible member of the *Cryptovirus* group subgroup A. Boccardo, G. *et al.* (1987) Adv. Virus Res. **32**, 171.

Michigan alfalfa virus. A *Luteovirus*. Johnstone, G.R. *et al.* (1984) Neth. J. Plant Path. **90**, 225.

microbial control. The use of micro-organisms and viruses as biological control agents for pests and diseases (e.g. the use of baculoviruses for the control of insect pests).

microbial pesticide. A micro-organism used as a biological control agent for an insect pest (*see* BACULOVIRUS).

microsomes. Fraction of a cell homogenate obtained by ultracentrifugation and comprising ribosomes and fragments (16-150 nm. in diameter) of rough endoplasmic reticulum. Has protein synthetic capacity from run-off of polysomes translating mRNAs.

Microviridae. (Greek 'micro, micros' = small). A family of phages containing a single genus (*MICROVIRUS*) of isometric viruses with a (+)-sense circular ssDNA genome.

   100nm

Microvirus. ssDNA phages with isometric un-enveloped particles (morphotype D1; *see* PHAGE), about 27 nm. in diameter (12 capsomeres T = 1) and knob-like spikes on the vertices. Particles have a mw. of *c.* 6.7 x 10⁶, sediment at about 114S and have a buoyant density in CsCl of 1.40 g/cc. Virions contain 60 copies of two major structural proteins (mw. 20 x 10³ and 50 x 10³) and at least two other protein species. The genome is circular

(+)-sense ssDNA (mw. 1.7 x 10^6). The viruses infect enterobacteria following adsorption to bacterial cell wall receptors. Phage DNA is converted to a circular ds replicative form (RF) by host enzymes. The genome codes for at least nine proteins, some of which are coded for by different reading frames on the same section of the genome. Lysis of the host commences shortly after the appearance of mature phage. The type species is ΦX174 PHAGE and the *Microvirus* genus is the only one presently classified within the *Microviridae*. Other members are listed in Appendix F.

Denhardt, D.T. (1977) **In** Comprehensive Virology. Vol. 7. p. 1. ed. H. Fraenkel-Conrat and R.R. Wagner. Plenum Press: New York.

Ritchie, D.A. (1983) **In** Topley & Wilson's Principles of Bacteriology, Virology and Immunity. Vol. 1. p. 177. ed. G. Wilson, A. Miles and M.T. Parker. Edward Arnold: London.

Middelburg virus. Family *Togaviridae*, genus *Alphavirus*. Isolated from mosquitoes in many countries in Africa. Infects sheep.

milk-vetch dwarf virus. A probable *Luteovirus*. Francki, R.I.B. *et al.* (1985) **In** Atlas of Plant Viruses. Vol. 1. p. 137. CRC Press: Boca Raton, Florida.

milker's node virus. Family *Orthopoxviridae*, subfamily *Chordopoxvirinae*, genus *Parapoxvirus*. Causes red papules on the udders of cows and the hands of milkers. Can be grown in human cell cultures.

millet red leaf virus. A possible *Luteovirus*. Yu, T.F. *et al.* (1958) Acta Phytopath. Sinica **4**, 1. Matthews, R.E.F. (1982) Intervirology **17**, 141.

minatitlan virus. Family *Bunyaviridae*, genus *Bunyavirus*. Isolated from a sentinel hamster in Mexico.

minimal essential medium (MEM). A medium for the culture of vertebrate cells. It differs from EAGLE'S MEDIUM mainly in its increased concentration of essential amino acids.
Eagle, H. (1959) Science **130**, 432.

minivirus. A term which has been applied to isometric virus particles (<18nm. in diameter) observed in some insect virus infections, such as crystalline array virus, cloudy wing particle and some putative satellite viruses (e.g. *Antheraea*

satellite virus; bee chronic paralysis virus associate).
Longworth, J.F. (1978) Adv. Virus Res. **23**, 103.

mink enteritis virus. Family *Parvoviridae*, genus *Parvovirus*. A virus very similar to feline panleucopenia virus.

minus strand. *See* NEGATIVE-SENSE STRAND 1.

minute virus of canines. *See* CANINE PARVOVIRUS.

minute virus of mice virus. Family *Parvoviridae*, genus *Parvovirus*. A natural and silent infection of mice. Grows when injected into newborn mice, rats and hamsters, causing retarded growth in mice but only a silent infection in rats. Replicates in mouse or rat embryo cell cultures.

Mirabilis mosaic virus. A *Caulimovirus*.
Francki, R.I.B. *et al.* (1985) **In** Atlas of Plant Viruses. Vol. 1. p. 17. CRC Press: Boca Raton, Florida.

Mirim virus. Family *Bunyaviridae*, genus *Bunyavirus* Isolated from sentinel monkeys and from mosquitoes in Brazil.

miscanthus streak virus. A probable *Geminivirus*, subgroup A.
Harrison, B.D. (1985) Ann. Rev. Phytopath. **23**, 55.

Mitchell River virus. Family *Reoviridae*, genus *Orbivirus*. Isolated from mosquitoes in Queensland, Australia. Antibodies found in cattle, wallabies and kangaroos.

mitochondrion. A membrane-surrounded organelle in the cytoplasm of eukaryotic cells. It contains the enzyme systems required for the citric acid cycle, electron transport and oxidative phosphorylation.

mitomycin C. Antitumour antibiotic which is also active against bacteria. It is a selective inhibitor of DNA synthesis. It induces lysogenic phage development as it does not inhibit viral DNA synthesis. Therefore, it is used as a tool for studying viral DNA synthesis in the absence of host cell DNA synthesis.

Mitsuhashi and Maramorosch Medium. A cell-culture medium designed for the propagation of mosquito (Diptera) and leafhopper (Homop-

tera) cells.
Mitsuhashi J. and Maramorosch, K. (1964) Contrib. Boyce Thompson Inst. 22, 435.

MNPV. Synonym: BUNDLE VIRION. Abbreviation used to describe the subtype of NUCLEAR POLYHEDROSIS VIRUSES where the majority of enveloped virions contain more than one nucleocapsid.

mock infection. Inoculation of cells or an organism with a solution not containing virus particles. Used as a control in virus infection experiments to ascertain any possible effects of materials in the inoculum other than infectious particles.

modal length. The most common length of particles in a preparation of a virus with rod-shaped particles. With plant viruses it can be used to distinguish groups of viruses.
Brandes, J. and Wetter, C. (1959) Virology **8**, 99.

Modoc virus. Family *Flaviviridae*, genus *Flavivirus*. Isolated from the deer mouse. Found in several states in Western USA.

Moju virus. Family *Bunyaviridae*, genus *Bunyavirus*. Isolated from sentinel mice, forest rats and mosquitoes in Brazil.

Mokola virus. Family *Rhabdoviridae*, genus *Lyssavirus*. Isolated from children in Nigeria with disease of central nervous system and from shrews. Shrews, mice, dogs and monkeys can be infected experimentally.

molecular radius (M$_r$). The radius of the space occupied by a molecule. Usually given as relative to the M$_r$ of a hydrogen atom. Considered to be a more correct term for macromolecules than MOLECULAR WEIGHT as it allows for hydration. *See* DALTON.

molecular weight (mw.) (mol.wt.). The sum of the atomic weights of all the atoms in a molecule. *See* DALTON.

molinia streak virus. A possible *Sobemovirus*. Hull, R. (1988) In The Plant Viruses. Vol. 3. p. 113. ed. R. Koenig. Plenum Press: New York.

molluscum contagiosum virus. Family *Poxviridae*, subfamily *Chordopoxvirinae*, unassigned to genus. Causes a disease in man which is confined to the skin. Pimples develop into very small nodules, which can persist for several months. Can be grown in WI-38 and human amnion cells.

Moloney leukaemia virus. Family *Retroviridae*, subfamily *Oncovirinae*, genus *Type C Oncovirus*. Isolated from a mouse sarcoma. Produces lymphoid leukaemia experimentally in mice and rats.

Monisarmio virus. Commercial preparation of the nuclear polyhedrosis virus of the Pine Sawfly, *Neodiprion sertifer*, produced by Kemira Oy Co. in Finland, for biological control of the homologous host.

monkey poxvirus. Family *Poxviridae*, subfamily *Chordopoxvirinae*, genus *Orthopoxvirus*. Causes disease similar to that caused by variola virus in monkeys and in man. The disease in man can be fatal. Lesions are produced by inoculation of rabbit skin. Produces pocks on chorioallantoic membrane of eggs and grows in a variety of cell cultures.

Mono Lake virus. Family *Reoviridae*, genus *Orbivirus*. Isolated from ticks in California.

monocistronic. A nucleic acid molecule coding for one CISTRON.

monoclonal antibody. An antibody preparation which contains only a single type of antibody molecule. They are synthesised and secreted by clonal populations of hybrid cells (hybridoma) prepared by the fusion of individual B lymphocyte cells from an immunized animal (usually a mouse or rat) with individual cells from a lymphocytic tumour (e.g. MYELOMA).

monovalent. A term to define the antigenic specificity of a vaccine, e.g. POLIOVIRUS type 1.

Montana myotis leukoencephalitis virus. Family *Flaviviridae*, genus *Flavivirus*. Isolated from a bat in Montana.

MOPS. 3-(N-morpholino)propane sulphonic acid (mw. 209.3). A biological buffer, pK$_a$ 7.20, with a pH range 6.2-8.2.
Good, N. *et al.* (1966) Biochem. **5**, 467.

Morator virus. A genus of viruses (including bee sacbrood virus) established in 1947 in one of the earliest classifications of insect viruses.

Morbillivirus. (Latin 'morbillus' = disease, hence measles.) A genus in the family *Paramyxoviridae*. Contains the viruses causing measles in man, rinderpest in cattle, canine distemper in dogs and peste des petits ruminants in goats. All are very closely related serologically.

Moriche virus. Family *Bunyaviridae*, genus *Bunyavirus*. Isolated from mosquitoes in Trinidad.

Moroccan pepper virus. A *Tombusvirus.*
Fischer, H.U. and Lockhart, B.E.L. (1977) Phytopath. **67**, 1352.
Gallitelli, D. and Russo, M. (1987) J. Phytopath. **119**, 106.
Martelli, G.P. *et al.* (1988) **In** The Plant Viruses. Vol. 3. p.13. ed. R. Koenig. Plenum Press: New York.

morphogenesis. The changes in form during the growth and differentiation of cells and tissue.

morphological subunit. The structural subunit of a virus particle as seen under the ELECTRON MICROSCOPE. These are often clusters of protein subunits (CAPSOMERES) especially in ISOMETRIC particles.

mosaic. A common symptom induced in leaves by some plant virus infections in which there is a pattern of dark green, light green and sometimes chlorotic areas. This pattern is often associated with the distribution of veins in the leaf. In monocotyledonous leaves it shows as stripes.

mosquito iridescent virus. Type species of the CHLORIRIDOVIRUS genus (*Iridoviridae*) isolated from the mosquito, *Aedes taeniorhynchus* (Diptera: Culicidae); originally referred to as insect iridescent virus type 3. Two strains of the virus have been reported; the 'regular' (R) and 'turquoise' (T) strain, the latter appearing spontaneously in a single larva during laboratory studies with the R strain. Virions of the R strain are isometric, *c.* 180 nm. (compared with *c.* 120 nm. for the small iridescent virus, IRIDOVIRUS, group). They sediment at about 4450S with a buoyant density in CsCl of 1.32-1.35 g/cc. Nine structural polypeptides have been detected (mw. 15-98 x 10^3) and the genome is a linear dsDNA (mw. 243 x 10^6). It has been reported that the R strain virion contains two identical DNA molecules. Virus infection probably occurs through cannibalism and by transovarial transmission.

Hall, D.W. (1985) **In** Viral Insecticides for Biological Control. p. 163. ed. K. Maramorosch and K.E. Sherman. Academic Press: New York.

Mossuril virus. Family *Rhabdoviridae*, unassigned to genus. Isolated from mosquitoes in several countries in Africa.

mottle. A diffuse form of the MOSAIC symptom in plant leaves in which the dark and light green are less sharply defined. This term is frequently used interchangeably with mosaic.

Mount Elgon bat virus. Family *Rhabdoviridae*, unassigned to genus. Isolated from a bat in Kenya. Will grow in brains of mice inoculated i.c.

mouse cytomegalovirus. Synonym: MURID BETAHERPESVIRUS 1. Family *Herpesviridae*, subfamily *Betaherpesvirinae*, genus *murine cytomegalovirus group*. Silent infection of wild mice. Young mice can be readily infected, with the virus localising in the salivary glands. Large doses of virus i.p. will kill mice in a few days. Small doses produce hepatitis. Virus can be grown in primary mouse fibroblasts.

mouse encephalomyelitis virus. *See* THEILER'S VIRUS.

mouse hepatitis virus. Family *Coronaviridae*, genus *Coronavirus*. Often causes a silent infection of laboratory mice but it can be activated by passage of other viruses or by other infectious agents, producing hepatitis. Neurotropic strains also exist.

mouse mammary tumour virus. Family *Retroviridae*, subfamily *Oncovirinae*, genus *Type B Oncovirus* group. Causes tumours by transmission in milk. Can be grown in cultures of mammary tumour tissue.

mouse poliovirus. *See* THEILER'S VIRUS.

mouse pox virus. *See* ECTROMELIA VIRUS.

mouse sarcoma virus. Family *Retroviridae*, subfamily *Oncovirinae*, genus *Type C Oncovirus* group. Several strains of virus exist. All rapidly induce sarcomas in mice. They transform fibroblasts in cell culture but do not produce progeny virus unless a helper virus such as mouse leukaemia virus is present.

Mozambique virus. *See* LASSA FEVER VIRUS.

MP variant. *See* MANY POLYHEDRA VARIANT.

M'Poko virus. Family *Bunyaviridae*, genus *Bunyavirus*. Isolated from mosquitoes in Central African Republic.

mRNA. Abbreviation for MESSENGER RNA.

MS2 phage. LEVIVIRUS type species isolated in San Francisco. Virions are unenveloped, 23 nm. in diameter, and probably contain 32 capsomeres arranged in T=3 symmetry. Particles have a mw. of 3.6×10^6, sediment at 79-80S, with a buoyant density of 1.46 g/cc. in CsCl. The capsid contains 180 copies of coat protein (129 amino acids) and one copy of 'A protein' (mw. 44×10^3). The (+)-sense linear ssRNA is 3569 nucleotides long. MS2 is closely related to other 'group I' leviviruses including phages f2 and R17, though it can be distinguished by serological and other criteria from other groups of structurally-similar phages including Qβ (*see* LEVIVIRUS).
Fiers, W. (1979) **In** Comprehensive Virology. Vol. 13. p. 69. ed. H. Fraenkel-Conrat and R.R. Wagner. Plenum Press: New York.

Mu phage. ('Mu' from 'mutator phage'). A temperate DNA-containing tailed phage (*Myoviridae*) which appears able to integrate at any site on the chromosome of bacterial hosts, thereby destroying bacterial gene functions at the site of integration. The PROPHAGE state is an essential stage in its replication. DNA replication apparently occurs by repeated transpositions to new sites on the host chromosome producing a series of copies of the Mu genome integrated at different sites. It undergoes up to 100 cycles of DNA transposition per hour during the lytic phase of its replication cycle. Maturation occurs with phage DNA excised from chromosomal sites and packaged into virions. The packaged DNA includes short sequences of host DNA at either end of the genome. Because of this method of replication, Mu phage is considered as a TRANSPOSON. A 'Mini-Mu' derivative of the phage DNA, containing the left and right hand ends of the genome, is used as a cloning vector.
Toussaint, A. and Résibois, A. (1983) **In** Mobile Genetic Elements. p. 105. ed. J.A. Shapiro. Academic Press: New York.

Mucambo virus. Family *Togaviridae*, genus *Alphavirus*. Isolated from man, rodents, birds and mosquitoes in Brazil, Trinidad, Surinam and French Guiana. Causes fever with headache and myalgia in man.

mucosal disease virus. Synonym: BOVINE VIRAL DIARRHOEA VIRUS. Family *Togaviridae*, genus *Pestivirus*. Causes mucosal disease with fever, diarrhoea and oral ulceration in several species, including cattle, sheep, pigs, buffaloes, moose and deer. If infection occurs during pregnancy, abortion or foetal abnormality often follows. Virus can be grown in rabbits and in cell cultures. Closely related antigenically to swine fever virus and border disease virus.

Mudjinbarry virus. Family *Reoviridae*, genus *Orbivirus*. Isolated from midges in Northern Territory, Australia. Antibodies have been found in man, wallabies, dingoes and domestic fowl but no disease reported.

mulberry latent virus. A *Carlavirus*.
Francki, R.I.B. *et al.* (1985) **In** Atlas of Plant Viruses. Vol. 2. p. 173. CRC Press: Boca Raton, Florida.

mulberry ringspot virus. A *Nepovirus*.
Tsuchizaki, T. (1975) CMI/AAB Descriptions of Plant Viruses No. 142.
Francki, R.I.B. *et al.* (1985) **In** Atlas of Plant Viruses. Vol. 2. p. 23. CRC Press: Boca Raton, Florida.

multi-hit kinetics. In virology the effect of concentration of a virus in which more than one particle is needed to initiate infection on the number of plaques or lesions induced. It is directly proportional to the $1/n^{th}$ power (where n = number of particles needed for infection) of the concentration of the inoculum. If the viral genome is divided between three particles the number of plaques or lesions will double when six times the concentration of virus is inoculated. *See* ONE-HIT KINETICS.

multicapsid NPV. *See* MNPV.

multicistronic messenger RNA. An mRNA which contains the coding sequences for two or more proteins. Used interchangeably with the term POLYCISTRONIC. *See* OPERON.

multicomponent virus. A virus in which the genome needed for full infection is divided between two or more particles. The partial genomes in separate particles may have some sequences in common but these are usually in non-coding regions. Examples are COWPEA MOSAIC VIRUS with two particle types and BROME MOSAIC VIRUS with three.

Reijnders, L. (1978) Adv. Virus Res. **23**, 79.

Fulton, R.. (1980) Ann. Rev. Phytopath. **18**, 131.

multigenic messenger RNA. *See* MULTI-CISTRONIC MESSENGER RNA.

multipartite genome. A viral genome divided between two or more nucleic acid molecules. These may be encapsidated in the same particle e.g. REOVIRUS, ORTHOMYXOVIRUS, or be in separate particles (e.g. COWPEA MOSAIC VIRUS) in which case they are termed MULTICOMPONENT.

multipartite virus. Synonym: MULTICOMPONENT VIRUS.

multiple nucleocapsid nuclear polyhedrosis virus. *See* MNPV.

multiplicity of infection (m.o.i.). Ratio of number of infectious virus particles added to a known number of cells in a culture.

multiplicity reactivation. A form of REASSORT-MENT or COMPLEMENTATION between two related viruses which have been INACTIVATED. The sites of inactivation must be in different parts of the genomes of the two viruses.

multiplier prefixes. The table gives multiplier prefixes for basic units:

Prefix	Abbreviation	Multiplier
exa-	E	10^{18}
penta-	P	10^{15}
tera-	T	10^{12}
giga-	G	10^9
mega-	M	10^6
kilo-	K	10^3
hecto-	h	$10^{2}*$
deka-	da	$10^{1}*$
deci-	d	$10^{-1}*$
centi-	c	$10^{-2}*$
milli-	m	10^{-3}
micro-	µ	10^{-6}
nano-	n	10^{-9}
pico-	p	10^{-12}
femto-	f	10^{-15}
atto-	a	10^{-18}

* used only as prefixes for metres

multiploid virus. A virus which comprises a population of particles which contain a varying number of genomes (0,1,2,...N), e.g. SENDAI VIRUS, NEWCASTLE DISEASE VIRUS.

Simon, E.H. (1972) Progr. Med. Virol. **14**, 36.

mumps virus. Family *Paramyxoviridae*, genus *Paramyxovirus*. A common infection of man, causing fever and parotitis and less frequently meningoencephalitis, orchitis and pancreatitis. Rhesus monkeys can be infected and show a disease similar to that seen in man. Hamsters, rats and mice can be infected. The virus grows in monkey kidney or human cell lines.

mungbean mottle virus. A possible *Potyvirus*. Francki, R.I.B. *et al.* (1985) **In** Atlas of Plant Viruses. Vol. 2. p. 183. CRC Press: Boca Raton, Florida.

mungbean yellow mosaic virus. A *Geminivirus*, subgroup B.

Honda, Y. and Ikegami, M. (1986) AAB Descriptions of Plant Viruses No. 323.

Harrison, B.D. (1985) Ann. Rev. Phytopath. **23**, 55.

Murray Valley encephalitis virus. Family *Flaviviridae*, genus *Flavivirus*. Occurs in Australia and New Guinea. Causes fever and sometimes encephalitis in man, often as epidemics. Natural host is probably a bird and the virus is transmitted by mosquitoes. Many species can be infected experimentally i.c., often with encephalitis. The virus can be grown in eggs, producing pocks on the chorioallantoic membrane.

Murutucu virus. Family *Bunyaviridae*, genus *Bunyavirus*. Causes fever in man. Isolated from a sentinel monkey, several rodent species and mosquitoes in Brazil.

mushroom virus 3. Synonym: AGARICUS BISPORUS VIRUS 3.

mushroom virus 4. Synonym: AGARICUS BISPORUS VIRUS 4.

muskmelon vein necrosis virus. A *Carlavirus*. Francki, R.I.B. *et al.* (1985) **In** Atlas of Plant Viruses. Vol. 2. p. 173. CRC Press: Boca Raton, Florida.

mutagen. An agent which raises the frequency of mutation above the spontaneous rate.

mutant. An individual containing a gene or allele which has undergone mutation and is often expressed in the phenotype.

mutation. An abrupt change in the genotype of an organism not resulting from recombination. In its simplest form it is the substitution of one nucleotide for another leading to changes in the structure of the protein coded for by a nucleotide sequence or modifying gene regulation sequences.

MV-L2 phage. Synonym for L2 PHAGE.

MV-L3 phage. Synonym: L3 PHAGE.

MVL51 phage. Proposed type species of PLECTROVIRUS genus (*Inoviridae*) isolated from *Acholeplasma laidlawii*. Virions are naked bullet-shaped particles (PHAGE morphotype F2) 70-90 nm. long and 13-16 nm. wide, with a buoyant density in CsCl of 1.37 g/cc. The particles contain four structural proteins (mw. 19, 30, 53 and 70 x 10^3). The genome is circular ssDNA (mw. 1.5 x 10^6; 4.5 kb). During replication, infecting viral DNA is converted into a replicative form. The virus is resistant to detergent and ether treatment. The host survives infection; virus particles are extruded through the cell membrane without lysis. Only *A. laidlawii* strains are susceptible to infection by virus particles, though MVL51 DNA will transfect *Mycoplasma gallisepticum*.
Maniloff, J. *et al.* (1982) Intervirology **18**, 177.

Mycogone perniciosa virus (MpV). Probable member of the *Totivirus* group.

mycoplasma virus type 1 phages. Vernacular name for *PLECTROVIRUS*.

mycoplasma virus type 2 phages. Vernacular name for *PLASMAVIRUS*.

mycoplasmaphages. *See* MYCOPLASMAVIRUSES.

mycoplasmaviruses. Phage-like viruses isolated from mycoplasmas (prokaryotes without cell walls). They include members of the *PODOVIRIDAE* (e.g. L3 phage) as well as the *PLECTROVIRUS* (e.g. MV-L51) and PLASMAVIRUS genera (e.g. MV-L2, renamed L2 phage) and several unclassified isolates.
Maniloff, J. *et al.* (1982) Intervirology **18**, 177.

mycovirus. A virus which replicates in cells of fungi.

myeloma. A tumour of the immune system.

Myoviridae. (Greek 'myos' = muscle). A family of phages with contractile tails (phage morphotypes A1-A3, *see* PHAGE), which contain linear dsDNA that may be circularly-permuted and/or terminally-redundant. The tail is long (80-455

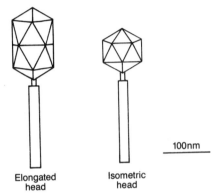

100nm

Elongated head   Isometric head

nm.) and complex, consisting of a central tube and a contractile sheath separated from the head by a neck. The family at present contains a single genus (the T-EVEN PHAGE GROUP) and the type species is T2 PHAGE. A large number of phages with these properties have been isolated from bacteria, cyanobacteria and mycoplasmas.
Ackermann, H.W. *et al.* (1984) Intervirology **22**, 181.

Myrobalan latent ringspot virus. A *Nepovirus*.
Dunez, J. *et al.* (1976) CMI/AAB Descriptions of Plant Viruses No. 160.
Francki, R.I.B. *et al.* (1985) **In** Atlas of Plant Viruses. Vol. 2. p. 23. CRC Press: Boca Raton, Florida.

myxoma virus. Family *Poxviridae*, subfamily *Chordopoxvirinae*, genus *Leporipoxvirus*. Isolated from rabbits, hares and squirrels. Exists as a silent infection of rabbits in Uruguay and Brazil. However, when introduced into rabbits in Australia, it caused a severe disease, usually fatal. The virus essentially eliminated the rabbit population in Britain in the 1950s. The signs included inflammation and swelling of the eyelids, nose, genitalia and anus. Transmission is by contact and by insects, mosquitoes and fleas. Virus can be grown in suckling mouse brain and in tissue cultures of cells from many species. Produces pocks on the chorioallantoic membrane.

N

NAD and NADP. Abbreviations for NICOTIN-AMIDE ADENINE DINUCLEOTIDE and its phosphate.

Nairobi sheep disease virus. Family *Bunyaviridae*, genus *Nairovirus*. Causes haemorrhagic gastro-enteritis in sheep and goats with high mortality. The spleen is enlarged and the female genital tract is also involved. In man, the virus can cause fever. Encephalitis ensues in mice inoculated i.c. Transmission is by tick. The virus will grow in cell cultures of lamb and goat tissues.

Nairovirus. A genus in the family *Bunyaviridae*, containing the important CRIMEAN-CONGO HAEMORRHAGIC FEVER and NAIROBI SHEEP DISEASE VIRUSES.

naked virus. A virus without a lipoprotein envelope, e.g. picornaviruses, many plant viruses.

Nandina mosaic virus. A possible *Potexvirus*.
Francki, R.I.B. *et al.* (1985) **In** Atlas of Plant Viruses. Vol. 2. p. 159. CRC Press: Boca Raton, Florida.

Nandina stem pitting virus. A possible member of the *Capillovirus* group.
Ahmed, N.A. *et al.* (1983) Phytopath **73**, 470.

narcissus degeneration virus. A *Potyvirus*.
Francki, R.I.B. *et al.* (1985) **In** Atlas of Plant Viruses. Vol. 2. p. 183. CRC Press: Boca Raton, Florida.

narcissus late season yellows virus. A possible *Potyvirus*.
Francki, R.I.B. *et al.* (1985) **In** Atlas of Plant Viruses. Vol. 2. p. 183. CRC Press: Boca Raton, Florida.

narcissus latent virus. A *Carlavirus*.
Brunt, A.A. (1976) CMI/AAB Descriptions of Plant Viruses No. 170.
Francki, R.I.B. *et al.* (1985) **In** Atlas of Plant Viruses. Vol. 2. p. 173. CRC Press: Boca Raton, Florida.

narcissus mosaic virus. A *Potexvirus*.
Mowat, W.P. (1971) CMI/AAB Descriptions of Plant Viruses No. 45.
Francki, R.I.B. *et al.* (1985) **In** Atlas of Plant Viruses. Vol. 2. p. 159. CRC Press: Boca Raton, Florida.

narcissus tip necrosis virus. A possible member of the *Carmovirus* group.
Mowat, W.P. (1976) CMI/AAB Descriptions of Plant Viruses No. 166.
Francki, R.I.B. *et al.* (1985) **In** Atlas of Plant Viruses. Vol. 2. p. 235. CRC Press: Boca Raton, Florida.
Morris, T.J. and Carrington, J.C. (1988) **In** The Plant Viruses. Vol. 3, p.73. ed. R. Koenig. Plenum Press: New York.

narcissus yellow stripe virus. A *Potyvirus*.
Brunt, A.A. (1971) CMI/AAB Descriptions of Plant Viruses No. 76.
Francki, R.I.B. *et al.* (1985) **In** Atlas of Plant Viruses. Vol. 2. p. 183. CRC Press: Boca Raton, Florida.

Nariva virus. Family *Paramyxoviridae*, genus *Paramyxovirus*. Isolated from forest rodents in Trinidad. Kills mice when injected i.c. Hamsters and guinea pigs can be infected without causing disease.

nascent cleavage. Proteolytic cleavage of a polypeptide while it is being synthesised by ribosomes.

nascent RNA. Oligoribonucleotides in the process of being transcribed from template DNA or RNA. REPLICATIVE INTERMEDIATES contain such RNA.

nasturtium mosaic virus. A possible *Carla-*

virus.
Francki, R.I.B. *et al.* (1985) **In** Atlas of Plant
Viruses. Vol. 2. p. 173. CRC Press: Boca Raton,
Florida.

Navarro virus. Family *Rhabdoviridae*, genus
unassigned. Isolated from turkey vulture in Colombia.

navel orange infectious mottling virus. An
unclassified plant virus with isometric particles.
Occurs in Japan.
Doi, Y. *Personal communication.*

Ndumu virus. Family *Togaviridae*, genus *Alpha
virus*. Isolated from mosquitoes in South Africa.
Kills new-born mice on injection.

nearest neighbour sequence analysis. A technique for studying the relationship between nucleic acids. The relative frequencies with which
pairs of the four bases occur in adjacent positions
is determined; this can distinguish between nucleic acids with identical or similar base compo-
sitions.

Nebraska calf diarrhoea virus. Family *Reoviridae*, genus *Rotavirus*. Causes gastro-enteritis
in calves. Can be grown in bovine embryo kidney
cell cultures.

neck. The region in TAILED PHAGE particles that
links the phage head to the tail. In some phages,
e.g. those which have contractile tails (MYO-
VIRIDAE) the neck is complex, including a connector, collar and whiskers (*see* T4 PHAGE).

Neckar river virus. A *Tombusvirus* isolated
from the water of the river Neckar in West Ger-
many.
Koenig, R. and Lesemann, D-E. (1985) Phyto-
path. Z. **112**, 105.
Gallitelli, D. and Russo, M. (1987) J. Phytopath.
119, 106.
Martelli, G.P. *et al.* (1988) **In** The Plant Viruses.
Vol. 3, p.13. ed. R. Koenig. Plenum Press: New
York.

necrosis (necrotic). The death of cells. This may
be due to virus infection or may occur in the
surrounding uninfected cells following the release of toxic materials from the infected cell. *See*
LOCAL LESION.

necrotic rhinitis virus. *See* BOVID HERPESVIRUS 1.

Necrovirus group. (necrosis). (Type member
TOBACCO NECROSIS VIRUS). Genus of plant viruses
with isometric particles, 28 nm. in diameter,
which sediment at 118S and band in CsCl at 1.40
g/cc. The capsid has icosahedral symmetry

   100nm

(T=3), the coat protein subunit having mw. = 22.6
x 10^3. Each particle contains one molecule of
linear (+)-sense ssRNA (mw. 1.3-1.6 x 10^6).
   The host range is wide. Virus particles are
found in most cell types. Necroviruses are easily
mechanically transmitted. They are transmitted
naturally by Chytrid fungi, *Olpidium* spp.
Matthews, R.E.F. (1982) Intervirology **17**, 146.
Francki, R.I.B. *et al.* (1985) **In** Atlas of Plant
Viruses. Vol. 1. p. 171. CRC Press: Boca Raton,
Florida.

Neethling virus. *See* LUMPY SKIN DISEASE VIRUS.

negative staining. A method of visualising virus
particles in the electron microscope. The electron
dense 'stain' dries down on the electron microscope grid, thus outlining the particles and reveal-
ing surface structures. It does not directly stain the
particles. *See* PHOSPHOTUNGSTIC ACID, URANYL
ACETATE.
Brenner, S. and Horne, R.W. (1959) Biochim.
Biophys. Acta **34**, 103.

negative-sense strand (negative strand). Nucleic acid complementary to the PLUS STRAND.
Some ssRNA viruses (e.g. PARAMYXOVIRUS,
ORTHOMYXOVIRUS, RHABDOVIRUS) package the
negative strand within the virus particle.

negative-strand virus. A virus whose genome is
NEGATIVE-SENSE RNA. Five families of negative-
strand viruses have been recognised: ARENA-
VIRIDAE, BUNYAVIRIDAE, ORTHOMYXOVIRIDAE,
PARAMYXOVIRIDAE AND RHABDOVIRIDAE.

Negishi virus. Family *Flaviviridae*, genus
Flavivirus. Isolated from cases of fatal encephalitis in man in Japan. Transmitted by ticks.

Negri body. Inclusion body found in the cyto-
plasm of brain cells of animals infected with
RABIES VIRUS. Used in diagnosis. Named after A.
Negri who first reported them in 1903.

negro coffee mosaic virus. A possible *Potexvirus.*
Francki, R.I.B. *et al.* (1985) **In** Atlas of Plant Viruses. Vol. 2. p. 159. CRC Press: Boca Raton, Florida.

Nelson Bay virus. Family *Reoviridae*, genus *Reovirus.* Isolated from blood of flying fox in Australia. Causes paralysis and death in mice inoculated i.c. Can be grown in pig kidney cells.

Neochek-S. Preparation of the nuclear polyhedrosis virus of the pine sawfly, *Neodiprion sertifer*, produced on an industrial scale by the US Forest Service for Pine Sawfly control.

Neodiprion sertifer nuclear polyhedrosis virus. A BACULOVIRUS (Subgroup A) of the SNPV type isolated from larvae of the pine sawfly, *N. sertifer.* It is host-specific and highly infectious for larvae of the homologous host. As with all known NPV infections of hymenopteran larvae, virus replication is restricted to cells of the gut epithelium. It has been available as a selective biological control agent for pine sawfly since 1974 in Finland and was registered for use in the USA in 1983 (as 'Neochek-S') and in the UK in 1985 (as 'Virox'). Comparative studies of *N. sertifer* NPV and NPVs isolated from lepidopteran hosts have revealed that the polyhedrins are distantly related.
Cunningham, J.C. and Entwistle, P.F. (1981) **In** Microbial Control of Pests and Plant Diseases 1979-1980. p. 379. ed. H.D. Burges. Academic Press: London.

neonatal calf diarrhoea virus. Family *Coronaviridae*, genus *Coronavirus.* Causes diarrhoea in very young calves. When serially transmitted, virus becomes attenuated. Can be grown in bovine kidney cell cultures.

neoplasm. An aberrant new growth of abnormal cells or tissues. *See* TUMOUR.

neotope. *See* EPITOPE.

Nepovirus group. (Sigla from nematode polyhedral). (Type member TOBACCO RINGSPOT VIRUS). Genus of MULTICOMPONENT plant viruses with isometric particles 28 nm. in diameter which sediment at 115-134S (bottom (B) component), 86-128S (middle (M) component) and 49-56S (top (T) component which lacks nucleic acid); CsCl banding densities (g/cc) are 1.51-1.53 (B),

1.43-1.48 (M) and 1.28 (T). The capsids are made up of subunits of a single polypeptide species (mw. $55\text{-}60 \times 10^3$) which is possibly an oligomer or may contain repeated amino acid sequences. The genome is two species of linear (+)-sense ssRNA, RNA-1 (mw. 2.8×10^6) and RNA-2 (mw. $1.3\text{-}2.4 \times 10^6$); the 5' terminus has a VPg (mw. 3.6×10^3), the 3' terminus is polyadenylated. M component particles contain a single molecule of RNA-2 and B component particles one molecule of RNA-1 or sometimes two molecules of RNA-2. RNA-1 codes for its replication; RNA-2 encodes the coat protein. The host ranges of nepoviruses are wide. Particles are found in most cell types. They are readily mechanically transmitted. Seed transmission is common. Most members are transmitted by soil-inhabiting nematodes of the longidorid genus.
Matthews, R.E.F. (1982) Intervirology **17**, 163.
Harrison, B.D. and Murant, A.F. (1977) CMI/AAB Descriptions of Plant Viruses No. 185.
Francki, R.I.B. *et al.* (1985) **In** Atlas of Plant Viruses. Vol. 2. p. 23. CRC Press: Boca Raton, Florida.

Nepuyo virus. Family *Bunyaviridae*, genus *Bunyavirus.* Isolated from bats and mosquitoes in several countries in the Caribbean and in Brazil.

Nerine latent virus. A *Carlavirus.*
Francki, R.I.B. *et al.* (1985) **In** Atlas of Plant Viruses. Vol. 2. p. 173. CRC Press: Boca Raton, Florida.

Nerine virus. A possible *Potyvirus.*
Francki, R.I.B. *et al.* (1985) **In** Atlas of Plant Viruses. Vol. 2. p. 183. CRC Press: Boca Raton, Florida.

Nerine virus X. A *Potexvirus.*
Francki, R.I.B. *et al.* (1985) **In** Atlas of Plant Viruses. Vol. 2. p. 159. CRC Press: Boca Raton, Florida.

neuraminic acid. Constituent of cell receptor recognised by orthomyxovirus haemagglutinin.

neuraminidase. A glycoprotein carried on the surface of members of the ortho- and paramyxoviruses. In orthomyxoviruses it forms a distinct spike, whereas in paramyxoviruses it forms a joint spike with the viral HAEMAG-

142 neutral red

GLUTININ. The detailed structure of influenza virus neuraminidase is known. It has enzymic activity which cleaves N-acetylneuraminic acid from the glycoprotein receptors on the cell membrane. In influenza virus it is serotype-specific as is haemagglutinin. Colman, P.M. *et al.* (1983) Nature **303**, 41.

neutral red. A photoreactive dye. It decreases the infectivity of REOVIRUSES.

neutralisation. The inactivation of infectious virus by reaction with its specific antibody, thereby blocking sites on the virus which normally adsorb to susceptible cells. However, other mechanisms have been proposed.
Dimmock, N.J. (1984) J. gen. Virol.**65**, 1015.
Dimmock, N.J. (1987) Trends in Biochem. Sci. **12**, 70.

neutralising antibody. An antibody which inhibits the infectivity of a virus.

New Minto virus. Family *Rhabdoviridae*, genus unassigned. Isolated from ticks in Alaska. Kills mice when injected i.c. Grows in Vero cells.

New South Wales kashmir bee virus. Unclassified small RNA virus (possible INSECT PICORNAVIRUS) isolated from the honey bee, *Apis mellifera*. The virus is physically indistinguishable from KASHMIR BEE VIRUS. Virions contain three major proteins (mw. 44.5, 35.4 and 25 x 10³). Serologically related to, but distinct from, KASHMIR BEE VIRUS, QUEENSLAND KASHMIR BEE VIRUS and SOUTH AUSTRALIA KASHMIR BEE VIRUS.
Bailey, L. *et al.* (1979). J. gen. Virol. **43**, 641.

Newcastle disease virus. Synonym: AVIAN PNEUMOENCEPHALITIS VIRUS, RANIKHET VIRUS. Family *Paramyxoviridae*, genus *Paramyxovirus*. Causes respiratory disease in many species of birds, with nasal discharge and diarrhoea. Occasionally, nervous system is involved with paralysis. Also infects man, causing conjunctivitis. Transmission in birds is by drinking water or dust. Encephalitis is produced when virus is inoculated into mice or hamsters. Can be grown in eggs or many types of cell culture.

nicked circular DNA. Synonym: OPEN CIRCULAR DNA.

Nicotiana velutina mosaic virus. A possible member of the *Furovirus* group.

Randles, J.W. (1978) CMI/AAB Descriptions of Plant Viruses No. 189.
Brunt, A.A. and Shikata, E. (1986) **In** The Plant Viruses. Vol. 2. p. 305. ed. M.H.V. van Regenmortel and H. Fraenkel-Conrat. Plenum Press: New York.

Nigerian horse virus. Family *Rhabdoviridae*, genus *Lyssavirus*. Isolated from the brain of a horse with staggers (meningo-encephalomyelitis). Infects mice when inoculated i.c.

NIH. National Institutes of Health. A group of medical research institutes in Bethesda, Maryland, U.S.A. and also the principal US Government agency for funding bio-medical research in universities.

Nique virus. Family *Bunyaviridae*, genus *Phlebovirus*. Isolated from *Lutzomyia panamensis* in Panama.

nitrocellulose. A nitrated derivative of cellulose which is used either as a powder or is made into membrane filters of defined porosity. Used to bind nucleic acids in NORTHERN BLOTTING and SOUTHERN BLOTTING procedures and proteins in WESTERN BLOTTING.

Noctua pronuba Cytoplasmic Polyhedrosis Virus. Cytoplasmic polyhedrosis virus (CPV) isolated in the United Kingdom from larvae of the cutworm, *Noctua (=Triphaena) pronuba* (Noctuidae, Lepidoptera). The virus is the type member of 'type 7' CPVs. Unrelated to *Bombyx mori* (type 1) CPV on the basis of RNA electropherotype. Viruses of similar electropherotype have been observed in other Lepidoptera (*see* APPENDIX B).
Payne, C.C. and Mertens, P.P.C. (1983) **In** The Reoviridae. p. 425. ed. W.K. Joklik. Plenum Press: New York.

Nodamura virus. Family *Nodaviridae*. Type species of the NODAVIRUS genus, isolated from the mosquito *Culex tritaeniorhynchus*, obtained in the village of Nodamura near Tokyo, Japan. Less-intensively studied than some other Nodaviruses, particularly BLACK BEETLE VIRUS. Unlike other Nodaviruses, it replicates not only in insects but is also pathogenic for suckling mice. It has a wide host range in insects being infectious for bees (*Apis mellifera*), Lepidoptera (e.g. *Galleria mellonella*) and Diptera (e.g. *Aedes aegypti*).

Moore, N.F. *et al.* (1985) J. gen. Virol. **66**, 647.

Nodaviridae. (Nodamura, village in Japan where virus first isolated.) A family of RNA viruses containing a single genus (NODAVIRUS). Virions are small unenveloped isometric particles *c*.30 nm. in diameter with T=3 symmetry; they sediment at 135-142S and have buoyant densities

   100nm

in CsCl of 1.30-1.35 g/cc. The particles contain one major polypeptide (mw. 39 x 10^3) and one minor species (43 x 10^3) derived by proteolytic cleavage of a 44 x 10^3 precursor protein which may also be present in small amounts in the mature particle. The (+)-sense ssRNA genome consists of two RNA molecules, one each of 1.1 x 10^6 (RNA1) and 0.48 x 10^6 (RNA2) in the same particle. Both RNAs are required for productive infection. The viruses replicate in the cytoplasm of susceptible cells. Infected cells contain three ssRNAs. RNA1 codes for a 104 x 10^3 mw. polymerase. RNA2 codes for the coat protein and RNA3 (0.15 x 10^6 and derived from RNA1) codes for a protein (mw. 10 x 10^3) of unknown function. RNAs 1 and 2 are required for the production of virions. Nodaviruses have only been naturally isolated from insects (Diptera, Coleoptera or Lepidoptera). Experimentally, most can be propagated in the wax moth, *Galleria mellonella* and in cultured *Drosophila* cells. However, Nodamura virus, unlike other members, grows in suckling mice but not in *Drosophila* cells. Moore, N.F. *et al.* (1985) J. gen. Virol. **66**, 647.

Nodavirus. The only genus of viruses currently classified within the *Nodaviridae*. The type species is NODAMURA VIRUS. Other known members are BLACK BEETLE VIRUS, BOOLARRA VIRUS, FLOCK HOUSE VIRUS, *LYMANTRIA DISPAR* NODAVIRUS, MANAWATU VIRUS; Arkansas, Boolarra and flock house viruses have similarities. All are serologically related. See also ENDOGENOUS *DROSOPHILA* LINE VIRUS.

Nola virus. Family *Bunyaviridae*, genus *Bunyavirus*. Isolated from mosquitoes in Central African Republic.

non-inclusion virus. *See* NON-OCCLUDED VIRUS.

non-ionic detergent. A detergent with no net surface charge, e.g. the TRITON series, NONIDET P40.

non-occluded Baculovirus. Proposed subgroup C of the *BACULOVIRUS* genus, characterised by virus particles containing rod-shaped nucleocapsids singly enveloped by a unit membrane, which replicate in the nucleus. Unlike productive infections with members of the other BACULOVIRUS subgroups (NUCLEAR POLYHEDROSIS and GRANULOSIS VIRUSES), no virus occlusion bodies are produced during infection. The type species is *ORYCTES RHINOCEROS* virus. Other morphologically-similar viruses include HZ-1 BACULOVIRUS and isolates from *Gyrinus natator*, *Hypera* sp., *Diabrotica undecimpunctata* (Insecta: Coleoptera), *Aphis* spp. (Insecta: Homoptera), *Gryllus campestris* (Insecta: Orthoptera), *Solenopsis* sp. (Insecta: Hymenoptera), *Bacillus rossius* (Insecta: Phasmatodea), *Chaoborus crystallinus* (Insecta: Diptera), *Panonychus* spp., *Pisaura mirabilis* (Arachnida), *Carcinus* spp. *Callinectes sapidus* and *Penaeus japonicus* (Crustacea). Some braconid parasitoid virus particles (e.g. *Mesoleius tenthredenis* virus) have similar properties. Brief descriptions of these viruses are given elsewhere in the Dictionary under the generic name of the host e.g. *GYRINUS* BACULOVIRUS. Granados, R.R. and Federici, B.A. (1986) The Biology of Baculoviruses. Vols. I and II. CRC Press: Boca Raton, Florida.

non-occluded rod-shaped nuclear viruses. *See* NON-OCCLUDED BACULOVIRUS.

non-occluded virus. Usually applied to describe insect-pathogenic viruses which do not produce proteinaceous OCCLUSION BODIES during infection (including NON-OCCLUDED BACULOVIRUS, DENSOVIRUS, IRIDOVIRUS, INSECT PICORNAVIRUS).

non-permissive cells. Cells in which a specific virus will not infect and replicate. The barrier may be lack of viral receptors or some cell factor or lack of cell factor which inhibits replication.

non-persistent transmission. The relationship between certain viruses (e.g. POTYVIRUSES) and their arthropod vectors which is characterised by a very short acquisition period after which the vector does not retain the virus for long after the initial transmission feed. The virus particles are thought to associate with the vector's mouth parts, those particles passing into the gut not being transmitted. *See* PERSISTENT TRANSMISSION.

non-producer cells. Cells carrying all or part of

a viral genome but not producing virus particles. They are usually transformed by the virus.

non-structural protein. Protein encoded by a viral genome but not involved in the structure of the virus particle. It is usually functional during replication.

Nonidet P40. A non-ionic detergent comprising octylphenol ethylene condensate. Used to disrupt cells and viral membranes.

nonsense codon. *See* STOP CODON.

nonsense mutant. Caused by a mutation that results in the premature termination of a polypeptide chain.

North American plum line pattern virus. An *Ilarvirus*.
Matthews, R.E.F. (1982) Intervirology **17**, 175.

northern blotting. A procedure analogous to SOUTHERN BLOTTING but involving the transfer of RNA on to nitrocellulose or activated paper sheets.

northern cereal mosaic virus. A plant *Rhabdovirus*, subgroup 1; transmitted by planthoppers.
Toriyama, S. (1986) AAB descriptions of Plant Viruses No. 322.
Francki, R.I.B. *et al.* (1985) **In** Atlas of Plant Viruses. Vol. 1. p. 73. CRC Press: Boca Raton, Florida.

Northway virus. Family *Bunyaviridae*, genus *Bunyavirus*. Isolated from mosquitoes in Alaska.

Norwalk agent. Unclassified but probably a *Calicivirus*. Causes acute gastro-enteritis in man.

Nothoscordum mosaic virus. A *Potyvirus*.
Francki, R.I.B. *et al.* (1985) **In** Atlas of Plant Viruses. Vol. 2. p. 183. CRC Press: Boca Raton, Florida.

notifiable disease. A disease which must be notified to the medical or agricultural authorities in a given country, e.g. in the UK, measles, foot-and-mouth disease, beet rhizomania.

NOV. Abbreviation sometimes used for 'non-occluded virus', often referring to those BACULOVIRUS particles which are not incorporated within occlusion bodies during infection.

Novobiocin. An antibiotic whose primary action is to inhibit DNA synthesis. Probably acts on DNA GYRASE.

NPV. Common abbreviation for NUCLEAR POLYHEDROSIS VIRUS.

Ntaya virus. Family *Flaviviridae*, genus *Flavivirus*. Isolated from mosquitoes in several countries in Africa.

nuclear magnetic resonance (NMR) spectroscopy. The measurement of absorption of energy by magnetic atomic nuclei placed in a strong magnetic field. Can give atomic detail of small macromolecules.

nuclear membrane. The membrane which surrounds the nucleus of a eukaryotic cell. It is really two membranes, the inner and the outer; the outer is contiguous with the endoplasmic reticulum.

nuclear polyhedrosis virus (NPV). Subgroup A of the *Baculovirus* genus. Virus particles may consist of one rod-shaped nucleocapsid surrounded by an envelope (SNPV morphotype), or two or more nucleocapsids surrounded by a common envelope (MNPV morphotype). An NPV from the silkworm, *Bombyx mori*, is characteristic of the SNPV type while the type species of NPVs, *Autographa californica* NPV, is of the MNPV type. During infection of susceptible insect hosts, many NPV virus particles are occluded in large numbers within large (0.5 to 15 μm) proteinaceous occlusion bodies which help maintain virus viability for many years outside the insect host. Most of the biochemical properties of the virus particles are shared with members of the baculovirus subgroups B and C (*see* BACULOVIRUS, GRANULOSIS VIRUS, NON-OCCLUDED BACULOVIRUS), including a large circular supercoiled dsDNA genome, about 25 structural polypeptides and an occlusion body matrix protein (*see* POLYHEDRIN), mw. about 29 x 10^3. The structure and replication of the type species has been intensively studied (*see* AUTOGRAPHA CALIFORNICA NPV); this virus and several other NPV isolates from Lepidoptera have been propagated successfully in insect cell culture. The majority of NPV isolates have been obtained from Lepidoptera, though NPVs have also been reported in Hymenoptera, Diptera, Neuroptera, Coleoptera, Trichoptera, Crustacea and mites. In lepidopteran larvae, virus infection predominates in the fat body tissue and is generally lethal. In Hymenop-

tera, the virus replicates only in the gut epithelial cells. Some NPV isolates (e.g. *Neodiprion sertifer* NPV) have very high host specificity, infecting only one insect species or a group of closely-related species. Others (e.g. *A. californica* NPV and *Mamestra brassicae* NPV) have a broader host range within one order of insects. No NPV infections have been recorded outside the Arthropoda. There is no formal nomenclature for each virus isolate. Viruses have usually been named after the host from which they were first isolated; names are often abbreviated in the literature e.g. AcNPV for *A. californica* NPV. Many NPVs have high pathogenicity for their insect hosts and several have been used on a commercial or semi-commercial scale as biological pest control agents. These include the NPVs of the cotton bollworm, *Heliothis zea*; gypsy moth, *Lymantria dispar*, pine sawfly, *Neodiprion sertifer*; Douglas fir tussock moth, *Orgyia pseudotsugata*, and cabbage moth, *Mamestra brassicae*. A list of insect hosts in which NPV infections have been recorded is given in Appendix A.
Granados, R.R. and Federici, B.A. (1986) The Biology of Baculoviruses. Vols. I and II. CRC Press: Boca Raton, Florida.

nuclease. An enzyme which can hydrolyse the internucleotide linkages of a nucleic acid.

nucleic acid. A long chain of repeating units built up from nucleotides. When the sugar is ribose the nucleic acid is ribonucleic acid (RNA); when deoxyribose it is deoxyribonucleic acid (DNA).

nucleocapsid. Viral nucleic acid enclosed by a protein capsid. Thus it is the particle of rod-shaped and simple isometric viruses. In more complex viruses, e.g. RHABDOVIRUSES, BACULOVIRUSES, the nucleocapsid or core is enclosed in a membranous structure or envelope.

nucleoid. The electron-dense central region observed in certain viruses, e.g. C-TYPE VIRUSES, in the electron microscope.

nucleopolyhedrosis virus. Synonym: NUCLEAR POLYHEDROSIS VIRUS.

nucleoprotein. A complex of nucleic acid with protein.

nucleoside. A constituent of nucleic acids comprising a sugar (ribose or deoxyribose) joined to a base. *See* NUCLEIC ACID.

nucleosome. Structures found in large DNA genomes and chromosomes which comprise DNA and histones (CHROMATIN). These structures alternate with protein-free stretches of nucleic acid.

nucleotide. A constituent of nucleic acid comprising a phosphate group joined to a sugar (ribose or deoxyribose) which, in turn, is joined to a base. *See* NUCLEIC ACID.

nucleotide phosphohydrolase. An enzyme which removes a phosphate group from the triphosphate end of nucleotides. Found in the virions of REOVIRUSES where it may be involved in the inhibition of host cell DNA synthesis or in the formation of the 5′ terminal CAP sequence on mRNA.

nucleotide sequence. *See* SEQUENCE.

nucleus. In eukaryotes, a membrane enclosed organelle that contains the chromosomes.

Nudaurelia β virus. Type species of the TETRAVIRUS genus, isolated from the pine emperor moth, *Nudaurelia cytherea capensis* (Saturniidae; Lepidoptera), in South Africa. Virus particles are isometric, 35-38 nm. in diameter, sediment at 213S and have a buoyant density of 1.30 g/cc. Particles contain a single structural polypeptide with a mw. of about 60×10^3. The capsid contains 240 copies of this protein arranged in a T=4 symmetry. The genome is a (+)-sense ssRNA, mw. 1.8×10^6. Several other related viruses have been identified by their serological reactions with antiserum against *Nudaurelia β* virus.
Moore, N.F. *et al.* (1985) J. gen. Virol. **66**, 647.

Nudaurelia β virus group. Vernacular name for the *Tetraviridae*.

Nudaurelia ε virus. Unclassified (probable RNA-containing) virus isolated from larvae of the pine emperor moth, *Nudaurelia cytherea capensis*. Particles are isometric, 40 nm. in diameter, sediment at 217S and have a buoyant density of 1.28 g/cc. (cf. Nudaurelia β virus; 213S; 1.30 g/cc.). The virus is serologically distinct from Nudaurelia ω virus.
Juckes, I.R.M. (1979) J. gen. Virol. **42**, 89.

Nudaurelia ω virus. Unclassified small RNA-containing virus isolated from larvae of the pine

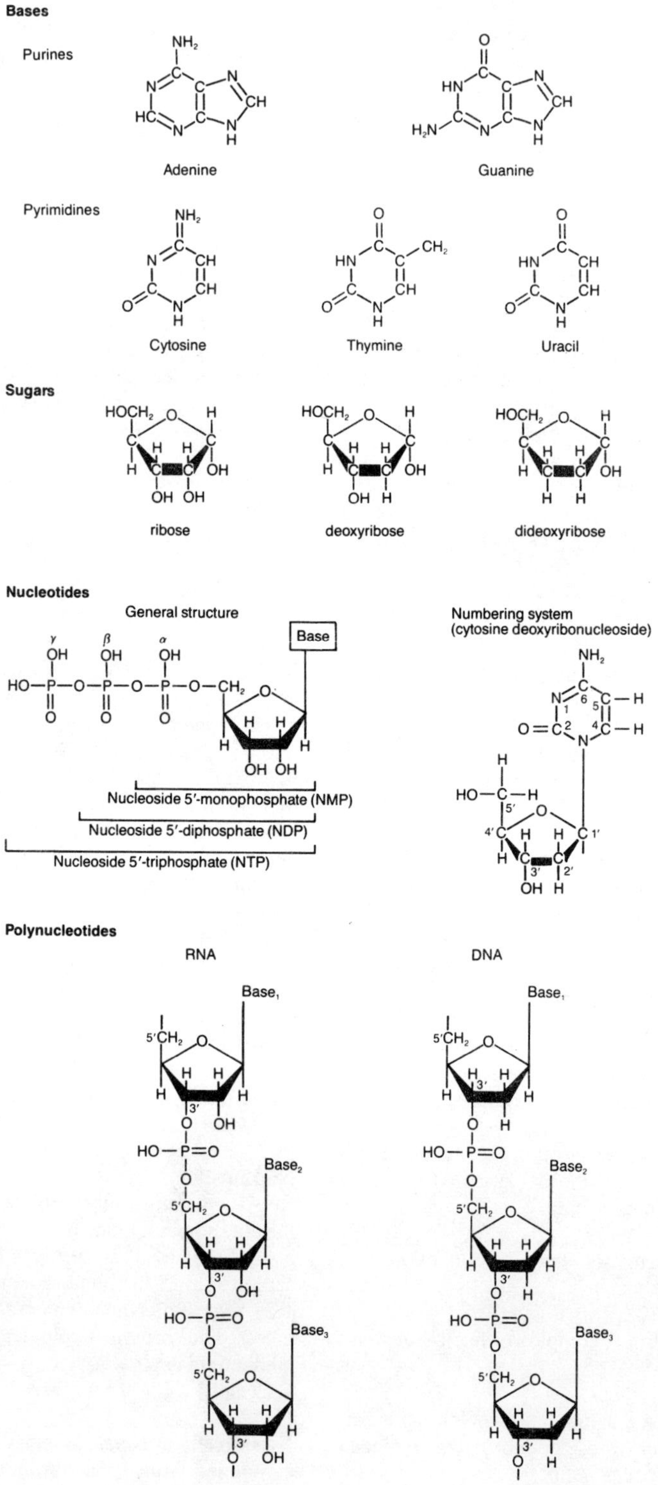
Bases
Purines
Adenine
Guanine
Pyrimidines
Cytosine
Thymine
Uracil
Sugars
ribose
deoxyribose
dideoxyribose
Nucleotides
General structure
Base
Nucleoside 5′-monophosphate (NMP)
Nucleoside 5′-diphosphate (NDP)
Nucleoside 5′-triphosphate (NTP)
Numbering system
(cytosine deoxyribonucleoside)
Polynucleotides
RNA
DNA
Base1
Base2
Base3

Nucleic acid

CONSTITUENTS OF NUCLEIC ACIDS

BASE		NUCLEOSIDE	NUCLEOTIDE		
Name	Abbreviation	Name	Abbreviation	Nucleoside monophosphate	Nucleoside diphosphate
PURINE					
Adenine	A	Adenosine	Ado	AMP*	ADP
Guanine	G	Guanosine	Guo	GMP	GDP
		2-Deoxyguanosine	dGuo	dGMP	dGDP
PYRIMIDINE					
Cytosine	C	Cytidine	Cyd	CMP	CDP
		2-Deoxycytidine	dCyd	dCMP	dCDP
Thymine	T	2-Deoxythymidine	dThd	dTMP	dTDP
Uracil	U	Uridine	Urd	UMP	UTP

*Full names:

AMP Adenosine 5′-monophosphate; Adenosine 5′-phosphoric acid; Adenylic acid.
dAMP Deoxyadenosine 5′-monophosphate; Deoxyadenosine 5′-phosphoric acid; Deoxyadenylic acid.
ADP Adenosine 5′-diphosphate; Adenosine 5′-pyrophosphoric acid.
dADP Deoxyadenosine 5′-diphosphate; Deoxyadenosine 5′-pyrophosphoric acid.
ATP Adenosine 5′-triphosphate; Adenosine 5′-triphosphoric acid.
dATP Deoxyadenosine 5′-triphosphate; Deoxyadenosine 5′-triphosphoric acid.

PROPERTIES OF NUCLEOSIDE TRIPHOSPHATES.

Abbreviation	mw. pH 7.0 (nm.)	λ_{max} x 10^{-3}	η_{max}
ATP	507.2	259	15.4
dATP	491.2	259	15.2
GTP	523.2	252	13.7
dGTP	507.2	253	13.7
CTP	483.2	271	9.1
dCTP	467.2	271	9.1
dTTP	482.2	267	9.6
UTP	484.2	262	10.0

Abbreviations: *See* Table
$L\lambda_{max}$ Wavelength of maximum ABSORPTION
λ_{max} SPECIFIC ABSORPTION at λ_{MAX}

Nucleic acid

emperor moth, *Nudaurelia cytherea capensis.* Particles are isometric, 40 nm. in diameter, and have a buoyant density in CsCl of 1.285 g/cc. There is a single major structural protein (mw. 65 x 10³; slightly larger than the protein of *Nudaurelia* β virus). Preliminary studies suggest that the RNA genome may exist as two molecules (mw. 0.9 x 10⁶ and 1.8 x 10⁶). Virus particles are serologically distinct from *Nudaurelia* ε virus, *Nudaurelia* β virus and tetraviruses isolated from *Antheraea eucalypti* and *Dasychira pudibunda.*
Hendry, D. *et al.* (1985) J. gen Virol. **66,** 627.

nugget virus. Family *Reoviridae,* genus *Orbivirus.* Isolated from nymphs of *Ixodes uriae* on Macquarie Island. Antibodies found in penguins.

Nyabira virus. Family *Reoviridae,* genus *Orbivirus.*

Nyamanini virus. Unclassified. Isolated from birds and ticks in Africa and Egypt. Kills newborn mice on injection.

Nyando virus. Unclassified isolated from man and mosquitoes in Kenya and Central African Republic. No disease reported.

O

O virus. A ROTAVIRUS isolated from water in which the intestines of sheep and cattle had been washed – hence 'O' for offal.

Malherbe, H.H. and Strickland-Cholmley, M. (1967) Archiv. ges. Virusforsch. **22**, 235.

oat blue dwarf virus. A possible member of the *Marafivirus* group.

Banttari, E.E. and Zeyen, R.J. (1973) CMI/AAB Descriptions of Plant Viruses No. 123.

oat chlorotic stripe virus. A possible *Geminivirus*, transmitted by aphids.

Harrison, B.D. (1985) Ann. Rev. Phytopath. **23**, 55.

oat golden stripe virus. A possible member of the *Furovirus* group.

Brunt, A.A. and Shikata, E. (1986) **In** The Plant Viruses. Vol. 2. p.305. ed. M.H.V. van Regenmortel and H. Fraenkel-Conrat. Plenum Press: New York.

oat mosaic virus. A member of the *Barley Yellow Mosaic Virus* group.

Hebert, T.T. and Panizo, C.H. (1975) CMI/AAB Descriptions of Plant Viruses No. 145.

oat necrotic mottle virus. A member of the *Ryegrass mosaic virus* group.

Gill, C.C. (1976) CMI/AAB Descriptions of Plant Viruses No. 169.

oat sterile dwarf virus. A *Phytoreovirus*.

Boccardo, G. and Milne, R.G. (1980) CMI/AAB Descriptions of Plant Viruses No. 217.

Francki, R.I.B. *et al.* (1985) **In** Atlas of Plant Viruses. Vol. 1. p. 47. CRC Press: Boca Raton, Florida.

oat striate virus. A plant *Rhabdovirus*, subgroup 2; transmitted by leafhopper.

Francki, R.I.B. *et al.* (1985) **In** Atlas of Plant Viruses. Vol. 1. p. 73. CRC Press: Boca Raton, Florida.

Obadhiang virus. Family *Rhabdoviridae*, genus unassigned. Isolated from mosquitoes in Sudan and Ethiopia.

occluded baculovirus. A BACULOVIRUS that produces proteinaceous virion-containing OCCLUSION BODIES during infection (i.e. NUCLEAR POLYHEDROSIS and GRANULOSIS VIRUSES).

occluded virus. A term applied to insect-pathogenic viruses that produce proteinaceous, virion-containing OCCLUSION BODIES during infection (i.e. NUCLEAR POLYHEDROSIS, GRANULOSIS, CYTOPLASMIC POLYHEDROSIS and ENTOMOPOX-VIRUSES).

occlusion body. A term applied to the large proteinaceous crystals which contain (occlude) virus particles, that are observed in infections of insects, mites and Crustacea by NUCLEAR POLYHEDROSIS, CYTOPLASMIC POLYHEDROSIS, GRANULOSIS and ENTOMOPOX-VIRUSES. They are often used to diagnose infection by light microscopy. The occlusion body serves to stabilize the virus for prolonged periods outside the host, being resistant to desiccation, normal temperature fluctuations and many proteases. The occlusion bodies range in size from about 0.5 to 24 µm, contain virions, a crystalline matrix protein (*see* POLYHEDRIN) and sometimes a structurally differentiated outer 'membrane' (which may be composed predominantly of carbohydrates). Virus particles are released *in vivo* following ingestion of the occlusion body by invertebrate hosts and solubilisation of the matrix protein by combined effects of pH and gut enzymes. Treatment of occlusion bodies *in vitro* with alkaline solutions >pH 10 (and sometimes a reducing agent) also releases the virions.

Payne, C.C. and Kelly, D.C. (1981) **In** Microbial Control of Insect Pests and Plant Diseases 1970-1980, p. 61. ed. H.D. Burges. Academic Press: London.

occult virus. Synonym: LATENT VIRUS. Often

used to describe inapparent virus infection.

Oceanside virus. Family *Bunyaviridae*, genus *Uukuvirus*.

ochre codon. *See* STOP CODON.

ochre mutant. Mutation resulting in the polypeptide chain termination codon UAA.

odontoglossum ringspot virus. A *Tobamovirus*.
Paul, H.L. (1975) CMI/AAB Descriptions of Plant Viruses No. 155.
Edwardson, J.R. and Zettler, F.W. (1986) **In** The Plant Viruses. Vol. 2. p. 233. ed. M.H.V. van Regenmortel and H. Fraenkel-Conrat. Plenum Press: New York.

Oita 293 virus. Family *Rhabdoviridae*, genus unassigned.

Okazaki fragments. Short pieces of DNA, with RNA PRIMERS attached, produced during DNA synthesis. Subsequently the primers are replaced by DNA and the fragments are ligated together. Named after the Japanese scientist who first described them. *See* SEMICONSERVATIVE REPLICATION.

Okhotskiy virus. Family *Reoviridae*, genus *Orbivirus*. Isolated from ticks in the USSR. Antibodies are found in guillemots, fulmars and cormorants.

okra mosaic virus. A *Tymovirus*.
Givord, L. and Koenig. R. (1974) CMI/AAB Descriptions of Plant Viruses No. 128.
Francki, R.I.B. *et al.* (1985) **In** Atlas of Plant Viruses. Vol. 1. p. 117. CRC Press: Boca Raton, Florida.
Hirth, L. and Girard, L. (1988) **In** The Plant Viruses. Vol. 3. p. 163. ed. R. Koenig. Plenum Press: New York.

Olifantsvlei virus. Family *Bunyaviridae*, genus *Bunyavirus*. Isolated from mosquitoes at Olifantsvlei, near Johannesburg, South Africa, and in the Sudan and Ethiopia.

oligonucleotide. A short chain nucleic acid molecule.

oligopeptide. A short chain of amino acids.

olive fruit fly viruses. At least two distinct viruses have been isolated from the olive fruit fly,

Dacus oleae. One is a putative REOVIRUS 60 nm. in diameter containing ten segments of double-stranded RNA. The other is an unclassified small RNA virus (probable INSECT PICORNAVIRUS). The latter virus has isometric particles, 35 nm. in diameter and an ssRNA genome (mw. 2.8 x 10^6). Both viruses appear to replicate in a cell line of the Mediterranean fruit fly, *Ceratitis capitata*.
Manousis, T. *et al.* (1987) Microbios **51**, 81.

olive latent ringspot virus. A *Nepovirus*.
Gallitelli, D. *et al.* (1985) AAB Descriptions of Plant Viruses No. 301.
Francki, R.I.B. *et al.* (1985) **In** Atlas of Plant Viruses. Vol. 2. p. 23. CRC Press: Boca Raton, Florida.

olive latent virus 1. A possible *Sobemovirus*.
Hull, R. (1988) **In** The Plant Viruses. Vol. 3. p. 113. ed. R. Koenig. Plenum Press: New York.

Omsk haemorrhagic fever virus. Family *Flaviviridae*, genus *Flavivirus*. Transmitted by ticks. Causes fever, enlargement of lymph nodes, gastro-intestinal symptoms and haemorrhages in man. Mortality is about 1%. Found in the USSR. Causes fever in Rhesus monkeys when injected i.p.

O'Nyong-Nyong virus. Family *Togaviridae*, genus *Alphavirus*. Occurs in several countries in Africa. Causes fever, lymphadenitis, joint pains and rash in man. Virus is transmitted by mosquitoes. Suckling mice can be infected i.c. The survivors are stunted and have alopecia. Virus can be grown in chick embryo fibroblasts.

oncogenic. Tumour-producing.

Oncovirinae. (Greek 'onkos' = tumour.) Subfamily in the family *Retroviridae*. Consists of three genera, *B, C* and *D Oncoviruses* which infect a wide range of animal species, including the important feline leukaemia and mouse mammary tumour viruses.

one- (single-) hit kinetics. In virology the effect of concentration of a virus in which one particle can initiate infection on the number of plaques or lesions induced. It is directly proportional to the first power of the concentration of the inoculum and so if the concentration is doubled the number of plaques or lesions is doubled. *See* MULTI-HIT KINETICS.

one-step growth curve. A single cycle of replication of a virus in a synchronously infected cell culture.

onion yellow dwarf virus. A *Potyvirus*.
Bos, L. (1976) CMI/AAB Descriptions of Plant Viruses No. 158.
Francki, R.I.B. *et al.* (1985) **In** Atlas of Plant Viruses. Vol. 2. p. 183. CRC Press: Boca Raton, Florida.

Ononis yellow mosaic virus. A *Tymovirus*.
Francki, R.I.B. *et al.* (1985) **In** Atlas of Plant Viruses. Vol. 1. p. 117. CRC Press: Boca Raton, Florida.
Hirth, L. and Girard, L. (1988) **In** The Plant Viruses. Vol. 3. p. 163. ed. R. Koenig. Plenum Press: New York.

opal codon. *See* STOP CODON.

opal mutant. Mutation resulting in the polypeptide terminating codon, UGA.

open reading frame (ORF). A set of codons for amino acids uninterrupted by STOP CODONS. Usually encodes a protein.

open-circular DNA. One of the three forms that dsDNA can take. It is circular DNA in which one or both strands are not covalently closed. In CsCl-ethidium bromide isopycnic gradients it has the same density as linear DNA. Also termed relaxed circular DNA.

operon. A set of two or more genes which are translated from a single mRNA molecule transcribed from a single promoter. Many bacterial genes are organised into operons but this form of gene organisation is rare or absent in eukaryotes.

optical density. *See* ABSORBANCE.

optical diffraction. The bending of light waves around the edges of an obstacle. Used to determine regular structures in electron micrographs as the amount of bending is inversely proportional to the repeat distance in a regular structure.

optical rotary dispersion. Optical rotation is the rotation of the plane of polarisation of plane-polarised light or of the major axis of the polarisation ellipse of elliptically polarised light by transmission through a substance or medium. Optical rotary dispersion (ORD) is the specific

rotation considered as a function of wavelength. Used in studying the structure and properties of nucleic acids which have characteristic ORD spectra.

oral vaccine. A vaccine applied by mouth, for example Sabin attenuated poliovirus vaccine.

orchid fleck virus. A possible plant *Rhabdovirus*; non-enveloped particles.
Doi, Y. *et al.* (1977) CMI/AAB Descriptions of Plant Viruses No. 183.

orcin, orcinol. A compound (3,5-dihydroxytoluene, $CH_3C_6H_3(OH)_2H_2O$) extracted from lichens which is used to test for pentoses, lignin, beet sugar, saccharoses, arabinose and diastase. It is specific in assaying for RNA as it does not react with DNA. *See* DIPHENYLAMINE.
Militzer, W.E. (1946) Archiv. Biochem. **9**, 85.

Oregon sockeye disease virus. Synonym: SOCKEYE SALMON VIRUS. Family *Rhabdoviridae*, genus unassigned. Causes necrosis of haematopoietic tissues of spleen and kidneys in trout and salmon. There is abdominal swelling and haemorrhages at base of fins. Occurs as epidemics in fish hatcheries. Can be grown in fat head minnow and other fish tissue cell lines. Closely related to infectious haematopoietic necrosis virus.

ORF. Abbreviation for OPEN READING FRAME.

Orf virus. Synonym: CONTAGIOUS PUSTULAR DERMATITIS OF SHEEP VIRUS, SCABBY MOUTH VIRUS, SORE MOUTH VIRUS. Family *Poxviridae*, subfamily *Chordopoxvirinae*, genus *Parapoxvirus*. Causes disease in lambs and kids. There are vesicles on the lips and nose, followed by pustules, ulcers and scabs. Disease can be fatal. Man can be infected by contact with animals. Cattle, horses, dogs, rabbits and monkeys can be infected experimentally. Virus can be grown in ovine and bovine cell cultures. The virus has a very unusual appearance in the electron microscope, resembling a ball of wool.

organ culture. A form of TISSUE CULTURE in which the organisation of the tissue is maintained.

Orgyia pseudotsugata nuclear polyhedrosis virus. The Douglas fir tussock moth, *Orgyia pseudotsugata*, is susceptible to infection by at least two distinct NPVs: an MNPV and an SNPV. The MNPV is more virulent for larvae of the homologous host and was registered for use as a biological control agent in 1976 by the US Forest Service under the name TM-Biocontrol-I. In 1983 the virus received temporary registration in Canada as 'Virtuss'. Although polyhedrin genes of the SNPV and MNPV strains are closely-related (76% nucleotide sequence homology) there is less than 1% total sequence homology between the viral DNAs. Limited homology (1%) has also been recorded between *O.pseudotsugata* MNPV (OpMNPV) and the prototype BACULOVIRUS *Autographa californica* MNPV (AcMNPV) though the overall organisation (colinearity) of the two viral genomes is similar. The repeated sequences (hr) observed in the AcMNPV genome have not been found in OpMNPV.

Leisy, D. *et al*. (1986) J. gen. Virol. **67**, 1073.

Oriboca virus. Family *Bunyaviridae*, genus *Bunyavirus*. Isolated from the opossum and rodents in Brazil, Surinam, French Guiana and Trinidad. Causes fever in man. Transmitted by mosquitoes.

Ornithogalum mosaic virus. A possible *Potyvirus*.
Francki, R.I.B. *et al*. (1985) **In** Atlas of Plant Viruses. Vol. 2. p. 183. CRC Press: Boca Raton, Florida.

Oropouche virus. Family *Bunyaviridae*, genus *Bunyavirus*. Isolated from a patient with fever in Trinidad. Has caused an epidemic of fever in Brazil, accompanied by myalgia, anthralgia and leucopenia. Transmitted by mosquitoes.

orphan virus. A virus which has not been identified as the cause of a disease, e.g. ECHO (ENTERIC CYTOPATHIC HUMAN ORPHAN) VIRUS, REO (RESPIRATORY ENTERIC ORPHAN) VIRUS.

Orthomyxoviridae. (Greek 'orthos' = correct, straight and 'myxa' = mucus.) A family of RNA viruses with pleomorphic particles, 80-120 nm. in diameter and filamentous forms up to several μm. in length. The particles have a lipid envelope and surface projections 8 nm. long. Nucleoproteins of

100nm

different size classes (50-130 nm. long) can be extracted from virus particles. There are eight proteins: haemagglutinin (HA) and neuraminidase (NA) are the surface projections and are glycosylated; the nucleoprotein (NP) is closely associated with the RNA and the matrix protein (M) is situated on the inside of the lipid bilayer; a non-structural protein (NS) and three polymerase proteins (P1, P2 and P3) are also present. The RNA is composed of eight segments of linear (–)-sense ssRNA, mw. 0.2-1.0 x 10^6.

Virus attaches to host cell receptors containing sialic acid and enters by fusion of virus envelope with plasma membrane. The RNA is transcribed to complementary RNA which functions as template for viral-specific protein synthesis and for replication of virus particle RNA.

There is a wide host range, including man, pigs, horses and birds. Other species are involved less frequently. The viruses can be grown in eggs or a variety of tissue cultures. Transmission is by contact or by aerosol and waterborne. There are three major serotypes, A, B and C.

Matthews, R.E.F. (1982) Intervirology **17**, 106.

Orthopoxvirus. (Greek 'orthos' = correct, straight.) A genus in the subfamily *Chordopoxvirinae* containing those viruses which infect mammals: buffalopox, camelpox, cowpox, ectromelia, monkeypox, rabbitpox, variola viruses.

Orungo virus. Family *Reoviridae*, genus *Orbivirus*. Isolated from man and mosquitoes in many countries in Africa. Causes fever, headache, conjunctivitis, myalgia, vomiting and rash in man. Antibodies found in man, other primates, sheep and cows in Nigeria. Can be passaged in mice i.c. Grows in BHK 21 cells.

Oryctes rhinoceros baculovirus. Type species of proposed subgroup C BACULOVIRUS genus, first discovered in the palm rhinoceros beetle (*O. rhinoceros*) in Malaysia. Virions are singly-enveloped nucleocapsids (220 x 120 nm.). The nucleocapsid carries a tail-like appendage (250 nm. long) when the envelope is removed. Particles sediment at 1640S, have a buoyant density in CsCl of 1.28 g/cc. (nucleocapsids; 1.47 g/cc.) and contain about 27 proteins (mw. 9.5-215 x 10^3) of which 14 are envelope proteins. The genome is a single molecule of circular supercoiled dsDNA (127 kbp) containing some repeated sequences. Virus replicates in the nuclei of larval fat body cells and in mid gut epithelium of larvae and adult beetles. In adults, infected gut epithelial cells proliferate, leading to the voiding of large quantities of virus in the faeces. It is generally believed that no occlusion bodies (*sensu stricto*) are formed. The virus has been successfully used as a self-perpetuating biological control agent for *O. rhinoceros* in many South Pacific countries by the introduction of infected adult beetles (*see* NON-OCCLUDED BACULOVIRUS).

Bedford, G.O. (1981) In Microbial Control of Pests and Plant Diseases 1970-1980. p. 409. ed. H.D. Burges. Academic Press: London.

Crawford, A.M. *et al.* (1985) J. gen. Virol. **66**, 2649.

Ossa virus.   Family *Bunyaviridae*, genus *Bunyavirus*. Isolated from man, rats and mosquitoes. Causes fever in man.

osteopetrosis virus.   Family *Retroviridae*, subfamily *Oncovirinae*, genus *Type C Oncovirus* group, sub-genus *Avian Type C Oncovirus*. Causes osteopetrosis in fowls, characterised by enlargement of the leg bones. It can be transmitted by inoculation of day old chicks or chick embryos or turkeys.

ouabain.   A glycoside obtained from the wood of *Acokanthera ouabio* or *A. schimperi* or from seeds of *Strophanthus gratus*. An inhibitor of Na$^+$- and K$^+$-dependent adenosine triphosphatase which reduces the rate of growth of cells and the synthesis of cellular proteins. Used as a selective agent in the production of HYBRIDOMAS.

Kozbov, D. *et al.* (1982) Proc. Natl. Acad. Sci. USA **79**, 6651.

Oubangui virus.   Unclassified arthropod-borne virus. Isolated from mosquitoes in Central African Republic.

Ouchterlony gel diffusion test.   Serological test in which ANTIBODY and ANTIGEN are placed in wells in a slab of agar and allowed to diffuse towards each other. A positive reaction shows as a band of precipitate between the antibody and antigen wells. Serological relationships can be detected by the interaction of the bands with each other. *See* SPUR FORMATION.

Ouchterlony, O. (1968) Handbook of Immunodiffusion and Immunoelectrophoresis. Prog. Allergy.   Ann   Arbor   Science   Publishers: Michigan.

OV.   Abbreviation sometimes used for 'occluded virus', referring to those particles which have been incorporated within proteinaceous occlusion bodies during some insect virus infections (e.g. some baculoviruses).

ovine adenovirus.   Family *Adenoviridae*, genus *Mastadenovirus*. Causes pneumoenteritis in lambs. Neutralising antibodies found in the majority of sheep. Multiplies in sheep kidney cell cultures.

ovine dependovirus.   Family *Parvoviridae*, genus *Dependovirus*.

ovine encephalomyelitis virus. Synonym: LOUP-ING-ILL   VIRUS,   OVINE   ADENO-ASSOCIATED VIRUS.

P

13p2 virus. Synonym: AMERICAN OYSTER REO-LIKE VIRUS.

P1 phage. Temperate tailed phage (*MYOVIRIDAE*) which does not integrate into the host chromosome. Instead the PROPHAGE resembles a plasmid and replicates autonomously at a low number of copies per cell.
Ritchie, D.A. (1983) **In** Topley and Wilson's Principles of Bacteriology, Virology and Immunity. Vol. 1. p. 177. ed. G. Wilson, A. Miles and M.T. Parker. Edward Arnold: London.

ø6 phage. Type species of CYSTOVIRUS genus (*CYSTOVIRIDAE*) isolated from *Pseudomonas phaseolicola* HB10Y. Virions are unique among phage isolates as they are composed of an isometric capsid (50 nm. in diameter) surrounded by a lipid-containing envelope and contain dsRNA. The intact particles are amorphous in shape, 60-70 nm. in diameter, sediment at 446S and have a buoyant density in CsCl of 1.27 g/cc. They contain at least ten polypeptides (mw. 6-82 x 10^3) and a genome composed of three segments of dsRNA (mw. 2.3, 3.1 and 5 x 10^6). The virion contains a transcriptase which produces complete ssRNA transcripts of the genome segments. Virus particles adsorb to pili of the host and infection results in lysis. The virus is the only described isolate within the *Cystoviridae*.
Mindich, L. (1978) **In** Comprehensive Virology. Vol. 12. p. 271. ed. H. Fraenkel-Conrat and R.R. Wagner. Plenum Press: New York.

øX phage group. Vernacular name for *MICROVIRUS* genus.

øX174 phage. MICROVIRUS type species infecting Enterobacteria. Particles are isometric, 27-30 nm. in diameter, with spikes at the vertices. They have a mw. of about 6.2 x 10^6, sediment at 114-120S and a buoyant density in CsCl of 1.4 g/cc. Virions contain 60 copies of a 46.4 x 10^3 mw. protein, 30-50 copies of a 4.2 x 10^3 mw. protein, as well as two protein species associated with the spikes (60 copies 19.1 x 10^3 mw. and 12 copies 35.8 x 10^3 mw.(H protein)). The genome is circular (+)-sense ssDNA (5386 nucleotides) and has been sequenced and genetically and physically mapped. It contains genetic information for at least ten different proteins. Five of these genes share their nucleotide sequence with another gene (overlapping genes), three of which read the base sequence in a different reading frame. The H protein facilitates adsorption to the host cell wall and is also required for DNA entry and the initiation of DNA replication. Production of replicative form (RF) DNA is primed by RNA and occurs during the first minute of infection. Only the minus strand of the RF is transcribed into mRNA. The RF is also used as the template for synthesis of further RF molecules by a semi-conservative process probably involving a rolling circle-like process without concatemeric DNA formation. Twenty minutes post infection, the RF gives rise to progeny ssDNA by a similar mechanism but because the (+)-sense (viral) strand is simultaneously encapsidated it cannot act as template for minus strand and RF formation. About 60 progeny RF molecules are produced per cell and about 500 progeny viral ssDNA strands. The replication cycle is complete within 25-30 minutes and the host cell is lysed.
Denhardt, D.T. (1977) **In** Comprehensive Virology. Vol. 7. p. 1. ed. H. Fraenkel-Conrat and R.R. Wagner. Plenum Press: New York.
Ritchie, D.A. (1983) **In** Topley and Wilson's Principles of Bacteriology, Virology and Immunity. Vol. 1. p. 177. ed. G. Wilson, A. Miles and M.T. Parker. Edward Arnold: London.
Weisbeek, P. (1984) **In** Genetic Maps 1984. Vol. 3. p. 22. ed. S.J. O'Brien. Cold Spring Harbor Laboratory: New York.

Pacora virus. Unclassified arthropod-borne virus. Isolated from mosquitoes in Panama.

Pacui virus. Family *Bunyaviridae*, genus

Phlebovirus. Isolated from rodents in Brazil and Trinidad.

Pahayokee virus. Family *Bunyaviridae*, genus *Bunyavirus.* Isolated from mosquitoes in Florida.

Palestina virus. Family *Bunyaviridae*, genus *Bunyavirus.*

palm mosaic virus. A possible *Potyvirus.*
Francki, R.I.B. *et al.* (1985) **In** Atlas of Plant Viruses. Vol. 2. p. 183. CRC Press: Boca Raton, Florida.

Palyam virus. Family *Reoviridae*, genus *Orbivirus.* Isolated from mosquitoes in India. Australia and Nigeria.

pangola stunt virus. A *Phytoreovirus.*
Milne, R.G. (1977) CMI/AAB Descriptions of Plant Viruses No. 175.
Francki, R.I.B. *et al.* (1985) **In** Atlas of Plant Viruses. Vol. 1. p. 47. CRC Press: Boca Raton, Florida.

Panicum mosaic virus. A possible *Sobemovirus.*
Niblett, C.L. *et al.* (1977) CMI/AAB Descriptions of Plant Viruses No. 177.
Hull, R. (1988) **In** The Plant Viruses Vol. 3. p. 113. ed. R. Koenig. Plenum Press: New York.

Panonychus 'baculovirus'. Unclassified viruses, morphologically similar to NON-OC-CLUDED BACULOVIRUSES have been observed in the mid gut epithelium of the European fruit tree red spider mite (*Panonychus ulmi*) and the citrus red mite (*Panonychus citri*). *P. citri* virus particles are enveloped (81 x 206 nm.) and contain a rod-shaped nucleocapsid (58 x 194 nm.). The virus occurs naturally in *P. citri* in California and Arizona where it significantly reduces mite populations.
Reed, D.K. (1981) **In** Microbial Control of Pests and Plant Diseases 1970-1980. p. 427. ed. H.D. Burges. Academic Press: London.

panzootic. Affecting many animals of different species over a wide area.

papaya mosaic virus. A *Potexvirus.*
Purcifull, D.E. and Hiebert, E. (1971) CMI/AAB Descriptions of Plant Viruses No. 56.
Francki, R.I.B. *et al.* (1985) **In** Atlas of Plant Viruses. Vol. 2. p. 159. CRC Press: Boca Raton, Florida.

papaya ringspot virus. A *Potyvirus.*
Purcifull, D.E. *et al.* (1984) CMI/AAB Descriptions of Plant Viruses No. 292.
Francki, R.I.B. *et al.* (1985) **In** Atlas of Plant Viruses. Vol. 2. p. 183. CRC Press: Boca Raton, Florida.

Papillomavirus. (Latin 'papilla' = nipple, pustule and Greek suffix '-oma'.) A genus in the family *Papovaviridae.* Consists of members isolated from man (nine types), cow (five types), deer, dog, goat, horse, rat and sheep.

Papovaviridae. (Sigla from PApilloma, POlyoma and VAcuolating agent (Latin 'papilla' = nipple, pustule; Greek suffix '-oma', used to form noun, denoting tumours and from Greek 'poly' = many.) A family of DNA viruses with non-enveloped icosahedral particles, 45-55 nm. in diameter which sediment at 240-300S and band in CsCl at 1.32 g/cc. There are 72 capsomeres in

    100nm

skew arrangement in the capsid, which consists of 5-7 proteins with mw. $10\text{-}75 \times 10^3$ and also contains host cell histones. Each capsid contains one molecule of ds circular DNA, mw. $3\text{-}5 \times 10^6$. Replication is in the nucleus. Expression is divided into early and late events and host cell histones are incorporated into virions during maturation in nucleus. Viruses of each genus share a common antigen but each virus has a distinct surface antigen. Viral DNA of polyomaviruses which integrates in cellular chromosomes of transformed cells. There are two genera, *PAPILLOMAVIRUS* and *POLYOMAVIRUS.*

Papillomaviruses cause papillomas in several species including man, cattle, dogs, rabbits, horses, monkeys, sheep, goats and birds and are generally species-specific. Polyomaviruses usually cause silent infection in their natural hosts but most are oncogenic when injected into new-born animals.
Matthews, R.E.F. (1982) Intervirology **17**, 62.

papular stomatitis of cattle virus. *See* BOVINE PUSTULAR STOMATITIS VIRUS.

parainfluenza virus. Family *Paramyxoviridae*, genus *Paramyxovirus.* There are several different parainfluenza viruses (types 1-5) of human, simian, bovine, canine, ovine and avian origin, causing febrile respiratory illness. Most can be grown in a variety of cell cultures.

parainfluenza virus type 1 murine. *See* SENDAI VIRUS.

Paramaribo virus. A strain of Venezuelan equine encephalomyelitis virus.

Paramushir virus. Unclassified arthropod-borne virus. Isolated from ticks in Eastern USSR. infesting guillemots and cormorants. Causes illness in suckling mice when injected i.c. Replicates in chick embryo fibroblasts and BHK cells.

Paramyxoviridae. (Greek 'para' = by the side of and 'myxa' = mucus.) A family of RNA viruses with three genera, PARAMYXOVIRUS, MORBILLIVIRUS (measles, rinderpest, canine distemper) and PNEUMOVIRUS (respiratory syncytial virus). The particles are pleomorphic, roughly spherical and are 150 nm. or greater in diameter. They have a lipid envelope with surface projections, 8 nm. long, and a nucleoprotein with well defined helical symmetry, 12-17 nm. in diameter and up to 1µm in length. The virus particle contains 5-7

100nm

proteins, mw. $35\text{-}200 \times 10^3$ and a variety of enzymes. The RNA is generally (–)-sense ssRNA, mw. $5\text{-}7 \times 10^6$ but some particles contain (+)-sense RNA. The virus enters the host cell by fusion of the viral envelope with the cell surface membrane. The RNA is transcribed to mRNAs on the nucleoprotein as the functional template. Replication of the RNA also occurs on nucleoprotein template. The host range of the family as a whole is wide but individual viruses may have narrow host ranges. Transmission is horizontal, mainly airborne.
Matthews, R.E.F. (1982) Intervirology **17**, 104.

Paramyxovirus. (Greek 'para' = by the side of and 'myxa' = mucus.) A genus in the family *Paramyxoviridae* containing members which infect man, birds, mice, cattle, sheep, monkeys and dogs.

Parana virus. Family *Arenaviridae*, genus *Arenavirus*. Isolated from rodents in Paraguay.

Parapoxvirus. (Greek 'para' = by the side of.) A genus in the subfamily *Chordopoxvirinae*. Consists of viruses of ungulates: bovine pustular stomatitis, chamois contagious ecthyma, milker's node and orf viruses.

paravaccinia virus. Synonym: MILKER'S NODE VIRUS.

Paroo River orbivirus. Family *Reovirus*, genus *Orbivirus*.

parrot paramyxovirus. Family *Paramyxoviridae*, genus *Paramyxovirus*.

Parry Creek rhabdovirus. Family *Rhabdoviridae*, unassigned to genus.

parsley latent virus. A probable plant *Rhabdovirus*, subgroup 1; aphid transmitted.
Francki, R.I.B. *et al.* (1985) **In** Atlas of Plant Viruses. Vol. 1. p. 73. CRC Press: Boca Raton, Florida.

parsley virus 5. A possible *Potexvirus*.
Francki, R.I.B. *et al.* (1985) **In** Atlas of Plant Viruses. Vol. 2. p. 159. CRC Press: Boca Raton, Florida.

parsnip mosaic virus. A *Potyvirus*.
Murant, A.F. (1972) CMI/AAB Descriptions of Plant Viruses No. 91.
Francki, R.I.B. *et al.* (1985) **In** Atlas of Plant Viruses. Vol. 2. p. 183. CRC Press: Boca Raton, Florida.

parsnip virus 3. A possible *Potexvirus*.
Francki, R.I.B. *et al.* (1985) **In** Atlas of Plant Viruses. Vol. 2. p. 159. CRC Press: Boca Raton, Florida.

parsnip yellow fleck virus. The parsnip strain is the type member of the *Parsnip Yellow Fleck Virus* group.
Murant, A.F. (1974) CMI/AAB Descriptions of Plant Viruses No. 129.
Francki, R.I.B. *et al.* (1985) **In** Atlas of Plant Viruses. Vol. 2. p. 235. CRC Press: Boca Raton, Florida.
Murant, A.F. (1988) **In** The Plant Viruses. Vol. 3. p. 273. ed. R. Koenig. Plenum Press: New York

parsnip yellow fleck virus group. (Named after type member). Genus of plant viruses with isometric particles about 30 nm. in diameter which sediment as two components, 152S (bottom (B) component containing 42% RNA) and 60S (top (T) component which lacks nucleic acid); CsCl banding densities are 1.52 g/cc for B and 1.30 g/cc for T component. The capsids are made up of

100nm

three major polypeptide species (mw. 31, 26 and 22.5 x 10^3). B component particles each contain one molecule of linear (+)-sense ssRNA of mw. 3.5 x 10^6. Replication is thought to occur in large cytoplasmic inclusion bodies adjacent to the nucleus. The host ranges are narrow. Viruses in this group are mechanically transmissible. In nature they are transmitted by aphids in the SEMI-PERSISTENT TRANSMISSION manner only in association with a helper virus.

Murant, A.F. (1988) In The Plant Viruses. Vol. 3. p. 273. ed. R. Koenig. Plenum Press: New York.

partial specific volume. The volume displaced when 1g solute (or 1kg in SI units) is added to an infinite volume of solvent. The symbol is v* which is formally given as:-

$$\bar{v} = (\frac{\delta V}{\delta g}) T_\rho$$

where V is volume of solution, g is mass of solute added at constant temperature and pressure. Measured in a pycnometer at constant pressure and temperature or can be estimated approximately if the % RNA of the virus is known. This estimate uses a $\bar{v}$ for RNA of 0.55 cm^{3}g^{-1} and for protein approximately 0.74 cm^{3}g^{-1} ; if the amino acid composition is known, the protein value can be derived more accurately from the weighted sum of $\bar{v}$ of individual amino acids *(see* AMINO ACID). $\bar{v}$ is also used in the calculation of molecular weight using the SVEDBERG EQUATION.

partition coefficient. The constant ratio of the concentration of a solute in the upper phase to its concentration in the lower phase when the solute is in equilibrium distribution between two liquid phases.

Partitiviridae. (from Latin 'partitus' = divided). Family of fungal viruses with divided dsRNA

genomes containing two genera, the PARTITIVIRUS GROUP and the PENICILLIUM CHRYSOGENUM VIRUS GROUP.

Partitivirus group. (Type member *Gaeumannomyces graminis* virus 019/6-A). Genus of PARTITIVIRIDAE with bipartite dsRNA genome. Formed by the amalgamation of the *Gaeumannomyces graminis* virus groups I and II and the *Penicillium stoloniferum* virus group. The particles are isometric, 30-35 nm. in diameter, sediment at 101-145S and band in CsCl at 1.35-1.36 g/cc. The CAPSIDS are composed of a single coat protein species (mw. 42-73 x 10^3). The genome comprises two species of dsRNA (mw. 0.9-1.6 x 10^6 for each segment) encapsidated in separate particles. One RNA species encodes the coat protein, the other a polypeptide which is probably the virus polymerase.

Buck, K.W. (1986) In Fungal Virology. p. 1. ed. K.W. Buck. CRC Press: Boca Raton, Florida.

parts per million (p.p.m). Concentration of a solute in a solvent. It is equivalent to μg solute per g solvent.

Parvoviridae. (Latin 'parvus' = small.) A family of unenveloped DNA viruses comprising three genera, PARVOVIRUS, DEPENDOVIRUS and DENSOVIRUS. They have isometric particles 18-26 nm. in diameter, possessing icosahedral symmetry (and probably 32 capsomeres), sedimenting at 110-122S and banding at 1.39-1.42 g/cc in CsCl. The

100nm

particles have a core with a diameter of 14-17 nm. The capsid consists of three proteins and surrounds a single molecule of ssDNA, mw. 1.5-2.0 x 10^6. In some members the single strands are complementary and after extraction these form dsDNA. Replication is in the nucleus. One or more cellular functions are necessary for replication. Members of *Parvovirus* and *Dependovirus* genera have a wide host range but the host range of the insect virus genus (*Densovirus*) is restricted to *Lepidoptera* and probably *Diptera* and *Orthoptera*.

Matthews, R.E.F. (1982) Intervirology **17**, 72.

Parvovirus. (Latin 'parvus' = small.) A genus in the family *Parvoviridae*. Contains viruses infecting many different species. The most important are those infecting cats (FELINE PANLEUCOPENIA VIRUS), mink (MINK ENTERITIS VIRUS) and dogs (CANINE PARVOVIRUS). Many molecular studies made on latent rat virus and minute virus of mice virus.

Paspalum striate mosaic virus. A possible *Geminivirus*.
Francki, R.I.B. *et al.* (1985) **In** Atlas of Plant Viruses. Vol. 1. p. 33. CRC Press: Boca Raton, Florida.

passage. The infection of a host with a virus or a mixture of viruses and the subsequent recovery of the virus from that host (usually after one infection cycle). It is a procedure by which for example, a specific virus can be separated from a mixture of viruses or (through a series of infection cycles) a virus can be adapted to grow well in a host in which it originally grew poorly.

passiflora latent virus. A *Carlavirus*.
Francki, R.I.B. *et al.* (1985) **In** Atlas of Plant Viruses. Vol. 2. p. 173. CRC Press: Boca Raton, Florida.

passiflora yellow mosaic virus. A possible *Tymovirus*.
Hirth, L. and Girard, L. (1988) **In** The Plant Viruses. Vol. 3. p. 163. ed. R. Koenig. Plenum Press: New York.

passionfruit ringspot virus. A possible *Potyvirus*.
Francki, R.I.B. *et al.* (1985) **In** Atlas of Plant Viruses. Vol. 2. p. 183. CRC Press: Boca Raton, Florida.

passionfruit woodiness virus. A *Potyvirus*.
Taylor, R.H. and Greber, R.S. (1973) CMI/AAB Descriptions of Plant Viruses No. 122.
Francki, R.I.B. *et al.* (1985) **In** Atlas of Plant Viruses. Vol. 2. p. 183. CRC Press: Boca Raton, Florida.

passive haemagglutination. A serological test which can be used to detect a virus-specific ANTI-BODY by coating red blood cells with viral antigen. If viral antibody is present in test samples, the red blood cells will agglutinate. *See* REVERSE PASSIVE HAEMAGGLUTINATION.
Herbert, W.J. (1967) **In** Handbook of Experimental Immunology. p. 720. ed. D.M. Weir. Blackwell Scientific Publications: Oxford.

Pata virus. Family *Reoviridae*, genus *Orbivirus*. Isolated from mosquitoes in Central African Republic.

Patas virus. *See* CERCOPITHECID HERPESVIRUS 5.

patchouli (Pogostemon patchouli) mottle virus. A possible plant *Rhabdovirus*.
Francki, R.I.B. *et al.* (1985) **In** Atlas of Plant Viruses. Vol. 1. p. 73. CRC Press: Boca Raton, Florida.

pathogen. An organism or virus which causes a disease.

pathogenesis related (PR) proteins Proteins which accumulate intracellularly in plant tissues in reaction to the hypersensitive response to viral or fungal infection or to certain chemical treatments.
Bol, J.F. *et al.* (1987) **In** Plant Resistance to Viruses. p. 72. Ciba Foundation Symposium No. 133. Wiley: Chichester.

pathogenicity. The ability of a pathogen to cause disease.

pathology. The study of diseases. For history *see* Dictionary of the History of Science (1981) ed. W.F. Bynum *et al.* Macmillan Press Ltd.: London.

Pathum Thani virus. Family *Bunyaviridae*, genus *Nairovirus*. Isolated from a tick in Thailand.

Patois virus. Family *Bunyaviridae*, genus *Bunyavirus*. Isolated from cotton rats and mosquitoes in Panama and Mexico.

PB viruses. Viruses isolated from the fungus *Penicillium brevicompactum* and other *Penicillium* spp.; reported also to grow in bacteria (e.g. *E. coli*). They include phage-like viruses, morphologically similar to members of the *MYOVIRIDAE*, *SIPHOVIRIDAE*, *PODOVIRIDAE* and *MICROVIRIDAE*.
Tikchonenko, T.I. (1978) **In** Comprehensive Virology. Vol. 12. p. 235. ed. H. Fraenkel-Conrat and R.R. Wagner. Plenum Press: New York.

PBCV-1 A virus from the green *Chlorella*-like alga *Paramecium bursaria*. It has isometric particles 150-190 nm. in diameter with the capsid comprising many protein species (mw. 10-135 x 10^3) containing dsDNA of about 330kbp. Particles

contain a lipid component essential for infectivity.
van Etten, J.L. *et al.* (1985) Virology **140**, 135.

PBS.   Abbreviation for PHOSPHATE-BUFFERED SALINE.

pea early browning virus. A *Tobravirus*.
Harrison, B.D. (1973) CMI/AAB Descriptions of Plant Viruses No. 120.
Francki, R.I.B. *et al.* (1985) **In** Atlas of Plant Viruses. Vol. 2. p. 147. CRC Press: Boca Raton, Florida.

pea enation mosaic virus. Type member of the *Pea Enation Mosaic Virus* group.
Peters, D. (1982) CMI/AAB Descriptions of Plant Viruses No. 257.

Pea enation mosaic virus group. (Named after the type virus). Monotypic genus comprising a MULTICOMPONENT plant virus with isometric particles, 28 nm. in diameter, which sediment as two components, bottom (B) 112S and middle (M) 99S but form one band in CsCl (after fixation) at 1.42 g/cc. The capsid of B component has 180

    ___100nm___

subunits of coat protein mw. 22 x 10^3, that of M component 150 subunits. The components differ in stability, the particles of B being relatively stable, those of M being disrupted in neutral chloride salts. The genome comprises two species of linear, (+)-sense ssRNA, B component containing RNA-1 (mw. 1.7 x 10^6) and M component RNA-2 (mw. 1.3 x 10^6). RNA-1 contains the information for coat protein. The virus is transmitted by aphids in the PERSISTENT TRANSMISSION manner. It is readily mechanically transmitted but this often leads to loss of aphid transmissibility.
Matthews, R.E.F. (1982) Intervirology **17**, 166.
Peters, D. (1982) CMI/AAB Descriptions of Plant Viruses No. 257.
Francki, R.I.B. *et al.* (1985) **In** Atlas of Plant Viruses. Vol. 2, p. 39. CRC Press: Boca Raton, Florida.

pea leaf roll virus. *See* BEAN LEAF ROLL VIRUS.

pea mild mosaic virus. A *Comovirus*.
Francki, R.I.B. *et al.* (1985) **In** Atlas of Plant Viruses. Vol. 2. p. 1. CRC Press: Boca Raton, Florida.

pea mosaic virus. A *Potyvirus*, closely related to *Bean Yellow Mosaic* and *Bean Common Mosaic Viruses*.
Taylor, R.H. and Smith, P.R. (1968) Aust. J. Biol. Sci. **21**, 429.

pea necrosis virus. *See* CLOVER YELLOW VEIN VIRUS.

pea seed-borne mosaic virus. A *Potyvirus*.
Hampton, R.O. and Mink, G.I. (1975) CMI/AAB Descriptions of Plant Viruses No. 146.
Francki, R.I.B. *et al.* (1985) **In** Atlas of Plant Viruses. Vol. 2. p. 183. CRC Press: Boca Raton, Florida.

pea streak virus. A *Carlavirus*. Synonym: ALFALFA LATENT VIRUS.
Bos, L. (1973) CMI/AAB Descriptions of Plant Viruses No. 112.
Francki, R.I.B. *et al.* (1985) **In** Atlas of Plant Viruses. Vol. 2. p. 173. CRC Press: Boca Raton, Florida.

peach enation virus. An unclassified plant virus with isometric particles; occurs in Japan.
Doi, Y. *Personal communication.*

peach rosette mosaic virus. A *Nepovirus*.
Dias, H.F. (1975) CMI/AAB Descriptions of Plant Viruses No. 150.
Francki, R.I.B. *et al.* (1985) **In** Atlas of Plant Viruses. Vol. 2. p. 23. CRC Press: Boca Raton, Florida.

peach yellow leaf virus. A *Closterovirus*, occurs in Japan.
Doi, Y. *Personal communication.*

peach yellow mosaic virus. An unclassified plant virus with isometric particles; occurs in Japan.
Doi, Y. *Personal communication.*

peanut chlorotic streak virus. A possible *Caulimovirus*.
Iizuka, N. and Reddy, D.V.R. (1986) Tech. Bull. Trop. Ag. Res. Center, Japan, **21**, 164.

peanut clump virus. A *Furovirus*.
Thouvenal, J-C. and Fauquet, C. (1981) CMI/AAB Descriptions of Plant Viruses No. 235.
Brunt, A.A. and Shikata, E. (1986) **In** The Plant Viruses. Vol. 2. p. 305. ed. M.H.V. van Regenmortel and H. Fraenkel-Conrat. Plenum

Press: New York.

peanut green mosaic virus. A *Potyvirus.*
Sreenivasulu, P. *et al.* (1981) Ann. Appl. Biol. **98**, 255.

peanut mottle virus. A *Potyvirus.*
Bock, K.R. and Kuhn, C.W. (1975) CMI/AAB Descriptions of Plant Viruses No. 141.
Francki, R.I.B. *et al.* (1985) **In** Atlas of Plant Viruses. Vol. 2. p. 183. CRC Press: Boca Raton, Florida.

peanut stripe virus. A *Potyvirus.*
Denski, J.W. and Lovell, G.R. (1985) Plant Disease **69**, 734.

peanut stunt virus. A *Cucumovirus.*
Mink, G.I. (1972) CMI/AAB Descriptions of Plant Viruses No. 92.
Francki, R.I.B. (1985) **In** The Plant Viruses. Vol. 1. p. 1. ed. R.I.B. Francki. Plenum Press: New York.

peanut yellow mottle virus. A *Tymovirus.*
Francki, R.I.B. *et al.* (1985) **In** Atlas of Plant Viruses. Vol. 1. p. 117. CRC Press: Boca Raton, Florida.
Hirth, L. and Girard, L. (1988) **In** The Plant Viruses. Vol. 3. p. 163. ed. R. Koenig. Plenum Press: New York.

pear ringspot virus. An *Ilarvirus*, occurs in Japan.
Doi, Y. *Personal communication.*

peaton virus. Family *Bunyaviridae*, genus *Bunyavirus.*

Pecicilis mosaic virus. A *Potyvirus*, occurs in Japan.
Doi, Y. *Personal communication.*

PEG. Abbreviation for POLYETHYLENE GLYCOL.

pelargonium flower-break virus. A member of the *Carmovirus* group.
Hollings, M. and Stone, O.M. (1974) CMI/AAB Descriptions of Plant Viruses No. 130.
Francki, R.I.B. *et al.* (1985) **In** Atlas of Plant Viruses. Vol. 2. p. 235. CRC Press: Boca Raton, Florida.
Morris, T.J. and Carrington, J.C. (1988) **In** The Plant Viruses. Vol. 3. p. 73. ed. R. Koenig. Plenum Press: New York.

pelargonium leafcurl virus. A *Tombusvirus.*
Francki, R.I.B. *et al.* (1985) **In** Atlas of Plant Viruses. Vol. 1. p. 181. CRC Press: Boca Raton, Florida.
Martelli, G.P. *et al.* (1988) **In** The Plant Viruses. Vol. 3. p. 13. ed. R. Koenig. Plenum Press: New York.

pelargonium line pattern virus. A member of the *Carmovirus* group.
Morris, T.J. and Carrington, J.C. (1988) **In** The Plant Viruses. Vol. 3. p. 73. ed. R. Koenig. Plenum Press: New York.

pelargonium vein clearing virus. A plant *Rhabdovirus*, subgroup 2.
Francki, R.I.B. *et al.* (1985) **In** Atlas of Plant Viruses. Vol. 1. p. 73. CRC Press: Boca Raton, Florida.

pelargonium zonate spot virus. An unclassified plant virus with isometric particles, 25-30 nm. in diameter, which sediment as three components (118, 90 and 80S) and band in CsCl at 1.35 g/cc. The capsids are composed of a single coat protein species (mw. 23×10^3) and contain two RNA species (mw. 1.25 and 0.95×10^6).
Gallitelli, D. *et al.* (1983) CMI/AAB Descriptions of Plant Viruses No. 272.

pellet. The material concentrated at the bottom of a centrifuge tube after centrifugation. *See* SUPERNATANT.

Penaeus 'baculovirus'. An occluded baculovirus (NUCLEAR POLYHEDROSIS VIRUS) isolated from the pink shrimp, *Penaeus duorarum* in the Gulf of Mexico. Virions consist of a single nucleocapsid (270 x 50 nm.) surrounded by an envelope. Particles are occluded within pyramidal-shaped occlusion bodies 0.5-20 μm in size. The polyhedrin has a mw. of 53×10^3 and cross-reacts immunologically with NPV polyhedrin from Lepidoptera, despite being almost twice the mw. Another occluded NPV which produces amorphous occlusion bodies has been detected in *P. monodon*. A putative NON-OCCLUDED BACULOVIRUS (310 x 72 nm.) was isolated from nuclei of infected mid gut tissue of *Penaeus japonicus* in Japan. The virus is a serious pathogen in intensive shrimp farming.
Couch, J.A. (1974) J. Invertebr. Pathol. **24**, 311.
Rohrmann, G.F. (1986) **In** The Biology of Baculoviruses. Vol. I. p. 203. ed. R.R. Granados and B.A. Federici. CRC Press: Boca Raton,

Florida.
Sano, T. *et al.* (1981). Abstr. 5th Internat. Congr. Virol. p. 297.

penetration. The second step in the initiation of infection in which the virus particle penetrates the cell surface. For viruses of animals it is an energy-dependent step and involves either translocation of the entire virus particle across the plasma membrane (e.g. POLIOVIRUS), PINOCYTOSIS or fusion of the virion envelope with the plasma membrane (e.g. ORTHO- and PARA-MYXOVIRUSES). Plant viruses probably enter through sites of mechanical damage. Some phages actively inject their nucleic acid through the host cell wall.

Penicillium brevicompactum virus. (PbV). A member of the *Penicillium chrysogenum virus group*.

Penicillium chrysogenum virus (PcV). Type member of the *Penicillium chrysogenum virus group*.

Penicillium chrysogenum virus group. (Named after the type member). A possible genus of the *PARTITIVIRIDAE*. Particles are isometric, 35-40 nm. in diameter and sediment at 145-150S. The capsid is composed of a single coat protein species (mw. 125 x 10³). Three dsRNA species with mw. in the range 1.9-2.4 x 10⁶ are encapsidated separately. The number of RNA segments required for replication is not known.
Buck, K.W. (1986) In Fungal Virology. p. 1. ed. K.W. Buck. CRC Press: Boca Raton, Florida

Penicillium cyaneo-fulvum virus (Pc-fV). A member of the *Penicillium chrysogenum virus group*.

Penicillium stoloniferum virus F (PsV-F). Possible member of the *Partitivirus group*.

Penicillium stoloniferum virus S (PsV-S). Member of the *Partitivirus group*.

pentamer. A group of five protein subunits which in the virus capsid form the vertices of an ICOSAHEDRON. *See* HEXAMER.

penton. Complex structures in the coat of ADENO-VIRUSES comprising a PENTAMER base, a fibre and a knob.

Pepino latent virus. A *Carlavirus*.

Francki, R.I.B. *et al.* (1985) In Atlas of Plant Viruses. Vol. 2. p. 173. CRC Press: Boca Raton. Florida.

Pepino mosaic virus. A *Potexvirus*.
Francki, R.I.B. *et al.* (1985) In Atlas of Plant Viruses. Vol. 2. p. 159. CRC Press: Boca Raton. Florida.

peplomer. Large, well-separated, petal- or knob-shaped glycoprotein SPIKES found, e.g. on the surface of CORONAVIRUS particles. They serve as antireceptors which bind the virion to the cell-surface receptors.

pepper mild mosaic virus. A *Potyvirus*.
Francki, R.I.B. *et al.* (1985) In Atlas of Plant Viruses. Vol. 2. p. 183. CRC Press: Boca Raton. Florida.

pepper mild mottle virus. A *Tobamovirus*.
Brunt, A.A. (1986) In The Plant Viruses. Vol. 2 p. 283. ed. M.H.V. van Regenmortel and H. Fraenkel-Conrat. Plenum Press: New York.

pepper mottle virus. A *Potyvirus*.
Nelson, M.R. and Zitter, T.A. (1982) CMI/AAB Descriptions of Plant Viruses No. 253.
Francki, R.I.B. *et al.* (1985) In Atlas of Plant Viruses. Vol. 2. p. 183. CRC Press: Boca Raton. Florida.

pepper ringspot virus. A *Tobravirus*, originally known as the CAM (Campina) strain of TOBACCO RATTLE VIRUS.
Harrison, B.D. and Robinson, D.J. (1986) In The Plant Viruses. Vol. 2. p. 339. ed. M.H.V. van Regenmortel and H. Fraenkel-Conrat. Plenum Press: New York.

pepper severe mosaic virus. A *Potyvirus*.
Francki, R.I.B. *et al.* (1985) In Atlas of Plant Viruses. Vol. 2. p. 183. CRC Press: Boca Raton. Florida.

pepper veinal mottle virus. A *Potyvirus*.
Brunt, A.A. and Kenton, R.H. (1972) CMI/AAB Descriptions of Plant Viruses No. 104.
Francki, R.I.B. *et al.* (1985) In Atlas of Plant Viruses. Vol. 2. p. 183. CRC Press: Boca Raton. Florida.

peptide. A compound of two or more amino acids joined by peptide bonds.

peptide bond. A chemical link formed by the reaction between the carboxylic acid group of one amino acid and the amino group of another thus uniting the carbon atom of one with the nitrogen atom of the other. *See* AMINO ACID for figure.

per os. (Latin = by, through the mouth). Usually refers to the method of transmission of a virus.

Perilla mottle virus. A *Potyvirus*.
Francki, R.I.B. *et al.* (1985) **In** Atlas of Plant Viruses. Vol. 2. p. 183. CRC Press: Boca Raton, Florida.

perinuclear. Situated in the region between the two membranes of the nucleus or close to the outer membrane. Certain RHABDOVIRUSES which are thought to bud through one nuclear membrane accumulate in the perinuclear space; members of the REOVIRIDAE replicate in the cytoplasm near the outer nuclear membrane.

periodic acid-Schiff's (PAS) reagent. A sensitive stain for glycoproteins. Frequently used for the detection of glycoproteins after SDS-gel electrophoresis.
Gordon, A.H. (1975) Electrophoresis of Proteins in Polyacrylamide and Starch Gels. North Holland/American Elsevier: Amsterdam.

permissive cells. Cells in which infection results in the production of infectious progeny virus.

Peromyscus virus. Family *Paramyxoviridae*, genus *Paramyxovirus*. Isolated from white-footed mice in Virginia, USA Kills suckling mice and hamsters injected i.c. and i.p. Grows in embryonated eggs and many different cell cultures.

peroxidase. An enzyme that catalyses reactions in which hydrogen peroxide is an electron acceptor. Used as a reporter molecule in diagnostic techniques such as ELISA.

persistence. The capacity of a virus to persist in a host for an extended time without inducing lysis.

persistent infection. An infection in which the virus does not induce lysis but reaches a form of equilibrium with the host. It is often associated with integration of the viral genome into the host genome. Three types of persistence have been distinguished: 1) the virus persists within the cell

but at no time is shed to the outside. This is rare as it is an evolutionary dead-end for the virus; it has been shown for VISNAVIRUS in sheep and SSPE (MEASLES VIRUS) in humans; 2) the virus is shed to the outside sporadically, e.g. as a result of reactivation of HUMAN PAPOVAVIRUS in the kidney; 3) the virus is continually shed to the outside but with no lysis, e.g. LYMPHOCYTIC CHORIOMENINGITIS. *See* LATENT INFECTIONS.
Virus Persistence (1982) ed. B.W.J. Mahy *et al.* Cambridge University Press.

persistent transmission. The relationship between a virus or plant pathogenic mycoplasm and its arthropod (or nematode) vector which is characterised by the need for long acquisition feeds, frequently a latent period before the vector is able to transmit and then the vector being able to transmit for long periods (days to weeks). The virus has to pass through the vector's gut wall to the haemocoel and then accumulates in the salivary glands where it may or may not multiply. Also termed circulative transmission. *See* NON-PERSISTENT TRANSMISSION, SEMIPERSISTENT TRANSMISSION.

Peru tomato virus. Synonym: TOMATO (PERU) VIRUS.

peste des petits ruminants. Synonyms: KATA VIRUS, PSEUDORINDERPEST VIRUS. Family *Paramyxoviridae*, genus *Morbillivirus*. Causes a disease in sheep and goats in West Africa similar to rinderpest in cattle. However, cattle are not affected by this virus. There is pyrexia with nasal and ocular discharge, necrotic stomatitis, severe enteritis and pneumonia. There is high mortality. Replicates in a variety of cell cultures. Closely related structurally and antigenically to canine distemper, measles and rinderpest viruses. In fact, rinderpest vaccine will protect against the disease and the virus grows well enough in cattle to protect them against rinderpest.

Pestivirus. (Latin 'pestis' = plague.) A genus in the family *Togaviridae*. Contains several important disease agents including those causing mucosal disease of cattle, swine fever (hog cholera) and Border disease of sheep.

pestivirus diarrhoea virus. Synonym: BOVINE DIARRHOEA VIRUS, MUCOSAL DISEASE VIRUS.

Petri dish. A shallow glass or plastic dish with a loosely fitting overlapping cover used for bacterial plate cultures and plant and animal tissue cultures. Named after Richard J. Petri (assistant to Dr. R. Koch) who invented the dish in 1887.

petunia asteroid mosaic virus. A *Tombusvirus.*
Francki, R.I.B. *et al.* (1985) **In** Atlas of Plant Viruses. Vol. 1. p. 181. CRC Press: Boca Raton, Florida.
Martelli, G.P. *et al.* (1988) **In** The Plant Viruses. Vol. 3. p. 13. ed. R. Koenig. Plenum Press: New York.

petunia vein clearing virus.   A possible *Caulimovirus.*
Francki, R.I.B. *et al.* (1985) **In** Atlas of Plant Viruses. Vol. 1. p. 17. CRC Press: Boca Raton, Florida.

phage. A general term used for viruses isolated from prokaryotes including bacteria, blue-green algae (cyanobacteria) and mollicutes (mycoplasma and spiroplasma). The viruses from these different host groups are termed bacteriophages, cyanophages and mycoplasmaphages respectively. The phages include a highly heterogeneous range of viruses which fall into four main morphological groups: tailed, cubic, rod-shaped (filamentous) and pleomorphic (*see* figure). These groups are further subdivided according to morphological and biochemical properties of the viruses as follows:

TAILED PHAGES:

Morphotypes A1 to A3:	Tails long and contractile.	*Myoviridae* (type species T2 PHAGE)
Morphotypes B1 to B3:	Tails long and non-contractile.	*Siphoviridae* (type species λ PHAGE)
Morphotypes C1 to C3:	Tails short.	*Podoviridae* (type species T7 PHAGE)

CUBIC PHAGES: (Phages with cubic symmetry)

Morphotype D1:	Small unenveloped phages containing ssDNA	*Microviridae* (type species φx174 PHAGE)
Morphotype D2:	Large unenveloped phages containing dsDNA.	Unclassified (e.g. SiI PHAGE)
Morphotype D3:	Large phages containing dsDNA and an internal lipid layer	*Corticoviridae* (type species PM2 PHAGE)
Morphotype D4:	Large phages containing dsDNA; double capsid structure and internal lipid.	*Tectiviridae* (type species PRD1 PHAGE)
Morphotype E1:	Small naked phages containing ssRNA.	*Leviviridae* type species MS2 PHAGE)
Morphotype E2:	Large isometric phages with segmented dsRNA genome and lipid envelope.	*Cystoviridae* (type φ6 PHAGE)

ROD-SHAPED OR FILAMENTOUS PHAGES (*Inoviridae*)

Morphotype F1:	Long flexible rods containing ssDNA.	INOVIRUS (type species fd PHAGE)
Morphotype F2:	Short straight rods containing ssDNA.	PLECTROVIRUS (type species MVL51 PHAGE)
Morphotype F3:	Enveloped rigid rods of variable length containing dsDNA.	Unclassified (e.g.TTV1 PHAGE)

PLEOMORPHIC PHAGES (*Plasmaviridae*)

Morphotype G:	Rounded phages with flexible envelope.	PLASMAVIRUS (type species L2 PHAGE)

Particles not to scale

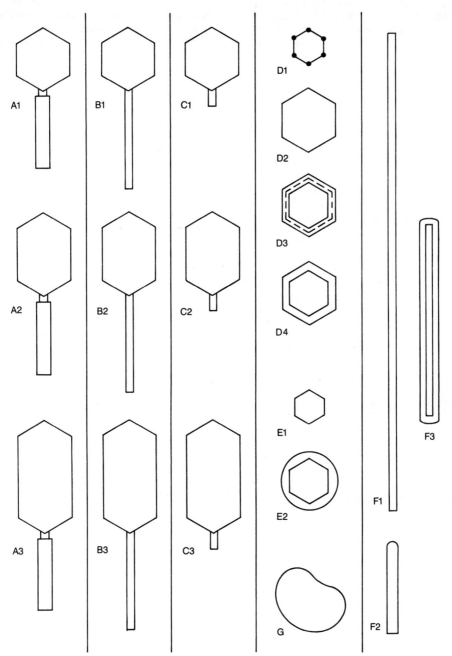

A given phage type infects a limited range of susceptible hosts and the host range usually follows the established taxonomic divisions among prokaryotes. While some phages attack only one or a few closely-related host strains, others may infect an entire species, or several related species and occasionally genera (e.g. among the enterobacteria). Where a phage adsorbs to a pilus rather than the cell wall, the susceptibility of the cell is governed by the availability of pili. Some pili are determined by plasmid-coded functions (e.g. F-plasmid codes for F pilus). Where these plasmids have a broad host-range, the phage host range is equally broad.

Unfortunately, there has been a marked tendency to give each new phage isolate a name regardless of the properties of the virus. At present there is no widespread agreement on nomenclature and this has led to a proliferation of names including many identical names for different phages. Descriptions of phage families, genera and type species are given in alphabetic order in Appendix F which includes approximately 2000 phage isolates from bacteria, cyanobacteria and mycoplasma that have been studied in detail, and/or listed in recent attempts to classify the viruses.
Ackermann, H.W. *et al.* (1978) Adv. Virus Res. **23**, 2.
Reanney, D.C. and Ackermann, H.W. (1982) Adv. Virus Res. **27**, 206.
Ackermann, H.W. and Du Bow, M.S. (1987) Viruses of Prokaryotes. CRC Press: Boca Raton, Florida.

phage typing. Subdivision of bacteria into 'phage types' according to their susceptibility to host-range specific phages. Phage typing is a useful epidemiological tool for tracing the origin of infection in epidemics by further subdividing the biotypes or serotypes of the bacterial hosts, e.g. in outbreaks of food poisoning by Staphylococci, *Salmonella* etc.
Kasatiya, S.S. and Nicolle, P. (1978) In Handbook of Microbiology. Vol. 2. p. 699. ed. Laskin, A.I. and Lechevalier, M.E. CRC Press: Boca Raton, Florida.

phagocyte. An amoeboid cell that engulfs foreign material. Also known as macrophage. They are part of the defence system of vertebrates against infection.

phagocytosis. A mechanism by which certain animal cells such as protozoans and phagocytes engulf and carry particles into the cytoplasm. One

of the mechanisms by which viruses can enter cells.

Phalaenopsis chlorotic spot virus. A possible plant *Rhabdovirus;* nonenveloped particles.
Francki, R.I.B. et al. (1985) In Atlas of Plant Viruses. Vol. 1. p. 73. CRC Press: Boca Raton, Florida.

phase-contrast microscope. A compound microscope that has an annular diaphragm in the front focal plane of the substage condenser and a phase plate at the rear focal plane of the objective to make visible, differences in phase or optical path in transparent or reflecting media.

phenotype. The outward, observable characteristics of an organism determined by its genotype but modified by the environment. For history see Dictionary of the History of Science (1981) ed. W.F. Bynum et al. Macmillan Press Ltd.: London.

phenotypic mixing. Synonym: genomic masking. A process by which an individual progeny from a mixed viral infection contains structural proteins derived from both viruses or has the genome of one virus encapsidated in the structural proteins of the other; this latter situation is termed genomic masking or transcapsidation.
Boettiger, D. (1979) Prog. med. Virol. 25, 37.
Zavada, J. (1976) Arch. Virol. 50, 1.

Phialophora sp. with lobed hyphobia (Phialophora radicola var. radicola) virus 2-2-A. (PrV-2-2-A). A probable member of the *Partitivirus group.*

phlebotomus fever virus. Synonym: sandfly FEVER VIRUS, HUNDSKRANKHEIT VIRUS. Family *Bunyaviridae*, genus *Phlebovirus*. Causes fever in man with pain in the eyes, head, back and limbs. There is also gastrointestinal involvement and leucopenia. Isolated in several parts of the world including Italy, Egypt, Iran and Pakistan. The sandfly is the vector. Grows in human, mouse and hamster kidney cell cultures. Occurs as at least two distinct antigenic types.

Phlebovirus. (Phlebotomine vectors: from Greek 'phlebos' = vein.) A genus in the family *Bunyaviridae* containing many viruses, of which the best known are RIFT VALLEY FEVER VIRUS and SANDFLY FEVER VIRUS.

phleum mottle virus. Synonym: COCKSFOOT MILD MOSAIC VIRUS.
Hull, R. (1988) In The Plant Viruses. Vol. 3. p. 113. ed. R. Koenig. Plenum Press: New York.

Phnom-Penh bat virus. Family *Flaviviridae*, genus *Flavivirus*. Isolated from bats in Kampuchia.

phosphatase. An enzyme that catalyses the hydrolysis and synthesis of phosphoric acid esters and the transfer of phosphate groups from phosphoric acid to other compounds. Used in various gene cloning techniques to remove the 5'-terminal phosphate from a nucleic acid.

phosphate-buffered saline. A solution used in cell culture and serology. It consists of 8.00g NaCl, 0.20g KCl, 0.20g KH_2PO_4 0.15g Na_2HPO_4 per litre (pH 7.2).

phosphodiester bond. Link formed between the nucleotides of polynucleotide chains by covalent bonding of the phosphoric acid with the 3'-hydroxyl group of one ribose or deoxyribose molecule and the 5'-hydroxyl group of the next ribose or deoxyribose ring. *See* NUCLEIC ACID.

phospholipase. An enzyme that catalyses the hydrolysis of a phospholipid, e.g. lecithinase that acts on lecithin. Used in the study of viral membranes.

phosphoprotein. A protein that has had one or more amino acids phosphorylated by a PROTEIN KINASE. The amino acids most commonly phosphorylated are serine, threonine and tyrosine. Examples of phosphoproteins are the GAG GENE PRODUCT in many RETROVIRUSES and the coat protein of CAULIFLOWER MOSAIC VIRUS.

phosphorylation. The esterification of a compound with phosphoric acid.

phosphotungstic acid. A NEGATIVE STAIN used for electron microscopy. It consists of dodecaungstophosphoric acid dissolved in water to give a 1-2% solution and adjusted to about pH 7 with NaOH.

photo-Shootur virus. *See* CAMEL POX VIURUS.

photon correlation spectroscopy. A technique which allows the study of movements of macromolecules, organelles and cells from their scatter-

ing of laser light. Can be used in the estimation of DIFFUSION COEFFICIENTS of virus particles.
Steer, M.W. *et al.* (1985) In Advances in Botanical Research, Vol. 11, p. 1. ed. J.A. Callow and H.W. Woolhouse. Academic Press: London.

photoreactivation. The enzymic repair of DNA damaged by UV-light. The cellular enzymes involved are activated by exposure to long wavelength light.

Phthorimaea operculella granulosis virus. BACULOVIRUS (Subgroup B) isolated from the potato tuber moth, *P. operculella*, and evaluated in field trials in Australia as a selective biological control agent for the homologous host. Laboratory tests have demonstrated that larvae can develop resistance to the virus under high selection pressure.
Briese, D.T. and Mende, H.A. (1983) Bull. ent. Res. **73**, 1.

phycovirus. Synonym: CYANOPHAGE.

Physalis mild chlorosis virus. A possible *Luteovirus*.
Francki, R.I.B. *et al.* (1985) In Atlas of Plant Viruses. Vol. 1. p. 137. CRC Press: Boca Raton, Florida.

Physalis mosaic virus. A *Tymovirus*.
Francki, R.I.B. *et al.* (1985) In Atlas of Plant Viruses. Vol. 1. p. 117. CRC Press: Boca Raton, Florida.
Hirth, L. and Girard, L. (1988) In The Plant Viruses. Vol. 3. p. 163. ed. R. Koenig. Plenum Press: New York.

Physalis vein blotch virus. A possible *Luteovirus*.
Francki, R.I.B. *et al.* (1985) In Atlas of Plant Viruses. Vol. 1. p. 137. CRC Press: Boca Raton, Florida.

Phytocryptic virus 'group'. Synonym: CRYPTOVIRUS 'GROUP'.

Phytoreovirus group. (Greek 'phytos' = plant and reovirus). (Type member WOUND TUMOUR VIRUS). One of the two genera of plant viruses in the family REOVIRIDAE; also known as plant reovirus subgroup 1. The particles are isometric, with an outer shell 70 nm. in diameter and an inner core 59 nm. in diameter; they sediment at 510S. The outer shell is made up of four polypeptide

100nm

species mw. 130, 96, 36 and 35 x 10^3 and the inner core of three polypeptides mw. 160, 118 and 58 x 10^3. Particles contain 12 species of dsRNA with mw. ranging from 3.0 to 0.3 x 10^6. Host range in plants varies between members from wide to narrow. Virus particles are found in most cell types; there are characteristic cytoplasmic inclusion bodies (VIROPLASMS). Phytoreoviruses also multiply in their insect vectors, cicadellid leafhoppers, which transmit them in the PERSISTENT TRANSMISSION manner. They are not transmitted mechanically.

Matthews, R.E.F. (1982) Intervirology **17**, 85.

PIB. Abbreviation sometimes used for 'polyhedral inclusion bodies' referring to the polyhedra of nuclear and cytoplasmic polyhedrosis viruses.

Pichinde virus. Family *Arenaviridae*, genus *Arenavirus*. Isolated from rodents, mosquitoes and mites in Colombia.

Picornaviridae. (Acronym from Old Spanish 'pico' = micro micro, RNA.) A family of four genera of viruses with small isometric particles 22-30 nm. in diameter which sediment between 140-165S and band in CsCl at 1.33-1.45 g/cc (depending on genus). The four genera, *APHTHO-*

100nm

VIRUS, CARDIOVIRUS, ENTEROVIRUS and *RHINOVIRUS* are distinguished by the stability of virions below pH7, their CsCl banding density, possession of a polycytidylic acid tract, base composition of RNA and clinical manifestations in susceptible hosts. They have no envelope or core, nor any surface projections. The capsids are T=1, the structural subunit comprising one molecule of each of the four major capsid polypeptides; these are derived by cleavage of the structural protein precursor (mw. 80-100 x 10^3). Each capsid contains one molecule of infectious (+)-sense ssRNA, mw. c.2.5 x 10^6, c.7,500 nucleotides and has a 5′ VPg (mw. c.2,400) and 3′ polyadenylic acid sequence. The cardio- and aphtho-viruses have a polycytidylic acid tract near the 5′ end. Most picornaviruses can be grown in cell culture.

Replication is in the cytoplasm and is via two distinct partially double-stranded replicative intermediates, one using (+)-RNA and the other (–)-RNA RNA as template; the polymerase is virus-coded. Virus mRNA is translated to give a large (mw. c.210,000) precursor polyprotein which is cleaved to give the functional proteins. The natural host range of most picornaviruses is host-species specific except for Coxsackie B5 virus, EMC virus and aphthoviruses.

Matthews, R.E.F. (1982) Intervirology **17**, 129.

Pieris spp. granulosis virus. A large number of closely-related genotypic variants of this BACULOVIRUS (GV) have been isolated from larvae of the cabbage white butterflies *Pieris* (= *Artogeia*) *rapae* and *P. brassicae*. The virus has been extensively tested as a selective control agent for these pests; large quantities of the virus are used for this purpose in the People's Republic of China for *P. rapae* control. Some protein sequence homology (53%) has been detected between the granulin component of *P. brassicae* GV and the polyhedrin of the prototype baculovirus, *Autographa californica* nuclear polyhedrosis virus.

Crook, N.E. (1986) J. gen. Virol. **67**, 781.

pig cytomegalovirus. Family *Herpesviridae*, subfamily *Betaherpesvirinae*, genus *Murine Cytomegalovirus* group. Causes rhinitis and sneezing in young piglets, often leading to death. Can be grown in pig cell cultures.

pig poxvirus. *See* SWINE POXVIRUS.

pigeon pea (Cajanus cajan) proliferation virus. A possible plant *Rhabdovirus*.

Francki, R.I.B. *et al.* (1985) **In** Atlas of Plant Viruses. Vol. 1. p. 73. CRC Press: Boca Raton, Florida.

pike fry disease rhabdovirus. *See* RED DISEASE OF PIKE VIRUS.

pilus. Filamentous non-motile appendage found on many gram-negative bacilli. Somatic pili (several hundred per cell) function in bacterial adherence, while conjugation between bacteria depends on the one or two sex pili on male cells. Sex pili are coded for by conjugative plasmids (*see* F PLASMID, I PLASMID, R PLASMID). Certain bacteriophages adsorb to specific sex pili and their host range is therefore restricted to bacteria containing the appropriate plasmid.

Hardy, K. (1986) Bacterial Plasmids. 2nd. ed.

Thomas Nelson: Walton-upon-Thames.

pine sawfly nuclear polyhedrosis virus. *See* NEODIPRION SERTIFER NUCLEAR POLYHEDROSIS VIRUS.

pineapple chlorotic leaf streak virus. A possible plant *Rhabdovirus*, subgroup 2.
Francki, R.I.B. *et al.* (1985) **In** Atlas of Plant Viruses. Vol. 1. p. 73. CRC Press: Boca Raton, Florida.

Pinellia mosaic virus. A *Potyvirus*, occurs in Japan.
Doi, Y. *Personal communication.*

pink eye virus. *See* EQUINE ARTERITIS VIRUS.

pinocytosis. A form of active transport of water and dissolved or suspended molecules across a cell membrane involving the internalisation of a fluid-filled vacuole. A common mode of entry for many animal viruses into cells. *See* VIROPEXIS.

pinwheel inclusions. CYTOPLASMIC INCLUSION BODIES comprising sheets of virus-coded protein which form characteristic structures in cells of plants infected with POTYVIRUSES.
Christie, R.G. and Edwardson, J.R. (1977) Light and Electron Microscopy of Plant Virus Inclusions. Florida Ag. Expt. Sta. Monograph, No. 9.

PIPES. Piperazine-N,N′-bis-2-ethanesulphonic acid (mw. 302.36). A biological buffer, $pK_a = 6.8$, used in the pH range 6.2 - 8.2.
Good, N. *et al.* (1966) Biochemistry **5**, 467.

Piry virus. Family *Rhabdoviridae*, genus *Vesiculovirus*. Isolated from an opposum in Brazil. Laboratory infection of man is common, resulting in fever with myalgia, anthralgia and soreness of abdomen. Very similar structurally to vesicular stomatitis virus but the antigenic relationship is tenuous.

Pisaura viruses. Unclassified viruses observed in the nuclei of hepatopancreatic cells of the spider *P. mirabilis* including particles resembling a NON-OCCLUDED BACULOVIRUS (nucleocapsids 300 x 30 nm., enveloped), an adeno-like virus (80 nm. in diameter, DNA mw. 12×10^6) and a small isometric virus (40-45 nm. in diameter).
Bergoin, M. *et al.* (1982) **In** Invertebrate Pathology and Microbial Control. p. 523. Proc. IIIrd Internat. Colloq. Invertebr. Pathol.

Pisum rhabdovirus. A plant *Rhabdovirus*, subgroup 1.
Francki, R.I.B. *et al.* (1985) **In** Atlas of Plant Viruses. Vol. 1. p. 73. CRC Press: Boca Raton, Florida.

pittosporum vein yellowing virus. A plant *Rhabdovirus*, subgroup 2.
Francki, R.I.B. *et al.* (1985) **In** Atlas of Plant Viruses. Vol. 1. p. 73. CRC Press: Boca Raton, Florida.

Pixuna virus. Family *Togaviridae*, genus *Alphavirus*. Isolated from rodents and mosquitoes in Brazil. Related antigenically to Venezuelan equine encephalomyelitis virus.

plant reovirus group. Term sometimes used collectively for the PHYTOREOVIRUS and FIJIVIRUS groups.
Boccardo, G. and Milne, R.G. (1984) CMI/AAB Descriptions of Plant Viruses No. 294.
Francki, R.I.B. *et al.* (1985) **In** Atlas of Plant Viruses. Vol. 1, p. 47. CRC Press: Boca Raton, Florida.

plant rhabdovirus group. A group of the *Rhabdoviridae* family consisting of plant viruses with characteristic BACILLIFORM or bullet-shaped particles which have many similarities to rhabdoviruses from vertebrates and insects; many of them replicate in their insect vectors. The basic shape, structure and molecular biology, where known, resemble closely the vertebrate rhabdoviruses. However, there are some members which appear to lack the outer membrane. The group is divided into two subgroups. Members of subgroup 1 (type member LETTUCE NECROTIC YELLOWS VIRUS) mature in association with the endoplasmic reticulum and accumulate in vesicles in the cytoplasm. Members of subgroup 2 (type member POTATO YELLOW DWARF VIRUS) tend to have wider particles than those of subgroup 1; they bud at the inner nuclear membrane and accumulate in the perinuclear space. Most plant rhabdoviruses have a narrow host range. However, as most are not sap-transmissible and for many no vector is known, it is uncertain that all the named plant rhabdoviruses are distinct. The natural vectors are plant-sucking arthropods (aphids, leafhoppers, plant bugs, mites, etc). Virus-vector relationships are very specific and in all cases are of the PERSISTENT TRANSMISSION type. For several members of the group there is direct evidence for virus multiplication in the vector.

Peters, D. (1981) CMI/AAB Descriptions of Plant Viruses No. 244.

Plantago mottle virus. A *Tymovirus*.
Francki, R.I.B. *et al*. (1985) **In** Atlas of Plant Viruses. Vol. 1. p. 117. CRC Press: Boca Raton, Florida.
Hirth, L. and Girard, L. (1988) **In** The Plant Viruses. Vol. 3. p. 163. ed. R. Koenig. Plenum Press: New York.

Plantago severe mottle virus. A *Potexvirus*.
Francki, R.I.B. *et al*. (1985) **In** Atlas of Plant Viruses. Vol. 2. p. 159. CRC Press: Boca Raton, Florida.

Plantago virus 4. A possible *Caulimovirus*.
Francki, R.I.B. *et al*. (1985) **In** Atlas of Plant Viruses. Vol. 1. p. 17. CRC Press: Boca Raton, Florida.

Plantago virus X. A *Potexvirus*.
Hammond, J. and Hull, R. (1983) CMI/AAB Descriptions of Plant Viruses No. 266.
Francki, R.I.B. *et al*. (1985) **In** Atlas of Plant Viruses. Vol. 2. p. 159. CRC Press: Boca Raton, Florida.

plantain 6 virus. A possible member of the *Carmovirus* group.
Hammond, J. (1981) Plant Path. **30**, 237.
Morris, T.J. and Carrington, J.C. (1988) **In** The Plant Viruses. Vol. 3. p. 73. ed. R. Koenig. Plenum Press: New York.

plantain (Plantago lanceolata) mottle virus. A possible plant *Rhabdovirus*.
Francki, R.I.B. *et al*. (1985) **In** Atlas of Plant Viruses. Vol. 1. p. 73. CRC Press: Boca Raton, Florida.

plaque. The area or 'hole' formed in a lawn of cells due to infection with a virus. In infections with bacteriophage VIRULENT PHAGE form clear plaques in bacterial lawns whereas TEMPERATE PHAGE form turbid plaques due to the survival and division of bacteria in which the phage has become LYSOGENIC.

plaque assay. An assay in which the concentration of infective particles in a virus solution is recorded as the number of plaques induced on a lawn of bacteria or cells. It can also be used to distinguish between different strains or distinct viruses by the features of the plaque.

plaque mutants. Mutants which produce PLAQUES differing in size or appearance from those produced by the wild-type virus.

plaque neutralisation test (plaque reduction test). A method for either identifying a virus (or serotype) or for titrating an antiserum by analysing the inhibitory effect of antibodies on the infectivity of the virus using the plaque assay.

plaque picking. The selection of individual plaques which are considered to be formed by a single infection event. Clones of a virus can thus be selected and further studied.

plaque-forming units (pfu). The number of PLAQUES formed per unit of volume or weight of a virus suspension.

Plasmaviridae. (Greek 'plasma' = shaped product). A family of phages containing a single genus (PLASMAVIRUS) of pleomorphic viruses with a circular supercoiled dsDNA genome (PHAGE morphotype G).

Plasmavirus. dsDNA phages with pleomorphic enveloped particles which have a broad size range (52-125 nm. in diameter). The virion is a nucleoprotein condensation bounded by a lipoprotein membrane. The viral genome is a molecule of circular supercoiled dsDNA (mw. 7.8×10^6). Virus infection is non-lytic; mature virus is released by budding and the genome is maintained as a PROPHAGE in the host chromosome. The type species is L2 PHAGE isolated from the mycoplasma *Acholeplasma laidlawii*.
Maniloff, J. *et al*. (1982) Intervirology **18**, 177.

plasmid. An extrachromosomal DNA element (usually of a bacterium) which is capable of independent replication. Plasmids vary in size from fewer than one kbp to more than 300kbp and in copy number from one to more than 100 per cell. Most plasmids are covalently closed dsDNA circles but some are linear molecules. The information that plasmids carry includes antibiotic resistance and toxin expression.

plasmid vector. A plasmid which is used for CLONING 'foreign' DNA. The plasmid is often manipulated to contain desirable features such as resistance to two or more antibiotics, ability to produce multiple copies, single-cutting restriction enzyme sites and strong promoters.

plating efficiency (efficiency of plating). The number of PLAQUES divided by the total number of virions in the inoculum. Used to quantify relative efficiencies with which cells can be infected and support viral replication.
Ellis, E.L. and Delbruck, M. (1939) J. gen. Physiol. **22**, 365.

Playas virus. Family *Bunyaviridae*, genus *Bunyavirus*.

Plectrovirus. Genus of phages (*Inoviridae*) characterised by rod-shaped virions with one rounded end (bullet-shaped) *c*.84 x 14 nm. from *Acholeplasma* or 250 x 13 nm. from *Spiroplasma*. The genome is a molecule of circular ssDNA (mw. 1.5 x 10^6). During infection, phages are extruded through the host membrane and the host survives infection. The type species is MVL51 PHAGE, isolated from *Acholeplasma laidlawii*. Other members and possible members are listed in Appendix F.
Maniloff, J. *et al.* (1982) Intervirology **18**, 177.

pleioblastus mosaic virus. A *Potyvirus*, occurs in Japan.
Doi, Y. *Personal communication.*

pleomorphic phages. Enveloped, dsDNA-containing phages without apparent capsid structure (PLASMAVIRUS).

Plodia interpunctella granulosis virus. BACULOVIRUS (Subgroup B) isolated from the Indian meal moth, *P. interpunctella*. The biochemical properties of the virus resemble those of other baculoviruses (*see* BACULOVIRUS, GRANULOSIS VIRUS, NUCLEAR POLYHEDROSIS VIRUS). The virus has been used with some success in trials as a selective biological control agent for its insect host.
Tweeten, K.A. *et al.* (1981) Microbial. Rev. **45**, 379.

plum line pattern virus *See* AMERICAN PLUM LINE PATTERN VIRUS, DANISH PLUM LINE PATTERN VIRUS.

plum pox virus. A *Potyvirus*.
Kegler, H. and Schade, C. (1971) CMI/AAB Descriptions of Plant Viruses No. 70.
Francki, R.I.B. *et al.* (1985) **In** Atlas of Plant

Viruses. Vol. 2. p. 183. CRC Press: Boca Raton, Florida.

plus strand. *See* POSITIVE-SENSE STRAND (POSITIVE STRAND).

PM2 phage. Type species of the *CORTICOVIRUS* genus, isolated from *Alteromonas espejiana*. Virus particles are icosahedral, about 60 nm. in diameter, non-enveloped and have brush-like spikes on the vertices. The particles are double-shelled, the internal and external shells separated by a lipid bilayer (predominantly phosphatidyl glycerol). Virions have a mw. of about 50 x 10^6, sediment at 230S and have a buoyant density in CsCl of 1.28 g/cc. Infectivity is sensitive to lipid solvents. There are four structural proteins, the spike protein (mw. 43 x 10^3), the outer shell protein (mw. 26 x 10^3), and two internal proteins (mw. 12.5 and 4.7 x 10^3) which behave as proteolipids. The virus contains polynucleotide pyrophosphorylase. The genome is a single molecule of circular supercoiled DNA (mw. 5.8 x 10^6). Infection causes lysis of the host.
Mindich, L. (1978) **In** Comprehensive Virology. Vol. 12. p. 271. ed. H. Fraenkel-Conrat and R.R. Wagner. Plenum Press: New York.
Ackermann, H.W. and Dubow, M.S. (1987). Viruses of Prokaryotes. CRC Press: Boca Raton, Florida.

PM2 phage group. Vernacular name for *CORTICOVIRUS*.

PMSF (phenyl methanesulphonyl fluoride, α-toluenesulphonyl fluoride). An inhibitor of PROTEASES. Used in studies on viral proteins where proteolysis is to be avoided.

pneumonia of mice virus. Family *Paramyxoviridae*, genus *Pneumovirus*. Causes a latent or mild respiratory infection of laboratory mice. Serial passage of lung tissue in mice activates the virus to cause accumulation of mononuclear cells around the bronchi and blood vessels. Grows in hamster kidney cell cultures and in BHK21 cells.

Pneumovirus. (Greek 'pneuma' = breath.) A genus in the family *Paramyxoviridae*. Consists of viruses causing respiratory syncytial disease in man and pneumonia in cattle, turkeys and mice.

Poa semi-latent virus. A *Hordeivirus*.
Francki, R.I.B. *et al.* (1985) **In** Atlas of Plant

Viruses. Vol. 2. p. 133. CRC Press: Boca Raton, Florida.

pock assay. An assay in which the concentration of infective particles in a virus solution is estimated by the number of lesions (pocks) it induces on the chicken egg allantoic membrane. Used for the assay of e.g. VACCINIA VIRUS.

Podoviridae. (Greek 'pous, podos' = foot). Family of virulent or temperate phages with short (c.20 nm. non-contractile tails (phage morphotypes C1-C3; *see* PHAGE) which contain linear dsDNA. The short tail may carry a base plate and tail fibres (dependent on virus strain). A large

    100nm

number of phage isolates with these properties have been obtained from bacteria, cyanobacteria and mycoplasmas (*see* Appendix F). The family at present contains a single genus (the T7 PHAGE GROUP), although other isolates will probably be assigned to several genera in future. The type species is T7 PHAGE.
Ackermann, H.W. *et al.* (1984) Intervirology **22**, 181.

Poikilovirus. A proposed genus in the family *Herpesviridae*. It consists of SUID HERPESVIRUS 1 (PSEUDORABIES VIRUS), equid herpesvirus 1 and HUMAN HERPESVIRUS 3.

poinsettia mosaic virus. A possible *Tymovirus*.
Koenig. R. *et al.* (1986) AAB Descriptions of Plant Viruses No. 311.
Hirth, L. and Girard, L. (1988) **In** The Plant Viruses. Vol. 3. p..163. ed. R. Koenig. Plenum Press: New York.

pokeweed mosaic virus. A *Potyvirus*.
Shepherd, R.J. (1972) CMI/AAB Descriptions of Plant Viruses No. 97.
Francki, R.I.B. *et al.* (1985) **In** Atlas of Plant Viruses. Vol. 2. p. 183. CRC Press: Boca Raton, Florida.

Pol gene. Gene which codes for a POLYMERASE.

PolA mutant. A mutant of a bacterium which affects the DNA repair enzyme, DNA-dependent DNA polymerase II.

polar mutant. A mutant having effects on genes transcribed downstream, e.g. NONSENSE MUTANT, TRANSPOSON insertions.

poliomyelitis virus. Family *Picornaviridae*, genus *Enterovirus*. A common infection of the gastro-intestinal tract which is usually silent. When the central nervous system is infected there is damage to the anterior horn cells and motor paralysis which can be severe. Causes similar disease in chimpanzees and monkeys when injected. Grows in monkey kidney and HeLa cells. Both the Salk (inactivated) and Sabin (attenuated) vaccines have been successful in controlling the disease in developed countries.

poly(A). Polyadenylic acid. Most eukaryotic messenger RNAs have a stretch of up to 300 nucleotides of adenosine at the 3′ end. The mRNAs of many eukaryotic viruses are polyadenylated as are the encapsidated RNAs of some viruses (e.g. PICORNAVIRUSES, COMOVIRUSES).

poly(A) polymerase. An enzyme which adds adenylate residues to the 3′ end of RNA.

polyacrylamide gel electrophoresis (PAGE). Electrophoresis in a gel composed of polyacrylamide which is made by cross-linking acrylamide usually with N,N′-methylene-bis-acrylamide; other cross-linkers can be used, e.g. bis acrylylcystamine which can be broken down by reducing agents thus solubilising the gel. Polymerisation is activated by the addition of a source of free radicals (usually ammonium persulphate or riboflavin) which, in turn, activates the TEMED. The TEMED acts as an electron carrier to activate the acrylamide monomer which then reacts with an unactivated monomer to begin the polymer chain elongation. The elongating polymer chains are randomly crosslinked by bis, resulting in a complex 'web' polymer with a characteristic porosity which depends upon the polymerisation conditions and the monomer concentrations. Polyacrylamide gel electrophoresis is used for the separation of nucleic acid or protein molecules according to their molecular size and thus in estimating relative molecular weights. In the case of proteins an ionic detergent, such as SODIUM DODECYL SULPHATE, is usually added to denature the protein and provide a uniform charge per unit molecular weight. The pore size is varied by changing the concentration of acrylamide and the extent of cross-linking.

polyadenylation. The addition of adenylate residues usually to the 3´ end of RNA molecules.

poly(C). Polycytidylic acid. Some PICORNA-VIRUSES have a stretch of about 100 nucleotides of cytosine in the genomic RNA.

polycistronic. A nucleic acid coding for more than one CISTRON.

polyclonal antibody. A preparation containing antibodies against more than one EPITOPE of an antigen. When an antigen is injected into a warm-blooded animal, antibodies are produced by different cell types against the various antigenic sites, the relative amounts being controlled by the antigenicity of the sites. *See* MONOCLONAL ANTI-BODY.

Polydnaviridae. (Greek 'poly' = many, DNA virus.) A family containing a single genus, *POLYDNAVIRUS*. The particles contain multiple supercoiled dsDNAs of variable mw. ranging from approximately 2 to 25 x 10^6 with an aggregate genome mw. of about 80-170 x 10^6. Earlier classified as subgroup D of *Baculovirus* genus. Stoltz, D.B. *et al.* (1984) Intervirology **21**, 1.

made up of a large number of molecules ranging in size from about 1.7 to 6.2 x 10^6. Viruses of this type have been found predominantly in parasitoids of the family Ichneumonidae. The type species is *CAMPOLETIS SONORENSIS* VIRUS. Another well-studied member is *HYPOSOTER EXIGUAE* VIRUS. Subgroup B includes viruses with rod-shaped nucleocapsids, variable in length (30-100 nm.), usually carrying tail appendages. A single envelope surrounds one or more nucleocapsids. Particles contain at least 18 polypeptides and a polydisperse dsDNA genome with as many as 15 different molecules ranging in size from 2-25 x 10^6. Viruses of this type have been found predominantly in parasitoids of the family *Braconidae*. No type species has been defined, though *APANTELES MELANOSCELUS* VIRUS is representative of the group. In both subgroups each virus isolate (as defined by the host wasp species) contains a distinct spectrum of DNA molecules which is constant for one individual wasp species. It is not yet clear how the different-sized DNA molecules are distributed among the virus particles. Cross-reacting antigenic determinants are present on viruses within each subgroup but no serological cross-reactions have been detected between subgroups. All viruses of this unusual group have

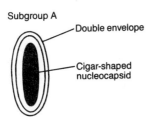

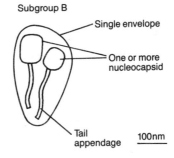

Polydnavirus. Genus of enveloped DNA-containing viruses with rod-shaped nucleocapsids isolated from parasitic hymenopteran insects (parasitoids) of the families Ichneumonidae and Braconidae. Two subgroups have been proposed based on morphological and other properties. Subgroup A includes viruses with fusiform (cigar-shaped) nucleocapsids of uniform length. Each nucleocapsid is surrounded by two unit-membrane envelopes, though in some members (e.g. a virus from *Glypta* sp.), more than one nucleocapsid may be enclosed within the envelopes. Virions are structurally complex, containing 20-30 polypeptides. The genome is multipartite (polydisperse) circular supercoiled dsDNA

been associated with parasitoid Hymenoptera. They replicate in the ovaries of female parasitoids and are injected (during oviposition) into the host of the parasitoid. The viruses appear to function by preventing encapsulation of the parasitoid egg and hence in promoting successful parasitism. Association of these viruses with female parasitoid Hymenoptera is probably the rule rather than the exception. As there may be >100,000 species of parasitoid Hymenoptera, this implies the existence of huge numbers of polydnavirus strains. There is no formal nomenclature for each virus although, conventionally, they have been named after the parasitoid host from which they were first isolated.

Stoltz, D.B. *et al.* (1984) Intervirology **21**, 1.
Stoltz, D.B. and Vinson S.B. (1979) Adv. Virus Res. **24**, 125.

polyethylene glycol (PEG). A polymer with the general formula $HOCH_2(CH_2OCH_2)_xCH_2OH$ which is available in molecular weights ranging from 200 to 20,000 (abbreviated PEG 200, PEG 20,000). PEG is a coacervate and will bind water; thus it can be used to concentrate solutions by withdrawing water from them. PEG 4000 and 6000 are commonly used to promote cell or protoplast fusion and to facilitate the uptake of DNA in transformation, especially of yeast. PEG 6000 is also used in precipitating viruses or DNA.

polyheads. Phage heads of normal diameter but exaggerated length due to mutations (e.g. in T-EVEN PHAGE, the lack of glycoprotein 20) or to growing the phage in the presence of certain amino acid analogues, e.g.L-canavanine.
Cummins, D.J. and Bolin, R.W. (1976) Bact. Rev. **40**, 314.
Showe, M.H. and Kellenberger, E. (1975) Symp. Soc. Gen. Microbiol. **25**, 407.

polyhedral protein. *See* POLYHEDRIN.

polyhedrin. The matrix protein which comprises the major component of the occlusion bodies produced during nuclear polyhedrosis (NPV) and cytoplasmic polyhedrosis (CPV) virus infections. Sometimes used as a generic term to include the matrix protein of granulosis (GV) virus occlusion bodies (*see* GRANULIN). NPV polyhedrin is a protein of mw. about 29×10^3 and is alkali soluble. During NPV virus infection, polyhedrin is produced in larger amounts than any other viral protein. Because of the high level of expression, the polyhedrin gene region has been used for the construction and high-level expression of foreign genes (*see* BACULOVIRUS EXPRESSION VECTOR). 5′ flanking sequences for polyhedrin genes contain a highly conserved sequence of 12 nucleotides with the consensus sequence AATAAGTATTTT beginning 20-70 nucleotides upstream from the protein initiation codon, followed by an AT-rich sequence believed to be important in the regulation of high-level expression. Polyhedrin is synthesised late in virus infection when it occludes virus particles within the large para-crystalline occlusion bodies. NPV polyhedrins and GV granulins are highly conserved and contain related sequences reflected in antigenic properties and amino acid sequences.

Lepidopteran NPV polyhedrins appear closely-related to one another and demonstrate 85-90% amino acid homology. Granulins are related but include additional cysteine residues and an N-terminal insertion not found in lepidopteran NPV polyhedrin. Polyhedrins from hymenopteran NPVs appear inter-related but have only limited relatedness to lepidopteran polyhedrins. Polyhedrin from a dipteran NPV (*Tipula paludosa* NPV), though similar in size, shows no serological or N-terminal amino acid sequence relatedness. CPV polyhedrins are of similar size to NPV polyhedrins but are unrelated to them as judged by serological comparisons or N-terminal amino acid sequence. Nonetheless, the protein serves the same function in all occluded viruses, forming a protective crystal around virus particles, allowing the virus to remain viable for many years outside the insect host (*see* also SPHEROIDIN).
Rohrmann, G.F. (1986) J. gen. Virol. **67**, 1499.

polyhedron. (plural polyhedra) OCCLUSION BODY produced during infection with NUCLEAR POLYHEDROSIS or CYTOPLASMIC POLYHEDROSIS VIRUSES, containing numerous virions surrounded by crystalline protein matrix (POLYHEDRIN).

polyhedrosis. Virus disease of invertebrates (generally insects) larvae characterised by breakdown of tissues and the presence of polyhedral OCCLUSION BODIES. *See* NUCLEAR POLYHEDROSIS VIRUS, CYTOPLASMIC POLYHEDROSIS VIRUS.

poly(I):poly(C). A synthetic duplex of polyriboinosinic acid and polycytidylic acid. Induces INTERFERON.

poly L-lysine. *See* POLY L-ORNITHINE.

poly L-ornithine. A polycation used to induce the uptake of virus or nucleic acid by protoplasts or cells. The function is considered to be twofold, to give the inoculum the correct charge for association with the plasma membrane and to damage the plasma membrane in such a way as to allow virus entry. Poly L-lysine is used for this purpose less frequently.
Watts, J.W. *et al.* (1981) Ciba Found. Symp. **80**, 56.

polylysogeny. A condition where bacterial strains are LYSOGENIC for several different phages. *See* LYSOGEN.

polymerase. An enzyme which catalyses the formation of RNA or DNA by the addition of ribonucleotide or deoxyribonucleotide triphosphates. *See* DNA-DEPENDENT DNA POLYMERASE, DNA-DEPENDENT RNA POLYMERASE, RNA-DEPENDENT RNA POLYMERASE, REVERSE TRANSCRIPTASE.

polymerase chain return (PCR). The selective amplification of DNA by repeated cycles of a) heat denaturation of the DNA, b) annealing of two oligonucleotide primers that flank the DNA segment to be amplified and c) the extension of the annealed primers with the heat-insensitive *Tag* DNA polymerase. Can be used to produce probes for virus diagnosis and in the amplification of low copy number sequences.
Saibi, R.K. *et al.* (1988) Science **239**, 487.

polynucleotide kinase. An enzyme (e.g. encoded by PHAGE T4) which phosphorylates the 5'-OH termini of RNA or DNA polynucleotide chains. Used to label nucleic acids at the 5' terminus.

polynucleotide ligase. Generic term for enzymes which catalyse the linking or repair of either DNA or RNA strands. *See* DNA LIGASE, RNA LIGASE.

polyoma virus. Family *Papovaviridae*, genus *Polyomavirus*. Causes a natural infection of laboratory mice and rats without any signs of disease. However, when injected into new-born mice or hamsters it is highly oncogenic. Grows in mouse embryo cell cultures and transforms hamster cell cultures.

Polyomavirus. (Greek 'poly' = many and suffix '-oma'.) A genus in the family *Papovaviridae*. Members have been isolated from man, mouse, rabbit, hamster and several species of monkey.

polypeptide. A chain of amino acids linked together by peptide bonds obtained by synthesis or by partial hydrolysis of a protein. Can also refer to the primary structure of a protein, e.g. polypeptide chain.

polyploid virus. A virus, the particles of which contain a variable number of genomes. The number of genomes per particle depends on factors such as host cell and cultural conditions but is independent of the number of genomes in the infecting particles. Polyploidy is found in viruses

of various groups including MYXOVIRUSES, HERPES VIRUS, VISNAVIRUS and NUCLEAR POLYHEDROSIS VIRUSES.
Simon, E.H. (1972) Prog. med. Virol. **14**, 36.

polyprotein. A large precursor protein subsequently cleaved to give two or more functional proteins, e.g. the primary translation product of PICORNAVIRUS RNA is a polyprotein which is cleaved to give the structural and non-structural proteins.

polyribosome. A structure comprising several ribosomes attached to messenger RNA during the translation of that RNA.

polysome. *See* POLYRIBOSOME.

Pongola virus. Family *Bunyaviridae*, genus *Bunyavirus*. Isolated from mosquitoes in several countries in Africa. Kills new-born mice on injection. Sheep, cattle and donkeys are the natural hosts. Antibodies are found in man but no disease is reported.

Ponteves virus. Family *Bunyaviridae*, genus *Bunyavirus*. Isolated from a tick in Southern France.

Poovoot virus. Family *Reoviridae*, genus *Orbivirus*.

poplar mosaic virus. A *Carlavirus*.
Biddle, P.G. and Tinsley, T.W. (1971) CMI/AAB Descriptions of Plant Viruses No. 75.
Francki, R.I.B. *et al.* (1985) **In** Atlas of Plant Viruses. Vol. 2. p. 173. CRC Press: Boca Raton, Florida.

POPOP. 1,4-bis-2-(5-phenyloxazolyl)benzene. A secondary solute used in SCINTILLATION FLUIDS to shift wavelength to enhance radioisotope detection by liquid scintillation counting.

porcine adenovirus. Family *Adenoviridae*, genus *Mastadenovirus*. Found in digestive tract of pigs but apparently causes no disease. Grows in a wide range of cell cultures.

porcine enterovirus. Synonym: TESCHEN DISEASE VIRUS, TALFAN DISEASE VIRUS, PORCINE POLIOMYELITIS VIRUS. Family *Picornaviridae*, genus *Enterovirus*. Usually present in intestinal tract without causing disease. However, virulent strains can cause diarrhoea or fever, convulsions, encephalomyelitis and paralysis. There are several antigenic variants. The virus grows in pig kidney cell cultures.

porcine haemagglutinating encephalitis virus. Synonym: VOMITING AND WASTING DISEASE VIRUS. Family *Coronaviridae*, genus *Coronavirus*. Causes encephalomyelitis in pigs with high mortality in young animals and stunted growth in older animals. Virus grows in primary pig kidney cell cultures.

porcine parvovirus. Family *Parvoviridae*, genus *Parvovirus*. A natural infection of pigs. May be the cause of prenatal death without maternal disease. Has been grown in cultures of kidney tissue from normal animals.

porcine transmissible gastroenteritis virus. Family *Coronaviridae*, genus *Coronavirus*. Occurs worldwide, causing a fatal disease of young pigs, following diarrhoea, vomiting and dehydration. Older pigs have diarrhoea but survive. Replicates in pig kidney cell cultures.

porcine type C oncovirus. Family *Retroviridae*, subfamily *Oncovirinae*, genus *Type C Oncovirus* group, sub-genus *Mammalian Type C Oncovirus*. Isolated from a cell line grown from a lymph node of a leukaemic pig.

Portillo virus. Family *Arenaviridae*, genus *Arenavirus*. Isolated from children in Buenos Aires with haemolytic-uraemic disease. Closely related to Junin virus.

positive-sense strand (positive strand). For RNA it is the strand which functions as the messenger (mRNA). For DNA it is the strand with the same sequence as the mRNA.

post-transcriptional cleavage. Cleavage of RNA into functional units, usually MONOCISTRONIC mRNAS. *See* SUBGENOMIC RNA.

post-transcriptional processing. Alterations to the structure of a mRNA after it has been transcribed from either DNA or RNA. These include CAPPING, addition of POLY(A) to the 3′ end, SPLICING and METHYLATION.

post-translational cleavage. Cleavage of a POLYPROTEIN into functional proteins. It can be effected by either virus-coded or host proteases.

potassium phosphotungstate. A NEGATIVE STAIN for electron microscope samples. Usually made up as a 2% solution of phosphotungstic acid with the pH adjusted to 6.8 with potassium hydroxide.
Brenner, S. and Horne, R.W. (1953) Biochim. Biophys. Acta **34**, 103-110.

potato 14R virus. A *Tobamovirus*.
Brunt, A.A. (1986) **In** The Plant Viruses. Vol. 2. p.183. ed. M.H.V. van Regenmortel and H.Fraenkel-Conrat. Plenum Press: New York.

potato aucuba mosaic virus. A possible *Potexvirus*.
Kassanis, B. and Govier, D.A. (1972) CMI/AAB Descriptions of Plant Viruses No. 98.
Francki, R.I.B. *et al.* (1985) **In** Atlas of Plant Viruses. Vol. 2. p. 159. CRC Press: Boca Raton, Florida.

potato black ringspot virus. A *Nepovirus*.
Salazar, L.F. and Harrison, B.D. (1979) CMI/AAB Descriptions of Plant Viruses No. 206.
Francki, R.I.B. *et al.* (1985) **In** Atlas of Plant Viruses. Vol. 2. p. 23. CRC Press: Boca Raton, Florida.

potato leaf roll virus (PLRV). A *Luteovirus*. Probably originated in South America and now found everywhere potatoes are grown. Causes leafrolling and stunting. It is one of the most important viruses of potatoes.
Harrison, B.D. (1984) CMI/AAB Descriptions of Plant Viruses No. 291.
Francki, R.I.B. *et al.* (1985) **In** Atlas of Plant Viruses. Vol. 1. p. 137. CRC Press: Boca Raton, Florida.
Casper, R. (1988) **In** The Plant Viruses. Vol. 3. p. 235. ed. R. Koenig. Plenum Press: New York.

potato moptop virus. A *Furovirus*.
Harrison, B.D. (1974) CMI/AAB Descriptions of Plant Viruses No. 138.
Brunt, A.A and Shikata, E. (1986) **In** The Plant Viruses. Vol. 2. p. 305. ed. M.H.V. van Regenmortel and H. Fraenkel-Conrat. Plenum Press: New York.

potato spindle tuber viroid. A VIROID, 359 nucleotides.
Diener. T.O. and Raymer, W.B. (1971) CMI/AAB Descriptions of Plant Viruses No. 66.
Diener. T.O. (1987) **In** The Viroids. ed. T.O. Diener. p. 221. Plenum Press: New York.

potato virus A. A *Potyvirus*.
Bartels, R. (1971) CMI/AAB Descriptions of Plant Viruses No. 54.
Francki, R.I.B. *et al.* (1985) **In** Atlas of Plant Viruses. Vol. 2. p. 183. CRC Press: Boca Raton, Florida.

potato virus M. A *Carlavirus*.
Wetter, C. (1972) CMI/AAB Descriptions of Plant Viruses No. 87.
Francki, R.I.B. *et al.* (1985) **In** Atlas of Plant Viruses. Vol. 2. p. 173. CRC Press: Boca Raton, Florida.

potato virus S. A *Carlavirus*.
Wetter, C. (1971) CMI/AAB Descriptions of Plant Viruses No. 60.
Francki, R.I.B. *et al.* (1985) **In** Atlas of Plant Viruses. Vol. 2. p. 173. CRC Press: Boca Raton, Florida.

potato virus T. A member of the *Capillovirus* group.
Salazar, L.F. and Harrison, B.D. (1978) CMI/AAB Descriptions of Plant Viruses No. 187.

potato virus V. A *Potyvirus*.
Jones, R.A.C. and Fribourg, C.E. (1986) AAB Descriptions of Plant Viruses No. 316.

potato virus X. Type member of the *Potexvirus* group.
Bercks, R. (1970) CMI/AAB Descriptions of Plant Viruses No. 4.
Francki, R.I.B. *et al.* (1985) **In** Atlas of Plant Viruses. Vol. 2. p. 159. CRC Press: Boca Raton, Florida.

potato virus Y. Type member of the *Potyvirus* group. One of the most important viruses of potatoes.
de Bokx, J.A. and Huttinga, H. (1981) CMI/AAB Descriptions of Plant Viruses No. 242.
Francki, R.I.B. *et al.* (1985) **In** Atlas of Plant Viruses. Vol. 2. p. 183. CRC Press: Boca Raton, Florida.

potato yellow dwarf virus. Type member of subgroup 2 of the plant *Rhabdoviruses*; transmitted by leafhopper.
Black, L.M. (1970) CMI/AAB Descriptions of Plant Viruses No. 35.
Francki, R.I.B. *et al.* (1985) **In** Atlas of Plant Viruses. Vol. 1. p. 73. CRC Press: Boca Raton, Florida.

Potexvirus group. (Sigla of potato X from the type member POTATO VIRUS X). Genus of plant viruses with flexuous rod-shaped particles, 470-580 nm. long and 13 nm. wide which sediment at 115-130S and band in CsCl at 1.31 g/cc. The coat

100nm

protein subunits (mw. $18\text{-}23 \times 10^3$) are arranged in a helix with a pitch of 3.4 nm. Each particle contains one molecule of linear, (+)-sense ssRNA (mw. 2.1×10^6) which has a CAP at the 5′ terminus; in some members SUB-GENOMIC RNAS may be encapsidated. The host range of individual members is narrow. Virus particles are found in most cell types where they aggregate to form cytoplasmic inclusion bodies. Potexviruses are readily transmitted mechanically and by contact. There are no known natural vectors.
Matthews, R.E.F. (1982) Intervirology **17**, 156.
Koenig. R. and Lesemann, D-E. (1978) CMI/AAB Descriptions of Plant Viruses No. 200.
Francki, R.I.B. *et al.* (1985) **In** Atlas of Plant Viruses. Vol. 2. p. 159. CRC Press: Boca Raton, Florida.

Potyvirus group. (Sigla of potato Y from the type member POTATO VIRUS Y). Genus of plant viruses with flexuous particles, 680-900 nm. long and 11 nm. wide which sediment at 150-160S and band in CsCl at 1.31 g/cc. The coat protein subunits

100nm

(mw. $32\text{-}36 \times 10^3$) are arranged in a helix with pitch of 3.4 nm. Each particle contains one molecule of linear (+)-sense ssRNA (mw. $3.0\text{-}3.5 \times 10^6$); the 3′ end of the RNA of some viruses has a poly(A) tract. The host range of individual members is narrow. Potyviruses occur in most cell types where they are characterised by cylindrical or conical proteinaceous inclusions (virus coded, mw. 70×10^3) which appear as pinwheels in transverse section. Nuclear inclusions (virus coded, mw. 49×10^3) are induced by some members. Potyviruses are easily mechanically transmitted. Most are transmitted by aphids in the NON-PERSISTENT TRANSMISSION manner requiring a virus-coded transmission factor (mw. $53\text{-}58 \times 10^3$). Some possible members are transmitted by whiteflies, mites or fungi.
Matthews, R.E.F. (1982) Intervirology **17**, 152.

Hollings, M. and Brunt, A.A. (1981) CMI/AAB Descriptions of Plant Viruses No. 245.
Francki, R.I.B. *et al.* (1985) **In** Atlas of Plant Viruses. Vol. 2. p. 183. CRC Press: Boca Raton, Florida.

Powassan virus. Family *Flaviviridae*, genus *Flavivirus*. Isolated from a human case of fatal encephalitis in Ontario, Canada, and from ticks in some parts of the USA. New-born mice can be infected experimentally. Antibodies are present in squirrels and chipmunks in Ontario.

Poxviridae. (Old English 'poc, pocc' = pustule, ulcer.) A family of several subfamilies and genera of viruses with large, brick-shaped or ovoid particles, 300-450 x 170-260 nm. with an external coat containing lipid and tubular or globular protein structures and a core, which contains the genome. There are more than 30 structural pro-

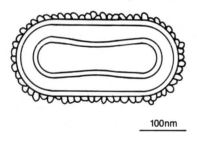

100nm

teins and several viral enzymes concerned with nucleic acid synthesis and processing. There is about 3-4% by weight of lipid and carbohydrate. Each particle contains one molecule of dsDNA, mw. 85-240 x 10^6. Replication occurs in the cytoplasm with inclusion bodies. Genetic recombination occurs within genera; non-genetic reactivation occurs both within and between genera. The natural host range is narrow. Transmission is by several routes including airborne, fomites, direct contact and mechanically by arthropods.
Matthews, R.E.F. (1982) Intervirology **17**, 42.

PPO. 2,5-diphenyloxazole. A primary solute (scintillator) used in SCINTILLATION FLUIDS. Its optimum concentration in toluene is 7g/l.

PRD1 phage. TECTIVIRUS type species (*Tectiviridae*) isolated from *Pseudomonas*. VIRIONS are unenveloped icosahedral, 65 nm. in diameter (PHAGE morphotype D4). Particles are characterised by a double capsid structure. The outer capsid is a rigid shell, 3 nm. thick, surrounding a

flexible inner coat, 5-6 nm. thick, containing lipid. Particles sediment at about 390S and have a buoyant density in CsCl of 1.28 g/cc. 16-18 proteins have been detected in the virion and the genome is a single piece of linear dsDNA (14.7 kbp) with long inverted terminal repeats and covalently bound at its 5′ ends to a protein. Infectivity is sensitive to organic solvents. Virions adsorb to pili of bacteria harbouring certain wide host-range drug-resistant plasmids and hence the virus host range includes a wide range of gram-negative bacteria. The host cell is lysed by infection.
Granados, D.D.and Ito, J. (1987) J. Virol. **61**, 594.

PRD1 phage group. Vernacular name for TECTIVIRUS.

precipitation reaction. The formation of a visible precipitate when adequate quantities of antigen and antibody are allowed to combine. It is the basic reaction which has contributed to the development of immunochemistry as a quantitative science. The precipitate is formed by a lattice of antigen and antibody molecules which grows in size. The precipitate will only form when the concentrations of antigen and antibody are within a certain range of ratios; it can be inhibited by excess of either. *See* AGGLUTINATION, RADIAL IMMUNODIFFUSION, RADIOIMMUNOPRECIPITATION, SINGLE RADIAL DIFFUSION TEST.

precipitin test. *See* IMMUNOPRECIPITATION, PRECIPITATION REACTION.

Preserve. Trials product of the nuclear polyhedrosis virus of the pine sawfly, *Neodiprion sertifer*, produced in 1986 by MicroGeneSys, West Haven, Connecticut, USA.

Pretoria virus. Family *Bunyaviridae*, genus *Nairovirus*. Isolated from a tick in South Africa.

Pribnow box. A nucleotide sequence found in prokaryotic promoters about 10 bp upstream of the start of transcription. It has the consensus sequence:

5′T A   T A A T G 3′
-12 -11 -10 -9 -8 -7 -6
nucleotides to the start of
transcription.

See TATA BOX.

primary culture. The establishment of cells in culture from tissue. The organised tissue needs dissociation into single cells usually by various hydrolytic enzymes (e.g. pronase or trypsin). It does not include cultures started from explants of tumours developed by injecting cultured cells into animals.

Kruse, P.F. and Patterson, M.K. (1973) Tissue Culture, Methods and Applications. Academic Press: New York.

primase. An enzyme which synthesises the RNA primers for DNA synthesis on a DNA template. *See* OKAZAKI FRAGMENTS.

primer. Small fragment of nucleic acid with a free 3´-hydroxyl group necessary for the initiation of DNA and, sometimes, RNA synthesis. *See* PRIMASE.

primula mosaic virus. A possible *Potyvirus*. Francki, R.I.B. *et al.* (1985) **In** Atlas of Plant Viruses. Vol. 2. p. 183. CRC Press: Boca Raton, Florida.

prion. (Sigla from proteinaceous infectious (particle)). Term used for the pathogens which induce some neurological diseases of vertebrates, e.g. scrapie disease of sheep and goats. It was considered at the time the term was introduced that the infectious agent was a protein, a very controversial concept. There is considerable uncertainty as to the causal agent of these diseases.

Diener, T.O. (1987) Cell **49**, 719.

probe. A specific sequence of DNA or RNA used to detect complementary sequences by hybridisation. The probe has reporter groups, e.g. radioactive LABEL, BIOTIN, by which successful hybridisation can be detected. Also used as a verb to describe the act of hybridisation.

probit analysis. A statistical method which involves the transformation of dose-response data represented as percentages or proportions to probits. This has the effect of changing the normal sigmoid curve characteristic of this type of data into a straight line by stretching the linear scale on which the percentages or proportions·are measured. Probits are often used in the analysis of biological assays and to make changes in series of percentages or proportions more readily comparable, e.g. a reduction in percent mortality from 99% to 98% indicates a doubling of the number of survivors while a drop from 51% to 50% is less spectacular and likely to be less important. Tables are available which give the probit transformation directly.

Fisher, R.A. and Yates, F. (1963) Statistical Tables for Biological, Agricultural and Medical Research. Oliver and Boyd Ltd.: Edinburgh.

Finney, D.J. (1971) Probit Analysis. 3rd edition. Cambridge University Press.

procapsid. A viral capsid without nucleic acid considered to be a stage in virion formation. Found, for example, in the synthesis of POLIO-VIRUS particles.

Jacobson, M. and Baltimore, D. (1968) J. mol. Biol. **33**, 369.

productive infection. Infection of a cell in which complete virus particles are formed.

progressive pneumonia virus. Family *Retroviridae*, subfamily *Lentivirinae*. Causes a chronic pulmonary disease in sheep, with gradual loss of weight. There is a proliferation of lymphocytes in the lungs and thickening of interalveolar septa. Very similar to Maedi virus. Spread by contact.

prokaryote, prokaryotic. Organisms which have a genome of a single circular dsDNA free in the cytoplasm and not in an organelle. e.g. nucleus. Prokaryotes are divided into the true bacteria (eubacteria), the archaebacteria and the blue-green algae (cyanobacteria).

promoter. A region of DNA, usually upstream of a coding sequence which binds RNA polymerase and directs the enzyme to the correct transcriptional start site. As well as the defined sequences, e.g. PRIBNOW BOX, TATA BOX, there are upstream and sometimes downstream sequences which attenuate or modulate transcription.

pronase. A non-specific proteolytic enzyme isolated from the bacterium *Streptomyces griseus*.

prophage. The genome of a phage which is perpetuated in the host cell by integration into the host chromosome or by plasmid formation. *See* TEMPERATE PHAGE.

protease. A generic term for an enzyme which cleaves a polypeptide chain.

protein. (Greek 'protos' = first, as chief constituent of living matter.) A class of high molecular weight polymer compounds composed of a variety of α-amino acids joined by peptide bonds.

protein A. A protein, found on the surface of the cells of certain *Staphylococcus aureus* strains, which has the ability to bind strongly to the Fc portion of an antibody molecule when that antibody is bound to an antigen. When protein A is labelled, e.g. with ^{125}I or a fluorescent molecule, it can be used to detect antibody-antigen complexes in various immunological tests.

protein kinase. Enzyme which catalyses the PHOSPHORYLATION of proteins usually in the presence of cyclic AMP or cyclic GMP. There are three main groups, protein-serine, protein-threonine and protein-tyrosine kinases which esterify phosphate to different amino acids.

proteolytic cleavage. The cleavage of a protein at specific site(s), e.g. the cleavage of a polyprotein to yield structural proteins.

protomers. Protein subunits which form a CAPSOMERE.

protoplast. A microbial or plant cell from which the cell wall has been removed, usually enzymatically. They are used for transformation and, in the case of plant cells, for studying the synchronous replication of viruses.

provirus. A virus genome integrated into the cell genome or into a plasmid in which it replicates. The viral DNA is thus passively replicated by the host machinery.

prune dwarf virus. An *Ilarvirus*.
Fulton, R.W. (1970) CMI/AAB Descriptions of Plant Viruses No. 19.
Francki, R.I.B. (1985) In The Plant Viruses. Vol. 1. p. 1. ed. R.I.B. Francki. Plenum Press: New York.

Prunus necrotic ringspot virus. An *Ilarvirus*.
Fulton, R.W. (1970) CMI/AAB Descriptions of Plant Viruses No. 5.
Francki, R.I.B. (1985) In The Plant Viruses. Vol. 1. p. 1. ed. R.I.B. Francki. Plenum Press: New York.

Prunus virus S. A *Carlavirus*, occurs in Japan.
Doi, Y. *Personal communication.*

Pseudoplusia includens icosahedral virus. Unclassified RNA virus isolated from *P. in-*

cludens; some physicochemical similarities with viruses of the *Tetraviridae* (*Nudaurelia* β virus group). Virions are isometric, 40 nm. in diameter; they sediment at 190S and have a buoyant density in CsCl of 1.33 g/cc. The viral genome is ssRNA (mw. 1.9×10^6) and particles contain a single structural polypeptide (mw. 55×10^3) somewhat smaller than the protein size recorded for acknowledged members of the TETRAVIRIDAE. The virus is not serologically related to two tetraviruses isolated from *Trichoplusia ni* and *Antheraea eucalypti*.
Chao, Y.C. *et al.* (1983) J. gen. Virol. **64**, 1835.

Pseudoplusia includens picornavirus. Unclassified small RNA virus (a possible INSECT PICORNAVIRUS) isolated from the armyworm, *Pseudoplusia includens*. Virions are isometric, 25 nm. in diameter, sediment at 178S and contain three major structural proteins (mw. 30, 31 and 34 x 10^3). The genome is ssRNA (mw. 3.3×10^6). Serologically-related to CRICKET PARALYSIS VIRUS.
Chao, Y.C., *et al.* (1986) J. Invertebr. Pathol. **47**, 247.

pseudorabies virus. *See* AUJESZKY'S DISEASE VIRUS.

pseudorecombinant. A virus produced by the *in vitro* mixing of segments of nucleic acid of two (closely related) viruses with fragmented or MULTIPARTITE GENOMES. Used as a method for mapping the genes encoded on the nucleic acid segments of plant viruses, e.g. BROMOVIRUSES. *See* REASSORTMENT.

pseudorinderpest virus. *See* AUJESZKY'S DISEASE VIRUS.

pseudotype. The genome of one virus enclosed in the capsid or outer coat of another resulting from a mixed infection.

PTA. Abbreviation for PHOSPHOTUNGSTIC ACID.

Puchong virus. Unclassified arthropod-borne virus. Isolated from mosquitoes in Malaysia.

Pueblo Viejo virus. Family *Bunyaviridae*, genus *Bunyavirus*.

puffinosis. A disease of possible viral aetiology causing blistering of the feet of the Manx Shearwater (*Puffinus puffinus*). Epizootic among fledglings on Skomer and Skokholm Islands off the coast of Wales.

pulmonary adenomatosis virus of sheep.
Family *Retroviridae*, subfamily *Oncovirinae*.
Causes a pulmonary disease of sheep in which
there are adenocarcinomas which may
metastasise.

pulse-chase. An experimental method in which
a radioactively-labelled product is added to cells
or a cell extract for a short period (pulse) and
followed by a large excess of unlabelled product
(chase) to prevent further significant incorporation of radioactivity. The course of metabolism of
the labelled product is then studied.

Punta Salinas virus. Family *Bunyaviridae*,
genus *Phlebovirus*. Isolated from man in Panama
and Colombia. It causes fever with myalgia and
enlarged liver and spleen.

purine. A heterocyclic compound containing
fused pyrimidine and imidazole rings; adenine
and guanine are the purine components of nucleic
acids. *See* NUCLEIC ACID.

puromycin. A broad spectrum antibiotic produced by a strain of *Streptomyces*. It has the
composition $C_{22}H_{29}O_5N_7$.

pustular dermatitis of camels virus. Family
Poxviridae, subfamily *Chordopoxvirinae*, genus
Parapoxvirus. Causes pustular dermatitis in
camels.

pyrimidine. A heterocyclic organic compound
containing nitrogen atoms at positions 1 and 3.
See NUCLEIC ACID.

Q

Qβ phage. A (LEVIVIRUS) ssRNA phage which has been used extensively in molecular biology studies of RNA phage and bacterial cell functions. Virions (mw. 4.2×10^6) are isometric, about 23 nm. in diameter, sediment at 83S and have a buoyant density in CsCl of 1.47 g/cc. Particles contain 180 molecules of the major coat protein (132 amino acid residues) and one molecule each of two 'A' proteins (mw. $c.40 \times 10^3$). The RNA genome is a (+)-sense ssRNA (mw. 1.5×10^6). The virus shows almost no immunological cross-reactivity with MS2 PHAGE, the LEVIVIRUS type species. The virus is specific for enterobacteria containing the F plasmid.
Fiers, W. (1978) **In** Comprehensive Virology. Vol. 13. p. 69, ed. H. Fraenkel-Conrat and R.R. Wagner. Plenum Press: New York.
Ritchie, D.A. (1983) **In** Topley and Wilson's Principles of Bacteriology, Virology and Immunity. Vol. 1. p. 177. ed. G. Wilson, A. Miles and M.T. Edward Arnold: London.

Qβ replicase. Replicase (RNA-dependent RNA polymerase) complex isolated in bacterial infections with Qβ phage. It is composed of four non-identical subunits: a) the 30S ribosomal protein S1 (mw. 70×10^3); b) the phage-coded replicase subunit (mw. 65×10^3); c) and d) protein synthesis elongation factors EF-TU (mw. 45×10^3) and EF-TS (35×10^3). The complex binds avidly and rather specifically to Qβ phage RNA.
Fiers, W. (1978) **In** Comprehensive Virology. Vol. 13. p. 69. ed. H. Fraenkel-Conrat and R.R. Wagner. Plenum Press: New York.

Qalyub virus. Family *Bunyaviridae*, genus *Nairovirus*. Isolated from the tick *Ornithodoros erraticus* in Egypt. Not reported to cause disease.

quail adenovirus. Synonym: QUAIL BRONCHITIS VIRUS. Family *Adenoviridae*, genus *Aviadenovirus*. Causes a respiratory disease in quail which is accompanied by a sudden fall in egg production and is often fatal. Chickens and turkeys are also susceptible. Embryonated chicken eggs are killed by inoculation of the virus.

quail avipoxvirus. Family *Poxviridae*, genus *Avipoxvirus*. Member of a group of antigenically related viruses which infect a number of different hosts, e.g. canary, pigeon, sparrow.

quail bronchitis virus. *See* QUAIL ADENOVIRUS.

quail pea mosaic virus. A *Comovirus*.
Moore, B.J. and Scott, H.A. (1981) CMI/AAB Descriptions of Plant Viruses No. 238.
Francki, R.I.B. *et al.* (1985) **In** Atlas of Plant Viruses. Vol. 2. p. 1. CRC Press: Boca Raton, Florida.

quakka poxvirus. Synonym: MARSUPIAL PAPILLOMA VIRUS. Probably belongs to the family *Poxviridae*. Isolated from the cytoplasm of cells in the dorsum of the tails of the quakka *Setonix brachyurus*.
Papadimitriou J.M. & Ashman R.B. (1972) J. gen. Virol. **16**, 87.

quaranfil virus. An unclassified arbovirus. Isolated from the cattle egret *Bubuleus ibis* and from pigeons but it is probably tick-borne. Associated with a febrile illness in man.

quasi-equivalence theory. A theory proposed by Caspar and Klug to solve the dilemma of the fact that the identical chemical subunits in an icosahedral virus particle are not arranged in a strictly mathematical equivalent manner. They suggested that the bonds between the subunits are the same throughout the particle and that they have to be slightly deformed in different ways in different parts of the structure.
Caspar, D.L.D. and Klug, A. (1962) Cold Spring Harbor Symp. Quant. Biol. **27**, 1.

Queensland Kashmir bee virus. Unclassified small RNA virus (possible INSECT PICORNAVIRUS)

183

isolated from the honey bee, *Apis mellifera*, in Queensland, Australia. The virus is physically indistinguishable from KASHMIR BEE VIRUS. Two strains have been identified, differing in their structural proteins (mw. 39.8, 33 and 24.8 x 10^3) and antigenic properties. Serologically related to, but distinct from KASHMIR BEE VIRUS, SOUTH AUSTRALIA KASHMIR BEE VIRUS and NEW SOUTH WALES KASHMIR BEE VIRUS.

Bailey, L. *et al.* (1979) J. gen. Virol. **43**, 641.

R

R plasmids. Plasmids containing genes for drug resistance that are particularly common in enterobacteria but also present in other gram-negative and gram-positive bacteria. Some code for sex pili and can influence phage host range. Thus the broad host range plasmid RP1 provides phage PRR1 (which adsorbs specifically to the RP1 pilus) with an equally broad host range.
Hardy, K. (1986) Bacterial Plasmids. 2nd ed. Thomas Nelson: Walton-on-Thames.
Ritchie, D.A. (1983) In Topley and Wilson's Principles of Bacteriology, Virology and Immunity. Vol. 1. p. 177, ed. G. S. Wilson, A. A. Miles and M.T. Parker. Edward Arnold: London.

R-loop mapping. A technique in which ssRNA is annealed to the complementary strand of partially denatured dsDNA. The formation of the RNA:DNA hybrid displaces the opposite DNA strand as a loop which can be visualised under the electron microscope. The region of the DNA complementary to the RNA can then be mapped.

rabbit fibroma virus. Synonym: SHOPE FIBROMA VIRUS. Family *Poxviridae*, genus *Leporipoxvirus*. Causes benign fibromas in the natural host, the wild cotton-tail rabbit. The extracted virus causes fibromas when injected into wild and domestic rabbits. Can be grown in cultures of rabbit, guinea pig, rat and human cells with cpe. See BERRY-DEDRICK phenomenon.
Berry G.P. and Dedrick H.M. (1936) J. Bact. **31**, 50.

rabbit herpesvirus. Family *Herpesviridae*, genus *Gammaherpesvirus*.

rabbit kidney vacuolating virus. See RABBIT VACUOLATING VIRUS

rabbit papilloma virus. Synonym: SHOPE PAPILLOMA VIRUS. Family *Papovaviridae*, genus *Papillomavirus*. Infects cotton-tail rabbits *Sylvilagus floridanus* naturally. Domestic rabbits can be infected by scarification. Skin warts, which may become malignant, are produced.
Stevens, J.G. and Wettstein, F.O. (1979) J. Virol. **30**, 891.

rabbit parvovirus. Family *Parvoviridae*, genus *Parvovirus*. Isolated from faeces; probably a nonpathogenic organism. Agglutinates human blood group O erythrocytes.

rabbit poxvirus. Family *Poxviridae*, subfamily *Chordopoxvirinae*, genus *Orthopoxvirus*. Similar to vaccinia virus. Causes generalised disease when given by the respiratory route.

rabbit reticulocyte lysate. A cell-free system prepared from lysed rabbit reticulocytes which is used for the translation of eukaryotic mRNAs. The rabbit is made anaemic by injection with acetyl phenylhydrazine and the reticulocytes extracted. After lysis the endogenous mRNAs are destroyed using micrococcal nuclease which is then inactivated by the addition of ETHYLENE-GLYCOLBIS(AMINOETHYLETHER)TETRA-ACETIC ACID. Translation in the extract is then totally dependent upon added mRNA. See WHEAT GERM EXTRACT.
Pelham, H.R.B. and Jackson, R.J. (1976) Eur. J. Biochem. **67**, 247.

rabbit syncytium virus. Probable member of the family *Reoviridae*, genus *Orbivirus*.

rabbit type C endogenous virus. Family *Retroviridae*, genus *Type C oncovirus*. Causes lymphosarcoma. Primary lymphosarcoma cell cultures from WH/J rabbits produce C type virus particles when treated with iododeoxyuridine. The particles contain RNA-dependent DNA polymerase.
Bedigian H.G. *et al*. (1978) J. Virol. **27**, 313.

rabbit vacuolating virus. Synonym: RABBIT KIDNEY VACUOLATING VIRUS. Family *Papovaviridae*, genus *Polyomavirus*. Infects cotton-tail rabbits without causing disease. Replicates in do-

185

mestic and cotton-tail rabbit kidney cell cultures producing vacuolation.

rabies virus. Synonym: HYDROPHOBIA VIRUS, LYSSAVIRUS, RAGE VIRUS. Family *Rhabdoviridae*, genus *Lyssavirus*. Can infect almost all warm-blooded animals, usually leading to death. Recovery is rare. Two forms, furious rabies and dumb rabies are seen in some species. The natural reservoir of infection varies: in Europe the fox is the most important but in the USA skunks, foxes, bats and raccoons spread the disease. Other species are involved in S. African (mongoose), India (jackal), South and Central America (vampire bat). The virus can be grown in tissue culture and in the brains of several species. Vaccines are available.
Turner, G.S. (1977). Recent Advances in Clinical Virology **1**, 79.

raccoon poxvirus. Family *Poxviridae*, subfamily *Chordopoxvirinae*, genus *Orthopoxvirus*. Isolated from raccoons in the USA. Experimental infection of raccoons results in a silent infection with the production of antibody. Replicates in chlorioallantoic membrane, producing white pocks, and in Vero cells, giving cpe.

Rachiplusia ou nuclear polyhedrosis virus. BACULOVIRUS (subgroup A) isolated from the looper, *R. ou*. Regarded as a genotypic variant of the prototype BACULOVIRUS, *AUTOGRAPHA CALIFORNICA* MNPV; the two viruses are very closely related and form stable recombinants in mixed infections.
Croizier, G. *et al.* (1988) J. gen. Virol. **69**, 177.

radial immunodiffusion. A serological test in which the antigen, placed in wells in agar gel containing antibody, diffuses radially into the agar and the resulting antigen-antibody complex forms a halo or ring of precipitate around the well. The diffusion continues until there is no more antigen available; the concentration of the antigen can be estimated by comparison of the diameter of the ring with those produced by standards of known concentration.

radioactivity. A particular type of radiation emitted by a radioactive substance. There are three major types of radioactivity, α decay in which an energetic helium ion is ejected, β decay in which an energetic negative ion is emitted and γ which involves the ejection of photons. The SI units are becquerel (Bq), $1 Bq = 1$ nuclear transformation per second; curies (Ci) are commonly

used as units, $1 Ci = 3.7 \times 10^{10}$ Bq. The radioisotopes most commonly used in virology are:

Isotope	Half life	Emission
^{14}C	5730y	β
^{3}H	12.43y	β
^{35}S	87.4d	β
^{32}P	14.3d	β
^{125}I	60.0d	γ

radioimmunoassay. An immunoassay in which a radioisotope (frequently ^{125}I) is attached as a reporter molecule to the antibody or antigen. *See* RADIOIMMUNOPRECIPITATION.

radioimmunoprecipitation. A serological test in which one of the reactants, usually the antibody, is labelled with a radioisotope; ^{125}I is the most commonly used radioisotope. As the amount of isotope precipitated with the antigen-antibody complex can be measured, it is a very accurate and sensitive method for quantifying immunoprecipitation. The amount of antibody in an immunoprecipitate can also be measured using radiolabelled PROTEIN A.

radish mosaic virus. A *Comovirus*.
Campbell, R.N. (1973) CMI/AAB Descriptions of Plant Viruses No. 121.
Francki, R.I.B. *et al.* (1985) **In** Atlas of Plant Viruses. Vol. 2. p. 1. CRC Press: Boca Raton, Florida.

radish yellow edge virus. A possible member of the *Cryptovirus* group, subgroup A.
Boccardo, G. *et al.* (1987) Adv. Virus Res. **32**, 171.

radish yellows virus. A *Luteovirus*; considered to be a strain of BEET WESTERN YELLOWS VIRUS.
Casper, R. (1988) **In** The Plant Viruses. Vol. 3. p. 235. ed. R. Koenig. Plenum Press: New York.

rage virus. *See* RABIES VIRUS.

Ranavirus. (Latin 'rana' = frog). Genus in the family *Iridoviridae*.

ranid herpesvirus 1. Synonym: LUCKÉ VIRUS. Family *Herpesviridae*. Causes renal carcinoma in *Rana pipiens* in N., Central and N.E. parts of the USA and adjacent parts of Canada. Replicates in frog cell cultures.
Granoff, A. (1973). **In**: The Herpesviruses, ed. A.S. Kaplan, Academic Press.

ranid herpesvirus 2. Synonym: FROG VIRUS 4. Family *Herpesviridae*. Isolated from urine of tumour-bearing frogs but genetically and antigenically distinct from ranid herpesvirus 1.

Ranikhet disease virus. *See* NEWCASTLE DISEASE VIRUS.

Ranunculus repens symptomless virus. A possible plant *Rhabdovirus*.
Francki, R.I.B. *et al.* (1985) **In** Atlas of Plant Viruses. Vol. 1. p. 73. CRC Press: Boca Raton, Florida.

Raphanus rhabdovirus. A probable plant *Rhabdovirus*.
Francki, R.I.B. *et al.* (1985) **In** Atlas of Plant Viruses. Vol. 1. p. 73. CRC Press: Boca Raton, Florida.

raspberry bushy dwarf virus. Unclassified, isometric particles, 33 nm. in diameter, which sediment at 115S. They contain three species of ssRNA (mw. 2.0, 0.8 and 0.3 x 10^6) and one protein species (mw. 29 x 10^3).
Murant, A.F.(1976) CMI/AAB Descriptions of Plant Viruses No. 165.

raspberry leaf curl virus. A possible *Luteovirus*.
Francki, R.I.B. *et al.* (1985) **In** Atlas of Plant Viruses. Vol. 1. p. 137. CRC Press: Boca Raton, Florida.
Casper, R. (1988) **In** The Plant Viruses. Vol. 3. p. 235. ed. R. Koenig. Plenum Press: New York.

raspberry ringspot virus. A *Nepovirus*.
Murant, A.F. (1978) CMI/AAB Descriptions of Plant Viruses No. 198.
Francki, R.I.B. *et al.* (1985) **In** Atlas of Plant Viruses. Vol. 2. p. 23. CRC Press: Boca Raton, Florida.

raspberry vein chlorosis virus. A plant *Rhabdovirus*, subgroup 1; aphid transmitted.
Jones, A.T. (1977) CMI/AAB Descriptions of Plant Viruses No. 174.
Francki, R.I.B. *et al.* (1985) **In** Atlas of Plant Viruses. Vol. 1. p. 73. CRC Press: Boca Raton, Florida.

rat coronavirus. Family *Coronaviridae*, genus *Coronavirus*. Causes silent pulmonary infection of laboratory rats. New-born rats given intranasal inoculation contract pneumonia and usually die. Older rats are resistant but develop antibodies.

Replicates in primary rat kidney cell cultures and gives cpe. Antigenically related to mouse hepatitis virus.

rat cytomegalovirus. Synonym: RAT SUBMAXILLARY GLAND VIRUS. Family *Herpesviridae*, subfamily *Betaherpesvirinae*, genus *Murine cytomegalovirus*. Present in the salivary glands of rats. Can cause abortion in laboratory rats.

rat leukaemia virus. Synonym: RAT SARCOMA VIRUS. Family *Retroviridae*, subfamily *Oncovirinae*, genus *Type C oncovirus*. A stable transforming virus.
Rasheed, S. *et al.* (1978) Proc. Nat. Acad. Sci. USA. **75**, 2972.

rat parvovirus. *See* KILHAM RAT VIRUS.

rat sarcoma virus. *See* RAT LEUKAEMIA VIRUS.

rat submaxillary gland virus. *See* RAT CYTOMEGALOVIRUS.

rate zonal gradient. A technique in which a sample containing macromolecules is centrifuged through a gradient of an inert material (e.g. sucrose, glycerol) and in which the constituents of the sample are separated on their SEDIMENTATION RATES as bands. The rate of sedimentation is influenced by the mw., size and shape of the macromolecule and not significantly by its density. *See* SUCROSE DENSITY GRADIENT, ISOPYCNIC GRADIENT.

Rauscher leukaemia virus. Family *Retroviridae*, subfamily *Oncovirinae*, genus *Type C oncovirus*. Similar to Friend leukaemia virus. A defective virus requiring a helper leukaemia virus for replication. Its genome contains specific leukaemia-producing sequences.

RAV (Rous associated virus). Family *Retroviridae*, subfamily *Oncovirinae*, genus *Type C oncovirus*. Acts as a helper for defective Rous sarcoma virus, providing the coat protein genes and thus controlling surface antigens and host range.

razavirus. Family *Bunyaviridae*, genus *Nairovirus*.

Razdan virus. Family *Bunyaviridae*, unassigned to any genus. Isolated from pool of female ticks *Dermacentor marginatus* collected from sheep in

USSR.

reactivation. The activation of a virus from a latent stage, e.g. HERPES VIRUS, or the 'rescue' of a defective virus. *See* RESCUE.

reading frame. A sequence of codons in RNA or DNA beginning with the initiation codon AUG. *See* OPEN READING FRAME.

readthrough. The reading of an mRNA through a STOP CODON. A suppressor tRNA causes the insertion of an amino acid into a growing polypeptide chain in response to the stop codon. The readthrough protein is thus longer than the usual polypeptide, e.g. with TOBACCO MOSAIC VIRUS, P126 is read through a stop codon to give P183.

reannealing. The linking by hydrogen bonds of complementary strands of nucleic acid after melting. *See* HYBRIDISATION.

reassortment. Synonym: GENETIC REASSORTMENT. The production of a hybrid virus which contains parts derived from the genomes of two viruses in a mixed infection. The process is known to occur with viruses which have segmented or multipartite genomes, e.g. ORTHOMYXOVIRUSES, REOVIRUSES. *See* PSEUDORECOMBINATION.

receptor. A site or structure in a cell which combines with a biological substance (e.g. virus) or drug.

recombinant. A term which has two uses. In classical genetics it is an organism which contains a combination of alleles different from either of its parents. In molecular biology it is a molecule containing a new combination of DNA or RNA sequences, e.g. a plasmid with 'foreign' DNA cloned into it.

recombinant DNA. DNA molecules in which sequences, not normally contiguous, have been placed next to each other by *in vitro* methods. *See* GENETIC ENGINEERING.

recombination. The exchange of genetic material from two or more virus particles into recombinant progeny virus during a mixed infection. It can be either intramolecular, within a single species of nucleic acid, or intermolecular, by exchange of nucleic acid species. *See* REASSORTMENT.

red clover cryptic virus 1. A member of the *Cryftovirus* group, subgroup A.
Boccardo, G. *et al.* (1987) Adv. Virus Res. **32**, 171.

red clover cryptic virus 2. A member of the *Cryptovirus* group, subgroup B.
Boccardo, G. *et al.* (1987) Adv. Virus Res. **32**, 171.

red clover enation mosaic virus. An unclassified plant virus with isometric particles, 28 nm. in diameter.
Musil, M. and Leskova, O. (1980) Ochr. Rost. **6**, 33.

red clover mosaic virus. A possible plant *Rhabdovirus*, subgroup 2.
Francki, R.I.B. *et al.* (1985) **In** Atlas of Plant Viruses. Vol. 1. p. 73. CRC Press: Boca Raton, Florida.

red clover mottle virus. A *Comovirus*.
Valenta, V. and Marcinka, K. (1971) CMI/AAB Descriptions of Plant Viruses No. 74.
Francki, R.I.B. *et al.* (1985) **In** Atlas of Plant Viruses. Vol. 2. p. 1. CRC Press: Boca Raton, Florida.

red clover necrotic mosaic virus. A *Dianthovirus*.
Hollings, M. and Stone, O.M. (1977) CMI/AAB Descriptions of Plant Viruses No. 181.

red clover vein mosaic virus. A *Carlavirus*.
Varma, A. (1970) CMI/AAB Descriptions of Plant Viruses No. 22.
Francki, R.I.B. *et al.* (1985) **In** Atlas of Plant Viruses. Vol. 2. p. 173. CRC Press: Boca Raton, Florida.

red disease of pike virus. Synonym: PIKE FRY DISEASE RHABDOVIRUS. Family *Rhabdoviridae*, unassigned genus. Causes haemorrhages in the muscles and kidneys of young pike, *Esox lucius*. Disease presents as red swollen areas on the trunk and is fatal. Virus can be grown in FHM cells, giving cpe.
Hill B.J. *et al.* (1975) J. gen. Virol. **27**, 369.

reed canary mosaic virus. A possible *Potyvirus*.
Francki, R.I.B. *et al.* (1985) **In** Atlas of Plant Viruses. Vol. 2. p. 183. CRC Press: Boca Raton, Florida.

refractive index. The ratio of the phase velocity of light in a vacuum to that in a specific medium. Measured using a refractometer. The symbol for refractive index is n; when measured using the sodium line at 25°C it is given as n_D^{25}. In virology it is mainly used to measure the concentration of solutions of caesium salts or sucrose; densities (ρ) can be calculated using the following formulae:

For CsCl $\rho = 10.8601 \times n_D^{25} - 13.4974$
For Cs_2SO_4 $\rho = 12.1200 \times n_D^{25} - 15.1662$
(density range 1.15-1.40 g/cc)
or $= 13.6986 \times n_D^{25} - 17.3233$ (density range 1.40-1.70 g/cc)

Vinograd, J. and Hearst, J.E. (1962) Progr. Chem. Org. Natur. Prod. **20**, 373.

Rehmannia virus X. A *Potexvirus*, occurs in Japan.
Doi, Y. *Personal communication.*

reiterated sequence. Nucleotide sequence which occurs many times in a nucleic acid.

relative centrifugal force (RCF). The centrifugal force induced by centrifugation relative to the force of gravity. It can be calculated from:
$$RCF (g) = 1118 \times 10^{-8} \times r \times N^2$$
or
$$RCF (g) = 284 \times 10^{-7} \times R \times N^2$$
where N = r.p.m.
R = distance from centre of rotation in inches.
r = distance from centre of rotation in cms.

Reoviridae. (from **R**espiratory **E**nteric **O**rphan viruses). A family containing six genera of viruses and several unassigned isolates, infecting vertebrate, invertebrate and plant hosts. Virus particles are unenveloped and icosahedral, 60-80 nm. in diameter and sedimenting between 400-730S with a buoyant density in CsCl of 1.36-1.44 g/cc. (depending on genus). Most genera have particles with two protein coats, the internal coat (or core; about 40 nm. in diameter) having 12 spikes arranged icosahedrally. The spatial inter-relationships of capsomeres have not been defined with certainty. Virus particles contain five to ten polypeptides (mw. 15-155 x 10³) of which five to six are present in the core. The RNA genome consists of ten to twelve segments of linear dsRNA (total mw. 12-20 x 10⁶) all packaged within a single virion, each segment representing a single gene. During infection, transcription of the genome segments into mRNA occurs by a conservative process, using transcriptase, nucleotide phosphohydrolase and RNA capping

enzymes present in the viral core. The viral mRNA is not polyadenylated. Virus replication occurs in the cytoplasm and is characterised by the development of perinuclear virus inclusions. Genetic recombination has been observed within some genera where it occurs readily by genome segment reassortment. The experimental host range of many isolates is wide. Genera isolated from vertebrates include REOVIRUS, ORBIVIRUS and ROTAVIRUS. Isolates infecting plants include the genera PHYTOREOVIRUS and FIJIVIRUS. Members of the genera ORBIVIRUS, PHYTOREOVIRUS and FIJI-VIRUS have insect vectors. CYTOPLASMIC POLY-HEDROSIS VIRUSES (sometimes referred to as CYPOVIRUS) have only been isolated from invertebrates. The different genera are readily distinguished by virus particle morphology and the number of RNA genome segments. Unclassified Reovirus-like agents include LEAFHOPPER A VIRUS, *DROSOPHILA* F VIRUS, *CERATITIS* REOVIRUS I, HOUSEFLY VIRUS and *MACROPIPUS* PARALYSIS VIRUS.
Joklik, W.K. (1983) The *Reoviridae*. Plenum Press: New York.

 100nm

repeated sequence. A nucleotide sequence which occurs more than once in a DNA or RNA sequence either in the same (direct repeats) or opposite (inverted repeats) orientation.

replicase. The enzyme involved in the replication of viral genomic nucleic acid. *See* DNA-DEPENDENT DNA POLYMERASE, DNA-DEPENDENT RNA POLYMERASE, RNA-DEPENDENT RNA POLYMERASE, REVERSE TRANSCRIPTASE, RNA REPLICASE.

replication. The duplication of the genomic DNA or RNA of a virus.

replicative form. The intracellular form of viral nucleic acid which is active in replication. For viruses with ss genomes it is usually the ds form of that nucleic acid.

replicative intermediate. Form of nucleic acid produced during the replication of a viral genome. For ssRNA viruses it is partially ds and partially ss and is formed by the simultaneous synthesis of one or more strands from a single complementary strand.

replicon. A portion of DNA that is able to

replicate from a single origin. Most viruses have a single origin of replication and thus the entire genome is a replicon.

repressor. A protein which prevents RNA polymerase from starting RNA synthesis by binding to a specific DNA sequence upstream of the transcription initiation site.

Reptilian type C Oncovirus. Family *Retroviridae*, subfamily *Oncovirinae*, genus *Type C oncovirus*.

rescue. The REACTIVATION of a defective virus either by RECOMBINATION or by COMPLEMENTATION of the defective functions.

resistance. The ability of a host to minimise the effects of a virus (or other pathogen) infection. There are various forms of resistance ranging from tolerance in which symptom expression is suppressed, through HYPERSENSITIVITY in which infection is restricted to a few cells around the originally infected cell, to immunity in which the virus does not multiply in the host. In animals resistance can also be effected by an immune response and/or INTERFERON production.

respiratory syncytial virus. Family *Paramyxoviridae*, genus *Pneumovirus*. Important cause of respiratory tract infection in infants. Causes colds in chimpanzees. Can be serially passed in ferrets without causing any signs of illness. Replicates in human cell lines such as HeLa and Hep2. Bovine strain causes mild respiratory disease in cattle. Replicates in bovine kidney and lung cell cultures, causing syncytia. Cross-reacts in serological tests with the human strain.

Restan virus. Family *Bunyaviridae*, genus *Bunyavirus*, serological group C. Isolated from *Culex* sp. in Trinidad and Surinam. Can cause fever in man.

restriction endonuclease. An enzyme, usually from a bacterium, which recognises a specific sequence of bases in dsDNA and cuts the DNA at or near that position. Type I enzymes bind to the recognition site but usually cut the DNA at random sites. Type II enzymes bind at the recognition site and cut either at or close to that site. The cuts in the two DNA strands may be either opposite one another, generating BLUNT ENDS, or not opposite one another, giving STICKY ENDS with 3′ or 5′ overlaps. The type II enzymes are used extensively in recombinant DNA technology. Some commonly used type II enzymes are shown in the table. A full list of restriction endonucleases is published annually in Nucleic Acids Research.

Enzyme	Cutting site ↓	End
Ava 1	5′ CPyCGPuG 3′ 3′ GPuGCPyC 5′ ↓ ↑	5′ sticky
Bam H1	5′ GGATCC 3′ 3′ CCTAGG 5′ ↓ ↑	5′ sticky
Bgl II	5′ AGATCT 3′ 3′ TCTAGA 5′ ↓ ↑	5′ sticky
Eco R1	5′ GAATTC 3′ 3′ CTTAAG 5′ ↓ ↑	5′ sticky
Hae III	5′ GGCC 3′ 3′ CCGG 5′ ↑↓	Blunt
Hha I	5′ GCGC 3′ 3′ CGCG 5′ ↕	3′ sticky
Hind III	5′ AAGCTT 3′ 3′ TTCGAA 5′ ↓ ↑	5′ sticky
Hpa I	5′ GTTAAC 3′ 3′ CAATTG 57 ↑ ↓	Blunt
Kpn I	5′ GGTACC 3′ 3′ CCATGG 5′ ↑ ↓	3′ sticky
Pst I	5′ CTGCAG 3′ 3′ GACGTC 5′ ↓↑	3′ sticky
Sau 3A	5′ GATC 3′ 3′ CTAG 5′ ↓↑	5′ sticky
Sma I	5′ CCCGGG 3′ 3′ GGGCCC 5′ ↑	Blunt

restriction endonuclease map. The cutting sites for restriction endonucleases marked on a linear or circular representation of a dsDNA molecule.

restriction enzyme. Synonym: RESTRICTION ENDONUCLEASE.

restriction fragment. A polynucleotide fragment produced by cutting dsDNA with a RESTRICTION ENDONUCLEASE.

reticulocyte lysate. *See* RABBIT RETICULOCYTE LYSATE.

Retroviridae. (Latin 'retro' = backwards, referring to reverse transcription during replication.) A family of spherical enveloped viruses comprising three sub-families, ONCOVIRINAE, SPUMAVIRINAE and LENTIVIRINAE and several genera and subgenera. The particles are 80-100 nm. in diameter with glycoprotein surface projections of 8 nm. There is an icosahedral capsid containing a helical ribonu-

100nm

cleoprotein. The central nucleoid is located acentrically in the B-TYPE PARTICLES but concentrically in C-TYPE PARTICLES. There are at least four internal non-glycosylated structural proteins (GAG), two glycosylated envelope (ENV) proteins and reverse transcriptase (POL). The RNA is an inverted dimer of linear (+)-sense. Defective viruses contain monomeric RNA, mw. c.3 x 10^6, polyadenylated at the 3' end and capped at the 5' end. The RNA is not infectious. A cellular tRNA, serving as primer for reverse transcription, is bound by base pairing to a specific primer attachment site about 100 residues from the 5' end of the genome. Entry of the virus into the host cell is probably by fusion of the viral envelope to the plasma membrane. The mechanism of uncoating is not known. Replication starts with reverse transcription of virus RNA into DNA which becomes integrated into the chromosomal DNA of the host. Integrated provirus is transcribed into virus RNA and mRNAs. Most primary translational products are polyproteins which require proteolytic cleavage before becoming functional. Endogenous oncoviruses occur widely among vertebrates and are associated with many diseases. Transmission is both vertical and horizontal.
Matthews, R.E.F. (1982) Intervirology **17**, 124.

rev./min. *See* REVOLUTIONS PER MINUTE

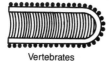

Vertebrates
and
invertebrates

reverse passive haemagglutination. A sensitive serological test in which red blood cells are coated with virus-specific antibody and used to test for the presence of antigen. If virus antigen is present, the red blood cells are agglutinated. *See* PASSIVE HAEMAGGLUTINATION.
Herbert, W.J. (1967) **In** Handbook of Experimental Immunology. p. 720. ed. D.M. Weir. Blackwell Scientific Publications: Oxford.

reverse transcriptase. An enzyme coded by RETROVIRUSES and probably HEPADNAVIRUSES and CAULIMOVIRUSES which makes a DNA copy of a primed RNA molecule. The enzyme is also capable of synthesising DNA from a DNA template. As well as the RNA-dependent DNA polymerase activity, reverse transcriptase carries RNaseH activity which digests the RNA moiety of an RNA-DNA duplex.

reverse transcription. The process of transferring the genetic information from RNA to DNA.

revolutions per minute. A measure of the speed of a centrifuge. *See* RELATIVE CENTRIFUGAL FORCE.

RF. Abbreviation for REPLICATIVE FORM.

RF virus. Family *Papovaviridae*, genus *Polyomavirus*. Isolated from the urine of a renal transplant patient. Indistinguishable from BK VIRUS antigenically but DNAs have 88% homology. Oncogenic in hamsters *in vivo* and transforms hamster cells *in vitro*. Grows in human and monkey cells.

Rhabdionvirus oryctes. Former name used for the virus disease of the coconut rhinoceros beetle, *Oryctes rhinoceros* which is now classified as a member of the NON-OCCLUDED BACULOVIRUS subgroup of the BACULOVIRUS genus (*see* ORYCTES RHINOCEROS BACULOVIRUS).
Huger, A.M. (1966) J. Invertebr. Pathol. 8, 38.

Rhabdoviridae. (Greek 'rhabdos' = a rod.) A large family of bullet or bacilliform-shaped viruses measuring 130-380 x 50-95 nm. with a lipid envelope and surface projections 5-10 nm. long and c.3 nm. in diameter. In thin section a

100nm

Plants

central axial channel is seen. Characteristic cross-striations (spacing 4.5 to 5.0 nm.) are seen in

negatively stained and thin-sectioned particles. Truncated particles ranging from one-tenth to two-thirds of the length of the virus are common except in members infecting plants. Abnormally long and double-length particles and tandem forms are sometimes seen. The virus contains one molecule of non-infectious (–) sense ssRNA (mw. $c.4 \times 10^6$) and five major proteins L (large), G (glycoprotein), N (nucleoprotein), NS (non-structural) and M (matrix). The surface projections consist of the G protein and the inner nucleocapsid, about 50 nm. in diameter, with helical symmetry, consists of an RNA-N protein complex together with the L and NS proteins, surrounded by the M protein. The nucleocapsid is infectious. The virus RNA is transcribed by the virion transcriptase into several (+)-sense RNA molecules corresponding to the five proteins. The natural host range is very wide, including vertebrates, invertebrates, arthropods and plants (*see* PLANT RHABDOVIRUS GROUP). Biological and mechanical vectors include mosquitoes, sandflies, tabanids, midges, mites, leafhoppers and aphids. Some of the viruses have been allotted to genera (e.g. *VESICULOVIRUS* for the vesicular stomatitis virus group and *LYSSAVIRUS* for the rabies virus group) but most have not been classified in this way.
Matthews, R.E.F. (1982) Intervirology **17**, 109.

rhabdovirus carpio. *See* SPRING VIRAEMIA OF CARP VIRUS.

rhabdovirus entamoeba. Family *Rhabdoviridae*, unassigned genus.

rhabdovirus of blue crab. Family *Rhabdoviridae*, unassigned genus.

rhabdovirus of eels. Family *Rhabdoviridae*, unassigned genus.
Hill, B.J. *et al.* (1980) Intervirology **14**, 208.

rhabdovirus of grass carp. Family *Rhabdoviridae*, unassigned genus. Causes disease of grass carp.

Rhadinovirus. A genus in the family *Herpesviridae*, subfamily *Gammaherpesvirinae* with ateline herpesvirus 2 as the type species and ateline herpesvirus 3 and saimirine herpesvirus 2 as other members.

rhinoceros pox virus. Family *Poxviridae*, subfamily *Chordopoxvirinae*, unassigned genus.

Causes disease in rhinoceros. Can be grown in chick embryos.

Rhinovirus. (Greek 'rhis, rhinos' = nose.) Family *Picornaviridae*. A genus comprising the common cold viruses. Distinguished from the other genera by its buoyant density in CsCl (1.40 g/cc) which is considerably higher than that of the enteroviruses, its base composition and by the instability of its members below pH5. There are many different serotypes (on the basis of neutralisation tests) which infect man but many of these may be related antigenically on the basis of priming experiments. There are two serotypes which infect horses. There is no vaccine.

Rhizidiomyces mycovirus. Type member of the RHIZIDIOVIRUS GROUP.

Rhizidiovirus group. (Type member *Rhizidiomyces* mycovirus). A monotypic genus of a fungal virus which has isometric particles, 60 nm. in diameter, which sediment at 625S and band in CsCl at 1.314 g/cc. The capsid comprises many protein species (mw. $26\text{-}84 \times 10^3$) and a single species of dsDNA (mw. 16.8×10^6). Virus particles are found in all stages of fungal development and are especially prevalent in the nuclei of developing sporangia. The virus is transmitted vertically. This genus is considered to have possible affinities with the ADENOVIRIDAE.
Dawe, V.H. and Kuhn, C.W. (1983) Virology **130**, 21.

rhizobiophage. PHAGE isolated from *Rhizobium* spp.

rhizomania. A disease of sugar beet caused by BEET NECROTIC YELLOW VEIN VIRUS and characterised by proliferation of rootlets.

rhododendron necrotic ringspot virus. A possible *Potexvirus*.
Francki, R.I.B. *et al.* (1985) **In** Atlas of Plant Viruses. Vol. 2. p. 159. CRC Press: Boca Raton, Florida.

Rhoeo discolor virus. A strain of TOBACCO MOSAIC VIRUS.
Thompson, S.M. and Corbett, M.K. (1985) Plant Disease **69**, 356.

Rhopalosiphum padi virus. Unclassified small RNA virus (possible INSECT PICORNAVIRUS) isolated from the birdcherry oat aphid, *R. padi*.

Virions are isometric, 27 nm. in diameter, sediment at 162S and have a buoyant density in CsCl of 1.37 g/cc. Particles contain three major proteins (mw. 31, 29.6 and 28.1 x 10^3) and a polyadenylated ssRNA genome which sediments at 31S. The virus is unrelated to barley yellow dwarf virus for which *R. padi* is a common vector (*see* APHID LETHAL PARALYSIS VIRUS).
D'Arcy, C.J. *et al.* (1981) Virology **112**, 346.

rhubarb temperate virus. A possible member of the *Cryptovirus* group, subgroup A.
Boccardo, G. *et al.* (1987) Adv. Virus Res. **32**, 171.

rhubarb virus 1. A possible *Potexvirus*.
Francki, R.I.B. *et al.* (1985) **In** Atlas of Plant Viruses. Vol. 2. p. 159. CRC Press: Boca Raton, Florida.

RI. Abbreviation for REPLICATIVE INTERMEDIATE.

ribavirin. 1-β-D-ribofuranosyl-1,2,4-triazole-3-carboxamide. A synthetic nucleoside analogue of guanosine which interferes with the synthesis of guanylic acid and thus has antiviral properties. It is active against both DNA and RNA viruses, e.g. INFLUENZA, MEASLES, HEPATITIS B, LASSA FEVER VIRUSES *in vitro* but as clinical trials have been inconclusive it does not have any widespread use in prevention or therapy.
Sidwell, R.W. *et al.* (1972) Science **177**, 705.
Streeter, D.G. *et al.* (1973) Proc. Natl. Acad. Sci. USA. **70**, 1174.

ribgrass mosaic virus. A *Tobamovirus*.
Oshima, N. and Harrison, B.D. (1975) CMI/AAB Descriptions of Plant Viruses No. 152.
Wetter, C. (1986) **In** The Plant Viruses. Vol. 2. p. 221. ed. M.H.V. van Regenmortel and H. Fraenkel-Conrat. Plenum Press: New York.

ribonuclease. An enzyme which hydrolyses RNA. There are various ribonucleases with different activities which can be used in the characterisation and sequencing of RNA.

B. cereus RNase: cleaves RNA on the 3' side of pyrimidine residues. Isolated from *Bacillus cereus*.

RNase A: cleaves RNA on the 3' side of pyrimidines to give oligonucleotides ending in pyrimidine 2',3'-cyclic phosphates. Much greater activity on ssRNA than on dsRNA under moderate salt concentrations. Isolated from bovine pancreas where it is the major RNase.

RNase C: cleaves preferentially on the 5' side of cytidine residues; isolated from human placenta.

RNase CL3: cleaves RNA predominantly on the 3' side of cytidine residues and most efficiently in urea at 50°C; cleaves at adenine and guanine residues less frequently and at uridine residues very infrequently. Isolated from chicken liver.

RNase H: *see* REVERSE TRANSCRIPTASE.

RNase P: removes the 5' terminal nucleotides from *E. coli* tRNA precursor molecules to generate the 5' termini of mature tRNA molecules. It comprises a catalytic RNA subunit and a protein subunit.

RNase Phy 1: cleaves RNA on the 3' side of adenine, guanine and uridine residues. Isolated from *Physarum polycephalum*, a slime mould.

RNase Phy M: cleaves RNA on the 3' side of adenine and uridine residues. Also isolated from *Physarum polycephalum*.

RNase S: derived by cleaving the main polypeptide chain of RNase A with the protease subtilisin. Has a similar enzymic activity to RNase A.

RNase T1: cleaves RNA on the 3' side of guanine residues. Isolated from the fungus *Aspergillus oryzae*.

RNase T2: cleaves RNA on the 3' side of all four nucleotides but shows a preference for adenine residues. Also isolated from *Aspergillus oryzae*.

RNase U2: cleaves RNA on the 3′ side of purine residues; the cleavage at guanine is suppressed in urea at pH 3.5, thus making it specific for adenine. Isolated from the fungus *Ustilago sphaerogena*.

RNase III: cleaves dsRNA and also ssRNA at specific sites. Isolated from *E. coli*.

ribonucleic acid. A linear polymer of ribonucleotides. Exists either as ss or ds molecules which are usually linear but can be covalently closed (*see* VIROID). Proteins are translated from messenger (m) RNA which is termed plus-strand RNA. Virus genomes can be ssRNA of the (+)- or (–)-sense or dsRNA. *See* NUCLEIC ACID.

ribonucleoprotein. A complex comprising ribonucleic acid and protein usually linked by electrostatic bonds, e.g. the core of ORTHO-MYXOVIRUSES.

ribonucleoside. A purine or pyrimidine base covalently bound to a ribose sugar molecule. *See* NUCLEIC ACID.

ribonucleotide. A ribonucleoside with one or more phosphate groups esterified to the 5′ position of the sugar moiety. *See* NUCLEIC ACID.

ribose. The sugar of ribonucleotides. *See* NUCLEIC ACID.

ribosomal RNA. The RNA species found in ribosomes; they are known by their sedimentation values (S). Prokaryotic ribosomes contain one 16S rRNA molecule (for *E. coli*, 1541 nucleotides) in the small subunit, one 23S rRNA molecule (for *E. coli*, 2904 nucleotides) and a 5S rRNA (for *E.coli*, 120 nucleotides) in the large subunit. Eukaryotic cytoplasmic ribosomes contain four rRNA species. The large subunit contains a 28S, a 5.8S and a 5S rRNA; the small subunit contains an 18S rRNA.

ribosome. A subcellular particle composed of two subunits (the large and the small) each comprising one or more RNA molecules and several proteins. Ribosomes are responsible for protein synthesis, the small subunit binding to mRNA and then being joined by the large subunit which accepts the aminoacy-tRNA and performs peptide bond formation.

ribosome binding site. A sequence of nucleotides in mRNA to which ribosomes will bind. In prokaryotes this is termed a Shine-Dalgarno sequence (after the Australians who first recognised it) and is a sequence complementary to the 3′ end of the 16S rRNA. These ribosomal binding sites are 3-9 bases long and precede the translation start codon by 3-12 bases. The consensus Shine-Dalgarno sequence is:

$$5' \text{ AAGGAGGU } 3' \text{ mRNA}$$
$$\circ \circ \circ \circ \circ \circ \circ \circ$$
$$3' \text{ AUUCCUCCACUAG } 5' \text{ } E. \text{ coli}$$
$$16S$$

rRNA 3′ end

In eukaryotes the ribosomes are thought to bind to the 5′ end of the mRNA.

rice black streaked dwarf virus. A *Phytoreovirus*.
Shikata, E. (1974) CMI/AAB Descriptions of Plant Viruses No. 135.
Francki, R.I.B. *et al.* (1985) **In** Atlas of Plant Viruses. Vol. 1. p. 47. CRC Press: Boca Raton, Florida.

rice dwarf virus. A *Fijivirus*.
Iida, T.T. *et al.* (1972) CMI/AAB Descriptions of Plant Viruses No. 102.
Francki, R.I.B. *et al.* (1985) **In** Atlas of Plant Viruses. Vol. 1. p. 47. CRC Press: Boca Raton, Florida.

rice gall dwarf virus. A possible member of the *Phytoreovirus* group.
Omura, T. and Inoue, H. (1985) AAB Descriptions of Plant Viruses No. 296.
Francki, R.I.B. *et al.* (1985) **In** Atlas of Plant Viruses. Vol. 1. p. 47. CRC Press: Boca Raton, Florida.

rice giallume virus. A *Luteovirus;* considered to be a strain of BEET WESTERN YELLOWS VIRUS.
Casper, R. (1988) **In** The Plant Viruses. Vol. 3. p. 235. ed. R. Koenig. Plenum Press: New York.

rice grassy stunt virus. A possible member of the *Tenuivirus* group.
Hibino, H. (1986) AAB Descriptions of Plant Viruses No. 320.

rice hoja blanca virus. A possible member of the *Tenuivirus* group.
Morales, F.J. and Niessen, A.I. (1985) AAB Descriptions of Plant Viruses No. 299.

rice necrosis mosaic virus. A member of the *Barley Yellow Mosaic Virus* group.
Inouye, T. and Fujii, S. (1977) CMI/AAB Descriptions of Plant Viruses No. 172.

rice ragged stunt virus. A *Fijivirus*.
Milne, R.G. *et al*. (1982) CMI/AAB Descriptions of Plant Viruses No. 248.
Francki, R.I.B. *et al*. (1985) **In** Atlas of Plant Viruses. Vol. 1. p. 47. CRC Press: Boca Raton, Florida.

rice stripe necrosis virus. A possible member of the *Furovirus* group.
Brunt, A.A. and Shikata, E. (1986) **In** The Plant Viruses, Vol. 2, p.305, ed. M.H.V. van Regenmortel and H. Fraenkel-Conrat. Plenum Press: New York.

rice stripe virus. Type member of the *Rice Stripe Virus* group.
Toriyama, S. (1983) CMI/AAB Descriptions of Plant Viruses No. 269.

rice stripe virus group. Synonym: TENUIVIRUS GROUP.

rice transitory yellowing virus. A plant *Rhabdovirus*, subgroup 2; transmitted by leafhoppers.
Shikata, E. (1972) CMI/AAB Descriptions of Plant Viruses No. 100.
Francki, R.I.B. *et al*. (1985) **In** Atlas of Plant Viruses. Vol. 1. p. 73. CRC Press: Boca Raton, Florida.

rice tungro bacilliform virus. A member of the *Cacao Swollen Shoot Virus* group; requires RICE TUNGRO SPHERICAL VIRUS for leafhopper transmission.
Francki, R.I.B. *et al*. (1985) **In** Atlas of Plant Viruses. Vol. 2. p. 235. CRC Press: Boca Raton, Florida.

rice tungro spherical virus. A possible *Machlovirus*.
Gingery, R.E. (1988) **In** The Plant Viruses. Vol. 3, p.259. ed. R. Koenig. Plenum Press: New York.

rice waika virus. Synonym: RICE TUNGRO SPHERICAL VIRUS.
Galvez, G.E. (1971) CMI/AAB Descriptions of Plant Viruses No. 67.

rice yellow mottle virus. A possible *Sobemovirus*.
Bakker, G.E. (1971) CMI/AAB Descriptions of Plant Viruses No. 149.
Francki, R.I.B. *et al*. (1985) **In** Atlas of Plant Viruses. Vol. 1. p. 153. CRC Press: Boca Raton, Florida.
Hull, R. (1988) **In** The Plant Viruses. Vol. 3. p. 113. ed. R. Koenig. Plenum Press: New York.

rifampicin, rifamycin. An antibiotic which binds to a subunit of the RNA polymerase of susceptible strains of *E. coli*, thus inhibiting RNA synthesis. Does not inhibit eukaryotic cellular RNA polymerases, but 3-oxime derivatives have some inhibitory action. Inhibits replication of POXVIRUSES.

Rift Valley fever virus. Family *Bunyaviridae*, genus *Phlebovirus*. Found in many parts of Africa. Causes abortion and death in pregnant and new-born cattle, sheep and goats. Lambs develop fever, vomiting and diarrhoea but cattle are less seriously affected. Buffaloes, camels and antelopes can be infected and die. The disease in man is much milder. Infection is usually via mosquitoes but contact infection also occurs. Control is by protection against mosquito bites. Mice, guinea pigs, ferrets and dogs can be infected experimentally. The virus multiplies in chick, mouse, rat and human cells and on the chorioallantoic membrane.

Riley virus. Synonyms: LACTIC DEHYDROGENASE VIRUS, ENZYME ELEVATING VIRUS. Family *Togaviridae*, unassigned genus. Isolated from wild and laboratory mice in Australia, Europe and the USA. Can infect *Mus caroli* experimentally, causing a life-long infection with permanent viraemia but no disease. There are abnormally high levels of lactic dehydrogenase in the plasma. Antibodies are produced but the virus antibody complexes are infectious. Replication occurs in mouse cell cultures but there is no cpe.
Mahy B.W.J. & Rowson K.E.K. (1985) J. gen. Virol. **66**, 2297.

Rinderpest virus. Synonym: CATTLE PLAGUE VIRUS. Family *Paramyxoviridae*, genus *Morbillivirus*. Serious infection of many species including cattle, buffalos, sheep, goats, pigs, camels, hippopotami, warthogs, giraffes, yaks and zebus. The disease involves high fever and generalised infection of the alimentary tract leading to constipation followed by diarrhoea. Transmission is by contact. Control is by slaughter and by vaccination. There is an excellent attenuated vaccine. The virus is related serologically to measles, canine distemper and peste des petits ruminants viruses.

ring-necked pheasant leukosis virus. Family *Retroviridae*, subfamily *Oncovirinae*, genus *Type C oncovirus*, sub-genus *Avian type C oncovirus*. A strain of subgroup F chicken leukosis sarcoma virus. Present as an endogenous virus in normal ring-necked pheasant cells. Possesses the genetic information to confer group F host range specificity to virus particles produced with it as a helper virus.

ring test. A serological test in which the antigen solution is mixed with glycerol, placed in a tube and antibody is carefully layered on top. The antigen-antibody precipitate forms as a ring at the interface.

Ringer's solution. A solution used for culturing cells. Consists of 0.66g NaCl, 0.015g KCl, 0.015g CaCl$_2$ per 100ml distilled water with the pH adjusted to 7.8 by the dropwise addition of NaHCO$_3$.
Ringer, S. (1886) J. Physiol. **7**, 291.

ringspot. Symptom induced by certain plant viruses in some hosts, e.g. TOBACCO RINGSPOT VIRUS in tobacco, which consists of chlorotic or necrotic ring LESIONS with a normal green centre. The rings can spread and coalesce giving an 'oak leaf' pattern.

Rio Bravo virus. Synonym: BAT SALIVARY VIRUS. Family *Flaviviridae*, genus *Flavivirus*. Isolated from the salivary gland of a bat caught in California. Vector not known. Infects mice and has caused laboratory infections.

Rio Grande virus. Family *Bunyaviridae*, genus *Phlebovirus*. Isolated from pack rats, *Neotoma micropus* in USA. This species is probably the natural host. Mode of transmission not known. Has not been reported to cause disease. Kills suckling mice when injected. Replicates in Vero cells, giving cpe.

RKV (rabbit) polyoma virus. Family *Papovaviridae*, genus *Polyomavirus*.

RNA. Abbreviation for RIBONUCLEIC ACID.

RNA ligase. An enzyme which can join RNA molecules. The usual source is phage T4 which encodes a ligase which will join RNA to RNA, DNA to DNA or RNA to DNA; the 5' end needs to be a phosphate and the 3' end a hydroxyl group. The enzyme requires ATP.

RNA polymerase. An enzyme synthesising RNA from either a DNA or RNA template. *See* DNA-DEPENDENT RNA POLYMERASE, RNA-DEPENDENT RNA POLYMERASE.

RNA processing. The modification of RNA after transcription. *See* POST-TRANSCRIPTIONAL PROCESSING.

RNA replicase. Synonym: RNA-DEPENDENT RNA POLYMERASE.

RNA segment. A distinct piece of RNA. Frequently refers to the genome segments of segmented genome viruses, e.g. the dsRNAs of REOVIRUS or the ssRNAs of ORTHOMYXOVIRUS.

RNA transcriptase. Synonym: RNA- OR DNA-DEPENDENT RNA POLYMERASE.

RNA-dependent DNA polymerase. Synonym: REVERSE TRANSCRIPTASE.

RNA-dependent RNA polymerase. An enzyme which transcribes RNA from an RNA template. Most RNA viruses encode their version of this enzyme. It is frequently found encapsidated in viruses with minus-strand (e.g. MYXO- AND PARAMYXO-VIRUS) or ds (e.g. REOVIRUS) RNA genomes.

RNase. Abbreviation for RIBONUCLEASE.

robinia mosaic virus. A strain of PEANUT STUNT VIRUS.
Richter, J. *et al.* (1979) Acta Virol. **23**, 489.
Schmelzer, K. (1971) CMI/AAB Descriptions of Plant Viruses No. 65.

Rocio virus. Family *Flaviviridae*, genus *Flavivirus*. Isolated from brain of a patient with encephalitis in Brazil. Caused an epidemic of encephalitis in Sao Paulo in 1975. Possibly transmitted by mosquito. Kills mice and hamsters on injection. Replicates in Vero and BHK21 cells giving cpe.
Monath T.P. *et al.* (1978) Am. J. trop. Med. Hyg. **27**, 1251.

rocket immunoelectrophoresis. An immunological technique for the determination of a single constituent in a protein mixture in a number of samples. The diluted samples are applied side by side into circular wells in an agarose gel containing an antiserum to the protein of interest. Upon electrophoresis rocket-shaped precipitates are formed which identify the protein; the protein is quantified by the area beneath the precipitin line. Axelsen, N.H. (1973) Scand. J. Immunol. **2**, suppl.1, 7.

rod-shaped phage. *See* INOVIRIDAE.

rodent paramyxoviruses. Family *Paramyxoviridae*, genus *Paramyxovirus*. Several species have been described: parainfluenza virus type 1 murine, PNEUMONIA VIRUS OF MICE, PEROMYSCUS VIRUS, NARIVA VIRUS, Mossman virus.

rodent parvoviruses. Family *Parvoviridae*, genus *Parvovirus*. Several have been described: latent rat virus, H-3 virus, X-14 virus, L-S virus, haemorrhagic encephalopathy virus of rats, Kirk virus, H-1 virus, HT virus, HB virus, minute virus of mice.

rolling circle. A nucleic acid replication mechanism in which the template is a circular molecule. The newly synthesised nucleic acid either copies as one circle (often displacing a second strand) or continues on round several times giving a CONCATEMERIC molecule. This mechanism is found in the DNA replication of phage λ and in the RNA replication of VIROIDS.

rose mosaic virus. An *Ilarvirus*. Different isolates are regarded as being strains of either *Prunus Necrotic Ringspot Virus* or *Apple Mosaic Virus*. Francki, R.I.B. (1985) **In** The Plant Viruses. Vol. 1. p. 1. ed. R.I.B. Francki. Plenum Press: New York.

rose tobamovirus. A *Tobamovirus*. Brunt, A.A. (1986) **In** The Plant Viruses. Vol. 2. p. 283. ed. M.H.V. van Regenmortel and H. Fraenkel-Conrat. Plenum Press: New York.

Ross River virus. Family *Togaviridae*, genus *Alphavirus*. Isolated from birds and mosquitoes in Fiji and Australia. Causes fever in man. Transmitted by *Aedes vigilax* and *Culex annulirostris*. Antibodies are found in cattle, horses, sheep, dogs, rats, bats and kangaroos in Australia, New Guinea and Solomon Islands.

Aaskov J.G. & Davies, C.E.A. (1979) J. immun. Meth. **25**, 37.

Rotavirus. Synonym: STELLAVIRUS. Genus of family *Reoviridae*. The capsid has a clearly defined outer layer, like the rim of a wheel, with an inner layer which gives the appearance of spokes. These features suggested the name. The outer shell is often removed spontaneously so that single as well as double-shelled particles are found. The RNA is present as eleven segments. There are eight proteins. Rotaviruses are found in many species, particularly infants. Many particles are present in the faeces. Some isolates grow well in cell culture but most grow poorly. The viruses from different host species can be distinguished serologically but there are common antigens associated with the inner capsid proteins. Protection requires gut immunity. This is provided by maternal antibodies. McNulty M.S. (1978) J. gen. Virol. **40**, 1. Flewett T.H. and Woode G.N. (1978) Arch. Virol. **57**, 1.

rougeole virus. *See* MEASLES VIRUS.

rough membrane. Endoplasmic reticulum encrusted with ribosomes. mRNAs are translated by the ribosomes and the membrane isolates and transports the proteins. *See* SMOOTH MEMBRANE.

Rous sarcoma virus. Family *Retroviridae*, subfamily *Oncovirinae*, genus *Type C oncovirus* group. The first virus shown to cause a solid tumour. There are a number of strains varying in their oncogenicity and host range. They are usually defective, requiring a leukaemia virus to code for the virus coat, which determines host range. The viruses transform cells in culture which do not produce infective virus unless also infected with a leukaemia virus.

Roux flask. A glass (or plastic) flask for culturing cells. Named after Emile Roux, assistant to Louis Pasteur.

Rowson-Parr virus. Family *Retroviridae*, subfamily *Oncovirinae*, genus *Type C oncovirus* group. A strain of mouse leukaemia virus isolated from a Friend virus preparation by end-point dilution. Depresses the immune response and eventually gives rise to neoplastic lymphoid cells in the spleen and other lymphoid tissues. Rowson K.E.K. and Parr I.B. (1970) Int. J. Cancer **5**, 96.

Royal Farm virus. Family *Flaviviridae*, genus *Flavivirus*. Isolated from a tick *Argas hermanni* in Afghanistan.

rRNA. Abbreviation for RIBOSOMAL RNA.

RT virus. Family *Parvoviridae*, genus *Parvovirus*. Isolated from the RT line of rat fibroblasts.
Hallauer C. *et al.*, (1971) Arch. Virol. **35**, 80.

Rubarth's disease virus. *See* CANINE HEPATITIS VIRUS.

rubber policeman. A glass rod with a rubber tube on one end used for scraping cells off surfaces and for resuspending pellets.

Rubella virus. Synonym: GERMAN MEASLES VIRUS. Family *Togaviridae*, genus *Rubivirus*. Causes a mild illness in man with generalised rash and enlarged lymph nodes. However, infection during first three to four months of pregnancy often results in infection of the foetus and severe congenital abnormalities. Several species can be infected experimentally but show no disease except leucopenia and rash. Major epidemics occur periodically. The virus is grown in duck eggs and a wide range of cell cultures. A good attenuated vaccine is available and should be given to all girls before puberty.
Banatvala J.E. (1977) Recent Adv. clin. Virol. **1**, 171.

Rubeola virus. *See* MEASLES VIRUS.

Rubivirus. (Latin 'rubeus' = reddish.) A genus in the family *Togaviridae*. The only member is rubella virus, the virus causing congenital abnormalities if the mother becomes infected early in pregnancy.

Rubus Chinese seed-borne virus. A possible *Nepovirus*.
Barbara, D.J. *et al.* (1985) Ann. Appl. Biol. **107**, 45.

Rubus yellow net virus. A member of the *Cacao Swollen Shoot Virus* group.
Stace-Smith, R. and Jones, A.T. (1978) CMI/AAB Descriptions of Plant Viruses No. 188.
Francki, R.I.B. *et al.* (1985) **In** Atlas of Plant Viruses. Vol. 2. p. 235. CRC Press: Boca Raton, Florida.

Runde virus. Family *Coronaviridae*, genus

Coronavirus. Isolated from the tick *Ixodes uriae* in sea bird colonies at Runde, Norway. Kills newborn mice when injected but produces a persistent infection in older mice. Replicates with cpe in BHK cells.
Traavick T. (1979). Acta Path. Microbiol. Scand. B, **87**, 1.

Russian autumn encephalitis virus. *See* JAPANESE B ENCEPHALITIS VIRUS.

Russian spring-summer encephalitis virus. *See* TICK-BORNE ENCEPHALITIS VIRUS.

Russian winter wheat mosaic virus. A probable plant *Rhabdovirus*; transmitted by leafhopper.
Francki, R.I.B. *et al.* (1985) **In** Atlas of Plant Viruses. Vol. 1. p. 73. CRC Press: Boca Raton, Florida.

ryegrass chlorotic streak virus. A *Luteovirus*.
Toriyama, S. *et al.* (1983) Ann. Phytopath. Soc. Japan **49**, 610.

ryegrass cryptic virus. A member of the *Cryptovirus* group, subgroup A.
Boccardo, G. *et al.* (1987) Adv. Virus Res. **32**, 171.

ryegrass mosaic virus. Type member of the *Ryegrass mosaic virus* group.
Slykhuis, J.T. and Paliwal, Y.C. (1972) CMI/AAB Descriptions of Plant Viruses No. 86.

ryegrass mosaic virus 'group'. Unofficial group of plant viruses with flexuous rod-shaped particles, 700-720 nm. long and 11-15 nm. wide. Coat protein subunits (mw. 29×10^3) are arranged

100nm

in a helix with pitch 3.3 nm. Each particle contains a single molecule of linear (+)-sense ssRNA (mw. 2.7×10^6). Host range is narrow. Virus particles are found in most cell types. Members are easily mechanically transmitted. Natural transmission is by Eriophyid mites in the SEMIPERSISTENT TRANSMISSION manner.

ryegrass mottle virus. An unclassified plant virus with isometric particles.
Toriyama, S. *et al.* (1983) Ann. Phytopath. Soc. Japan **49**, 610.

S

S1Ar 126 virus. Family *Bunyaviridae*, genus *Bunyavirus*.

s$_w^{20}$. SEDIMENTATION COEFFICIENT expressed in Svedbergs and adjusted to 20°C and water.

SA6 virus. *See* CERCOPITHECID HERPESVIRUS 2.

SA8 virus. *See* CERCOPITHECID HERPESVIRUS 3.

SA11 virus. Family *Reoviridae*, genus *Rotavirus*.

SA12 virus. Family *Papovaviridae*, genus *Polyomavirus*. Virus isolated from baboons.

S-adenosyl-L-homocysteine (AdoHcy, SAH). An inhibitor of methylation, e.g. of nucleic acids, as it is an analogue of S-ADENOSYL-L-METHIONINE.

S-adenosyl-L-methionine (AdoMet, SAM). An intracellular source of activated methyl groups including those used for RNA or DNA methylation. Stimulates transcription in some viruses, e.g. CYTOPLASMIC POLYHEDROSIS VIRUSES. Also required by class I restriction endonucleases for their initial binding.

Sabo virus. Family *Bunyaviridae*, genus *Bunyavirus*. Isolated from cattle, goats and *Culicoides* sp. in Nigeria.

Saboya virus. Family *Flaviviridae*, genus *Flavivirus*. Isolated from Kenya's gerbil *Tatera kempi* in Senegal. Antibodies present in many species.

sacbrood virus. Unclassified small RNA virus (a possible INSECT PICORNAVIRUS) isolated from the European honey bee *Apis mellifera*. Larvae infected with the disease fail to pupate. Virions are isometric, 30 nm. in diameter, sediment at 160S and have a buoyant density in CsCl of 1.35 g/cc. Particles contain three structural proteins (mw.

31.5, 28 and 25 x 10³) and an ssRNA genome (mw. 2.6 x 10⁶). It is serologically related, though not identical, to THAI SACBROOD VIRUS.
Bailey, L. *et al.* (1982) J. Invertebr. Pathol. **39**, 264.
Moore, N.F. *et al.* (1985) J. gen. Virol. **66**, 647.

Saccharomyces cerevisiae virus L1 (ScV-L1). (Synonym: SCV-L1A). A member of the *Totivirus* group.

Saccharomyces cerevisiae virus L1 group. Now the TOTIVIRUS GROUP.

Saccharomyces cerevisiae virus La (ScV-La). (Synonym: ScV-LB/C). A possible member of the *Totivirus* group.

Sacramento River Chinook salmon disease. Synonym: CHINOOK SALMON VIRUS. Family *Rhabdoviridae*, unassigned genus. Causes necrosis of haematopoietic tissues of spleen and kidney in trout and salmon. The fish often have subdermal lesions at back of head. The virus causes epidemic disease in hatcheries. Replicates in FHM cells or other fish cell lines with cpe.

Sagiyama virus. Family *Togaviridae*, genus *Alphavirus*. Isolated from mosquitoes in Japan. Kills new-born mice on injection.

saguaro cactus virus. A member of the *Carmovirus* group.
Nelson, M.R. *et al.* (1975) CMI/AAB Descriptions of Plant Viruses No. 148.
Francki, R.I.B. *et al.* (1985) **In** Atlas of Plant Viruses. Vol. 2. p. 235. CRC Press: Boca Raton, Florida.
Morris, T.J. and Carrington, J.C. (1988) **In** The Plant Viruses. Vol. 3. p. 73. ed. R. Koenig. Plenum Press: New York.

SAH. Abbreviation for S-ADENOSYL-L-HOMO-CYSTEINE.

Saint Augustine decline virus.   A strain of
PANICUM MOSAIC VIRUS.

Saint Floris virus.   Family *Bunyaviridae*, genus
Phlebovirus. Isolated from a gerbil of *Tatera* sp.
in Central Africa.

Saint Louis encephalitis virus.   Family
Flaviviridae, genus *Flavivirus*. Occurs in Canada, USA , Central and S. America, causing fever
and occasionally encephalitis. Injection into mice
causes encephalitis. Virus can be grown in eggs
and in many cell cultures, causing cpe.

Saintpaulia leaf necrosis virus. A possible plant
Rhabdovirus.
Francki, R.I.B. *et al*. (1985) **In** Atlas of Plant
Viruses. Vol. 1. p. 73. CRC Press: Boca Raton,
Florida.

Sakhalin virus.   Family *Bunyaviridae*, genus
Nairovirus. First isolated from the tick *Ixodes
putens* on Sakhalin Island where it is associated
with guillemots. Also found on the Kuril, Commodore and Aleutian Islands and the north coasts
of Europe, Canada and the USA. Associated with
penguin colonies in the southern hemisphere.
Infects mice injected i.c. and replicates in experimentally infected *Culex modestus* mosquitoes.

Salanga virus.   Unclassified arthropod-borne
virus, isolated from the blood of a rat trapped in
Salanga (Central African Republic).

Salchabad virus.   Family *Bunyaviridae*, genus
Phlebovirus. Isolated from female sandflies
Phlebotomus sp. in Iran. Antibodies present in
man and sheep.

saline.   A solution of NaCl, usually at 0.15M,
either buffered or in water.

Salmonid rhabdoviruses.   Family *Rhabdoviridae*, unassigned genus. Viruses causing disease in young fish: INFECTIOUS HAEMATOPOIETIC
NECROSIS VIRUS, OREGON SOCKEYE DISEASE VIRUS,
SACRAMENTO RIVER CHINOOK SALMON DISEASE VIRUS
in USA and Canada, and EGTVED VIRUS in Europe.

salting out.   The precipitation of proteins or
nucleic acids in concentrated salt solutions.
Ammonium sulphate and sodium sulphate are
frequently used for the precipitation of proteins
and, since different proteins precipitate at different salt concentrations, the method can be used

for their fractionation and partial purification.
Similarly the precipitation of different nucleic
acids depends on the concentration of salt solution, e.g. LiCl. *See* SOLUBLE RNA.

SAM.   Abbreviation for S-ADENOSYL-L-METHIONINE.

sambucus vein clearing virus.   A possible plant
Rhabdovirus.
Francki, R.I.B. *et al*. (1985) **In** Atlas of Plant
Viruses. Vol. 1. p. 73. CRC Press: Boca Raton,
Florida.

Samford virus.   Family *Bunyaviridae*, genus
Bunyavirus. Isolated from *Culicoides brevitarsis*
in Queensland, Australia. Antibodies present in
horses.
Doherty R.L. (1972) Aust. vet. J. **48**, 81.

Sammon's opuntia virus.   A *Tobamovirus*.
Brunt, A.A. (1986) **In** The Plant Viruses. Vol. 2.
p. 285. ed. M.H.V. van Regenmortel and H.
Fraenkel-Conrat. Plenum Press: New York.

SAN 240I.   Experimental product of the nuclear
polyhedrosis virus of *Heliothis zea* later developed by Sandoz Inc. as ELCAR.

SAN 404I.   Experimental preparation of the
nuclear polyhedrosis virus of *Autographa californica* produced for commercial trials by Sandoz
Inc. during the early 1980s for the biological
control of the cabbage looper, *Trichoplusia ni* and
other susceptible lepidopteran hosts.

SAN 405I.   Experimental product of the nuclear
polyhedrosis virus of the cabbage looper,
Trichoplusia ni, produced by Sandoz Inc. during
the early 1980s for the biological control of the
homologous host.

SAN 406I.   Experimental product of the granulosis virus of the codling moth, *Cydia pomonella*
produced by Sandoz Inc. during the early 1980s
for the biological control of the codling moth.

San Angelo virus.   Family *Bunyaviridae*, genus
Bunyavirus. Isolated from mosquitoes in Texas.
Antibodies found in raccoons and opossums.

San Juan virus.   Family *Bunyaviridae*, genus
Bunyavirus.

San Miguel sea lion virus. Family *Caliciviridae*,

genus *Calicivirus*. First isolated from sea lions on San Miguel island off the coast of California. Since isolated from many other marine mammalian species living in the Pacific Ocean. Considered to be a cause of abortion in sea lions. Its close serological and structural relationship to vesicular exanthema virus suggests that marine mammals may be a reservoir of the virus for terrestrial animals. Pigs and African green monkeys can be infected experimentally. The virus grows well in several cells, giving cpe.
Burroughs J.N. *et al.* (1978) Intervirology **10**, 51.

sandfly fever virus. Synonym: HUNDS-KRANKHEITVIRUS. Family *Bunyaviridae*, genus *Phlebovirus*. Isolated from man in Italy, Egypt, Iran and Pakistan. The sandfly *Phlebotomus papatasi* is the vector. Causes fever, which may be recurrent. Replicates in human, mouse and hamster kidney cell cultures, with cpe.

Sandjimba virus. Family *Bunyaviridae*, genus *Bunyavirus*. Isolated from a bird *Acrocephalus schoenobaenus* in the Central African Republic.

Sanger method. The DIDEOXY METHOD for sequencing DNA. The DNA to be sequenced is in a single-stranded form usually by cloning in PHAGE M13. A primer comprising a sequence complementary to that immediately 5′ of the insert is annealed to the ssDNA and the reaction mixture is divided into four aliquots. To each are added the four deoxyribonucleotides (dNTPs) (one of which is labelled, usually radioactively, with α–^{32}P–ATP) plus a dideoxyribonucleotide (ddNTP) (*see* DIDEOXYRIBOSE) such that each aliquot has a different ddNTP. The KLENOW FRAGMENT of DNA polymerase is added and starts the synthesis of DNA complementary to the ssDNA template. When a ddNTP is incorporated into the newly synthesised strand no further dNTP can be added because the ddNTP lacks a 3′ hydroxyl group. The ratio of ddNTP:dNTP is adjusted so that the synthesis of the new DNA is terminated at all the positions at which the ddNTP can be inserted. Thus a nested set of fragments, based 5′ on the primer and ending in a ddNTP, is formed. These fragments are separated by electrophoresis in a polyacrylamide gel, the four reaction mixtures being placed in adjacent tracks. The fragments are detected as bands by autoradiography and the sequence is read directly from the band pattern or ladder.
Sanger, F. *et al.* (1977) Proc. Natl. Acad. Sci. USA. **74**, 5463.

Sango virus. Family *Bunyaviridae*, genus *Bunyavirus*. Isolated from cattle and from flies of *Culicoides* sp. in Nigeria and Kenya.

Santa Rosa virus. Family *Bunyaviridae*, genus *Nairovirus*.

santhosai temperate virus. A possible member of the *Cryptovirus* group, subgroup A.
Boccardo, G. *et al.* (1987) Adv. Virus Res. **32**, 171.

sarkosyl. Sodium lauroyl sarcosinate. An anionic detergent used to disrupt virus particles or cells.

Sarranacia purpurea rhabdovirus. A possible plant *Rhabdovirus*.
Francki, R.I.B. *et al.* (1985) **In** Atlas of Plant Viruses. Vol. 1. p. 73. CRC Press: Boca Raton, Florida.

satellite virus. A defective virus which requires a helper virus to provide functions necessary for replication. It may code for its own coat protein (*see* TOBACCO NECROSIS SATELLITE VIRUS) or various other products.

Sathuperi virus. Family *Bunyaviridae*, genus *Bunyavirus*. Isolated from cattle and mosquitoes in India and Nigeria.

satsuma dwarf virus. A member of the *Fabavirus* group.
Usugi, T. and Saito, Y. (1979) CMI/AAB Descriptions of Plant Viruses No. 208.

Saumarez Reef virus. Family *Flaviviridae*, genus *Flavivirus*. Isolated from sea-bird tick *Ornithodoros capensis*, from sooty terns on islands off the coast of Queensland, Australia, and *Ixodes endyptidis* from silver gulls in Tasmania. Paralyses new-born mice on injection.

Sawgrass virus. Family *Rhabdoviridae*, unassigned genus. Isolated from the eastern dog tick *Dermacentor variabilis* and the rabbit tick *Haemaphysalis leporis-palustris* in Florida.

Scabby mouth virus. *See* ORF VIRUS.

scanning electron microscope. A type of electron microscope in which a beam of electrons a few hundred angstroms in diameter systematically sweeps over the specimen; the intensity of

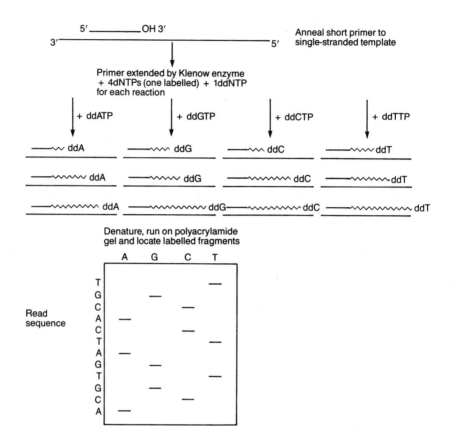

secondary electrons generated at the point of impact of the beam on the specimen is measured and the resulting signal is fed into a cathode-ray tube display which is scanned in synchrony with the scanning of the specimen. Used to give enlarged images of the surface of structures.

Scatchard plot. A method for analysing data to assess the interaction between small molecules which bind ions (ligands) and the ions themselves. Values of ϖ/M are plotted against ϖ where ϖ = moles ion bound/moles ligand and M = molarity of the ion. From the intercept on the ϖ axis the number of ions bound can be estimated; the slope gives the binding constant.
Scatchard, G. (1949) Ann. N.Y. Acad. Sci. **51**, 660.

Schiff's reagent. *See* FEULGEN STAIN.

Schlieren optics. An optical system which detects gradients of concentration of a solute in a solvent. In its simplest form light is collimated by a lens and focused on to a knife-edge by a second lens, the gradient is placed between the two lenses and the resultant diffraction pattern recorded. By the use of a phase plate a differentiated curve of the gradient can be obtained.

Schneider's Drosophila Medium. A cell-culture medium for the propagation of insect cells from *Drosophila* spp.
Schneider, I. (1966) J. Embryol. exp. Morph. **15**, 271.

scintillation counter. A device in which the scintillations produced in a fluorescent material by ionizing radiation are detected and counted by a multiplier phototube and associated circuitry.

scintillation fluid. A solvent, usually organic, containing one, two or three solutes which convert the kinetic energy of a nuclear particle traversing the medium into light photons at a maximum efficiency. The solutes, also known as scintillators, fluors or lumifluors are ranked according to their level of fluorescent energy, that with the highest being the primary solute and the others secondary solutes. The secondary solutes shift the emission spectrum of the system to longer wavelengths to match the spectral responses of the multiplier cathode. One of the most commonly used scintillation fluids is composed of PPO and POPOP dissolved in toluene. A third solute might be added to increase the solubilising power for the sample specimen, e.g. TRITON X-100 can be added to enhance water incorporation.

Peng, C.T. (1977) Sample Preparation in Liquid Scintillation Counting. The Radiochemical Centre, Amersham, England.

scrapie agent. Unclassified. This agent causes ataxia in sheep and goats which usually results in death. Mode of transmission is not known but is probably vertical. The disease can be transmitted experimentally in sheep and goats by injection of brain tissue from infected animals. Mice and hamsters can also be infected experimentally. The incubation period is several months. The agent is probably similar to that causing Creutzfeldt-Jakob disease. There has been much controversy regarding the nature of the agent and there are still no definite clues to its identity. *See* PRION.

Kimberlin, R.H. (1984). Principles of Bacteriology, Immunology, Vol. 4, ed. G.S. Wilson and F. Brown.

Scrophularia mottle virus. A *Tymovirus.*
Bercks, R. (1973) CMI/AAB Descriptions of Plant Viruses No. 113.
Francki, R.I.B. *et al.* (1985) **In** Atlas of Plant Viruses. Vol. 1. p. 117. CRC Press: Boca Raton, Florida.
Hirth, L. and Girard, L. (1988) **In** The Plant Viruses. Vol. 3. p. 163. ed. R. Koenig. Plenum Press: New York.

ScV-L1A . *See* SACCHAROMYCES CEREVISIAE VIRUS L1.

ScV-LB/C. *See* SACCHAROMYCES CEREVISIAE VIRUS LA.

SDI. Abbreviation for SEROLOGICAL DIFFERENTIATION INDEX.

SDS. Abbreviation for SODIUM DODECYL SULPHATE.

SDS-polyacrylamide gel electrophoresis. Electrophoresis in polyacrylamide gels of proteins denatured with the anionic detergent SODIUM DODECYL SULPHATE (SDS). The treatment with SDS usually gives proteins equal charge per unit mw. and thus the proteins are separated according to their mw.

Gordon, A.H. (1975) Electrophoresis of Proteins in Polyacrylamide and Starch Gels. North Holland/American Elsevier: Amsterdam.

sea lion pox virus. Family *Poxviridae*, subfamily *Chordopoxvirinae*, genus *Parapoxvirus.* Isolated from a captive California sea lion *Zalophus californianus.* Causes severe disease in sea lions.

seal morbillivirus. A MORILLIVIRUS which caused an epizootic of seals in the Baltic Sea and North Sea in 1988 leading to the death of many animals. Closely related to CANINE DISTEMPER and RINDERPEST VIRUSES.

Cosby, S.L. *et al.* Nature **336,** 115.

Sebokele virus. Unclassified. Isolated from an adult female rodent *Hylomyscus* sp. in the Central African Republic.

sedimentation coefficient. The rate at which a macromolecule sediments under a defined gravitational force. The basic unit is the Svedberg (S) which is 10^{-13} sec. The sedimentation coefficient (s), often measured in an ANALYTICAL ULTRACENTRIFUGE, is given by the equation:

$$s = \frac{1}{w^2 r} \times \frac{dr}{dt}$$

where r = radius (distance in cm. between the macromolecule and the centre of rotation)
w = the angular velocity of the centrifuge rotor in radians per second

$\dfrac{dr}{dt}$ = the rate of movement of the macromolecule in cm. per sec.

See SVEDBERG EQUATION.

sedimentation rate. The rate at which a macromolecule moves in a standard gravitational field. This rate is influenced by the mass, shape, hydration and PARTIAL SPECIFIC VOLUME of the macromolecules as well as the temperature and viscosity of the solvent. The measure of the rate in the standard gravitational field of 1 cm sec^{-2} is the SEDIMENTATION COEFFICIENT.

Seletar virus. Family *Reoviridae*, genus *Orbivirus*. Isolated from the tick *Boophilus microplus* in Singapore and Malaysia. Antibodies found in cattle, pigs and carabao.

Sembalum virus. Unclassified. Isolated from herons in India.

semiconservative replication. A model of DNA-DEPENDENT DNA REPLICATION in which the synthesis uses both strands of dsDNA as templates resulting in progeny each comprising one parent strand and one new strand. DNA polymerases only synthesise in the $3' \rightarrow 5'$ direction. Thus on circular molecules the synthesis will be bidirectional. On linear molecules synthesis usually starts at one end; it proceeds continuously in the $5' \rightarrow 3'$ direction on one template but discontinuously on the other. In the discontinuous strand short pieces of DNA are synthesised in the $5' \rightarrow 3'$ direction and are joined to each other by DNA ligase. The leading edge of synthesis, where the dsDNA becomes separated to give the ss templates, is termed the replication fork. It is considered that this is the mechanism by which DNA replication usually takes place in nature. *See* OKAZAKI FRAGMENTS.

semipersistent transmission. The relationship between a plant virus and its arthropod vector which is intermediate between NON-PERSISTENT TRANSMISSION and PERSISTENT TRANSMISSION. It has the features of short acquisition feed and no latent period found in non-persistent transmission, but the vector remains able to transmit the virus for periods of hours to days which is longer than the non-persistent transmission.

Semliki Forest virus. Family *Togaviridae*, genus *Alphavirus*. Natural host and vector are not known but antibodies are found in man and wild primates in many countries in Africa and in Borneo and Malaya. Multiplies in the mosquito *Aedes aegypti*. Not usually associated with disease in man but there is one well-documented case of fatal infection of a laboratory worker in Germany. Causes encephalitis in adult mice, guinea pigs, rabbits and Rhesus monkeys on injection. The virus grows well in eggs and in many cell cultures, causing cpe.

Sendai virus. Synonym: PARAINFLUENZA VIRUS TYPE 1 MURINE; JAPANESE HAEMAGGLUTINATING VIRUS. Family *Paramyxoviridae*, genus *Paramyxovirus*. Occurs as a latent infection of labora-

tory mice, rats and guinea pigs. Causes inapparent infection in ferrets, monkeys and pigs. Grows in many cell cultures and in embryonated eggs.
Ishida N. and Homma M. (1978) Adv. Virus Research **23**, 349.

Sepik virus. Family *Flaviviridae*, genus *Flavivirus*. Isolated from mosquitoes in New Guinea. Causes fever in man.

sequence. The order of nucleotides in RNA or DNA or of amino acids in a polypeptide.

sequencing. The determination of a sequence of a nucleic acid or protein. *See* SANGER METHOD, and MAXAM AND GILBERT METHOD for DNA, EDMAN DEGRADATION for proteins.

serological differentiation index. A measure of the degree of serological cross reactivity of two antigens. It is the number of two-fold dilution steps separating the homologous and heterologous TITRES. Because of the variation in the response of animals to antigens it is only reliable if it is the average of several measurements.
Van Regenmortel, M.H.V. and von Wechmar, M.B. (1970) Virology **41**, 330.

serology. The branch of science dealing with properties and reactions of sera and particularly the use of antibodies in the sera to examine the properties of antigens.

serotype. A group of viruses or microorganisms distinguished on the basis of their antigenic properties.

Serra de Navio virus. Family *Bunyaviridae*, genus *Bunyavirus*. Isolated from *Aedes fulvus* mosquitoes in Brazil. Not reported to cause disease.

serum hepatitis virus. *See* HEPATITIS B VIRUS.

serum neutralisation. The inhibition of the infectivity of a virus by antiserum.

serum-free medium. A culture medium used for growing viruses in animal tissues or cells. The medium does not contain foetal calf serum or other serum additives, but does contain essential salts, amino acids and an energy source (usually glucose).

SEV. Abbreviation sometimes used to describe the SNPV subtype of NUCLEAR POLYHEDROSIS VIRUSES where the enveloped virions contain predominantly single nucleocapsids. *See* MEV.

SF-Naples virus. Family *Bunyaviridae*, genus *Phlebovirus*.

SF-Sicilian virus. Family *Bunyaviridae*, genus *Phlebovirus*.

shallot latent virus. A *Carlavirus*.
Bos, L. (1982) CMI/AAB Descriptions of Plant Viruses No. 250.
Francki, R.I.B. *et al.* (1985) **In** Atlas of Plant Viruses. Vol. 2. p. 173. CRC Press: Boca Raton, Florida.

Shamonda virus. Family *Bunyaviridae*, genus *Bunyavirus*. Isolated from cattle and *Culicoides* sp. in Nigeria.

Shark River virus. Family *Bunyaviridae*, genus *Bunyavirus*. Isolated from mosquitoes in Florida, Mexico and Guatemala.

sheep capripox virus. Synonyms: CLAVELEE VIRUS, ISIOLO VIRUS, KEDONG VIRUS, VARIOLA OVINA VIRUS. Family *Poxviridae*, subfamily *Chlordopoxvirinae*, genus *Capripoxvirus*. Causes a generalised pock disease in sheep, often with tracheitis and involvement of the lungs. High mortality. Occurs in Africa, Asia, Middle East, Southern Europe and the Iberian Peninsula. Replicates in sheep, goat and calf cultures with cpe. An attenuated vaccine is used successfully. Singh I.P. *et al.* (1979) Vet. Bull. **49**, 145.

Shine-Dalgarno sequence. *See* RIBOSOME BINDING SITE.

shipping fever virus. Family *Paramyxoviridae*, genus *Paramyxovirus*. A bovine strain of parainfluenza virus type 3. Causes respiratory disease in cattle.

Shokwe virus. Family *Bunyaviridae*, genus *Bunyavirus*.

Shope fibroma virus. *See* RABBIT FIBROMA VIRUS.

Shope papilloma virus. *See* RABBIT PAPILLOMA VIRUS.

show fever virus. *See* FELINE PANLEUCOPENIA VIRUS.

Shuni virus. Family *Bunyaviridae*, genus *Bunyavirus*. Isolated from man, cattle and sheep in Nigeria and South Africa. Causes disease in man.

sialic acid. Synonym: N-ACETYLNEURAMINIC ACID. A constituent of many glycoproteins. Cleaved from glycoprotein virus receptors by NEURAMINIDASE.

sialodacryoadenitis virus. Family *Coronaviridae*, genus *Coronavirus*. Isolated from rats with sialodacryoadenitis, a disease affecting the submaxillary, salivary and Harderian glands. There is necrosis of the ductal epithelium with acute lymphocytic infiltration and gelatinous oedema. Staining of the fur due to porphyrin excretion in the tears occurs. Injection i.c. into new-born mice causes ataxia, paralysis and death. Older mice are resistant. Related antigenically to rat coronavirus and mouse hepatitis virus. Will replicate in primary rat kidney cells.

sigla. A name formed from letters or other characters taken from the words in a compound term. Virus group names are often sigla, e.g. POTYVIRUS from potato virus Y, PAPOVAVIRUS from papilloma, polyoma and vacuolating agent.

Sigmavirus. Synonym: CARBON DIOXIDE SENSITIVITY VIRUS. Probable member of the *Rhabdoviridae* not yet assigned to a genus. A disease of adult fruit flies (*Drosophila* spp.) which is congenitally transmitted. The presence of Sigmavirus in the fly appears to cause no harm unless the insect is exposed to pure CO_2, when even brief exposure to the gas causes death of the insect (hence, the alternative name). The virion proteins and nucleic acid resemble those of other rhabdoviruses, particularly vesicular stomatitis virus. Teninges, D. and Bras-Herreng, F. (1987) J. gen. Virol. **68**, 2625.

signal peptide. A short segment usually of 15-30 amino acids at the N-terminus of a secreted or exported protein which enables it to pass through the membrane of the cell or organelle. The signal peptide is then usually removed by a specific protease and therefore is not present in the mature protein.
Lingappa, V.R. *et al.* (1979) Nature **281**, 117.

signal sequence. a) Synonym: SIGNAL PEPTIDE or

b) the nucleotide sequence encoding a signal peptide.

SiI phage. Unclassified PHAGE isolated from *Aquaspirillum itersonii*. Particles are isometric, 63 nm. in diameter, unenveloped and contain linear dsDNA (mw. 26×10^6). Representative of phage morphotype D2 (*see* PHAGE).
Clark-Walker, G.D. and Primrose, S.B. (1971) J. gen. Virol. **11**, 139.

silent infection. An infection with no detectable symptoms. Can also be termed a latent infection.

silverwater virus. Family *Bunyaviridae*, genus *Bunyavirus*. Isolated from ticks removed from hares in Canada.

Simbu virus. Family *Bunyaviridae*, genus *Bunyavirus*. Isolated from mosquitoes in S. Africa, Central African Republic and Cameroons. Antibodies found in man but no disease has been reported. Gives name to group of viruses which are related serologically.

simian adenovirus. Family *Adenoviridae*, genus *Mastadenovirus*. Most infections are silent but some strains are associated with respiratory and enteric disease in baboons, Rhesus, *Erythrocebus* and *Cercopithecus* monkeys. Experimentally some strains are oncogenic in hamsters. Viruses cause cpe in monkey kidney cells.

simian enterovirus. Family *Picornaviridae*, genus *Enterovirus*. Isolated from monkey tissue or excreta. Can cause latent infection in monkey kidney cell cultures.
Hull, R.N. (1968). Virology Monographs No.2.

simian foamy virus. Family *Retroviridae*, subfamily *Spumavirinae*. Isolated from throat washings of several different species of monkey. Exists as several distinct antigenic types. Causes foamy degeneration and syncytia formation in cell culture.

simian haemorrhagic fever virus. *Family Flaviviridae*, genus *Flavivirus*. Causes high fever, facial oedema, splenomegaly and severe haemorrhagic diathesis in Rhesus monkeys. Non-pathogenic in mice. Replicates in Rhesus monkey cells with cpe.
Trousdale, M.D. *et al.* (1975). Proc. Soc. exp. Biol. Med. **150**, 707.

simian immunodeficiency virus (SIV). Family *Retroviridae*, subfamily *Lentivirinae*. Causes a disease in monkeys which is similar to AIDS in man. This suggests the monkey as a good model for studying AIDS.

simian papilloma virus. Family *Papovaviridae*, genus *Papillomavirus*. Causes papillomas in *Cebus* monkeys. The virus can be transmitted experimentally to both new and old world monkeys, causing thickening of the epidermis but there is no invasion of normal tissue.
Lucké, B. *et al.* (1950). Federation Proc. Am. Soc. Exp. Biol. **9**, 336.

simian sarcoma virus. *See* WOOLLY MONKEY SARCOMA VIRUS.

simian vacuolating virus. *See* SIMIAN VIRUS 40.

simian virus 40. Synonym: SIMIAN VACUOLATING VIRUS. Family *Papovaviridae*, genus *Polyomavirus*. A natural and silent infection of Rhesus, cynomolgus and cercopithecus monkeys. Often isolated from kidney cell cultures. Produces tumours on injection into new-born hamsters, grivets, baboons and Rhesus monkeys.

Simplexvirus. A genus in the family *Herpesviridae*, subfamily *Alphaherpesvirinae*. Herpes simplex 1 is the type member and the human herpes viruses types 1 and 2 and bovine herpesvirus 2 are members.

Sindbis virus. Family *Togaviridae*, genus *Alphavirus*. Isolated from birds but antibodies are found in man and domestic ungulates. Found in Egypt, South Africa, India, Malaya, the Philippines and Australia. Replicates in eggs, killing the embryo and in cell cultures of chick, human and monkey tissues with cpe. Kills suckling mice on injection and causes myositis and encephalitis in infant mice.

single nucleocapsid nuclear polyhedrosis virus. *See* SNPV.

single radial diffusion test. The same as RADIAL IMMUNODIFFUSION.

single-stranded. A term describing nucleic acid molecules consisting of only one polynucleotide chain. Because of intra-strand base-pairing they all contain ds regions and thus share some properties (e.g. susceptibility to certain enzymes) with ds nucleic acids. Examples of ssDNA are the genomes of PHAGE M13 and of GEMINIVIRUSES, and of ssRNA and mRNAs, tRNAs and the genomes of many viruses.

Siphoviridae. Family of TAILED PHAGES in which the tail is long and noncontractile; includes phage morphotypes B1 to B3 (*see* PHAGE). The name for this family was approved in 1984 by the ICTV and supersedes the name '*Styloviridae*' which was earlier proposed for this group. Members of the family have been isolated from bacteria, cyanobacteria and mycoplasma (*see* APPENDIX F) and are extremely variable in dimensions and physical properties. Phage heads may be isometric or elongated and attached to noncontractile tails, in which the capsomeres are helically organised and approximately 60-540 nm. in length. The tail may

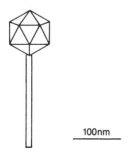

100nm

terminate with tail fibres, knobs with tail fibres attached or have no structural differentiation at the tail tip. Particles contain a linear dsDNA genome, and the phages may be VIRULENT or TEMPERATE. One genus has been defined (λ PHAGE GROUP), although other isolates will probably be assigned to several genera in future. λ PHAGE is the type species.
Ackermann, H.W. *et al.* (1984) Intervirology **22**, 181.

Sixgun City virus. Family *Reoviridae*, genus *Orbivirus*. Isolated from the tick *Argos cooleyi* in Texas and Colorado.

slow bee-paralysis virus. *See* BEE SLOW PARALYSIS VIRUS.

slow virus. A virus causing a slow progressive disease of the central nervous system, e.g. LENTIVIRUSES.

small iridescent insect virus. Vernacular name for *IRIDOVIRUS* genus.

smallpox virus. Synonym: VARIOLA VIRUS. Family *Poxviridae*, subfamily *Chordopoxvirinae*, genus *Orthopoxvirus*. Causes smallpox in man which was a severe disease with confluent rash, fever and prostration, frequently followed by death. The disease has now been eliminated by a concerted vaccination policy organised by the World Health Organisation.

Smith River virus. Family *Bunyaviridae*, genus *Bunyavirus*.

Smithiavirus. Former name for a member of the CYTOPLASMIC POLYHEDROSIS VIRUS group (after K.M. Smith).

smooth membrane. Endoplasmic reticulum not encrusted with ribosomes. *See* ROUGH MEMBRANE.

Snowshoe Hare virus. Family *Bunyaviridae*, genus *Bunyavirus*. Isolated from hares, lemmings and mosquitoes in northern USA, Canada and Alaska. Antibodies found in man and other animals. Causes subclinical infection in man.

SNPV. Abbreviation used to describe the subtype of NUCLEAR POLYHEDROSIS VIRUSES where the enveloped virions contain predominantly single nucleocapsids. *See* MNPV.

Sobemovirus group. (Sigla of SOUTHERN BEAN MOSAIC, the type member). Genus of plant viruses with isometric particles, 30 nm. in diameter, which sediment at 115S and band in CsCl at 1.36 g/cc. The particles are stabilised by pH-depend-

 100nm

ent and divalent cation mediated protein-protein interactions and protein-RNA links. Capsid structure is icosahedral (T=3), the coat protein subunits having mw. 30×10^3. Each particle contains one molecule of linear (+)-sense ssRNA (mw. 1.4×10^6; 4193 nucleotides). The host range of each member is narrow. Virus particles are found in the cytoplasm and nuclei of most cell types. Sobemoviruses are easily mechanically transmitted and by seed in several hosts and by beetles.
Matthews, R.E.F. (1982) Intervirology **17**, 144.
Francki, R.I.B. *et al.* (1985) **In** Atlas of Plant Viruses. Vol. 1. p. 153. CRC Press: Boca Raton, Florida.
Hull, R. (1988) **In** The Plant Viruses. Vol. 3. p.

113. ed. R. Koenig. Plenum Press: New York.

sockeye salmon virus. *See* OREGON SOCKEYE DISEASE VIRUS.

sodium dodecyl sulphate (SDS). Also known as sodium lauryl sulphate (SLS). An ANIONIC DETERGENT which has many uses including disruption of virus particles and cells, inhibiting nucleases during nucleic acid extractions and denaturing proteins for gel electrophoresis.

soil-borne wheat mosaic virus. Type member of the *Furovirus* group.
Brakke, M.K. (1971) CMI/AAB Description of Plant Viruses No. 77.
Brunt, A.A. and Shikata, E. (1986) **In** The Plant Viruses. Vol. 2. p. 305. ed. M.H.V. van Regenmortel and H. Fraenkel-Conrat. Plenum Press: New York.

Sokoluk virus. Family *Flaviviridae*, genus *Flavivirus*. Isolated from bats in USSR.

solanum apical leaf curling virus. A possible *Geminivirus*.
Harrison, B.D. (1985) Ann. Rev. Phytopath. **23**, 55.

solanum nodiflorum mottle virus. A *Sobemovirus*.
Greber, R.S. and Randles, J.W. (1986) AAB Descriptions of Plant Viruses No. 318.
Francki, R.I.B. *et al.* (1985) **In** Atlas of Plant Viruses. Vol. 1. p. 153. CRC Press: Boca Raton, Florida.
Hull, R. (1988) **In** The Plant Viruses. Vol. 3. p. 113. ed. R. Koenig. Plenum Press: New York.

solanum yellows virus. A *Luteovirus*; considered to be a strain of POTATO LEAF ROLL VIRUS.
Francki, R.I.B. *et al.* (1985) **In** Atlas of Plant Viruses. Vol. 1. p. 137. CRC Press: Boca Raton, Florida.
Casper, R. (1988) **In** The Plant Viruses. Vol. 3. p. 235. ed. R. Koenig. Plenum Press: New York.

Soldado virus. Family *Bunyaviridae*, genus *Nairovirus*. Isolated from (a) a pool of the *Ornithodoros* sp. in the Caribbean; (b) *Ornithodoros capensis* associated with sea birds in Ethiopia, Seychelles, USA. and Senegal and (c) *Ornithodoros maritimus* in France, Ireland and Wales. Kills suckling mice when injected. Does not cause disease in man.

Solenopsis 'baculovirus'. Unclassified virus morphologically resembling a NON-OCCLUDED BACULOVIRUS, observed in Brazilian samples of the fire ant, *Solenopsis* sp.
Avery, S.W. *et al.* (1977) Florida Entomol. **60**, 17.

soluble antigen. An antigen which is virus-specific but not the virion itself. It often comprises isolated structural subunits but can be non-structural virus-coded proteins.

soluble RNA. RNA which is soluble in strong salt solutions, e.g. 3M sodium acetate. Consists mainly of small species such as transfer RNA and ribosomal 5S RNA.

Sonchus rhabdovirus. A member of the plant *Rhabdovirus* subgroup 1.
Francki, R.I.B. *et al.* (1985) **In** Atlas of Plant Viruses. Vol. 1. p. 73. CRC Press: Boca Raton, Florida.

Sonchus yellow net virus. A member of the plant *Rhabdovirus,* subgroup 2; aphid transmitted.
Jackson, A.O. and Christie, S.R. (1979) CMI/AAB Descriptions of Plant Viruses No. 205.

sorghum stunt mosaic virus. A probable plant *Rhabdovirus*; transmitted by leafhopper.
Mayhew, D.E. and Flock, R.A. (1981) Plant Disease **65**, 84.

Sororoca virus. Family *Bunyaviridae*, genus *Bunyavirus*. Isolated from mosquitoes of *Sabethini* sp. in Brazil.

South Australia Kashmir bee virus. Unclassified small RNA virus (possible INSECT PICORNAVIRUS) isolated from the honey bee, *Apis mellifera* in South Australia. The virus is physically indistinguishable from KASHMIR BEE VIRUS. Virions contained three major proteins (mw. 44.3, 35.6 and 25 x 10³); serologically related to, but distinct from, KASHMIR BEE VIRUS, QUEENSLAND KASHMIR BEE VIRUS and NEW SOUTH WALES KASHMIR BEE VIRUS.
Bailey, L. *et al.* (1979) J. gen. Virol. **43**, 641.

southern bean mosaic virus. Type member of the *Sobemovirus* group.
Tremaine, J.H. and Hamilton, R.I. (1983) CMI/AAB Descriptions of Plant Viruses No. 274.
Francki, R.I.B. *et al.* (1985) **In** Atlas of Plant Viruses. Vol. 1. p. 153. CRC Press: Boca Raton,

Florida.
Hull, R. (1988) **In** The Plant Viruses. Vol. 3. p. 113. ed. R. Koenig. Plenum Press: New York.

Southern blotting. A technique, named after its inventor, E.M. Southern, in which DNA fragments, separated by gel electrophoresis, are immobilised on a membrane, usually nitrocellulose; more recently nylon-based membranes have been used. The immobilised DNA is available for hybridisation with a labelled ssDNA or RNA probe. The usual procedure is to separate the DNA fragments by electrophoresis in an agarose gel, and then to denature them by treatment with alkali (ssDNA binds much better to nitrocellulose than does dsDNA). The denatured DNA fragments are transferred to the nitrocellulose by drawing a concentrated salt solution (often 20 x SSC) through the gel and the nitrocellulose which has been placed upon it; the DNA fragments are then firmly linked to the nitrocellulose by baking (80°C under vacuum). The DNA fragments are detected by hybridisation with the probe which may be labelled either with radioactivity (e.g.^{32}P) or with a non-radioactive reporter group (e.g. biotin); before hybridisation non-specific sites are blocked by prehybridising with DENHARDT'S SOLUTION. Successful hybridisation is then detected by autoradiography or colour development of the biotinylated probe.
Southern, E.M. (1975) J. mol. Biol. **98**, 503.

sowbane mosaic virus. A possible *Sobemovirus*. Seed-transmitted at a high rate in *Chenopodium* spp.
Kado, C.I. (1971) CMI/AAB Descriptions of Plant Viruses No. 64.
Francki, R.I.B. *et al.* (1985) **In** Atlas of Plant Viruses. Vol. 1. p. 153. CRC Press: Boca Raton, Florida.
Hull, R. (1988) **In** The Plant Viruses. Vol. 3. p. 113. ed. R. Koenig. Plenum Press: New York.

sowthistle yellow vein virus. A member of the plant *Rhabdovirus*, subgroup 2; aphid transmitted.
Peters, D. (1971) CMI/AAB Descriptions of Plant Viruses No. 62.
Francki, R.I.B. *et al.* (1985) **In** Atlas of Plant Viruses. Vol. 1. p. 73. CRC Press: Boca Raton, Florida.

soybean chlorotic mottle virus. A *Caulimovirus*.
Vervier, J. *et al.* (1987) J. gen. Virol. **68**, 159.

soybean crinkle leaf virus. A probable *Geminivirus*, subgroup B.
Harrison, B.D. (1985) Ann. Rev. Phytopath. **23**, 55.

soybean dwarf virus. A *Luteovirus*.
Tamada, T. and Kojima, M. (1977) CMI/AAB Descriptions of Plant Viruses No. 179.
Francki, R.I.B. *et al.* (1985) **In** Atlas of Plant Viruses. Vol. 1. p. 137. CRC Press: Boca Raton, Florida.
Casper, R. (1988) **In** The Plant Viruses. Vol. 3. p. 235. ed. R. Koenig. Plenum Press: New York.

soybean mild mosaic virus. An unclassified plant virus with isometric particles 27-28 nm. in diameter. The virus is transmitted in seed and by aphids in the NON-PERSISTENT TRANSMISSION manner.
Takahashi, K. *et al.* (1980) Bull. Tohoku Natl. Agric. Expt. Sta. **62**, 1.

soybean mosaic virus. A *Potyvirus*.
Bos, L. (1972) CMI/AAB Descriptions of Plant Viruses No. 93.
Francki, R.I.B. *et al.* (1985) **In** Atlas of Plant Viruses. Vol. 2. p. 183. CRC Press: Boca Raton, Florida.

soybean stunt virus. A *Cucumovirus*.
Hanada, K. and Tochihara, H. (1982) Phytopath. **72**, 761.

Sp Ar 2317 virus. Family *Bunyaviridae*, genus *Bunyavirus*.

Spartina mottle virus. A possible *Potyvirus*.
Francki, R.I.B. *et al.* (1985) **In** Atlas of Plant Viruses. Vol. 2. p. 183. CRC Press: Boca Raton, Florida.

specific absorbance. The absorbance per unit mass of a substance usually at the wavelength of the maximum of the absorbance spectrum of that substance. The symbol is A and the value is usually quoted as A_n^x, where x is the concentration of the substance and n is the wavelength. The $A_{260\,nm}^{0.1\%}$ is 25 for ssDNA and RNA, and 20 for dsDNA and RNA; the $A_{280\,nm}^{0.1\%}$ for proteins is approximately 1.0 but the actual value depends on the proportion of aromatic amino acids (tyrosine and phenylalanine) present in the molecule.

specific activity. The activity of a radioisotope of an element, other labelled product, protein or enzyme per unit weight present in the sample.

specific infectivity. The infectivity of a virus expressed per unit weight or per unit number of particles. It can be measured by, for example, PLAQUE or POCK ASSAY or by LOCAL LESIONS.

spectrophotometry. A procedure to measure photometrically the wavelength range of radiant energy absorbed by a sample under analysis; can be by visible light, UV light or X-rays. For viral constituents it is usually by UV light in the range 220-330 nm.

spermidine. A triamine, composition $H_2N(CH_2)_3NH(CH_2)_4NH_2$, found in semen and animal and plant tissues. A constituent of some viruses, e.g. TURNIP YELLOW MOSAIC VIRUS, DENSO-VIRUSES.

sphaeroplast, spheroplast. A bacterial or plant cell from which the cell wall has been removed. The term is used interchangeably with protoplast but most frequently refers to bacterial cells. *See* PROTOPLAST for uses.

spheroid. Term applied to the crystalline protein OCCLUSION BODY which incorporates virus particles and is produced during ENTOMOPOXVIRUS infections.

spheroidin. Matrix protein of ENTOMOPOXVIRUS occlusion bodies (SPHEROIDS). Composed of a single polypeptide with high mw. (102-110 x 10^3), soluble at high pH in the presence of disulphide bond-reducing agents.

spheroidosis. *See* ENTOMOPOXVIRUS.

spike. Projections from the surface of a virus particle usually associated with binding of the particle to the cell surface. Composed of either protein or glycoprotein. Most viruses which have membrane-bound particles have spikes. Found, for example, in ORTHOMYXOVIRUSES which have glycoprotein spikes with NEURAMINIDASE and HAEMAGGLUTININ activities. Also found in viruses not possessing membranes, e.g. ADENOVIRUSES and REOVIRUSES.

spinach latent virus. An *Ilarvirus*.
Bos, L. (1984) CMI/AAB Descriptions of Plant Viruses No. 281.
Francki, R.I.B. (1985) In The Plant Viruses. Vol. 1. p. 1. ed. R.I.B. Francki. Plenum Press: New York.

spinach temperate virus. A member of the *Cryptovirus* group, subgroup A.
Boccardo, G. *et al.* (1987) Adv. Virus Res. **32**, 171.

spindle. A term often applied to the bipyramidal paracrystalline proteinaceous inclusion bodies (0.5-25 μm) devoid of virus particles which are produced in large numbers during infections by ENTOMOPOXVIRUSES in Lepidoptera and Coleoptera. They are sometimes occluded within SPHEROIDS. Similar inclusion bodies have been observed in some NUCLEAR POLYHEDROSIS VIRUS infections. The origin, function and significance of the spindles is unknown, though the protein components are distinct from SPHEROIDIN or POLYHEDRIN.
Arif, B.M. (1984) Adv. Virus Res. **29**, 195.
Naser, W.C. *et al.* (1987) J. gen. Virol. **68**, 1251.

spindle virosis. Term used for certain ENTOMOPOXVIRUS infections in which SPINDLE-shaped inclusion bodies are produced.

spiroplasma viruses. Phage-like viruses isolated from helical mycoplasmas (spiroplasmas). They include spv-1 and spv-2, isolated from the sex-ratio organism of *Drosophila*, which contain DNA and resemble phage morphotype C, being polyhedral with a short tail (*see* PHAGE). Isolate SV-C3 from *Spiroplasma citri* has similar morphology. All three isolates are regarded as members of the *Podoviridae*. SV-C2 virus, also from *S. citri* has a polyhedral head and a long tail (morphotype B). Isolate SV-C1 is a possible PLECTROVIRUS although its size (230-280 nm. x 10-15 nm.) is larger than the proposed type plectrovirus MVL51.
Maniloff, J. *et al.* (1978) Adv. Virus Res. **21**, 343.
Maniloff, J. *et al.* (1982) Intervirology **18**, 177.

splicing. The mechanism of RNA processing in which INTRONS are removed and the adjacent EXON sequences are re-ligated. The sequence 5' of the intron is called the splice donor and that 3' of the intron the splice acceptor. Consensus sequences proposed for splice donors are

GAG/GTAAGT
A   G

and for splice acceptors are
(T)ANCAG/G.
C   T

Mount, S.M. (1982) Nucl. Acids Res. **10**, 459.

Spodoptera exempta Cytoplasmic Polyhedrosis Virus. Cytoplasmic polyhedrosis viruses (CPV) of five distinct electropherotypes have been isolated from cultures of the mystery armyworm, *Spodoptera exempta* (*Noctuidae*, Lepidoptera) (*see* APPENDIX B). Two isolates, both from laboratory-reared cultures of *S. exempta*, are recognised as the type members of 'type 3' and 'type 12' CPVs. Type 3 CPV is unrelated to *Bombyx mori* (type 1) CPV, on the basis of RNA electropherotype and serological properties. Type 12 CPV differs from type 1 CPV in the electrophoretic mobility of at least three RNA genome segments and is distantly related as judged by serological properties and nucleic acid homology.
Payne, C.C. and Mertens, P.P.C. (1983) In The *Reoviridae* p. 425. ed. W.K. Joklik. Plenum Press: New York.

Spodoptera exempta nuclear polyhedrosis virus. BACULOVIRUS (subgroup A) isolated from the mystery armyworm, *S. exempta*. Reported to be very closely related to the prototype BACULOVIRUS, *AUTOGRAPHA CALIFORNICA* MNPV.
Brown, S.E. *et al.* (1985) J. gen. Virol. **66**, 2431.

Spodoptera exigua cytoplasmic polyhedrosis virus. Cytoplasmic polyhedrosis virus isolated in the UK from a laboratory culture of the beet armyworm, *Spodoptera exigua* (Noctuidae, Lepidoptera). The virus is the type member of 'type 11' CPVs. Unrelated to *Bombyx mori* (type 1) CPV on the basis of RNA electropherotype. Viruses of similar electropherotype have been observed in other Lepidoptera (*see* APPENDIX B).
Payne, C.C. and Mertens, P.P.C. (1983) In The *Reoviridae*. p. 425. ed. W.K. Joklik. Plenum Press: New York.

Spodoptera exigua nuclear polyhedrosis virus. BACULOVIRUS (subgroup A) isolated from the beet armyworm, *S. exigua*. The virus is of the MNPV type and is cross-infectious for some other noctuid species including the cabbage moth, *Mamestra brassicae*. It has been used in field trials conducted in the USA, Thailand and The Netherlands as a selective biological control agent for larvae of the homologous host.
Gelernter, W.D. *et al.* (1986) J. Econ. Entomol. **79**, 714.

Spodoptera frugiperda nuclear polyhedrosis virus. BACULOVIRUS (subgroup A) isolated from the fall armyworm, *S. frugiperda* which is also infectious and pathogenic for the beet armyworm, *S. exigua*. The virus is of the MNPV type, and a range of closely related genotypes have been isolated. The best studied of these shares 33-53% homology with the MNPV from *S. exempta*, including conservation of repeated sequences. The virus replicates in cell lines derived from *S. frugiperda*.
Hamm, J.J. and Styer, E.L. (1985) J. gen. Virol. **66**, 1249.

Spodoptera littoralis nuclear polyhedrosis virus. BACULOVIRUS (subgroup A) isolated from the cotton leafworm, *S. littoralis*. The virus is of the MNPV type. It has been extensively tested as a selective biological control agent for *S. littoralis* on cotton and is now produced in France on a commercial scale for the control of this pest on cotton and vegetables (*see* SPODOPTERIN). The virus is closely-related to an NPV isolated from *S. litura*, and also shares extensive DNA sequence homology with NPVs from *S. frugiperda* and *S. exigua*.
Kislev, N. (1985) Intervirology **24**, 50.

Spodopterin. Commercial product of the nuclear polyhedrosis virus of the cotton leafworm, *Spodoptera littoralis*, produced by Calliope, Béziers, France. Used for the control of *S. littoralis* on cotton and vegetable crops.

Spondweni virus. Family *Flaviviridae*, genus *Flavivirus*. Isolated from culicine mosquitoes in South Africa, Nigeria, Mozambique, Cameroons. Causes fever and hepatitis in man. Antibodies present in cattle, sheep and goats in which it causes disease.

spongiform encephalopathy agents. Unclassified. These are the agents causing scrapie, kuru, transmissible mink encephalopathy and Creutzfeldt-Jakob disease. All these diseases are neurological with a common pathological picture, leading to death. There is spongiform degeneration of the brain but absence of inflammatory reaction. The agents have not been characterised physicochemically.
Kimberlin, R.H. (1985). **In**: Principles of Bacteriology, Virology and Immunology, Vol. **4**. ed. G.S. Wilson and F. Brown. Edward Arnold: London.

spontaneous mutant. A mutation that occurs naturally.

spring beauty latent virus. A *Bromovirus.*
Valverde, R.A. (1985) Phytopath. **75**, 395.

spring viraemia of carp virus. Synonyms:
RHABDOVIRUS CARPIO, INFECTIOUS DROPSY OF CARP
VIRUS. Family *Rhabdoviridae*, unassigned genus.
Causes disease and death in fish farms in the USA
and Europe. Infects carp and pike. Replicates in
FHM cells, best at 20-22° but also at 31°. Proba-
bly same as SWIM BLADDER INFLAMMATION VIRUS.
Hill B.J. *et al.* (1975) J. gen. Virol. **27**,
369.

Spumavirinae. (Latin 'spuma' = foam.) Subfamily in the family *Retroviridae*. There are
members which infect a wide range of mammalian species, usually leading to persistent infection. Members include bovine and feline
syncytial viruses and human and simian foamy
viruses.

spur formation. Pattern formed by precipitates
in certain antigen-antibody interactions in
double-diffusion serological tests. It usually occurs when there is a partial relationship between
antigens in adjacent wells but can also form if two
identical antigens are present at very different
concentrations. *See* OUCHTERLONY GEL DIFFUSION
TEST.

SpV4 phage. Unclassified phage isolated from
Spiroplasma mellifera, with several characteristics of MICROVIRUS isolates from enterobacteria
(27 nm. diameter particles; circular ssDNA
genome 1.7 x 10⁶, 4.3 kb) but with larger capsid
protein (mw. 60 x 10³).
Ackermann, H.W. and DuBow, M.S. (1987)
Viruses of Prokaryotes. CRC Press: Boca Raton,
Florida.

squash leaf curl virus. A *Geminivirus*, transmitted by whitefly.
Harrison, B.D. (1985) Ann. Rev. Phytopath. **23**,
55.

squash mosaic virus. A *Comovirus.*
Campbell, R.N. (1971) CMI/AAB Descriptions
of Plant Viruses No. 43.
Francki, R.I.B. *et al.* (1985) **In** Atlas of Plant
Viruses. Vol. 2. p. 1. CRC Press: Boca Raton,
Florida.

squirrel fibroma virus. Family *Poxviridae*, subfamily *Chordopoxvirinae*, genus *Lepripoxvirus*.
Causes fibromas in grey squirrels *Sciurus caro-*

linensis in North America and in domestic rabbits.

ss. Abbreviation for SINGLE-STRANDED.

SSC. Abbreviation for standard saline citrate. 1
x SSC is 0.15M NaCl, 0.015M Na citrate. Various
concentrations of SSC are used in SOUTHERN BLOT-
TING and NORTHERN BLOTTING and hybridisation
experiments.

ssRNA phages. Vernacular name for PHAGES
classified within the LEVIVIRUS genus.

SSV1 phage. Unclassified virus-like particle
from the aerobic archaebacterium *Sulfolobus
acidocaldarius*. Particles are lemon-shaped
(*c.* 100 x 60 nm.) morphologically resembling bee
chronic paralysis virus and *Drosophila* RS virus.
The genome is a 15.5kb circular DNA.
Ackermann, H.W. and DuBow, M.S. (1987)
Viruses of Prokaryotes. CRC Press: Boca Raton,
Florida.

stainsall. 1-ethyl-2-(3-(1-ethylnaphtho(1,2d)
thiazolin-2-ylidene)-2-methylpropenyl)-
naphtho(1,2d)thiazolium bromide. Used for
staining proteins and nucleic acids in polyacrylamide gels. RNA stains bluish purple, DNA
stains blue and proteins stain red.
Dahlberg, A.E. *et al.* (1969) J. mol. Biol. **41**, 139.

standard deviation. This is a commonly used
measure of dispersion or spread of data values and
is given in the same units as the mean and thus of
the original observations. It can therefore be of
great value in the intuitive interpretation of the
data and is often used in the estimation of confi-
dence limits and significance tests. It is calculated
and further defined as the square root of the
variance and is given by:-
$$S.D. = \sqrt{\tau (x - \bar{x})_2}$$
Where n = no. of observations
  x = individual data values
  $\bar{x}$ = mean value for x
Statistics books contain details of its calculation
and use.

standard error. This is the STANDARD DEVIATION
of the mean values calculated for each of a number of samples taken from a given population. It
helps in the determination of the potential degree
of discrepancy between the sample mean and the
(usually) unknown population mean. It cannot be
computed exactly but is estimated by:-

$$S.E. = \frac{s}{\sqrt{n}}$$

Where: s = standard deviation
   n = number of observations or data
values
Statistics books contain details of its calculation
and use.

Staphylococcus aureus V8 protease. An enzyme cleaving polypeptides at the carboxyl side
of aspartate and glutamate residues.

starch lesion. Local accumulation of starch in
virus-infected leaf demonstrated by decolorising
the leaf and staining with iodine. Can be used as
an assay for virus concentration. *See* LOCAL
LESION.

starling pox virus. Family *Poxviridae*,
subfamily *Chordopoxvirinae*, genus *Avipoxvirus*.

start codon. The trinucleotide in an mRNA at
which ribosomes start the process of translation
and which sets the READING FRAME for the translation. In eukaryotes it is AUG which is decoded as
methionine. In prokaryotes AUG (giving N-
formylmethionine) is the most common start
codon but GUG (valine) is sometimes used.

statice virus Y. A possible *Potyvirus*.
Francki, R.I.B. *et al.* (1985) **In** Atlas of Plant
Viruses. Vol. 2. p. 183. CRC Press: Boca Raton,
Florida.

steady state infection. A situation where both
virus replication and cell multiplication occur
simultaneously in cell culture. The majority of the
cells are infected and virus is continuously released from them, but there is no CYTOPATHIC
EFFECT. Addition of antiviral antibody will not
cure such infections. A frequent occurrence
among viruses which mature by budding, e.g.
RETROVIRUSES, PARAMYXOVIRUSES, TOGAVIRUSES.

stellavirus. *See* ROTAVIRUS.

sticky ends. The single-stranded ends on DNA
produced by many type II RESTRICTION ENDONU-
CLEASES. They may be either 5′ or 3′ extensions of
the DNA molecule.

Stokes' radius. The radius of the spherical
volume occupied by a macromolecule in solution.

It takes account of hydration and any non-iso-
metric features of the macromolecule.

stop codon. The trinucleotide sequence at which
protein synthesis is terminated due to there being
no tRNA molecule to insert an amino acid into the
polypeptide chain at that site. There are three stop
codons: UAA (OCHRE), UAG (AMBER) and UGA
(OPAL). *See* SUPPRESSOR.

strain. An isolate of a virus which resembles the
type virus in the major properties which define the
type, but differs in minor properties such as vector
species specificity, symptoms induced, serological and genetic properties. In practice it is often
very difficult to differentiate the boundaries between variants, strains and distinct viruses. *See*
SEROTYPES.

strand polarity. The organisation of the base
sequence in a single-stranded nucleic acid. Positive (+) polarity refers to ss molecules that contain
the same base sequence as mRNA. Negative (−)
polarity molecules have a base sequence complementary to the (+)-sense strand.

Stratford virus. Family *Flaviviridae*, genus
Flavivirus. Isolated from *Aedes vigilax* in
Queensland, Australia. Not reported to cause
disease in any species.

strawberry crinkle virus. A plant *Rhabdovirus*,
subgroup 1; transmitted by aphids.
Sylvester, E.E. (1976) CMI/AAB Descriptions of
Plant Viruses No. 163.
Francki, R.I.B. *et al.* (1985) **In** Atlas of Plant
Viruses. Vol. 1. p. 73. CRC Press: Boca Raton,
Florida.

strawberry latent ringspot virus. A member of
the *Fabavirus* group.
Murant, A.F. (1974) CMI/AAB Descriptions of
Plant Viruses No. 126.

strawberry mild yellow edge virus. A possible
Luteovirus.
Francki, R.I.B. *et al.* (1985) **In** Atlas of Plant
Viruses. Vol. 1. p. 137. CRC Press: Boca Raton,
Florida.
Casper, R. (1988) **In** The Plant Viruses. Vol. 3. p.
235. ed. R. Koenig. Plenum Press: New York.

strawberry vein-banding virus. A *Caulimovirus*.
Frazier, N.H. and Converse, R.H. (1980) CMI/

AAB Descriptions of Plant Viruses No. 219. Francki, R.I.B. *et al.* (1985) **In** Atlas of Plant Viruses. Vol. 1. p. 17. CRC Press: Boca Raton, Florida.

streptavidin. A protein from *Streptomyces avidinii* which has a very high affinity for BIOTIN. Four biotin molecules bind to each streptavidin molecule. When coupled with an enzyme (e.g. horseradish peroxidase) it is used to detect biotin in biotinylated DNA used as a probe, or in ELISA. Hiller, Y. (1987) Biochem. J. **248**, 167.

stressors. Factors which induce a manifestation of infection from a LATENT condition. *See* INDUCTION.

strigid herpesvirus 1. Family *Herpesviridae*.

striposomes. *See* UNCOATING.

structural protein. A protein which forms part of the structure of a virus particle.

stylet-borne. *See* NON-PERSISTENT.

Styloviridae. Name proposed for the family of PHAGES with long noncontractile tails, which has now been superseded by the name '*SIPHOVIRIDAE*'.

subcutaneous injection. A route of injection of antigens between the skin and underlying tissue used in vaccination and in the preparation of antisera.

subgenomic RNA. A species of RNA less than genomic length found in infected cells and occasionally encapsidated; when it is encapsidated it is not involved in natural infection. The genomic RNA codes for several proteins but because of the constraints of eukaryotic ribosomes translating non-5' CISTRONS, all but the 5' open reading frame will effectively be closed for translation. The formation of subgenomic RNAs overcomes this problem as each species has a different cistron at its 5' end, thus opening it for translation.

subterranean clover mottle virus. A *Sobemovirus*.
Francki, R.I.B. *et al.* (1985) **In** Atlas of Plant Viruses. Vol. 1. p. 153. CRC Press: Boca Raton, Florida.
Hull, R. (1988) **In** The Plant Viruses. Vol. 3. p. 113. ed. R. Koenig. Plenum Press: New York.

subterranean clover red leaf virus. A *Luteovirus*; considered to be a strain of SOYBEAN DWARF VIRUS.
Francki, R.I.B. *et al.* (1985) **In** Atlas of Plant Viruses. Vol. 1. p. 137. CRC Press: Boca Raton, Florida.
Casper, R. (1988) **In** The Plant Viruses. Vol. 3. p. 235. ed. R. Koenig. Plenum Press: New York.

subterranean clover stunt virus. An unclassified plant virus originally thought to be a *Luteovirus*. Has small isometric particles 17-19 nm. in diameter which band in CsCl at 1.34 g/cc and contain a single species of coat protein (mw. 19×10^3) and three to four species of ssDNA of 850-880 nucleotides.
Francki, R.I.B. *et al.* (1985) **In** Atlas of Plant Viruses. Vol. 1. p. 137. CRC Press: Boca Raton, Florida.
Chu, P.W.G. and Helms, K. (1987) Abstr. R36.1 VII International Congress of Virology, Edmonton, Canada.

sucrose gradient, sucrose density gradient centrifugation. A technique in which particles of different SEDIMENTATION COEFFICIENTS are separated by centrifugation. A gradient of sucrose concentrations, increasing from top to bottom, is constructed in a centrifuge tube and the solution containing the particles is placed on top. On centrifugation in a swing-out rotor the various particle species sediment as bands at different rates. The particles usually do not reach an equilibrium density. *See* ISOPYCNIC CENTRIFUGATION, RATE ZONAL GRADIENT.

sugarcane Fiji disease virus. Synonym: FIJI DISEASE VIRUS. Type member of the *Fijivirus* group.
Hutchinson, P.B. and Francki, R.I.B. (1973) CMI/AAB Descriptions of Plant Viruses No. 119. Francki, R.I.B. *et al.* (1985) **In** Atlas of Plant Viruses. Vol. 1. p. 47. CRC Press: Boca Raton, Florida.

sugarcane mosaic virus. A *Potyvirus*.
Pirone, T.P. (1972) CMI/AAB Descriptions of Plant Viruses No. 88.
Francki, R.I.B. *et al.* (1985) **In** Atlas of Plant Viruses. Vol. 2. p. 183. CRC Press: Boca Raton, Florida.

suid alphaherpesvirus 1. *See* AUJESZKY'S DISEASE VIRUS.

suid herpesvirus 1. *See* AUJESZKY'S DISEASE VIRUS.

Suid herpesvirus 1 group. A genus in the subfamily *Alphaherpesvirinae*. Contains the pseudorabies virus and the equid virus causing abortion.

suid herpesvirus 2. Synonym: PIG CYTOMEGALO-VIRUS. Family *Herpesviridae*, subfamily *Betaherpesvirinae*, genus *Cytomegalovirus*. Causes rhinitis with distortion of the snout. Death is common in young piglets. Can be cultivated in primary pig cell cultures.

Suipoxvirus. (Latin 'sus' = pig.) A genus in the family *Poxviridae*, subfamily *Chordopoxvirinae*, consisting of swinepox virus.

Sunday Canyon virus. Family *Bunyaviridae*, unassigned genus. Isolated from the tick *Argas cooleyi* in the USA. Pathogenic for suckling mice.

sunnhemp mosaic virus. A *Tobamovirus*.
Kassanis, B. and Varma, A. (1975) CMI/AAB Descriptions of Plant Viruses No. 153.
Varma, A. (1986) In The Plant Viruses. Vol. 2. p. 249. ed. M.H.V. van Regenmortel and H. Fraenkel-Conrat. Plenum Press: New York.

supercoiled DNA. A conformation that a dsDNA molecule can adopt. When both strands of a ds molecule are covalently closed one of the strands becomes over- or under-wound in relation to the other. The torsional strain causes the molecule to coil into a characteristic shape.

superhelical DNA. *See* SUPERCOILED DNA.

superinfection. Attempt to infect a host with a second virus, usually a different strain of the first infecting virus. It can result in; 1) interference between the two viruses, the first virus preventing the replication of the second, (*see* CROSS-PROTEC-TION); 2) the two viruses replicating independently (*see* SYNERGISM); 3) recombination between the two viruses; 4) PHENOTYPIC MIXING.

superinfection exclusion. The restriction of growth (interference) of a second virus in a cell already infected with another virus. The term is usually applied to growth restrictions imposed on superinfecting phage in prokaryotic systems. A range of mechanisms is involved in the exclusion

process including DNA breakdown and phage-induced changes in the host cell membrane. *See* CROSS PROTECTION.

supernatant. The material remaining above the PELLET after centrifugation of a suspension of macromolecules.

suppressor. A gene which overcomes (suppresses) the effects of mutations in other, unlinked, genes. A common type involves mutation of tRNAs in which changes in the anticodon reverse the effects of stop codons; these are referred to as suppressor tRNAs. *See* READTHROUGH.

sus 1-4 mastadenoviruses. Family *Adenoviridae*, genus *Mastadenovirus*. Species of the genus affecting pigs.

susceptibility. A property of a host which enables a virus (or other pathogen) to replicate in it. The replication does not necessarily lead to overt symptoms, e.g. LATENT INFECTION. *See* RESISTANCE.

SV4 phage. Unclassified phage isolated from *Spiroplasma* with several chartacteristics of *Microvirus* isolated from the enterobacteria (27nm diameter particles: circular ssDNA genome 1.7 x 10⁶) but with a larger capsid protein (mw. 60 x 10³).
Ackermann, H.W. and DuBow, M.S. (1987) Viruses of Prokaryotes. CRC: Boca Raton, Florida.

SV40 virus. Abbreviation for SIMIAN VIRUS 40. Family *Papovaviridae*, genus *Polyomavirus*. Isolated from a case of progressive multifocal leucoencephalopathy with African green monkey kidney cells. Antibodies are often found in humans in contact with monkeys.

Svedberg equation. A formula for estimating the molecular weight of a macromolecule:

$$M = \frac{RTs}{D(1-\bar{v}\rho)}$$

Where M = weight of one mole of the macromolecule.

R = gas constant per mole (8.314 x 10⁷ erg mole⁻¹K⁻¹ in cgs units, or 8.134 J mole⁻¹K⁻¹ in SI units).

s = sedimentation coefficient.

D = diffusion coefficient.

$\bar{v}$ = partial specific volume.

ρ = solvent density.
T = absolute temperature in degrees Kelvin.

Svedberg unit. *See* SEDIMENTATION CO-EFFICIENT.

swamp fever virus. *See* INFECTIOUS EQUINE ANAEMIA VIRUS.

sweet clover necrotic mosaic virus. A *Dianthovirus.*
Hiruki, C. (1986) AAB Descriptions of Plant Viruses No. 321.

sweet potato feathery mottle virus. A *Potyvirus.*
Francki, R.I.B. *et al.* (1985) **In** Atlas of Plant Viruses. Vol. 2. p. 183. CRC Press: Boca Raton, Florida.

sweet potato mild mottle virus. A possible *Potyvirus.*
Holling, M. *et al.* (1976) CMI/AAB Descriptions of Plant Viruses No. 162.

sweet potato russet crack virus. A possible *Potyvirus.*
Francki, R.I.B. *et al.* (1985) **In** Atlas of Plant Viruses. Vol. 2. p. 183. CRC Press: Boca Raton, Florida.

sweet potato virus A. A possible *Potyvirus.*
Francki, R.I.B. *et al.* (1985) **In** Atlas of Plant Viruses. Vol. 2. p. 183. CRC Press: Boca Raton, Florida.

swim bladder inflammation virus. *See* SPRING VIRAEMIA OF CARP VIRUS.

swine fever virus. *See* HOG CHOLERA VIRUS.

swine influenza virus. *See* INFLUENZA VIRUS.

swine poxvirus. Synonyms: PIG POXVIRUS, VARIOLA SUILLA. Family *Poxviridae,* subfamily *Chordopoxvirinae,* genus *Suipoxvirus.* Type species of genus. Causes generalised disease in young pigs. Can be transmitted by lice. Injection into rabbits causes papular lesions. Replicates with cpe in pig kidney, testis, brain and embryo lung cell cultures.

swine vesicular disease virus. (SVDV). Family *Picornaviridae,* genus *Enterovirus.* Causes a condition in pigs similar to foot-and-mouth disease. Disease first described in Italy in 1966 but has since been found in many countries. Closely related serologically to Coxsackie B5 virus. Has caused meningitis and other Coxsackie virus-like symptoms in man. Grows with cpe in pig kidney and HeLa cells.

swing-out rotor. A centrifuge rotor having buckets which are free to swing out to the horizontal position when the rotor is spinning. The gravitational force is thus along the long axis of the tubes. *See* SUCROSE GRADIENT.

syndrome. A group of signs or symptoms which together characterise a disease.

synergism. An association of two biological organisms in which the effects or behaviour of one organism are enhanced by the presence of the other. With viruses it is the symptoms caused by the joint infection by two viruses being greater than the sum of those caused by the individual viruses.

synergistic factor (SF). A lipoprotein of mw. 93-126 x 10^3 isolated from the granulosis virus of *Pseudaletia unipuncta* which enhances BACULOVIRUS infections *in vivo* and *in vitro*. It is present in the matrix of the GV occlusion body and it enhances the attachment of enveloped baculoviruses to the plasma membrane of certain cells including those of the midgut epithelium of susceptible lepidopteran larvae.
Tanada, Y. *et al.* (1980) J. Invertebr. Pathol. **34,** 249.

systemic infection. Infection resulting from the spread of virus from the site of infection to all or most cells of an organism.

T

T2 phage. Type species of T-EVEN PHAGE GROUP (*Myoviridae*) isolated from *E. coli*. Virions are structurally complex with an elongated head (111 x 80 nm.), contractile tail (113 x 16 nm.) collar, six spikes and six kinked tail fibres. The particles have a mw. of 210 x 10^6, sediment at 1042S and have a buoyant density in CsCl of 1.49 g/cc. Virions contain at least 41 polypeptides ranging from mw. 8 to 155 x 10^3, with a major capsid protein of *c*.43 x 10^3. Particles also contain ATP (required for tail contraction) and a number of enzyme activities (e.g. thymidylate synthetase; dehydrofolate reductase). The genome is linear dsDNA (mw. 121 x 10^6) which is circularly permuted and terminally redundant. The number of genome permutations is very large. The DNA contains 5-hydroxymethylcytosine instead of cytosine and these residues are glycosylated in specific patterns (protecting viral DNA from degradation by host restriction enzymes). Virions adsorb to the cell wall of susceptible hosts and completely inhibit host nucleic acid and protein synthesis. Viral DNA replicates as a concatemer and is packaged by the HEADFUL PACKAGING mechanism. Virus heads, tails and tail fibres assemble independently in three different pathways. T2 is a virulent phage and lyses the cells of susceptible hosts (enterobacteria). The replication of T2 phage has been studied in less detail than T4 phage (*see* T4 PHAGE).

Ackermann, H.W. (1978) **In** Handbook of Microbiology. Vol. 2. p. 643, ed. A.I. Laskin and H.A. Lechevalier. CRC Press: Boca Raton, Florida.
Matthews, R.E.F. (1982) Intervirology **17**, 1.

T4 phage. Member of T-EVEN PHAGE GROUP (*Myoviridae*) and probably the best-studied and most complex of all the TAILED PHAGES; isolated from *E. coli*. The phage head is 95 nm. long, 65 nm. wide and the tail 80 x 16 nm. About 36 structural proteins have been observed including at least six virus-specified enzymes. Like the type species of the T-even phage group (T2), the DNA genome (mw. 130 x 10^6; 166 kbp) contains 5-

hydroxymethylcytosine instead of cytosine and is circularly permuted. Each molecule carries a 3% circularly permuted. Detailed genetic and physical maps of the genome are available. Particles adsorb to lipopolysaccharide receptors on the bacterial cell wall by the virus tail fibres. The baseplate is then brought close to the cell surface, accompanied by tail sheath contraction and DNA injection. Within a few minutes of DNA injection, bacterial DNA replication stops and the host DNA is degraded and used as precursor for T4 DNA. Expression of T4 genes is temporally-regulated at the level of transcription by changes in the ability of the host RNA polymerase to recognise promoters. 'Early' and 'mid' genes are transcribed from one DNA strand and 'late' genes from the other. T4 DNA synthesis is initiated at a specific origin, and produces concatemers which are packaged by the HEADFUL PACKAGING mechanism. Phages are assembled from four main subassembly reactions: 1) neck collar and whiskers; 2) DNA-filled head; 3) baseplate, tube and sheath; 4) tail fibres. Infection results in lysis of the host.

Mathews, C.K. *et al.* (1983) Bacteriophage T4. American Society for Microbiology: Washington.
Mosig, G. (1984) **In** Genetic Maps 1984. Vol. 3. p. 35. ed. S.J. O'Brien. Cold Spring Harbor Laboratory: New York.

T7 phage. Type species of the T7 PHAGE GROUP (*Podoviridae*). Virions have an isometric head (65 nm. in diameter) and a short tail (17 nm. long) with six short fibres. The particles have a mw. of 47 x 10^6, sediment at 507S and have a buoyant density in CsCl of 1.5 g/cc. Virions contain 10-12 proteins (mw. 14-150 x 10^3) including the major capsid protein (mw. 38 x 10^3). The genome is linear dsDNA containing the conventional bases, which (unlike T2 and T4 DNAs) is not circularly permuted, but contains short regions (*c*.260 bp) of terminal redundancy. Virus particles adsorb to cell walls of susceptible hosts (female entero-

bacteria). Host cell DNA is degraded by phage-coded enzymes to provide substrate for viral DNA synthesis. Virus codes for about 30 polypeptide gene products and requires several host factors for replication. Like T4 PHAGE, transcription is temporally regulated but only one DNA strand is used as template. T7 DNA synthesis is initiated at a specific origin and is bidirectional. concatemers are produced and packaged by a mechanism which ensures that the genomes have unique (non-permuted) sequences. Infection results in lysis of the host.
Ackermann, H.W. (1978) In Handbook of Microbiology. Vol. 2. p. 643. ed. A.I. Laskin and H.A. Lechevalier. CRC Press: Boca Raton, Florida.
Ritchie, D.A. (1983) In Topley and Wilson's Principles of Bacteriology, Virology and Immunity. Vol. 1. p. 177. ed. G. Wilson, A. Miles and M.T. Parker. Edward Arnold: London.

T7 phage group. Vernacular name for the only genus of phages currently defined within the *Podoviridae* (phages with short tails). The genus includes the type species, T7 phage as well as phages H, PTB, R, T3, Y, W31, φI and φII.
Ackermann, H.W. and Du Bow, M.S. (1987) Viruses of Prokaryotes. CRC Press: Boca Raton, Florida.

T-even phage group. Vernacular name for the only officially recognised genus of phages with contractile tails (*see MYOVIRIDAE*) which includes T2, T4 and T6 phages as well as >50 other isolates. T2 phage is the type species.
Ackermann, H.W. and Du Bow, M.S. (1987) Viruses of Prokaryotes. CRC Press: Boca Raton, Florida.

T antigens. Abbreviation for transforming or tumour antigens. Gene products of PAPOVA- and ADENO-VIRUSES which are essential for viral DNA replication and for cellular transformation.

Tacaiuma virus. Family *Bunyaviridae*, genus *Bunyavirus*. Isolated from sentinel monkeys and mosquitoes in Brazil.

Tacaribe virus. Family *Arenaviridae*, genus *Alphavirus*. Isolated from bats and mosquitoes in Trinidad. Guinea pigs can be silently infected.

Taggert virus. Family *Bunyaviridae*, genus *Nairovirus*. Isolated from ticks which infest penguins on Macquarie Island (S.E. of Tasmania). Antibodies found in penguin sera.

Tahyna virus. Family *Bunyaviridae*, genus *Bunyavirus*. Isolated in several countries in Europe. Causes fever in man. Antigenically indistinguishable from Lumbo virus.

Taiassui virus. Family *Bunyaviridae*, genus *Bunyavirus*.

tail. Structure associated with the most complex DNA phages that is involved in the attachment of the phage to specific receptors on the host bacterium and acts as the passageway through which the nucleic acid enters the cell. Some tails are contractile (*see* MYOVIRIDAE) facilitating DNA transport. Others are non-contractile and may be 'short' (c. 20 nm.; *see* PODOVIRIDAE) or 'long' (60-540 nm.; *see* SIPHOVIRIDAE). The tail probably originates from a specialised vertex of the phage head. Tails may carry neck, collar and whisker attachments near the connection of the tail and head sections of the particle (*see* T4 PHAGE). Contractile tails (e.g. T4) have a core and contractile sheath composed of helically-arranged capsomeres. The termini of phage tails may be undifferentiated or carry base plate, tail pins and tail fibres involved in virus attachment and DNA injection.
Eiserling, F.A. (1978) In Comprehensive Virology. Vol. 13. p.543. ed. H. Fraenkel-Conrat and R.R. Wagner. Plenum Press: New York.
Ritchie, D.A. (1983) In Topley and Wilson's Principles of Bacteriology, Virology and Immunity. Vol. 1 p.177. ed. G.S.Wilson, A.A. Miles and M.T. Parker. Edward Arnold: London.

tailed phages. Large group of unenveloped DNA-viruses isolated from prokaryotes, characterised by complex virus particles consisting of a cubic head (capsid) and helical tail (phage morphotypes A, B and C; *see* PHAGE). Several hundred isolates with these properties have been reported (*see* APPENDIX F) and are currently classified in three families: *Myoviridae* (contractile tails e.g. T2 PHAGE, T4 PHAGE); *Siphoviridae* (long non-contractile tails e.g. λPHAGE); *Podoviridae* (short non-contractile tails e.g. T7 PHAGE). The viruses may be temperate or virulent and have been isolated from more than 90 genera of bacteria, cyanobacteria and mycoplasma.
Ackermann, H.W. (1978) In Handbook of Microbiology. Vol. 2. p. 643. ed. A.I. Laskin and H.E. Lechevalier. CRC Press: Boca Raton, Florida.
Matthews, R.E.F. (1982) Intervirology **17**, 1.

tailing. A term describing the addition of

stretches of nucleotides to the 3′-termini of ds nucleic acid to facilitate cloning into a vector.

Talfan disease virus. Family *Picornaviridae*, genus *Enterovirus*. Causes mild form of Teschen disease. First isolated in England and Denmark in 1955.

tamarillo mosaic virus. A *Potyvirus*. Hollings, M. and Brunt, A.A. (1981) **In** Handbook of Plant Virus Infections. ed. E. Kurstak. p. 731. Elsevier: North Holland.

Tamdy virus. Family *Bunyaviridae*, genus *Bunyavirus*. Isolated from ticks infesting men working with sheep and camels in the desert regions of the USSR. Pathogenic for suckling and very young mice when injected i.c.

Tamiami virus. Family *Arenaviridae*, genus *Arenavirus*. Isolated from cotton-rats in Florida.

Tanapox virus. Family *Poxviridae*, subfamily *Chordopoxvirinae*, genus unassigned. Causes skin lesions in children in Kenya and lesions in monkeys, which in turn infect those handling them. Grows in monkey kidney cell cultures. Serologically related to Yaba virus and swine pox virus.

Tanga virus. Unclassified arthropod-borne virus. Isolated from mosquitoes in Tanzania and Uganda.

Tanjong Rabok virus. Unclassified arthropod-borne virus. Isolated from sentinel Rhesus monkeys in Malaysia. Antibodies found in man, birds, wild rodents and bats.

TATA box. A sequence of DNA in the region of a eukaryotic promoter which is considered analaogous to the Pribnow box of prokaryotes. It is also termed the Goldberg-Hogness box and has the canonical sequences:-

for animals $\text{TATATA}_{1\text{-}3}\text{CA}$
              A   T
for plants   TATATAT
              A A

Tataguine virus. Family *Bunyaviridae*, genus *Bunyavirus*. Isolated from man and mosquitoes in several countries in Africa.

tau virus. Unclassified virus isolated from the crab, *Carcinus mediterraneus*. It shares the general characteristics of the non-occluded baculoviruses but differs in its bow-shaped structure. Enveloped particles are 350 nm. long by 80 nm. wide.
Pappalardo, R. *et al.* (1986) J. Invertebr. Pathol. **47**, 361.

teasle mosaic virus. A possible *Potyvirus*. Francki, R.I.B. *et al.* (1985) **In** Atlas of Plant Viruses. Vol. 2. p. 183. CRC Press: Boca Raton, Florida.

Tectiviridae. (Latin 'tectus' = covered). A family of phages containing a single genus (TECTI-VIRUS) of large isometric viruses (65 nm. in diameter) with a double capsid structure and a linear dsDNA genome (PHAGE morphotype D4).

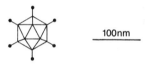

100nm

Tectivirus. dsDNA phages with unenveloped isometric particles, 65 nm. in diameter (PHAGE morphotype D4) and a double capsid structure with the inner layer containing lipid (comprising 10-20% by weight of the particle). The inner layer is flexible and able to transform itself, upon adsorption to a bacterium, into a tail-like tube. Some isolates have 20 nm. long spikes on the vertices. The genome is one piece of non-permuted linear dsDNA (mw. 9 x 10^6) covalently linked to protein. Virions infect *Bacillus* spp. and gram-negative bacteria which contain certain drug-resistant plasmids, adsorbing to the tips of plasmid-specified pili. The host is lysed by the infecting virus. The type species is PRD1 PHAGE and other members include L17, PR3, PR4, PR5, PR772, AP50, Bam35 and φNS11 (Appendix F).
Mindich, L. (1978) **In** Comprehensive Virology. Vol. 12. p. 271. ed. H. Fraenkel-Conrat and R.R. Wagner. Plenum Press: New York.
Matthews, R.E.F. (1982). Intervirology **17**, 1.
Ackermann, H.W. and Du Bow, M.S. (1987). Viruses of Prokaryotes. CRC Press: Boca Raton, Florida.

Tehran virus. Family *Bunyaviridae*, genus *Phlebovirus*.

Tembe virus. Unclassified arthropod-borne virus. Isolated from mosquitoes in Brazil.

Tembusuvirus. Family *Flaviviridae*, genus *Flavivirus*. Isolated from mosquitoes in Malaysia, Thailand and Sarawak.

TEMED. N,N,N′,N′-Tetramethylethylenediamine. A catalyst providing free radicals necessary for the polymerisation of monomer acrylamide and bisacrylamide to give the gel used in POLYACRYLAMIDE ELECTROPHORESIS.
Davis, B.J. (1964) Ann. N. Y. Acad. Sci. **121**, 404.

temperate phage. A phage which can grow in its prokaryotic host in two distinct ways; a LYTIC CYCLE where the virus replicates and the host cell is killed, or a LYSOGENIC state where the viral genome is integrated into the host chromosome (e.g.λ PHAGE) or a plasmid (e.g. P1 PHAGE). During this prophage stage replication is suppressed and the viral genome is transmitted to progeny cells with each cell division.

temperature-sensitive mutant. A mutant which is not 'viable' at as high a temperature as the WILD TYPE strain.

template. A nucleic acid molecule from which a complementary nucleic acid molecule is being synthesised.

Tensaw virus. Family *Bunyaviridae*, genus *Bunyavirus*. Causes fever and encephalitis in man. Virus has been isolated from dogs and marsh rabbits. Antibodies present in man, cows and chickens.

Tenuivirus group. (Latin 'tenuis' = thin or fine, referring to the virus particles). (Type member RICE STRIPE VIRUS). Genus of plant viruses with filamentous particles, 8 nm. in diameter and of varying length, occasionally branched. Particles

100nm

are composed of a supercoiled nucleoprotein, the polypeptide having mw. 32 x 10³. Particles contain four species of linear ssRNA (mw. 1.9, 1.4, 1.0 and 0.9 x 10⁶).
   Host ranges of members are wide. Mechanical transmission is difficult. Transmitted naturally by plant hoppers in the PERSISTENT TRANSMISSION manner.
Francki, R.I.B. *et al.* (1985) **In** Atlas of Plant Viruses. Vol. 2. p. 237. CRC Press: Boca Raton, Florida.
Toriyama, S. (1986) Microbiol. Sci. **3**, 347.

Tephrosia symptomless virus. A possible member of the *Carmovirus* group.
Bock, K.R. (1982) CMI/AAB Descriptions of Plant Viruses No. 256.
Morris, T.J. and Carrington, J.C. (1988) **In** The Plant Viruses. Vol 3, p.73, ed. R. Koenig. Plenum Press: New York.

teratoma. A neoplasm composed of bizarre and chaotically arranged tissues which are foreign embryologically as well as histologically to the area in which the tumour is found.

terminal redundancy. The presence of identical nucleotide sequences at both ends of a nucleic acid molecule which have linear chromosomes, e.g. T2, T4, T7 PHAGE, and in RETROVIRUS genomic RNA.

terminal repetition. The presence of nucleotide sequences at both ends of a nucleic acid molecule which are either identical (*see* TERMINAL REDUNDANCY) or are identical but inverted. Inverted terminal repeats are found in the genomic DNA of ADENOVIRUSES.

terminal transferase, deoxynucleotidyl transferase. An enzyme which will add deoxynucleotide triphosphates (dNTP) to the 3′ hydroxyl group of a DNA fragment. It is usually isolated from calf thymus. Used in 3′ end labelling of DNA and in tailing DNA fragments with complementary dNTPs to facilitate cloning.

termination codon. *See* STOP CODON.

termite paralysis virus. Unclassified virus (possible INSECT PICORNAVIRUS), isolated in Australia from the termites *Nasutitermes exitiosus*, *Coptotermes lacteus* and *Porotermes adamsoni*. Virus particles are isometric, 30 nm. in diameter, sediment at 155S and have a buoyant density in CsCl of 1.34 g/cc. The particles are stable at pH 2.5. The nucleic acid has not been analysed. Serologically unrelated to bee acute paralysis virus.
Moore, N.F. *et al.* (1985) J. gen. Virol. **66**, 647.

TES. N-tris(hydroxymethyl)methyl-2-amino-

ethanesulphonic acid (mw. 229.25). A biological buffer, pK_a 7.5, with a pH range 6.6-8.6.
Good, N. *et al.* (1966) Biochemistry **5**, 467.

Teschen disease virus. Synonym: PORCINE ENTEROVIRUS. Family *Picornaviridae*, genus *Enterovirus*. Causes fever, convulsions and paralysis with encephalomyelitis. Named after region in Czechoslovakia where first outbreaks occurred.

test plant. A plant species which is used for diagnosing, or evaluating infectivity of, a plant virus. *See* INDICATOR PLANT.

Tete virus. Family *Bunyaviridae*, genus *Bunyavirus*. Isolated from several species of birds in S. Africa and Nigeria.

tetracyclines. A group of broad spectrum antibiotics which are isolated from *Streptomyces* spp. They act by preventing charged transfer RNAs binding to ribosomes. Some plasmid cloning vectors, e.g. pBR322, confer tetracycline resistance to bacteria; this region of the plasmid is used as a selective marker for transformation.

Tetraviridae. A family of insect-pathogenic viruses containing a single genus, TETRAVIRUS. Virions are unenveloped, isometric, 35 nm. in diameter. The viral genome is composed of a

 100nm

single (+)-sense ssRNA molecule. The capsid contains one major polypeptide (mw. 60-70 x 10^3), arranged in T=4 symmetry. The name of the group derives from the particle symmetry. The family is also referred to as the *Nudaurelia* β virus group.

Tetravirus. The only genus of viruses currently classified within the *Tetraviridae (Nudaurelia* β virus group). The type species is *Nudaurelia* β virus. Virus particles are isometric, consisting of 240 copies of a single protein subunit arranged in a T=4 symmetry, unenveloped and 35 nm. in diameter. They sediment at 194-210S and have a buoyant density in CsCl of 1.29 g/cc. Particles are stable at pH3 and contain one major polypeptide (mw. 60-70 x 10^3). The (+)-sense ssRNA genome consists of a single molecule (mw. 1.8 x 10^6) which is not polyadenylated. The viruses replicate primarily in the cytoplasm of gut cells of

susceptible Lepidoptera. Detailed studies of replication have been hindered by the absence of a suitable permissive *in vitro* cell culture system. *In vitro* translation of viral RNA has revealed distinct products, the largest having a mw. in excess of 100 x 10^3. Other viruses with similar properties have been isolated from the lepidopterans, *Antheraea eucalypti, Samia (=Philosamia) ricini, Saturnia pavonia (Saturniidae; Darna trima, Setora nitens, Thosea asigna* (Limacodidae); *Dasychira pudibunda, Lymantria ninayi* (Lymantriidae); *Trichoplusia ni* (Noctuidae); *Acherontia atropas* (Sphingidae); *Eucocytis meeki* (Cocytiidae); *Hypocrita jacobeae* (Arctiidae); *Agraulis vanillae* (Nymphalidae); *Euploea corea* (Danaidae). Most of the isolates were identified on the basis of their serological reaction with antiserum raised against *Nudaurelia* β virus. Members are serologically interrelated but distinguishable. No infections have yet been achieved in cultured insect cells.
Moore, N.F. *et al.* (1985) J. gen. Virol. **66**, 647.

Thai sacbrood virus. Unclassified small RNA virus (a possible INSECT PICORNAVIRUS) first isolated from the Eastern honey bee, *Apis cerana*, in Thailand. Virus differs principally from SACBROOD VIRUS in the sizes and antigenic properties of the virus structural proteins (mw. 38.7, 34 and 30.2 x 10^3).
Bailey, L. *et al.* (1982) J. Invertebr. Pathol. **39**, 264.

Theiler's virus. Synonym: MOUSE ENCEPHALO-MYELITIS VIRUS, MOUSE POLIOMYELITIS VIRUS. Family *Picornaviridae*, provisionally in genus *Enterovirus*. First isolated from the central nervous system of mice with flaccid paralysis. Passaging in mice causes necrosis of motor neurones. Mice surviving infection develop demyelinating disease. Relationship with other enteroviruses is unclear. It may be more closely related to the Cardioviruses.

thermal inactivation point. A temperature at which virus infectivity is lost. With plant viruses, samples of sap from an infected plant are heated to various temperatures for ten minutes and then assessed for the infectivity of the virus. Inactivation points range from 45°C for TOMATO SPOTTED WILT VIRUS to 95°C for TOBACCO MOSAIC VIRUS.

thermolysin. An enzyme which catalyses the hydrolysis of peptide bonds on the N-terminal side of valine, leucine, isoleucine, phenylalanine, tyrosine, methionine, and sometimes alanine residues.

Ambler, R.P. and Medway, R.J. (1968) Biochem. J. **108**, 893.

thermotherapy. The curing of a host or cell line of a virus infection by heat treatment. One of the methods used for curing plants of virus infection. *See* CHEMOTHERAPY.

Thetalymphocryptovirus. A proposed genus in the family *Herpesviridae*. It contains GALLID HERPESVIRUS 2 (Marek's disease virus) as its type species and meleagrid herpesvirus 1 as a second member.

Thielaviopsis basicola viruses. Possible members of the *Totivirus* group.

Thimiri virus. Family *Bunyaviridae*, genus *Bunyavirus*. Isolated from the pond heron in India and the lesser white-throat in Egypt.

thin-layer chromatography. CHROMATOGRAPHY on thin layers of adsorbents rather than in columns; the adsorbents include alumina, silica gel, silicates, charcoals or cellulose.

thistle mottle virus. A *Caulimovirus*. Francki, R.I.B. *et al.* (1985) **In** Atlas of Plant Viruses. Vol. 1. p. 17. CRC Press: Boca Raton, Florida.

Thogoto virus. Family *Bunyaviridae*, genus *Bunyavirus*. Causes optic neuritis and meningo-encephalitis in man. Virus isolated from man, cattle, camels and ticks in Egypt, Kenya, Nigeria and Sicily. Antibodies found in cattle, sheep, goats and camels.

Thottapalayam virus. Unclassified arthropod-borne virus. Isolated from shrews in India.

thymidine kinase. An enzyme catalysing the phosphorylation of thymidine to thymidylic acid. It is induced in cells infected with DNA viruses; that of HERPESVIRUS is encoded by the viral DNA whereas in PAPOVAVIRUS-infected cells virus replication depends on a cell-encoded enzyme.

thymidine. A nucleoside of thymine and deoxyribose. *See* NUCLEIC ACID.

thymine. A constituent pyrimidine base of DNA. *See* NUCLEIC ACID.

tick-borne encephalitis virus. Synonyms: BIPHASIC MILK FEVER VIRUS, RUSSIAN SPRING-SUMMER ENCEPHALITIS VIRUS. Family *Flaviviridae*, genus *Flavivirus*. A severe infection of man, causing paralysis, meningitis and frequently death. Occurs in several central European countries but particularly in the USSR. The virus will infect mice, guinea pigs, monkeys, sheep, goats, and wild rodents experimentally. There are two well-defined subtypes. Disease can be controlled by elimination of ticks or by vaccination.

Tillamook virus. Family *Bunyaviridae*, genus *Nairovirus*.

Tilligerry virus. Family *Reoviridae*, genus *Orbivirus*. Isolated from mosquitoes in New South Wales, Australia.

Timbo virus. Family *Rhabdoviridae*, unassigned to genus. Isolated from lizards in Brazil. Infects mosquitoes experimentally and kills suckling mice. Can be grown in Vero cells.

Timboteua virus. Family *Bunyaviridae*, genus *Bunyavirus*.

Tinaroo virus. Family *Bunyaviridae*, genus *Bunyavirus*.

Tindholmur virus. *Family Reoviridae*, genus *Bunyavirus*.

Tipula paludosa nuclear polyhedrosis virus. BACULOVIRUS (Subgroup A) isolated from the leatherjacket, *T. paludosa*, a dipteran pest of grassland. Virus infection was probably first described in Scotland in 1923. The virus is of the SNPV type. Unlike NPV infections of lepidopteran hosts, the viral occlusion bodies produced are crescent-shaped and require the use of a reducing agent as well as high pH to dissolve the crystalline matrix protein (polyhedrin). The viral polyhedrin is unrelated both serologically and in N-terminal amino acid sequence to polyhedrins of NPVs isolated from Lepidoptera.

Guelpa, B. *et al.* (1977) C. R. hebd. Acad. Sci. Paris D **285**, 779.

tissue culture. The growth or maintenance of tissue cells or undifferentiated cells *in vitro* in either a liquid or soft gel medium. Tissue culture in the strict sense involves the maintenance of fragments of tissue or an undifferentiated callus. The term is also used for CELL CULTURE and ORGAN

CULTURE. *See* PROTOPLASTS.

titration. The measurement of TITRE.

titre. The concentration of virus present in a preparation measured by bioassay or a relative measure of the concentration of a specific antibody in antiserum.

TK. Abbreviation for THYMIDINE KINASE.

Tlacotalpan virus. Family *Bunyaviridae*, genus *Bunyavirus*. Isolated from mosquitoes in Mexico.

TLCK N-α-p-tosyl-L-lysine chloromethyl ketone HCl. An inhibitor of CHYMOTRYPSIN.

Tm. Abbreviation for MELTING TEMPERATURE.

TM-biocontrol-I. Preparation of a nuclear polyhedrosis virus of the Douglas fir tussock moth, *Orgyia pseudotsugata*, produced on an industrial scale and registered for use as a biological control agent for the tussock moth by the US Forest Service.

TN 368 cells. Insect ovarian cell line from the cabbage looper, *Trichoplusia ni,* susceptible to infection by AUTOGRAPHA CALIFORNICA NUCLEAR POLYHEDROSIS VIRUS and some other baculoviruses including Hz-1 virus.

TNM-FH medium. Standard medium for the culture of the insect cell line TN-368.

tobacco etch virus. A *Potyvirus*.
Purcifull, D.E. and Hiebert, E. (1982) CMI/AAB Descriptions of Plant Viruses No. 258.
Francki, R.I.B. *et al.* (1985) **In** Atlas of Plant Viruses. Vol. 2. p. 183. CRC Press: Boca Raton, Florida.

tobacco leafcurl virus. A *Geminivirus*, subgroup B.
Osaki, T. and Inouye, T. (1981) CMI/AAB Descriptions of Plant Viruses No. 232.
Harrison, B.D. (1985) Ann. Rev. Phytopath. **23**, 55.

tobacco mild green mosaic virus. A *Tobamovirus*.
Wetter, C. (1986) **In** The Plant Viruses. Vol. 2. p. 205. ed. M.H.V. van Regenmortel and H. Fraenkel-Conrat. Plenum Press: New York.

tobacco mosaic virus. Type member of the *Tobamovirus* group. This virus has been in the forefront of the development of concepts in virology. It was shown by Mayer (1886) to be transmissible by sap and by Iwanowski (1892) to be smaller than a bacterium as the infective agent passed through a Chamberland filter-candle. The significance of this latter observation was realised by Beijerinck (1898) who provided evidence that the pathogen was not a bacterium but a *contagium vivum fluidum*. The shape and size of the particles was inferred in the 1930s from studies on optical birefringence, sedimentation behaviour and X-ray diffraction of purified preparations of virus. In 1939 Kausche, Pfankuch and Ruska visualised the particles for the first time in the electron microscope. Helen Purdy (1929) showed that TMV particles contained protein and the RNA was first detected by Bawden and Pirie in 1937. It was in 1956 that Gierer and Schramm and Fraenkel-Conrat showed that the RNA was infectious. The virus is now understood in considerable detail including the structure of the particle at the atomic level, the sequence of the genomic RNA and the *in vivo* expression of the RNA. The host range of TMV is very wide and it causes a mosaic symptom in many species. However, it is latent in many other species including ornamentals in which it is prevalent. The virus has a very widespread distribution. It has several strains including the type strain (also named vulgare, U1 or OM) and the U5 and legume or cowpea strains.
Fraenkel-Conrat. H. (1986) **In** The Plant Viruses. Vol. 2. p. 5. ed. M.H.V. van Regenmortel and H. Fraenkel-Conrat. Plenum Press: New York.
Bloomer, A.C. and Butler, P.J.C. (1986) **In** The Plant Viruses. Vol. 2. p. 19. ed. M.H.V. van Regenmortel and H. Fraenkel-Conrat. Plenum Press: New York.
Palukaitis, P. and Zaitlin, M. (1986) **In** The Plant Viruses. Vol. 2. p. 105. ed. M.H.V. van Regenmortel and H. Fraenkel-Conrat. Plenum Press: New York.
Edwardson, J.R. and Christie, R.G. (1986) **In** The Plant Viruses. Vol. 2. p. 153. ed. M.H.V. van Regenmortel and H. Fraenkel-Conrat. Plenum Press: New York.

tobacco mosaic virus U2. A strain of *Tobacco Mild Green Mosaic Virus*.

tobacco mottle virus. Unencapsidated RNA, dependent on TOBACCO VEIN DISTORTION VIRUS for aphid transmission.

Adams, A.N. and Hull, R. (1972) Ann. Appl. Biol. **71**, 135.

tobacco necrosis satellite virus. An unclassified plant virus with isometric particles, 17 nm. in diameter which sediment at 49-58S. The capsid has icosahedral symmetry (T=1) and is composed of a single coat protein species (mw. 21.6 x 10³). Each particle contains a single molecule of linear (+)-sense ssRNA (mw. 0.34 x 10⁶; 1239 nucleotides) which codes for coat protein. The virus is dependent upon tobacco necrosis virus for replication.
Kassanis, B. (1970) CMI/AAB Descriptions of Plant Viruses No. 15.
Fraenkel-Conrat. H. (1988) In The Plant Viruses. Vol. 3. p. 147. ed. R. Koenig. Plenum Press: New York.

tobacco necrosis virus. Type member of the *Necrovirus* group.
Kassanis, B. (1970) CMI/AAB Descriptions of Plant Viruses No. 14.
Fraenkel-Conrat. H. (1988) In The Plant Viruses. Vol. 3. p. 147. ed. R. Koenig. Plenum Press: New York.

tobacco necrosis virus group. *See* NECROVIRUS GROUP.

tobacco necrotic dwarf virus. A *Luteovirus*; considered to be a strain of POTATO LEAFROLL VIRUS.
Kubo, S. (1981) CMI/AAB Descriptions of Plant Viruses No. 234.
Francki, R.I.B. *et al.* (1985) In Atlas of Plant Viruses. Vol. 1. p. 137. CRC Press: Boca Raton, Florida.
Casper, R. (1988) In The Plant Viruses. Vol. 3. p. 235. ed. R. Koenig. Plenum Press: New York.

tobacco rattle virus. (TRV). Type member of the *Tobravirus* group. Causes a range of symptoms including necrosis in potato tubers (known as spraing) systemic necrotic spots and lines on tobacco, mottle and necrosis of bulbous ornamentals (e.g. tulip, narcissus, gladiolus) and mosaics, blotches and ringspots on the leaves of many plants. The virus has a very wide host range and has been found naturally infecting more than 100 species. Isolates of the virus fall into two classes, stable (M-type) in which nucleoprotein particles containing both RNA species are found, and unstable (NM) which are caused by infections by RNA-1 alone and do not have coat protein.

Harrison, B.D. (1970) CMI/AAB Descriptions of Plant Viruses No. 12.
Francki, R.I.B. *et al.* (1985) In Atlas of Plant Viruses. Vol. 2. p. 147. CRC Press: Boca Raton, Florida.

tobacco ringspot virus. Type member of the *Nepovirus* group.
Stace-Smith, R. (1985) CMI/AAB Descriptions of Plant Viruses No. 305.
Francki, R.I.B. *et al.* (1985) In Atlas of Plant Viruses. Vol. 2. p. 23. CRC Press: Boca Raton, Florida.

tobacco streak virus. Type member of the *Ilarvirus* group.
Fulton, R.W. (1985) AAB Descriptions of Plant Viruses No. 307.
Francki, R.I.B. (1985) In The Plant Viruses. Vol. 1. p. 1. ed. R.I.B. Francki. Plenum Press: New York.

tobacco stunt virus. An unclassified plant virus with rod-shaped particles, 300-340 nm. long and 18 nm. wide. The particles are made up of a single species of coat protein (mw. 48 x 10³) and two species of dsRNA (mw. 4.5 and 4.0 x 10⁶). The virus is transmitted by the phycomycete fungus, *Olpidium brassicae*. These properties are similar to those of LETTUCE BIG VEIN VIRUS.
Kuwata, S. and Kubo, S. (1986) AAB Descriptions of Plant Viruses No. 313.

tobacco vein distorting virus. A possible *Luteovirus*.
Francki, R.I.B. *et al.* (1985) In Atlas of Plant Viruses. Vol. 1. p. 137. CRC Press: Boca Raton, Florida.

tobacco vein mottling virus. A *Potyvirus*.
Francki, R.I.B. *et al.* (1985) In Atlas of Plant Viruses. Vol. 2. p. 183. CRC Press: Boca Raton, Florida.

tobacco yellow dwarf virus. A *Geminivirus*, subgroup C.
Thomas, J.E. and Bowyer, J.W. (1984) CMI/AAB Descriptions of Plant Viruses No. 278.

tobacco yellow net virus. A possible *Luteovirus*.
Francki, R.I.B. *et al.* (1985) In Atlas of Plant Viruses. Vol. 1. p. 137. CRC Press: Boca Raton, Florida.

tobacco yellow vein assistor virus. A possible

Luteovirus.
Francki, R.I.B. *et al.* (1985) **In** Atlas of Plant Viruses. Vol. 1. p. 137. CRC Press: Boca Raton, Florida.
Casper, R. (1988) **In** The Plant Viruses. Vol. 3. p. 235. ed. R. Koenig. Plenum Press: New York.

tobacco yellow vein virus. Unencapsidated RNA, dependent on TOBACCO YELLOW VEIN VIRUS ASSISTOR VIRUS for aphid transmission.
Adams, A.N. and Hull, R. (1972) Ann. Appl. Biol. **71**, 135.

Tobamovirus group. (Sigla for TOBACCO MO-SAIC, the type member). Genus of plant viruses with rigid rod-shaped particles, 300 nm. long and 18 nm. in diameter, which sediment at 194S and band in CsCl at 1.325 g/cc. The coat protein

[▭]   100nm

subunits are arranged in a helix with pitch of 2.3 nm. Each particle contains one molecule of linear (+)-sense ssRNA (mw. 2.0 x 10^6; 6395 nucleo-tides) which has at the 5′ end a CAP and at the 3′ end a tRNA-like structure which accepts histidine, or valine in the cowpea strain of tobacco mosaic virus. Replication is in the cytoplasm and is via a REPLICATIVE INTERMEDIATE. The 5′ end of the genomic RNA translates to give products of mw. 126 and 183 x 10^3, the latter being a READTHROUGH product of the former. 3′-coterminal SUBGENOMIC RNAS translate to give proteins of mw. 30 x 10^3 and, from the 3′ end, the coat protein. Most members of this group have moderate host ranges. Virus particles are found in most cell types and often form large crystalline arrays. They are very easily transmitted mechanically. In nature they are transmitted by contact and some members by seed.
Matthews, R.E.F. (1982) Intervirology **17**, 158.
Gibbs, A.J. (1977) CMI/AAB Descriptions of Plant Viruses No. 184.
Francki, R.I.B. *et al.* (1985) **In** Atlas of Plant Viruses. Vol. 2. p. 103. CRC Press: Boca Raton, Florida.
Gibbs, A.J. (1986) **In** The Plant Viruses. Vol. 2. p. 168. ed. M.H.V. van Regenmortel and H. Fraenkel-Conrat. Plenum Press: New York.

Tobravirus group. (Sigla of TOBACCO RATTLE, the type member). Genus of MULTICOMPONENT plant viruses with rigid rod-shaped particles of two lengths. Long (L) particles are 180-215 nm.

long and short (S) particles are 46-114 nm. long depending on isolate; both are 21-23 nm. in dia-meter. The coat protein subunits (mw. 22 x 10^3)

[▭]   [▢]   100nm

are arranged in a helix with pitch of 2.5 nm. The genome comprises two species of linear (+)-sense ssRNA, the L particles containing RNA-1 (mw. 2.4 x 10^6) and the S particles RNA-2 (mw. 0.6-1.4 x 10^6); the size of the latter depends on isolate. RNA-1 can multiply and spread in the plant on its own but produces only unencapsidated nucleic acid. RNA-2 codes for the coat protein. The host range is wide. Virus particles are found in most cell types and often associate with mitochondria. Most members are easily mechanically transmit-ted. Natural transmission is mainly by nematodes (*Trichodorus* and *Paratrichodorus* spp.). Some are seed-transmitted.
Matthews, R.E.F. (1982) Intervirology **17**, 170.
Francki, R.I.B. *et al.* (1985) **In** Atlas of Plant Viruses. Vol. 2. p. 147. CRC Press: Boca Raton, Florida.
Harrison, B.D. and Robinson, D.J. (1986) **In** The Plant Viruses. Vol. 2. p. 339. ed. M.H.V. van Regenmortel and H. Fraenkel-Conrat. Plenum Press: New York.

Togaviridae. (Latin 'toga' = gown, cloak.) A family of enveloped RNA viruses with spherical particles 40-70 nm. in diameter sedimenting at 150-300S and having a buoyant density of less than 1.25 g/cc in CsCl. The viruses have a spheri-cal nucleocapsid 25-35 nm. in diameter and pos-sess an icosahedral symmetry. Surface projec-

   100nm

tions are clearly defined in most viruses. There are four genera, *ALPHAVIRUS, ARTERIVIRUS, RUBIVIRUS* and *PESTIVIRUS*. The viruses contain three to four proteins, one or more of which are glycosylated. Each capsid contains a single molecule of infec-tious (+)-sense ssRNA, mw. *c*.4 x 10^6. Replica-tion is in the cytoplasm in arthropods as well as in vertebrates. Members of the genera *Rubivirus* and *Pestivirus* are not arthropod-borne.
Matthews, R.E.F. (1982) Intervirology **17**, 97.

tomato (Peru) mosaic virus. A possible *Potyvirus.*

Fribourg, C.E. and Fernandez-Northcote, E.N. (1982) CMI/AAB Descriptions of Plant Viruses No. 255.
Francki, R.I.B. *et al.* (1985) **In** Atlas of Plant Viruses. Vol. 2. p. 183. CRC Press: Boca Raton, Florida.

tomato apical stunt viroid. A VIROID, 360 nucleotides.
Walter, B. (1987) **In** The Viroids. p. 321. ed. T.O. Diener, Plenum Press: New York.

tomato aspermy virus. A *Cucumovirus*.
Hollings, M. and Stone, O.M. (1971) CMI/AAB Descriptions of Plant Viruses No. 79.
Francki, R.I.B. (1985) **In** The Plant Viruses. Vol. 1. p. 1. ed. R.I.B. Francki. Plenum Press: New York.

tomato black ring virus. A *Nepovirus*.
Murant, A.F. (1970) CMI/AAB Descriptions of Plant Viruses No. 38.
Francki, R.I.B. *et al.* (1985) **In** Atlas of Plant Viruses. Vol. 2. p. 23. CRC Press: Boca Raton, Florida.

tomato bunchy top viroid. A VIROID.
Diener, T.O. (1987) **In** The Viroids. p. 329. ed. T.O. Diener. Plenum Press: New York.

tomato bushy stunt virus. Type member of the *Tombusvirus* group.
Martelli, G.P. *et al.* (1971) CMI/AAB Descriptions of Plant Viruses No. 69.
Francki, R.I.B. *et al.* (1985) **In** Atlas of Plant Viruses. Vol. 1. p. 181. CRC Press: Boca Raton, Florida.
Martelli, G.P. *et al.* (1988) **In** The Plant Viruses. Vol. 3, p. 13. ed. R. Koenig. Plenum Press: New York.

tomato golden mosaic virus. A *Geminivirus*, subgroup B. The genome consists of two species of ssDNA, 2588 and 2508 nucleotides.
Buck, K.W. and Coutts, R.H.A. (1985) AAB Descriptions of Plant Viruses No. 303.
Harrison, B.D. (1985) Ann. Rev. Phytopath. **23**, 55.

tomato leaf curl virus. A probable *Geminivirus*, subgroup B.
Harrison, B.D. (1985) Ann. Rev. Phytopath. **23**, 55.

tomato mosaic virus. A *Tobamovirus*.

Hollings, M. and Huttinga, H. (1976) CMI/AAB Descriptions of Plant Viruses No. 156.
Brunt, A.A. (1986) **In** The Plant Viruses. Vol. 2. p. 181. ed. M.H.V. van Regenmortel and H. Fraenkel-Conrat. Plenum Press: New York.

tomato planto macho viroid. A VIROID, 360 nucleotides.
Galindo, A.J. (1987) **In** The Viroids. p. 315. ed. T.O. Diener. Plenum Press: New York.

tomato pseudo-curly top virus. A possible *Geminivirus*, transmitted by treehoppers.
Harrison, B.D. (1985) Ann. Rev. Phytopath. **23**, 55.

tomato ringspot virus. A *Nepovirus*.
Stace-Smith, R. (1984) CMI/AAB Descriptions of Plant Viruses No. 290.
Francki, R.I.B. *et al.* (1985) **In** Atlas of Plant Viruses. Vol. 2. p. 23. CRC Press: Boca Raton, Florida.

tomato spotted wilt virus. Type member of the *Tomato Spotted Wilt Virus* group.
Ie, T.S. (1970) CMI/AAB Descriptions of Plant Viruses No. 39.
Francki, R.I.B. *et al.* (1985) **In** Atlas of Plant Viruses. Vol. 1. p. 101. CRC Press: Boca Raton, Florida.

Tomato spotted wilt virus group. (Named after the type virus). Monotypic genus of a plant virus with spherical particles, 82 nm. in diameter which sediment at 560S and band in sucrose at 1.21 g/cc. The particles have a lipid bilayer envelope surrounding an internal ribonucleoprotein core.

100nm

There are four major virion polypeptides (mw. 78, 58, 52 and 27 x 10^3). The particles contain ssRNA of uncertain polarity which occurs as three species (mw. 2.7, 1.7 and 1.1 x 10^6). The host range is very wide. Virus particles are found in most cell types. Infected cells contain characteristic granular inclusions (VIROPLASMS). The virus is readily transmitted mechanically. In nature it is transmitted by thrips in the PERSISTENT TRANSMISSION manner.

The structure and replication of this virus are considered to resemble those of BUNYAVIRUSES.
Matthews, R.E.F. (1982) Intervirology **17**, 123.

Ie, T.S. (1970) CMI/AAB Descriptions of Plant Viruses No. 39.
Francki, R.I.B. *et al.* (1985) **In** Atlas of Plant Viruses. Vol. 1. p. 101. CRC Press: Boca Raton, Florida.

tomato top necrosis virus.   A possible *Nepovirus.*
Francki, R.I.B. *et al.* (1985) **In** Atlas of Plant Viruses. Vol. 2. p. 23. CRC Press: Boca Raton, Florida.

tomato vein clearing virus.   A plant *Rhabdovirus* occurring in Japan.
Kano, T. *et al.* (1985) Ann. Phytopath. Soc. Japan **51**, 606.

tomato vein-yellowing virus. A plant *Rhabdovirus.*
El Maataoui, M. *et al.* (1985) Phytopath. **75**, 109.

tomato white necrosis virus. A possible *Tymovirus.*
Hirth, L. and Girard, L. (1988) **In** The Plant Viruses. Vol. 3. p. 163. ed. R. Koenig. Plenum Press: New York.

tomato yellow dwarf virus. Synonym: TOBACCO LEAFCURL VIRUS.

tomato yellow leaf curl virus. Synonym: TOMATO LEAF CURL VIRUS.

tomato yellow mosaic virus. Synonym: TOMATO GOLDEN MOSAIC VIRUS.

tomato yellow net virus. A possible *Luteovirus.*
Francki, R.I.B. *et al.* (1985) **In** Atlas of Plant Viruses. Vol. 1. p. 137. CRC Press: Boca Raton, Florida.

tomato yellow top virus. A *Luteovirus*, considered to be a strain of POTATO LEAFROLL VIRUS.
Francki, R.I.B. *et al.* (1985) **In** Atlas of Plant Viruses. Vol. 1. p. 137. CRC Press: Boca Raton, Florida.
Casper, R. (1988) **In** The Plant Viruses. Vol. 3. p. 235. ed. R. Koenig. Plenum Press: New York.

Tombusvirus group. (Sigla of TOMATO BUSHY STUNT, the type member). Genus of plant viruses with isometric particles, 30 nm. in diameter,

   100nm

which sediment at 149S and band in CsCl at 1.35 g/cc. The particles have icosahedral symmetry (T=3), the coat protein subunit having mw. 41 x 10^3. Each particle contains one molecule of linear (+)-sense ssRNA (mw. 1.5 x 10^6). The host ranges of members are wide. Virus particles are found in the cytoplasm and nuclei of most cell types; infections are characterised by compact membranous inclusions. Tombusviruses are readily mechanically transmitted. They are also transmitted through soil.
Matthews, R.E.F. (1982) Intervirology **17**, 142.
Francki, R.I.B. *et al.* (1985) **In** Atlas of Plant Viruses. Vol. 1. p. 181. CRC Press: Boca Raton, Florida.
Martelli, G.P. *et al.* (1988) **In** The Plant Viruses. Vol. 3. p. 13. ed. R. Koenig. Plenum Press: New York.

tombusvirus Neckar. *See* NECKAR RIVER VIRUS.

Tonate virus.   Family *Togaviridae*, genus *Alphavirus.*

Toroviridae. A family of enveloped, peplomer-bearing particles containing an elongated tubular nucleocapsid with helical symmetry. The capsid may bend into an open torus, conferring a biconcave disc or kidney-shaped morphology to the virion (120-140 nm.) or the capsid may be a rod-shaped particle (35-170 nm.). The virus genome is (+)-sense ss RNA, mw. *c.* 6.5 x 10^6 and has a polyadenylic acid tract at its 3′ end. The RNA is surrounded by a major nucleocapsid phosphoprotein, mw. *c.* 20 x 10^3 which in turn is enveloped by a membrane containing one major protein, mw. 22 x 10^3 and a phosphoprotein, mw. 37 x 10^3. The peplomers, *c.* 20 nm. in length, carry determinants for neutralisation and haemagglutination, and consist of a polydisperse N-glycosylated protein, mw. 75-100 x 10^3. Infected cells contain four subgenomic polyadenylated RNAs with mw. 3.0, 0.71, 0.46 and 0.26 x 10^6. All the toroviruses identified so far cause enteric infection. The viruses occurring in man, horse and cattle are serologically related.
Horzinek, M.C. *et al.* (1987) Intervirology **27**, 17.

Toscana virus.   Family *Bunyaviridae*, genus *Phlebovirus.*

Totiviridae.   (Latin 'totus' = whole or undivided). Family of fungal viruses with monopartite dsRNA containing a single genus, the TOTIVIRUS GROUP.

Totivirus group. (Type member *Saccharomyces cerevisiae* virus L1). Only genus of TOTIVIRIDAE. The particles are isometric, 40-43 nm. in diameter, sediment at 161-172S and band in CsCl at 1.40-1.42 g/cc. The capsid is composed of a single coat protein species (mw. 73-88 x 10^3). Each particle contains a single molecule of dsRNA (mw. 3.3-4.2 x 10^6) and also RNA polymerase activity. Some virus isolates also have satellite dsRNAs which encode 'killer' proteins and which are encapsidated separately in particles encoded by the helper virus genome.
Buck, K.W. (1986) **In** Fungal Virology. p. 1. ed. K.W. Buck. CRC Press: Boca Raton, Florida.

Toure virus. Unclassified arthropod-borne virus. Isolated from Kemp's gerbil in Senegal.

Tradescantia (Zebrina) virus. A possible *Potyvirus*.
Francki, R.I.B. *et al.* (1985) **In** Atlas of Plant Viruses. Vol. 2. p. 183. CRC Press: Boca Raton, Florida.

transcapsidation. *See* PHENOTYPIC MIXING.

transcript. The RNA molecule produced by TRANSCRIPTION. The primary transcript is often processed or modified to give the mature functional RNA, e.g. mRNA, rRNA or tRNA.

transcriptase. *See* DNA-DEPENDENT RNA POLYMERASE or RNA-DEPENDENT RNA POLYMERASE.

transcription. The process of RNA synthesis by polymerases from a nucleic acid template. *See* REVERSE TRANSCRIPTION.

transducing phage. Defective phage that contains host DNA sequences as well as, or instead of, viral genes. *See* TRANSDUCTION.

transduction. The transfer of genes from one host cell to another by a virus. The mechanism is best understood with phages but also occurs with other viruses which integrate into host DNA (e.g. RETROVIRUSES which transfer host DNA including tumour-inducing 'oncogenes'). Two mechanisms of transduction are known in phage: a) specialized transduction (restricted to TEMPERATE PHAGES) in which phages obtained from a lysogen carry specific regions of the host chromosome, usually close to the prophage integration site (following faulty excisions); b) generalised transduction, where all segments of the host chromosome may be transferred (by faulty packaging of DNA during virion formation). Specialised transducing phages include λ PHAGE. Faulty excision of the prophage produces molecules with both viral and host DNA, and occurs at a low frequency of about one in 10^5 (low frequency transduction). Co-infections of a defective transducing phage and a normal 'helper' phage can yield transducing phages at a frequency of one in two (high frequency transduction). Generalised transduction arises by encapsidation of cellular DNA by the headful packaging mechanism. Virulent or temperate phages (e.g. T1, T4, P1, P22) are able to form generalised transducing particles.
Bishop, J.M. (1984) **In** The Microbe 1984 I: Viruses. p. 121. ed. B.W.J. Mahy and J.R. Pattison. Cambridge University Press: Cambridge.
Ritchie, D.A. (1983) **In** Topley and Wilson's Principles of Bacteriology, Virology and Immunity. Vol. 1. p. 177. ed. G. Wilson, A. A. Miles and M.T. Parker. Edward Arnold: London.

transfection. The successful virus-infection of cells following their inoculation with viral nucleic acid.

transfer RNA. The RNA molecule responsible for decoding the genetic information in an mRNA. There are tRNA molecules specific for

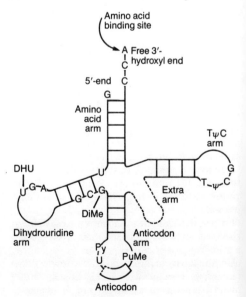

ψ, pseudouridine; Pu, purine; Py, pyrimidine; DiMe, dimethyl; DHU, dihydro-uridine.

each amino acid. The RNA is folded into a 'clover-leaf' secondary structure. The anticodon loop has a sequence complementary to the specific codon; the acceptor stem is charged with the specific amino acid, via its 3' adenosine residue, by the enzyme, aminoacyl-tRNA synthetase. *See* WOBBLE HYPOTHESIS.

transformation. Gene transfer by the uptake of DNA by competent cells. Following the uptake the input DNA may recombine with that of the host or may replicate independently as a plasmid or virus. The term is also used to describe the change of a cell from the normal to the tumorigenic state.

transformation assay. A test for the frequency with which cells are rendered tumorigenic by a virus. The effect of the virus on a monolayer of cells is observed; transformed cells grow in a manner different from that of normal cells, often forming plaques of heaped-up colonies.

translation. The process of synthesis of a protein by ribosomes moving along an mRNA.

transmission. The act of transfer of a virus from an infected organism to a non-infected one. Examples are: for certain viruses of vertebrates by droplet inhalation, by physical contact or by vector, often an arthropod (e.g. arboviruses); for plant viruses by contact (e.g. TOBACCO MOSAIC VIRUS), by arthropod vector (e.g. aphid transmission of POTYVIRUSES), by nematode vector (e.g. NEPOVIRUSES), by pollen (e.g. ILARVIRUSES) or by fungus (e.g. TOBACCO NECROSIS VIRUS); for insect viruses by mouth or by wounding by parasites; for fungal viruses by anastamosis. These forms of transmission are termed HORIZONTAL TRANSMISSION. VERTICAL TRANSMISSION is from mother to progeny, e.g. seed transmission of some plant viruses.

transmission electron microscope. An ELECTRON MICROSCOPE in which the electron beam is passed through a thin specimen, either a preparation of virus particles, NEGATIVELY STAINED or shadowed, or a thin section of tissue mounted on a grid. The image of the specimen is magnified using electrostatic lenses and observed on a fluorescent screen or photographed. The resolution of biological materials is about 2 nm. for negatively stained samples and 5 nm. for sectioned material.

transport protein. A protein involved in the transport of small molecules around the cell.

transposon. A discrete piece of DNA which can insert itself into other DNA sequences within the cell. The ends of the transposon DNA are usually inverted repeats. Transposons in bacteria often carry genes which confer antibiotic or heavy-metal resistance to the host cell. Transposons have been found in many organisms including bacteria, insects, humans, yeast and plants. They can be used to study gene function by examining their effect when inserted into different regions of the (viral) genome.

triangulation number. A description of a triangular face of an ICOSAHEDRON indicating the number of triangles into which each face is divided when forming an ICOSADELTAHEDRON. The triangulation number $T = Pf^2$, where P is any number of the series 1, 3, 7, 13, 19, 21, 31,... (=h^2 + hk + k^2, h and k are any integers with no common factor) and f is any integer. *See* ICOSAHEDRAL SYMMETRY, QUASI EQUIVALENCE THEORY.

f=1   f=2   f=3

Tribec virus. Family *Reoviridae*, genus *Orbivirus*. Isolated from ticks, mice and goats in Slovakia. Antibodies found in man.

trichloracetic acid (TCA). CCl_3COOH (mw. 163.39). A chemical used (5-10% solutions) for the quantitative precipitation of nucleic acids and proteins. It is frequently used when the radioactive counts incorporated into macromolecules are being measured. The precipitates are caught on glass fibre disks, the unincorporated radioactivity washed away, the disks dried, placed in scintillation fluid and counted in a scintillation counter.

Trichoplusia ni cytoplasmic polyhedrosis virus. Cytoplasmic polyhedrosis virus (CPV) isolated in the United Kingdom from laboratory cultures of the cabbage looper, *Trichoplusia ni* (Noctuidae, Lepidoptera). The virus is the type member of 'type 5' CPVs. Viruses of similar electropherotype have been observed in other Lepidoptera (Appendix B) and are unrelated to *Bombyx mori* (type 1) CPV on the basis of RNA electropherotype, serological properties and

{triangulation number}

Numbers of morphological and structural subunits
in particles with icosahedral symmetry.

P	f	T	Morphological subunits	Structural subunits	Example subunits
1	1	1	12	60	Phage øX Poliovirus Tobacco necrosis satellite virus
3	1	3	32	180	Turnip yellow mosaic virus
1	2	4	42	240	*Nudaurelia* β virus
7	1	7	72*	360	Polyoma virus
1	3	9	92	540	Reovirus
1	4	16	162	960	Herpes virus Varicella zoster virus
1	5	25	252	1500	Adenovirus

* Morphological subunits are 72 PENTAMERS.

nucleic acid homology.
Payne, C.C. and Mertens, P.P.C. (1983) **In** The *Reoviridae*. p. 425. ed. W.K. Joklik. Plenum Press: New York.

Trichoplusia ni granulosis virus. Type species of Subgroup B of the BACULOVIRUS genus. The basic structure and composition of the virus particles are very similar to those of other baculoviruses (*see* BACULOVIRUS, GRANULOSIS VIRUS, NUCLEAR POLYHEDROSIS VIRUS). The granulin amino acid sequence is also related to that of other GVs (e.g. *Pieris brassicae* GV; 77% amino acid homology) and NPVs (e.g. *Autographa californica* MNPV; 53% amino acid homology). Productive virus infection has been achieved in cell lines derived from the homologous host although susceptibility of the cells decreased during serial passage.
Granados, R.R. *et al*. (1986) Virology **152**, 472.

Trichoplusia ni nuclear polyhedrosis virus. The cabbage looper, *Trichoplusia ni*, is susceptible to infection by at least two distinct NPVs; an MNPV and an SNPV. The MNPV is closely related to, and regarded as a genotypic variant of, *Autographa californica* NPV. The SNPV is serologically related to some other SNPVs isolated from related insect species (e.g. *Pseudoplusia includens*). Both the MNPV and SNPV have been successfully cultivated *in vitro*.
Granados, R.R. *et al*. (1986) Virology **152**, 472.

tricine.   N-tris(hydroxymethyl)methylglycine (mw. 179.17). A biological buffer, pK$_a$ 8.15, used in the pH range 7.0 - 9.2. Differs from TRIS in that it has significant binding capacity for Mg^{2+}, Ca^{2+}, Mn^{2+} and Cu^{2+}.
Good, N. *et al*. (1966) Biochemistry **5**, 467.

Tricornaviridae. ('trico', referring to tripartite genome, and RNA). An unofficial family comprising the BROMOVIRUS, CUCUMOVIRUS, ALFALFA MOSAIC VIRUS and ILARVIRUS GROUPS. Characterised by having isometric or quasi-isometric particles containing ssRNA genomes comprising three species.
van Vloten-Doting, L. *et al*. (1981) Intervirology **15**, 198.

Triphaena pronuba Cytoplasmic Polyhedrosis Virus. *See* NOCTUA PRONUBA CYTOPLASMIC POLYHEDROSIS VIRUS.

tris. Tris(hydroxymethyl)methylamine (mw. 121.14). A widely used biological buffer, pK$_a$ 8.3, with a pH range 7.0-9.2.
Good, N. *et al*. (1966) Biochemistry **5**, 467.

Triticum aestivum chlorotic spot virus. A possible plant *Rhabdovirus*.
Francki, R.I.B. *et al*. (1985) **In** Atlas of Plant Viruses. Vol. 1. p. 73. CRC Press: Boca Raton, Florida.

Triton X-100.   Iso-octylphenoxypolyethoxyethanol, the polyethoxy chain containing approximately ten ethoxy units. A NON-IONIC DETERGENT with many uses e.g. disruption of cells, stabilising proteins, solubilising aqueous samples in scintillation fluids.

Trivittatus virus. Family *Bunyaviridae*, genus *Bunyavirus*. Isolated from mosquitoes in several states in mid-west of the USA.

tRNA. Abbreviation for TRANSFER RNA.

Trubanaman virus. Family *Bunyaviridae*, genus *Bunyavirus*. Isolated from mosquitoes. Antibodies found in man and several other species.

trypsin. An enzyme catalysing the hydrolysis of peptide bonds on the carboxyl side of arginine, lysine and amino-ethyl cysteine residues.

tryptic peptide A peptide formed from a protein by the action of trypsin. It has arginine, lysine or amino-ethyl cysteine at the C-terminus.

ts-mutant. Abbreviation for TEMPERATURE SENSITIVE MUTANT.

tsetse fly virus. An unclassified DNA-containing virus isolated from the tsetse fly, *Glossina pallidipes* (Glossinidae, Diptera), in Kenya. Virus particles morphologically resemble elongated BACULOVIRUS nucleocapsids, being rod-shaped, 57 nm. wide by 700-1300 nm. long. Particle lengths fall into two size classes with 'short' particles averaging 869 nm. and 'long' particles 1175 nm. Unlike baculoviruses, no fully enveloped virions have been found in purified preparations. There are at least twelve structural polypeptides (major component, mw. 39×10^3). The genome is a dsDNA which is linear and may be heterogeneous in size. Virus infection in adult tsetse flies is characterised by hypertrophied (enlarged) salivary glands and sterility in male flies. In this regard and in some structural properties the virus may be related to virus particles observed in the syrphid, *Merodon equestris* (Diptera).
Odindo, M.O. *et al.* (1986) J. gen. Virol. **67**, 527.

Tsuruse virus. Family *Bunyaviridae*, genus *Bunyavirus*. Isolated from a bird in Japan.

TTV1 phage. Unclassified phage isolated in Iceland from the anaerobic, thermophilic archaebacterium *Thermoproteus tenax*. Particles are rigid rods *c.*400 nm. long, consisting of a capsid surrounded by a lipid bilayer envelope (phage morphotype F3). The genome is DNA. Other isolates with particles of variable length (up to 2500 nm.) have been obtained from the same source (TTV2 to TTV4).
Ackermann, H.W. and Du Bow, M.S. (1987) Viruses of Prokaryotes. CRC Press: Boca Raton, Florida.

Tulare apple mosaic virus. *See* APPLE (TULARE) MOSAIC VIRUS.

tulip breaking virus. A *Potyvirus*.
van Slogteren, D.M.H. (1971) CMI/AAB Descriptions of Plant Viruses No. 71.
Francki, R.I.B. *et al.* (1985) **In** Atlas of Plant Viruses. Vol. 2. p. 183. CRC Press: Boca Raton, Florida.

tulip chlorotic blotch virus. A *Potyvirus*.
Mowat, W.P. (1985) Ann. Appl. Biol. **106**, 65.

tulip virus X. A *Potexvirus*.
Mowat, W.P. (1984) CMI/AAB Descriptions of Plant Viruses No. 276.
Francki, R.I.B. *et al.* (1985) **In** Atlas of Plant Viruses. Vol. 2. p. 159. CRC Press: Boca Raton, Florida.

tunicamycin. A compound which blocks the formation of N-glycosidic protein- carbohydrate linkages.

turbidimetric analysis. A procedure for measuring the light scattered by a turbid (cloudy, hazy) solution. Used e.g. for measuring the density of cells in solution.

turbot herpesvirus. *See* BOTHID HERPESVIRUS 1.

turkey avipoxvirus. *See* AVIPOXVIRUS.

turkey bluecomb disease virus. Synonym: TRANSMISSIBLE ENTERITIS OF TURKEYS VIRUS. Family *Coronaviridae*, genus *Coronavirus*. Causes diarrhoea in young birds. Grows in embryonated turkey eggs.

turkey herpesvirus. *See* GALLID HERPESVIRUS 2.

turkey meningo-encephalitis virus. Family *Flaviviridae*, genus *Flavivirus*. Causes a progressive and fatal paralysis of turkeys. Mice injected i.c. develop encephalitis. Virus grows in chick embryo fibroblasts and in eggs, killing the embryo.

Turlock virus. Family *Bunyaviridae*, genus *Bunyavirus*. Isolated from birds, rabbits and mosquitoes in Canada, USA, Caribbean and South America.

turnip crinkle virus. A member of the *Carmovirus* group. The virus is transmitted by flea-beetles.
Hollings, M. and Stone, O.M. (1972) CMI/AAB Descriptions of Plant Viruses No. 109.
Morris, T.J. and Carrington, J.C. (1988) **In** The Plant Viruses. Vol. 3. p. 73. ed. R. Koenig. Plenum Press: New York.

turnip mosaic virus. A *Potyvirus*.
Tomlinson, J.A. (1970) CMI/AAB Descriptions of Plant Viruses No. 8.
Francki, R.I.B. *et al.* (1985) **In** Atlas of Plant Viruses. Vol. 2. p. 183. CRC Press: Boca Raton, Florida.

turnip rosette virus. A *Sobemovirus*.
Hollings, M. and Stone, O.M. (1973) CMI/AAB Descriptions of Plant Viruses No. 125.
Francki, R.I.B. *et al.* (1985) **In** Atlas of Plant Viruses. Vol. 1. p. 153. CRC Press: Boca Raton, Florida.
Hull, R. (1988) **In** The Plant Viruses. Vol. 3. p. 113. ed. R. Koenig. Plenum Press: New York.

turnip yellow mosaic virus. Type member of the *Tymovirus* group.
Matthews, R.E.F. (1980) CMI/AAB Descriptions of Plant Viruses No. 230.
Francki, R.I.B. *et al.* (1985) **In** Atlas of Plant Viruses. Vol. 1. p. 117. CRC Press: Boca Raton, Florida.
Hirth, L. and Girard, L. (1988) **In** The Plant Viruses. Vol. 3. p. 163. ed. R. Koenig. Plenum Press: New York.

turnip yellows virus. A *Luteovirus*; considered to be a strain of BEET WESTERN YELLOWS VIRUS.
Francki, R.I.B. *et al.* (1985) **In** Atlas of Plant Viruses. Vol. 1. p. 137. CRC Press: Boca Raton, Florida.
Casper R. (1988) **In** The Plant Viruses. Vol. 3. p. 235. ed. R. Koenig. Plenum Press: New York.

TVX virus. Family *Parvoviridae*, genus *Parvovirus*. Isolated from a human amnion cell line.

two-dimensional electrophoresis. A technique in which the constituents of a sample are separated by electrophoresis in one dimension on one property and in a second dimension, usually at right angles to the first, on another property. It is used to resolve complex mixtures of molecules. Examples are: 1) the 2-D separation of proteins based, in the first dimension, on their ISOELECTRIC POINTS and, in the second dimension, on their denatured molecular size; 2) the separation of DNA in the first dimension as ds molecules in a neutral buffer and in the second dimension as ss molecules in a denaturing solution.
Favoloro, J.M. *et al.* (1980) Methods Enzymol. **65**, 718.

Tymovirus group. (Sigla of TURNIP YELLOW MOSAIC, the type member). Genus of plant viruses with isometric particles, 29 nm. in diameter, which sediment as two components at 115S (bottom (B)) and 54S (top (T) which is nucleic acid free) and band in CsCl at 1.42 and 1.29 g/cc respectively. The particles have icosahedral

   100nm

symmetry (T=3) and are composed of coat protein subunits of mw. 20×10^3. Each B component particle contains a single molecule of linear (+)-sense ssRNA (mw. 2.0×10^6) which has a CAP at the 5′ terminus and a tRNA-like structure which accepts valine at the 3′ end. Replication is probably in vesicles induced on the outside of the chloroplast membrane. The host ranges are narrow. Tymoviruses infect most cell types and cause characteristic clumping of the chloroplasts. They are easily mechanically transmitted. In nature they are transmitted by beetles.
Matthews, R.E.F. (1982) Intervirology **17**, 138.
Koenig, R. and Lesemann, D-E. (1979) CMI/AAB Descriptions of Plant Viruses No. 214.
Francki, R.I.B. *et al.* (1985) **In** Atlas of Plant Viruses. Vol. 1. p. 117. CRC Press: Boca Raton, Florida.
Hirth, L. and Givord, L. (1988) **In** The Plant Viruses. Vol. 3, p. 163. ed. R. Koenig. Plenum Press: New York.

Type B Oncovirus group. A genus in the subfamily *Oncovirinae*. The most important member is the MOUSE MAMMARY TUMOUR VIRUS.

Type C Oncovirus group. A genus in the subfamily *Oncovirinae*. Members cause leukaemia and sarcomas. They can be exogenous when the virus is shed by infected cells or endogenous when the genetic material is integrated into the cell genome.

type-specific antigen. An antigen specific to a certain type of virus. *See* GROUP-SPECIFIC ANTIGEN.

typing phages. Bacteriophages used in the subdivision of species, biotypes or serotypes of bacteria into 'phage types' based on their susceptibility to more or less host-range specific phages. *See* PHAGE TYPING.

Tyuleniy virus. Family *Flaviviridae*, genus *Flavivirus*. Isolated from ticks in USA and USSR. Kills young and old mice when injected i.c. Mosquitoes can be infected experimentally. Replicates in pig kidney cell cultures. Antibodies are found in humans, fur seals and birds.

U

U2-tobacco mosaic virus. *See* TOBACCO MOSAIC VIRUS U2.

UDP. Abbreviation for uridine diphosphate, the 5′ pyrophosphate of uridine.

Uganda S virus. Synonym: MAKONDE VIRUS. Family *Flaviviridae*, genus *Flavivirus*. Isolated from mosquitoes and birds in Uganda, Nigeria and the Central African Republic. Antibodies found in man.

UIV-SL-573 cells. Insect cell line from the cotton leafworm, *Spodoptera littoralis*, susceptible to infection by *S. LITTORALIS* NUCLEAR POLYHEDROSIS VIRUS.

ulcerative stomatitis of cattle virus. *See* BOVINE PUSTULAR STOMATITIS VIRUS.

ullucus mild mottle virus. A *Tobamovirus*.
Brunt, A.A. (1986). In The Plant Viruses. Vol. 2. p. 283. ed. M.H.V. van Regenmortel and H. Fraenkel-Conrat. Plenum Press: New York.

ullucus virus C. A *Comovirus*.
Brunt, A.A. (1984) CMI/AAB Descriptions of Plant Viruses No. 277.
Francki, R.I.B. *et al.* (1985) **In** Atlas of Plant Viruses. Vol. 2. p. 1. CRC Press: Boca Raton, Florida.

ultracentrifuge. An instrument which can spin a rotor at great speeds thus creating forces many times that of gravity. It is used to separate cells, virus particles or molecules according to their size or density. The preparative centrifuge is used to prepare and purify biological samples by sedimenting them to the bottom of tubes or as bands in gradients of various materials. In ANALYTICAL ULTRACENTRIFUGES the movement of particles under a gravitational field can be observed and the SEDIMENTATION COEFFICIENT or isopycnic banding density measured. These observations are usually made using either schlieren or UV absorption optics. *See* RATE ZONAL GRADIENT, ISOPYCNIC GRADIENT.

ultrastructure. The fine structure of a macromolecule or a cell usually resolved by an electron microscope.

ultraviolet radiation. Electromagnetic radiation in the wavelength range 40-400 nm.

Umatilla virus. Family *Reoviridae*, genus *Orbivirus*. Isolated from the sparrow and from mosquitoes in several states in the USA.

Umbre virus. Family *Bunyaviridae*, genus *Bunyavirus*.

Una virus. Family *Togaviridae*, genus *Alphavirus*. Isolated from mosquitoes in the Caribbean and South America.

uncoating. The removal of the outer layers of a virus particle on infection thus releasing the viral nucleic acid. With phage this takes place at the cell surface with the viral coat remaining outside the cell. In animal cells uncoating may occur at the cell membrane or in the cytoplasm. For some plant viruses and possibly some vertebrate viruses ribosomes are involved in uncoating the particles; this process, termed co-translational disassembly, gives rise to complexes of virus particles and ribosomes, structures which have been named 'striposomes'.
Wilson, T.M.A. (1985) J. gen. Virol. **66**, 1201.

unicapsid NPV. *See* SNPV.

Upolu virus. Unclassified arthropod-borne virus. Isolated from ticks in Australia. Antibodies present in cattle and kangaroos.

uracil. A pyrimidine base, one of two principal bases in RNA. Can be formed in DNA by the deamination of cytosine which in turn gives a point mutation as it is replicated to give adenine. *See* NUCLEIC ACID.

uranyl acetate. A NEGATIVE STAIN for electron microscope samples. Often used as a 1% solution with a natural pH of 4.0.
Huxley, H.E. and Zubay, G. (1961) J. Biophys. Biochem. Cytol. **11**, 273.

urd bean leaf crinkle virus. An unclassified plant virus with isometric particles 25-30 nm. in diameter which contain ssRNA. The virus is transmitted by seed and by beetles.
Bhardway, S.V. and Dubey, G.S. (1984) Ind. J. Plant Path. **2**, 64.

URF. Abbreviation for unidentified reading frame which is deduced from a DNA sequence but for which no protein or genetic function is known. *See* OPEN READING FRAME.

uridine. The nucleoside of uracil and ribose. *See* NUCLEIC ACID.

uridine 5′-triphosphate (UTP). A pyrimidine nucleotide, one of the four major constituents of RNA. The trisodium salt has a mw. of 586.12. *See* NUCLEIC ACID.

Uronema gigas 'virus'. Virus-like particles found in a green alga. The particles are 390 nm. in diameter with a tail of up to 1.0 μm. The capsid has one major protein species (mw. 45×10^3) and nine minor species (mw. $26\text{-}64 \times 10^3$) surrounding dsDNA (mw. $8\text{-}72 \times 10^6$).

Cole, A. *et al.* (1980) Virology **100**, 166.

Urucuri virus. Family *Bunyaviridae*, genus *Phlebovirus*. Isolated from an apparently normal rodent, *Proechimys guyannensis* in Brazil. Antibodies found in this species of rodent.

Uruma virus. Family *Togaviridae*, genus *Alphavirus*. Cause of an outbreak of disease in man in Bolivia. Chief symptoms were fever and headache.

Ustilago maydis virus P1-H1 (UmV-P1-H1) and related viruses **UmV-P4-H1** and **UmV-P6-H1**. Members of the *Totivirus* group.

Usutu virus. Family *Flaviviridae*, genus *Flavivirus*. Isolated from mosquitoes and birds in several African countries.

Utinga virus. Family *Bunyaviridae*, genus *Bunyavirus*.

Utive virus. Family *Bunyaviridae*, genus *Bunyavirus*.

Uukuniemi virus. Family *Bunyaviridae*, genus *Uukuvirus*. Isolated from a rodent, mosquitoes and birds in Finland, Czechoslovakia, Poland, Hungary, Lithuania, USSR and USA. Antibodies found in man.

Uukuvirus. (Uukuniemi, place in Finland where virus was isolated.) A genus in the family *Bunyaviridae* which is serologically unrelated to other genera in the family.

V

V8 protease. *See* STAPHYLOCOCCUS AUREUS V8 PROTEASE.

vaccine. A suspension of killed or attenuated viruses or subviral entities injected into vertebrates to produce active immunity.

vaccinia virus. Family *Poxviridae*, subfamily *Chordopoxvirinae*, genus *Orthopoxvirus*. Probably derived from cowpox virus, the first virus to be used as a vaccine, by Jesty in 1774 and Jenner in 1796. Introduction of the virus into the skin by scarification produces a local lesion in most cases but can cause generalised infection in people with impaired immune responsiveness. Long-lasting immunity is induced. The virus was the basis of a highly successful vaccine against smallpox. Similar local lesions are produced in calf, sheep, rabbit and guinea pigs. Virus grows in many cell culture systems and in eggs. The virus is now being used experimentally as a vector for genes expressing foreign protein antigens (e.g. rabies virus glycoprotein); this may enable the virus to be used as a vaccinating vehicle for various virus diseases.

vacuolating virus. A virus characterised by forming vacuoles in infected cells, e.g. species of *Polyomavirus*: SIMIAN VIRUS 40, RABBIT VACUOLATING VIRUS.

Vagoiavirus. Former name for ENTOMOPOXVIRUS (after C. Vago).

varicella zoster virus. Synonyms: CHICKENPOX VIRUS, HUMAN HERPESVIRUS 3. Family *Herpesviridae*, subfamily *Alphaherpesvirinae*, genus *Alpha herpesvirus 3*. Causes chickenpox, usually in the young, sometimes with encephalitis, on primary infections. Reactivation causes herpes zoster, a painful local condition with skin lesions. Virus can be grown in HeLa cells and monkey kidney cell lines but there is no animal model.

variola ovina virus. *See* SHEEP CAPRIPOX VIRUS.

variola suilla virus. *See* SWINEPOX VIRUS.

variola virus. Family *Poxviridae*, subfamily *Chordopoxvirinae*, genus *Orthopoxvirus*. Causative agent of smallpox, which was a severe and often fatal disease of man, involving an extensive rash, fever and prostration. The World Health Organisation eradicated the disease from the world in 1977 by a well-organised vaccination campaign. This was the first human infectious disease to be so eradicated. Two main types were recognised: variola major virus, which occurred in Asia and the Middle and Far East, caused a high level of mortality; variola minor virus (synonyms: ALASTRIM VIRUS, AMAAS VIRUS, KAFFIR POX VIRUS) which occurred in South America and West Africa, caused lower mortality. Both viruses produced pocks in the chorioallantoic membrane and killed chick embryos. The virus multiplied readily in suckling mice.

vector. This word has two meanings: a) in molecular biology it is a self-replicating DNA molecule into which fragments of DNA can be cloned. Most vectors are derived from bacterial plasmids or from viruses. A vector should contain one or more unique restriction endonuclease(s) site(s) and one or more selectable feature(s), e.g. antibiotic resistance, distinct plaque formation; b) in virus transmission it is the organism which transmits a virus from the infected to the uninfected host. *See* TRANSMISSION.

VEE virus. Abbreviation for VENEZUELAN EQUINE ENCEPHALOMYELITIS VIRUS.

vegetative cycle. *See* LYTIC CYCLE.

Vellore virus. Family *Togaviridae*, genus *Alphavirus*. Isolated from mosquitoes in Vellore, India.

velvet tobacco mottle virus. A *Sobemovirus.*
Randles, J.W. and Francki, R.I.B. (1986) AAB
Descriptions of Plant Viruses No. 317.
Francki, R.I.B. *et al.* (1985) **In** Atlas of Plant
Viruses. Vol. 1. p. 153. CRC Press: Boca Raton,
Florida.
Hull, R. (1988) **In** The Plant Viruses. Vol. 3. 113.
ed. R. Koenig. Plenum Press: New York.

Velvet tobacco mottle virus group. Now included in the SOBEMOVIRUS GROUP.

Venezuelan equine encephalitis virus. Family
Togaviridae, genus *Alphavirus.* Causes viscero-
tropic disease in horses, donkeys and man but
some strains are neurotropic. In horses and donkeys there is damage to blood vessels and many
organs, with fever, diarrhoea and general loss of
condition. In man there is fever, severe headache,
tremors and mortality approaching 1%. Laboratory infections are common. Horses, dogs, cats,
sheep and goats can be infected experimentally.
Mosquitoes are the chief vectors. Grows well in
several tissue culture cell lines.

Venkatapuram virus. Unclassified arthropod-
borne virus. Isolated from mosquitoes in
India.

Vero cell. Named from Esperanto for truth. A
heteroploid fibroblast cell line derived from the
kidney of a normal African green monkey *Cercopithecus aethiops.* Widely used in propagation of
viruses and study of their replication, e.g. SIMIAN
VIRUS 40, MEASLES, MUMPS, RUBELLA, VARICELLA
ZOSTER, POLIOMYELITIS VIRUS.
Yasumura, T. and Kawakita, Y. (1963) Nippon
Rinsho **21**, 1201.

vertical transmission. Synonym: GENETIC
TRANSMISSION. The passage of the viral genome
from one host generation to the next. This can be
either as a PROVIRUS (e.g. in RETROVIRUSES) or in
close association with the genome of the gamete,
e.g. in transovarial transmission of BUNYAVIRUSES
in their insect vectors or in the seed transmission
of plant viruses.

vervet monkey disease virus. Family
Papovaviridae, genus *Polyomavirus.* Isolated
from a monkey kidney cell culture in S. Africa.
Grows in primary Rhesus kidney cell cultures.

vesicular exanthema of swine virus. (VESV).
Type species of family *Caliciviridae*, genus

Calicivirus. First seen in California in 1932. Affects pigs and will also infect horses. Initially
mistaken for foot-and-mouth disease because of
similar clinical signs (lesions on tongue, feet,
snout and teats). Not seen outside North America.
Several distinct serotypes, easily distinguished
serologically. Grows well in a variety of cell
cultures. Similar morphologically to San Miguel
sea lion virus.

vesicular stomatitis virus. (VSV). Type species
of family *Rhabdoviridae*, genus *Vesiculovirus.*
Disease first seen in horses in the USA. Produces
blisters on tongue and feet of cattle, pigs, horses,
similar to those found with foot-and-mouth disease. The virus will also infect man, causing
influenza-like symptoms. Important epidemiol-
ogically because of clinical similarity to other
vesicular diseases. Seen only in the Americas.
Occurs as two major serotypes, Indiana and New
Jersey. Antibodies have been found in the turtle
and snake. Most species of animal can be infected
experimentally and the virus can be grown in
large amounts in a variety of tissue culture cells.
Crick, J, and Brown F. (1979). **In** Rhabdoviruses
Vol. 1, ed. D.H.L. Bishop. CRC Press: New York,
pp. 1-22.

Vesiculovirus. (Latin 'vesicula' = bladder, blis-
ter.) A genus in the family *Rhabdoviridae.*
Contains VESICULAR STOMATITIS VIRUS (Indiana
and New Jersey and the closely related Argentina,
Brazil, Cocal and Isfahan strains) and PIRY and
CHANDIPURA VIRUSES. The latter two viruses are
only distantly related to the vesicular stomatitis
viruses.

vicia cryptic virus. A member of the *Crypto-
virus* group, subgroup A.
Kenton, R. *et al.* (1978) Ann. Rep. Rothamsted
Expt. Sta. 1977, 222.

Vigna sinensis mosaic virus. A possible plant
Rhabdovirus.
Francki, R.I.B. *et al.* (1985) **In** Atlas of Plant
Viruses. Vol. 1. p. 73. CRC Press: Boca Raton,
Florida.

Vinces virus. Family *Bunyaviridae*, genus
Bunyavirus.

viola mottle virus. A *Potexvirus.*
Lisa, V. *et al.* (1982) CMI/AAB Descriptions of
Plant Viruses No. 247.
Francki, R.I.B. *et al.* (1985) **In** Atlas of Plant

Viruses. Vol. 2. p. 159. CRC Press: Boca Raton, Florida.

viraemia, viremia. The spread of virus particles through the blood stream. Primary viraemias, which frequently occur early in infection, cause the dissemination of the virus to other susceptible organs, e.g. liver and spleen. In secondary viraemia the virus moves to other target organs, e.g. the skin in the case of MOUSEPOX (ectromelia). Chronic viraemia, where the virus circulates, sometimes as virus-antibody complexes, is found in some infections, e.g. hepatitis in man.

viral haemorrhagic septicaemia virus. *See* EGTVED VIRUS.

viral hepatitis. Disease of the liver caused by a virus. In man it is most commonly caused by HEPATITIS VIRUSES A, B or NON-A, NON-B but can be caused by other viruses, e.g. MEASLES VIRUS after certain immunisation procedures, RUBELLA VIRUS or certain herpesviruses (α1 or 2, γ4).

Virazole. Trade name for RIBAVIRIN.

Virex. Former trials product of *Heliothis zea* nuclear polyhedrosis virus produced by Hayes-Sammons, USA, for the biological control of *Heliothis* spp; now discontinued.

Virgin river virus. Family *Bunyaviridae*, genus *Bunyavirus*.

Virin Diprion. Preparation of *NEODIPRION SERTIFER* NUCLEAR POLYHEDROSIS VIRUS produced on an industrial scale in the USSR for the biological control of pine sawfly.

Virin-ENSh. Preparation of *LYMANTRIA DISPAR* NUCLEAR POLYHEDROSIS VIRUS produced on an industrial scale in the USSR for the biological control of the gypsy moth.

virion. The infectious unit of a virus.

virogenic stroma. Term applied to describe the dense network of chromatin-like material and viroplasm which represents the nuclear site of viral DNA synthesis and nucleocapsid assembly in nuclear insect virus infections (e.g. baculovirus, densovirus). Also used to describe the sites of virus synthesis in the cytoplasm of insect cells infected with cytoplasmic polyhedrosis viruses.

viroid. A group of pathogens characterised by being unencapsidated circular ssRNA of low mw. (247-574 nucleotides). The nucleic acid has extensive secondary structure, making it very stable. Replication is in the nucleus and is thought to be by rolling circle producing oligomeric forms which are processed to unit length. The nucleic acid does not code for any proteins. Host ranges of viroids vary from wide to narrow. They are transmitted by mechanical damage.
Diener, T.O. (1979) Viroids and Viroid Diseases. John Wiley and Sons: New York.
Diener, T.O. ed. (1987) The Viroids. Plenum Press: New York.
Semancik, J.S. ed. (1987) Viroids and Viroid-Like Pathogens. CRC Press: Boca Raton, Florida.

virolysis. Lysis or holes produced in the membrane of enveloped viruses by reaction with antibodies or complement.
Berry, D.M. and Almeida, J.D. (1968) J. gen. Virol. **3**, 97.

Viron. Nuclear polyhedrosis virus preparations produced on a semi-commercial scale by International Minerals and Chemical Corp., USA; now discontinued. Viron-H was developed for the biological control of *Heliothis* spp.; Viron-P for *Prodenia* sp.; Viron-S for *Spodoptera* sp. and Viron-T for *Trichoplusia ni.*

viropexis. A form of PINOCYTOSIS in which virus particles are engulfed by the cell plasma membrane. It is an active process by the cell and is probably an important mechanism of penetration of viruses into animal cells. After viropexis the virus particles are still confined within vesicles in the cytoplasm; they have to be released before the virus can replicate.
Fazekas de St. Groth, S. (1948) Nature **162**, 294.

viroplasm. An amorphous cytoplasmic INCLUSION BODY associated with a virus infection. Usually proteinaceous and without a surrounding membrane. Found, for example, in plant cells infected with WOUND TUMOUR VIRUS or with CAULIFLOWER MOSAIC VIRUS.

virosis. Condition induced by a virus infection.

virosome. LIPOSOMES where viral proteins have been inserted at the surface, e.g. the removal of haemagglutinin and neuraminidase subunits from the envelope of influenza virus into such liposomes creates a structure with many of the antigenic properties of the viral envelope. Almeida, J.D. *et al.* (1975) Lancet **ii,** 899.

virostatic. Able to inhibit viral replication which can then resume when the virostatic agent is removed. Actinomycin D and 5-fluoro-deoxyuridine are virostatic to DNA-containing viruses and 5-fluorouracil, thiouracil and puromycin to RNA viruses. *See* BASE ANALOGUE.

Virox. Commercial preparation of the nuclear polyhedrosis virus of the pine sawfly, *Neodiprion sertifer,* produced during 1985 and 1986 by Microbial Resources Ltd., UK, for the biological control of the homologous host.

Virtuss. Preparation of NUCLEAR POLYHEDROSIS VIRUS of the Douglas fir tussock moth, *Orgyia pseudotsugata,* produced and provisionally registered for use in Canada by the Canadian Forest Pest Management Institute for the control of the homologous host.

virucidal. Causes inactivation of a virus, e.g. formalin, chlorine.

virulence. The disease-producing capacity of a micro-organism.

virulent phage. A phage which produces a LYTIC infection in its host and cannot induce a lysogenic state. *See* TEMPERATE PHAGE, LYSOGENY.

virus. (Latin = poison or slime.) Infectious units (obligate parasites) comprising either RNA or DNA enclosed in a protective coat. Their nucleic acid contains information necessary for their replication in a susceptible host cell. They contain no energy-producing enzyme systems, no functional ribosomes or other cellular organelles; these are supplied by the cell in which they replicate. The cell may also supply some of the enzymes necessary for viral replication. The host cell may or may not be destroyed in the process of viral replication and release.

virus assembly. The coming together of the constituent parts of a virus particle and the subsequent formation of that particle. The process can vary from the autoassembly of protein subunits around viral nucleic acid, a process which can be performed *in vitro* with e.g. TOBACCO MOSAIC VIRUS to the assembly of complex viruses at cell membranes, e.g. ORTHO- or PARA-MYXOVIRUSES, or the complicated assembly processes of T-EVEN PHAGES.

virus induction. *See* INDUCTION.

virus-like particle. Structure resembling a virus particle but which has not been demonstrated to be pathogenic, e.g. Syrian hamster intercisternal A- particle.

virus particle. The morphological form of a virus. In some virus types it consists of a genome surrounded by a protein CAPSID; others have additional structures (e.g. envelopes, tails, etc.).

virus replication. The process of forming progeny virus from input virus. It involves the expression and replication of the viral genomic nucleic acid and the assembly of progeny virus particles.

viscosity. The internal friction of a particle in a fluid. The units in cgs are poises (P); water has a viscoscity of 0.01P approximately. SI units are Pascal.secs ($1P = 10^{-1}PaS$). The symbol is η.

Visna virus. Family *Retroviridae,* subfamily *Lentivirinae.* Causes a slow demyelinating disease of the CNS in sheep. The first signs are abnormal carriage of the head, with lip tremor, followed by progressive paralysis and death. Transmission is horizontal but requires close contact. Antibodies are produced but the virus is not eliminated from the animal. Antigenic change occurs during the progression of the disease, probably by selection under pressure from the antibody. The virus will transform mouse cells *in vitro.*

voandzeia necrotic mosaic virus. A *Tymovirus.* Fauquet, C. *et al.* (1984) CMI/AAB Descriptions of Plant Viruses No. 279.
Hirth, L. and Girard, L. (1988) **In** The Plant Viruses. Vol. 3. p. 163. ed. R. Koenig. Plenum Press: New York.

vomiting and wasting disease virus. *See* PORCINE HAEMAGGLUTINATING ENCEPHALITIS VIRUS.

von Magnus phenomenon. The increase in the proportion of defective virus particles produced on the repeated passage of a virus at high multiplicity of infection. Originally observed with INFLUENZA VIRUS A. *See* DEFECTIVE VIRUS.
von Magnus, P. (1954) Adv. Virus Res. **2,** 59.

VPg. Abbreviation for virion protein, genome-linked. A small virus-coded protein attached through a phosphodiester linkage from an amino acid (e.g. to the phenolic hydroxyl group of a tyrosine residue in POLIOVIRUS) to the 5′ end of the virion nucleic acid of certain viruses, e.g. PICORNAVIRUSES, COMOVIRUSES.

W

W2 virus. Synonym: CALLINECTES W2 VIRUS. An unclassified reovirus-like particle, isolated from the Mediterranean shore crab, *Carcinus mediterraneus*. Virus particles are 65-70 nm. in diameter, with a buoyant density of 1.34 g/cc. in CsCl. The capsid consists of two protein shells and contains at least six polypeptides (mw. 120, 94, 76, 44, 32 and 24 x 10^3). The genome is dsRNA with at least nine segments in four different size classes. The virus infects the connective tissue of many organs in the crab.
Mari, J. and Bonami, J-R. (1988) J. gen. Virol. **69**, 561.

Wad Medani virus. Family *Reoviridae*, genus *Orbivirus*. Isolated from ticks in Sudan, Jamaica, India and Pakistan.

Wallal virus. Family *Reoviridae*, genus *Orbivirus*. Isolated from mosquitoes in Queensland, Australia. Antibodies found in wallabies and kangaroos.

Wanowrie virus. Unclassified arthropod-borne virus. Isolated from ticks and mosquitoes in India, Sri Lanka and Egypt. Also isolated from the brain of a patient with hepatitis and haemorrhagic disease of the gut. Pathogenic experimentally for new-born mice. Grows in BHK21 cells.

Warrego virus. Family *Reoviridae*, genus *Orbivirus*. Isolated from mosquitoes in Queensland, Australia. Antibodies found in cattle, kangaroos and wallabies.

wart hog virus. *See* AFRICAN SWINE FEVER VIRUS.

watermelon mosaic virus 1. A strain of PAPAYA RINGSPOT VIRUS.
van Regenmortel, M.H.V. (1971) CMI/AAB Descriptions of Plant Viruses No. 63.

watermelon mosaic virus 2. A *Potyvirus*. Purcifull, D.E. *et al.* (1984) CMI/AAB Descrip-tions of Plant Viruses No. 293.
Francki, R.I.B. *et al.* (1985) **In** Atlas of Plant Viruses. Vol. 2. p. 183. CRC Press: Boca Raton, Florida.

WEE virus. Abbreviation for WESTERN EQUINE ENCEPHALOMYELITIS VIRUS.

Wesselbron disease virus. Family *Flaviviridae*, genus *Flavivirus*. Causes abortion and death of lambs and pregnant ewes. Haemorrhages and jaundice occur in the ewes and meningo-encephalitis in the lambs. Infects man, causing fever and muscular pains. Found in Thailand and in many countries in Africa. Transmission is by mosquito. Can infect mice i.c. causing encephalitis, and guinea pigs and rabbits causing abortion. Grows in lamb kidney cell cultures and in eggs.

West Nile virus. Family *Flaviviridae*, genus *Flavivirus*. Causes fever in humans. Occurs in Egypt, Uganda, S. Africa, Israel, France and India. The virus is spread by mosquitoes. When inoculated i.c. into rodents, chickens and monkeys it causes encephalitis. The virus will grow in eggs and in many cell cultures, including mosquito cells.

Western blotting. The transfer of proteins which have been separated on a polyacrylamide gel to an immobilising matrix, commonly nitrocellulose. The proteins are frequently transferred using electrophoresis, a process termed electroblotting. The proteins on the matrix can be probed with suitable antibodies which are labelled or detected using radioactivity e.g. ^{125}I or an enzymic assay, e.g. with antibodies conjugated with peroxidase.
Towbin, H. *et al.* (1979) Proc. Natl. Acad. Sci. USA **76**, 4350.

Western equine encephalomyelitis virus. Family *Togaviridae*, genus *Alphavirus*. Causes a disease in horses and man similar to EASTERN EQUINE ENCEPHALOMYELITIS VIRUS but milder.

Mortality in horses is 20-30% and in man 10%. Found in most of the USA, Canada and S. America. Transmission is by mosquitoes. Maintained in the wild in birds, small rodents and reptiles. Inoculation i.c. into rodents, rabbits, pigs and monkeys causes meningoencephalomyelitis. The virus grows in eggs and a range of cell cultures.

Whataroa virus. Family *Togaviridae*, genus *Alphavirus*. Isolated from mosquitoes and birds in New Zealand.

wheat chlorotic streak mosaic virus. A plant *Rhabdovirus*, subgroup 1; leafhopper transmitted.
Francki, R.I.B. *et al.* (1985) **In** Atlas of Plant Viruses. Vol. 1. p. 73. CRC Press: Boca Raton, Florida.

wheat chlorotic streak virus. A strain of BARLEY YELLOW STRIATE MOSAIC VIRUS.
Milne, R.G. (1986) Intervirology **25**, 83.

wheat dwarf virus. A *Geminivirus*, subgroup A. The genome is a single species of ssDNA of 2749 nucleotides.
Harrison, B.D. (1985) Ann. Rev. Phytopath. **23**, 55.

wheat germ extract. Cell-free extract from wheat germ used for the translation of eukaryotic mRNAs. The extract is made from commercial wheat germ but it is important to obtain it from the right source; that from many sources contains inhibitors of translation. *See* RABBIT RETICULOCYTE LYSATE.
Davies, J.W. (1979) **In** Nucleic Acids in Plants. Vol. 2. p. 114. ed. T.C. Hall and J.W. Davies. CRC Press: Boca Raton, Florida.

wheat mottle dwarf virus. An unclassified plant virus with isometric particles, occurring in Japan.
Doi, Y. *Personal communication.*

wheat rosette stunt virus. A plant *Rhabdovirus*, subgroup 1; transmitted by leafhoppers.
Francki, R.I.B. *et al.* (1985) **In** Atlas of Plant Viruses. Vol. 1. p. 73. CRC Press: Boca Raton, Florida.

wheat spindle streak mosaic virus. A possible *Potyvirus*, transmitted by fungus.
Slykhuis, J.T. (1976) CMI/AAB Descriptions of Plant Viruses No. 167.

wheat spindle streak virus. A possible *Potyvirus*, transmitted by aphids.
Matthews, R.E.F. (1982) Intervirology **17**, 154.

wheat streak mosaic virus. A possible *Potyvirus*.
Brakke, M.K. (1971) CMI/AAB Descriptions of Plant Viruses No. 48.

wheat streak virus. A possible *Potyvirus*.
Francki, R.I.B. *et al.* (1985) **In** Atlas of Plant Viruses. Vol. 2. p. 183. CRC Press: Boca Raton, Florida.

wheat striate mosaic virus. A member of the plant *Rhabdovirus*, subgroup 1; transmitted by leafhoppers.
Francki, R.I.B. *et al.* (1985) **In** Atlas of Plant Viruses. Vol. 1. p. 73. CRC Press: Boca Raton, Florida.

wheat yellow leaf virus. A member of the *Closterovirus*, subgroup 2.
Inouye, T. (1976) CMI/AAB Descriptions of Plant Viruses No. 157.
Francki, R.I.B. *et al.* (1985) **In** Atlas of Plant Viruses. Vol. 2. p. 219. CRC Press: Boca Raton, Florida.

wheat yellow mosaic virus. A member of the *Barley Yellow Mosaic Virus* group.
Hibino, H. *et al.* (1981) Ann. Phytopath. Soc. Japan **47**, 510.

white bryony mosaic virus. A possible *Potyvirus*.
Francki, R.I.B. *et al.* (1985) **In** Atlas of Plant Viruses. Vol. 2. p. 183. CRC Press: Boca Raton, Florida.

white clover cryptic virus 1. Type member of the *Cryptovirus* group, subgroup A.
Boccardo, G. *et al.* (1987) Adv. Virus Res. **32**, 171.

white clover cryptic virus 2. Type member of the *Cryptovirus* group, subgroup B.
Boccardo, G. *et al.* (1987) Adv. Virus Res. **32**, 171.

white clover cryptic virus 3. A member of the *Cryptovirus* group.
Boccardo, G. *et al.* (1987) Adv. Virus Res. **32**, 171.

white clover mosaic virus. A *Potexvirus*.
Bercks, R. (1971) CMI/AAB Descriptions of Plant Viruses No. 41.
Francki, R.I.B. *et al*. (1985) **In** Atlas of Plant Viruses. Vol. 2. p. 159. CRC Press: Boca Raton, Florida.

white clover temperate virus. A mixture of three viruses, probably the same as WHITE CLOVER CRYPTIC VIRUSES 1, 2 and 3, occurring in Japan.
Natsuaki, K.T. *et al*. (1984) J. Ag. Sci. Tokyo Univ. of Agric. **29**, 49.

white pox virus. Family *Poxviridae*, subfamily *Chordopoxvirinae*, genus *Orthopoxvirus*. Isolated from cynomolgus monkeys, a chimpanzee, a sun squirrel and a multimammate mouse. Very similar to monkey pox virus. However, it produces white pocks on chorioallantoic membrane like variola virus, from which it is difficult to distinguish. Grows well in cell cultures.

wild cucumber mosaic virus. A *Tymovirus*.
van Regenmortel, M.H.V. (1972) CMI/AAB Descriptions of Plant Viruses No. 105.
Francki, R.I.B. *et al*. (1985) **In** Atlas of Plant Viruses. Vol. 1. p. 117. CRC Press: Boca Raton, Florida.
Hirth, L. and Girard, L. (1988) **In** The Plant Viruses. Vol. 3. p. 163. ed. R. Koenig. Plenum Press: New York.

wild potato mosaic virus. A possible *Potyvirus*.
Francki, R.I.B. *et al*. (1985) **In** Atlas of Plant Viruses. Vol. 2. p. 183. CRC Press: Boca Raton, Florida.

wild strain. A distinct virus isolate made from a natural infection. To be contrasted with LABORATORY STRAIN which has been selected under laboratory conditions.

wild type. The original parental strain of a virus. It is usually laboratory-adapted and is used as a basis for comparison of mutants.

wildebeest herpesvirus. *See* BOVID HERPESVIRUS 3.

wineberry latent virus. A *Potexvirus*.
Jones, A.T. (1985) AAB Descriptions of Plant Viruses No. 304.
Francki, R.I.B. *et al*. (1985) **In** Atlas of Plant Viruses. Vol. 2. p. 159. CRC Press: Boca Raton, Florida.

winter wheat mosaic virus. A possible member of the *Rice Stripe Virus* group.
Fedotina, V.L. *et al*. (1985) Archiv f. Phytopathologie u. Pflanzenschutz **21**, 111.

Wipfelkrankheit. Literally 'wilt disease'; a term first applied to the observation of disease in the larvae of the nun moth, *Lymantria monacha*, probably infected with a nuclear polyhedrosis virus.

wisteria vein mosaic virus. A *Potyvirus*.
Francki, R.I.B. *et al*. (1985) **In** Atlas of Plant Viruses. Vol. 2. p. 183. CRC Press: Boca Raton, Florida.

Witwatersrand virus. Family *Bunyaviridae*, genus *Bunyavirus*. Isolated from rodents and mosquitoes in Uganda, Mozambique and S. Africa. Kills mice on injection.

wobble hypothesis. The third base of the anticodon loop of a tRNA can bind to more than one base of the mRNA codon, e.g. the codons for arginine can be AGG or AGA and the 3' A or G will associate with 5' U on the anticodon. Thus the 'wobble' hypothesis states that a certain amount of variation or 'wobble' is tolerated in the third position of a codon.

Wongal virus. Family *Bunyaviridae*, genus *Bunyavirus*. Isolated from mosquitoes in Queensland, Australia.

Wongorr virus. Unclassified arthropod-borne virus. Isolated from mosquitoes in Queensland, Australia. Antibodies found in man, cattle and wallabies.

woolly monkey sarcoma virus. Synonym: SIMIAN SARCOMA VIRUS. Family *Retroviridae*, subfamily *Oncovirinae*, genus *Type C Oncovirus* group, sub-genus *Mammalian Type C Oncovirus*. Found in a fibrosarcoma of a woolly monkey. Causes sarcomas on injection into marmosets.

wound tumour virus. Type member of the *Phytoreovirus* group.
Black, L.M. (1970) CMI/AAB Descriptions of Plant Viruses No. 34.
Francki, R.I.B. *et al*. (1985) **In** Atlas of Plant Viruses. Vol. 1. p. 47. CRC Press: Boca Raton, Florida.

Wyeomyia virus. Family *Bunyaviridae*, genus

Bunyavirus. Causes fever in man. Isolated from birds and mosquitoes in Caribbean and some countries in S. America.

X

X bodies. INCLUSION BODIES found in TOBACCO MOSAIC VIRUS infected cells.

Xerosiavirus. Former name for NUCLEAR POLYHEDROSIS VIRUSES (NPVs) isolated from Diptera (Insecta), e.g. *Tipula paludosa* NPV, which produces orange segment-shaped occlusion bodies (after N. Xeros).

X-gal. 5-chloro-4-bromo-3-indolyl-β-D-galactoside. A colourless compound which, when hydrolysed by β-galactosidase, releases a bright blue non-diffusible dye. It is used as a colour indicator for bacteria able to use lactose and in cloning vectors involving β-galactosidase, e.g. the M13 mp phage and the pUC plasmid systems.

xylene cyanol FF. A dye which is used as a marker in polyacrylamide gel electrophoresis of DNA or RNA. In a 5% polyacrylamide gel it comigrates with DNA of about 250bp, in a 10% gel about 110bp and in a 20% gel about 45bp.

Y

Yaba 1 virus. Synonym: LEDNICE VIRUS. Family *Bunyaviridae*, genus *Bunyavirus*. Isolated from ticks in Nigeria and at a town, Lednice, in S. Moravia. Pathogenic for new-born mice. Grows in goose and duck embryo cells.

Yaba virus. Family *Poxviridae*, subfamily *Chordopoxvirinae*, unassigned to genus. Causes benign fibrous tumours of the head and limbs of monkeys. Disease often spreads to persons in contact with infected animals. Symptoms in man are fever and local lesions. Grows in the chorio-allantoic membrane and in human and monkey kidney cell cultures.

yam internal brown spot virus. A member of the *Cacao Swollen Shoot Virus* group.
Francki, R.I.B. *et al.* (1985) **In** Atlas of Plant Viruses. Vol. 2. p. 235. CRC Press: Boca Raton, Florida.

yam mosaic virus. Synonym: DIOSCOREA GREEN BANDING VIRUS.
Thouvenal, J-C. and Fauquet, C. (1986) AAB Descriptions of Plant Viruses No. 314.
Francki, R.I.B. *et al.* (1985) **In** Atlas of Plant Viruses. Vol. 2. p. 183. CRC Press: Boca Raton, Florida.

Yaquina Head virus. Family *Reoviridae*, genus *Orbivirus*. Isolated from ticks in Oregon and Alaska.

Yata virus. Family *Rhabdoviridae*, unassigned to genus. Isolated from mosquitoes in Central African Republic.

yellow fever virus. Family *Flaviviridae*, genus *Flavivirus*. Causes a fatal infection in man with high fever, albuminuria, jaundice, black vomit and haemorrhages. The disease is endemic in Africa and S. America. The reservoir is wild primates and the virus is spread to man by mosquitoes. Some species of monkey die after experimental infection. The virus will kill hedgehogs and mice experimentally. Grows in chicken and mouse embryo cells and in eggs. Theiler's 17D vaccine is famous historically as one of the first attenuated vaccines. It is effective and safe except in very young children.

Yogue virus. Unclassified arthropod-borne virus. Isolated from a bat in Senegal.

Yucaipa virus. Family *Paramyxoviridae*, genus *Paramyxovirus*. Isolated from tracheal exudate of chickens in which it causes a mild disease.

Yucca bacilliform virus. A member of the *Cacao Swollen Shoot Virus* group.
Milne, R.G. *et al.* (1985) Abstr. Assoc. Appl. Biol. Meeting on New Developments in Techniques for Virus Detection, Cambridge.

Z

Z-DNA. *See* DEOXYRIBONUCLEIC ACID.

Zaliv-Terpeniya virus. Family *Bunyaviridae*, genus *Uukuvirus*. Isolated from ticks in Tyuleniy Island in Patience Bay where there are sea bird colonies. Pathogenic for suckling mice.

Zaysan virus. Family *Togaviridae*, genus *Alphavirus*. Isolated from mosquitoes in the USSR.

Zea mais rhabdovirus. A possible plant *Rhabdovirus*, subgroup 1.
Francki, R.I.B. *et al.* (1985) **In** Atlas of Plant Viruses. Vol. 1. p. 73. CRC Press: Boca Raton, Florida.

Zegla virus. Family *Bunyaviridae*, genus *Bunyavirus*. Isolated from the cotton rat in Central America.

Zika virus. Family *Flaviviridae*, genus *Flavivirus*. Causes fever with rash in man. Isolated from man, monkeys and mosquitoes in several countries in Africa and in Malaya. Rhesus monkeys experimentally infected develop a fever.

Zinga virus. Unclassified arthropod-borne virus. Causes fever in man. Isolated from man and mosquitoes. Antibodies found in rodents, elephants, pigs, buffalo, hartebeest and birds.

Zingilamo virus. Unclassified arthropod-borne virus. Isolated from a bird in the Central African Republic.

Zirqa virus. Family *Bunyaviridae*, genus *Nairovirus*. Isolated from ticks on Zirqa Island in the Persian Gulf.

zonal centrifugation. Centrifugation in a cylindrical or bowl-shaped rotor in which the solutions are in a large central cavity rather than individual tubes or buckets. The zonal rotor is usually used for gradient centrifugation. The gradient is loaded through the core while the rotor is spinning slowly (about 2000 rev/min) and the sample is then loaded and the rotor taken up to speed. The macromolecules separate as concentric rings (zones) in the gradient. The gradient is then unloaded by displacement with a dense solution, usually at slow revolutions, through the core. Used for separating macromolecules in large quantities.

zoonosis. A disease or infection which is naturally transmitted between vertebrate animals and man.
World Health Organisation/FAO Expert Committee on Zoonoses (1959) Tech. Rept. Series. No. 169.

zoysia mosaic virus. A *Potyvirus*, occurring in Japan.
Doi, Y. *Personal communication.*

zucchini yellow mosaic virus. A *Potyvirus*.
Lisa, V. and Lecoq, H. (1984) CMI/AAB Descriptions of Plant Viruses No. 282.
Francki, R.I.B. *et al.* (1985) **In** Atlas of Plant Viruses. Vol. 2. p. 183. CRC Press: Boca Raton, Florida.

zwogerziekte virus. *See* MAEDI VIRUS.

zygocactus virus X. A possible *Potexvirus*.
Francki, R.I.B. *et al.* (1985) **In** Atlas of Plant Viruses. Vol. 2. p. 159. CRC Press: Boca Raton, Florida.

Appendix A

A list of insect species in which infections caused by occluded baculovirus (nuclear polyhedrosis (NPV) and granulosis (GV)) infections have been recorded

Genus	Species	Family	Order	Virus Type[1]
Abraxas	*grossulariata*	Geometridae	Lepidoptera	NPV
Acantholyda	*erythrocephala*	Pamphilidae	Hymenoptera	NPV
Achaea	*janata*	Noctuidae	Lepidoptera	GV, NPV
Achroia	*grisella*	Pyralidae	Lepidoptera	NPV
Acidalia	*carticcaria*	Geometridae	Lepidoptera	NPV
Acleris	*gloverana*	Tortricidae	Lepidoptera	NPV
Acleris	*variana*	Tortricidae	Lepidoptera	GV, NPV
Acronicta	*aceris*	Noctuidae	Lepidoptera	NPV
Actebia	*fennica*	Noctuidae	Lepidoptera	NPV
Actias	*selene*	Saturniidae	Lepidoptera	NPV
Adisura	*atkinsoni*	Noctuidae	Lepidoptera	NPV
Adoxophyes	*fasciata*	Tortricidae	Lepidoptera	GV
Adoxophyes	*orana*	Tortricidae	Lepidoptera	GV, NPV
Aedes	*aegypti*	Culicidae	Diptera	NPV
Aedes	*annandalei*	Culicidae	Diptera	NPV
Aedes	*atropalpus*	Culicidae	Diptera	NPV
Aedes	*epactius*	Culicidae	Diptera	NPV
Aedes	*nigromaculis*	Culicidae	Diptera	NPV
Aedes	*scutellaris*	Culicidae	Diptera	NPV
Aedes	sollicitans	Culicidae	Diptera	NPV
Aedes	*taeniorhynchus*	Culicidae	Diptera	NPV
Aedes	*tormentor*	Culicidae	Diptera	NPV
Aedes	*triseriatus*	Culicidae	Diptera	NPV
Aedia	*leucomelas*	Noctuidae	Lepidoptera	NPV
Aglais	*urticae*	Nymphalidae	Lepidoptera	NPV
Agraulis	*vanillae*	Nymphalidae	Lepidoptera	NPV
Agrotis	*exclamationis*	Noctuidae	Lepidoptera	GV, NPV
Agrotis	*ipsilon*	Noctuidae	Lepidoptera	GV, NPV
Agrotis	*segetum*	Noctuidae	Lepidoptera	GV, NPV
Agrotis	*tokionis*	Noctuidae	Lepidoptera	GV
Alabama	*argillacea*	Noctuidae	Lepidoptera	NPV
Aletia	*oxygala*	Noctuidae	Lepidoptera	NPV
Alphaea	*phasma*	Arctiidae	Lepidoptera	NPV
Alsophila	*pometaria*	Geometridae	Lepidoptera	GV, NPV
Amathes	*c-nigrum*	Noctuidae	Lepidoptera	GV, NPV
Amelia	*pallorana*	Tortricidae	Lepidoptera	GV
Amphelophaga	*rubiginosa*	Sphingidae	Lepidoptera	NPV
Amphidasis	*cognataria*	Geometridae	Lepidoptera	NPV
Amsacta	*albistriga*	Arctiidae	Lepidoptera	NPV

Genus	Species	Family	Order	Virus Type[1]
Amsacta	*lactinea*	Arctiidae	Lepidoptera	GV, NPV
Amsacta	*moorei*	Arctiidae	Lepidoptera	NPV
Amsacta	*sp.*	Arctiidae	Lepidoptera	NPV
Amyelois	*transitella*	Pyralidae	Lepidoptera	NPV
Anadevidia	*peponis*	Noctuidae	Lepidoptera	NPV
Anagasta	*kuehniella*	Pyralidae	Lepidoptera	NPV
Anagrapha	*falcifera*	Noctuidae	Lepidoptera	NPV
Anaitis	*plagiata*	Geometridae	Lepidoptera	NPV
Andraca	*bipunctata*	Eupterotidae	Lepidoptera	GV
Anisota	*senatoria*	Saturniidae	Lepidoptera	NPV
Anomis	*flava*	Noctuidae	Lepidoptera	NPV
Anomis	*sabulifera*	Noctuidae	Lepidoptera	NPV
Anomogyna	*elimata*	Noctuidae	Lepidoptera	NPV
Anopheles	*crucians*	Culicidae	Diptera	NPV
Anthela	*varia*	Anthelidae	Lepidoptera	NPV
Anthelia	*hyperborea*	Geometridae	Lepidoptera	NPV
Antheraea	*paphia*	Saturniidae	Lepidoptera	NPV
Antheraea	*pernyi*	Saturniidae	Lepidoptera	NPV
Antheraea	*polyphemus*	Saturniidae	Lepidoptera	NPV
Antheraea	*yamamai*	Saturniidae	Lepidoptera	NPV
Anthonomus	*grandis*	Curculionidae	Coleoptera	NPV
Anthrenus	*museorum*	Dermestidae	Coleoptera	NPV
Anticarsia	*gemmatalis**	Noctuidae	Lepidoptera	NPV*
Apamea	*anceps*	Noctuidae	Lepidoptera	GV, NPV
Apamea	*sordens*	Noctuidae	Lepidoptera	GV
Apocheima	*cinerarius*	Geometridae	Lepidoptera	NPV
Apocheima	*pilosaria*	Geometridae	Lepidoptera	NPV
Aporia	*crataegi*	Pieridae	Lepidoptera	NPV
Aproaerema	*modicella*	Gelechiidae	Lepidoptera	NPV
Araschnia	*levana*	Nymphalidae	Lepidoptera	NPV
Archippus	*breviplicanus*	Tortricidae	Lepidoptera	GV
Archippus	*packardianus*	Tortricidae	Lepidoptera	GV
Archips	*argyrospila*	Tortricidae	Lepidoptera	GV
Archips	*cerasivoranus*	Tortricidae	Lepidoptera	NPV
Archips	*longicellana*	Tortricidae	Lepidoptera	GV
Arctia	*caja*	Arctiidae	Lepidoptera	NPV
Artica	*villica*	Arctiidae	Lepidoptera	NPV
Arctonis	*alba*	Lymantriidae	Lepidoptera	
Ardices	glatignyi	Arctiidae	Lepidoptera	NPV
Arge	*pectoralis*	Argidae	Hymenoptera	NPV
Argynnis	*paphia*	Nymphalidae	Lepidoptera	NPV
Argyresthia	*conjugella*	Argyresthiidae	Lepidoptera	
Argyresthia	cupressella	Argyresthiidae	Lepidoptera	
Argyrogramma	basigera	Noctuidae	Lepidoptera	NPV
Argyrotaenia	*velutinana*	Tortricidae	Lepidoptera	GV
Artona	*funeralis*	Zygaenidae	Lepidoptera	GV
Asterocampa	*celtis*	Nymphalidae	Lepidoptera	NPV
Athetis	*albina*	Noctuidae	Lepidoptera	GV
Autographa	*biloba*	Noctuidae	Lepidoptera	NPV
Autographa	*bimaculata*	Noctuidae	Lepidoptera	NPV
Autographa	*californica**	Noctuidae	Lepidoptera	GV, NPV*
Autographa	*gamma*	Noctuidae	Lepidoptera	NPV

Genus	Species	Family	Order	Virus Type[1]
Autographa	*nigrisigna*	Noctuidae	Lepidoptera	NPV
Autographa	*precationis*	Noctuidae	Lepidoptera	NPV
Batocera	*lineolata*	Cerambycidae	Coleoptera	NPV
Bellura	*gortynoides*	Noctuidae	Lepidoptera	NPV
Bhima	*undulosa*	Lasiocampidae	Lepidoptera	NPV
Biston	*betularia*	Geometridae	Lepidoptera	NPV
Biston	*hirtaria*	Geometridae	Lepidoptera	NPV
Biston	*hispidaria*	Geometridae	Lepidoptera	NPV
Biston	*marginata*	Geometridae	Lepidoptera	NPV
Biston	*robustum*	Geometridae	Lepidoptera	NPV
Biston	*strataria*	Geometridae	Lepidoptera	NPV
Boarmia	*bistortata*	Geometridae	Lepidoptera	NPV
Boarmia	*obliqua*	Geometridae	Lepidoptera	NPV
Bombyx	*mori**	Bombycidae	Lepidoptera	NPV*
Bucculatrix	*thurberiella*	Lyonetiidae	Lepidoptera	NPV
Bupalus	*piniarius*	Geometridae	Lepidoptera	NPV
Buzura	*suppressaria*	Geometridae	Lepidoptera	NPV
Buzura	*thibtaria*	Geometridae	Lepidoptera	NPV
Cadra	*cautella*	Pyralidae	Lepidoptera	GV, NPV
Cadra	*figulilella*	Pyralidae	Lepidoptera	GV, NPV
Calliphora	*vomitoria*	Calliphoridae	Diptera	NPV
Calophasia	*lunula*	Noctuidae	Lepidoptera	NPV
Canephora	*asiatica*	Pyschidae	Lepidoptera	NPV
Caripeta	*divisata*	Geometridae	Lepidoptera	NPV
Carposina	*niponensis*	Carposinidae	Lepidoptera	GV, NPV
Catabena	*esula*	Noctuidae	Lepidoptera	NPV
Catocala	*conjuncta*	Noctuidae	Lepidoptera	NPV
Catocala	*nymphaea*	Noctuidae	Lepidoptera	NPV
Catocala	*nymphagoga*	Noctuidae	Lepidoptera	NPV
Catopsilia	*pomona*	Pieridae	Lepidoptera	NPV
Cephalcia	*abietis*	Pamphiliidae	Hymenoptera	NPV
Cephalcia	*fascipennis*	Pamphiliidae	Hymenoptera	GV
Ceramica	*picta*	Noctuidae	Lepidoptera	NPV
Ceramica	*pisi*	Noctuidae	Lepidoptera	NPV
Cerapteryx	*graminis*	Noctuidae	Lepidoptera	NPV
Cerura	*hermelina*	Notodontidae	Lepidoptera	NPV
Chilo	*infuscatellus*	Pyralidae	Lepidoptera	GV
Chilo	*sacchariphagus*	Pyralidae	Lepidoptera	GV
Chilo	*suppressalis*	Pyralidae	Lepidoptera	GV, NPV
Chironomus	*tentans*	Chironomidae	Diptera	NPV
Choristoneura	*conflictana*	Tortricidae	Lepidoptera	GV, NPV
Choristoneura	*diversana*	Tortricidae	Lepidoptera	NPV
Choristoneura	*fumiferana**	Tortricidae	Lepidoptera	GV, NPV*
Choristoneura	*murinana*	Tortricidae	Lepidoptera	GV, NPV
Choristoneura	*occidentalis*	Tortricidae	Lepidoptera	GV, NPV
Choristoneura	*pinus*	Tortricidae	Lepidoptera	NPV
Choristoneura	*retiniana*	Tortricidae	Lepidoptera	GV
Choristoneura	*rosaceana*	Tortricidae	Lepidoptera	NPV
Choristoneura	*viridis*	Tortricidae	Lepidoptera	GV
Chrysodeixis	*chalcites*	Noctuidae	Lepidoptera	NPV

Genus	Species	Family	Order	Virus Type[1]
Chrysodeixis	*eriosoma*	Noctuidae	Lepidoptera	NPV
Chrysopa	*perla*	Chrysopidae	Neuroptera	NPV
Cingilia	*caternaria*	Geometridae	Lepidoptera	NPV
Clepsis	*persicana*	Tortricidae	Lepidoptera	GV
Cnaphalocrocis	*medinalis*	Pyralidae	Lepidoptera	GV
Cnidocampa	*flavescens*	Limacodidae	Lepidoptera	GV, NPV
Coleophora	*laricella*	Coleophoridae	Lepidoptera	NPV
Coleotechnites	*milleri*	Gelechiidae	Lepidoptera	GV
Colias	*electo*	Pieridae	Lepidoptera	NPV
Colias	*eurytheme*	Pieridae	Lepidoptera	NPV
Colias	*lesbia*	Pieridae	Lepidoptera	NPV
Colias	*philodice*	Pieridae	Lepidoptera	NPV
Coloradia	*pandora*	Saturniidae	Lepidoptera	NPV
Corcyra	*cephalonica*	Pyralidae	Lepidoptera	NPV
Cosmotriche	*podatoria*	Lasiocampidae	Lepidoptera	NPV
Cossus	*cossus*	Cossidae	Lepidoptera	NPV
Cryptoblabes	*lariciana*	Pyralidae	Lepidoptera	NPV
Cryptophlebia	*leucotreta*	Tortricidae	Lepidoptera	GV
Cryptothelea	*junodi*	Pyschidae	Lepidoptera	NPV
Cryptothelea	*variegata*	Pyschidae	Lepidoptera	NPV
Culcula	*panterinaria*	Geometridae	Lepidoptera	NPV
Culex	*pipiens*	Culicidae	Diptera	NPV
Culex	*salinarius*	Culicidae	Diptera	NPV
Cyclophragma	*undans*	Lasiocampidae	Lepidoptera	NPV
Cyclophragma	*yamadai*	Lasiocampidae	Lepidoptera	NPV
Cydia	*nigricana*	Tortricidae	Lepidoptera	GV
Cydia	*pomonella**	Tortricidae	Lepidoptera	GV*, NPV
Darna	*trima*	Limacodidae	Lepidoptera	GV
Dasychira	*abietis*	Lymantriidae	Lepidoptera	NPV
Dasychira	*argentata*	Lymantriidae	Lepidoptera	NPV
Dasychira	*axutha*	Lymantriidae	Lepidoptera	NPV
Dasychira	*basiflava*	Lymantriidae	Lepidoptera	NPV
Dasychira	*confusa*	Lymantriidae	Lepidoptera	NPV
Dasychira	*glaucinoptera*	Lymantriidae	Lepidoptera	NPV
Dasychira	*locuples*	Lymantriidae	Lepidoptera	NPV
Dasychira	*mendosa*	Lymantriidae	Lepidoptera	NPV
Dasychira	*plagiata*	Lymantriidae	Lepidoptera	NPV
Dasychira	*pseudabietis*	Lymantriidae	Lepidoptera	NPV
Dasychira	*pudibunda*	Lymantriidae	Lepidoptera	NPV
Deilephila	*elpenor*	Sphingidae	Lepidoptera	NPV
Deileptenia	*ribeata*	Geometridae	Lepidoptera	NPV
Dendrolimus	*latipennis*	Lasiocampidae	Lepidoptera	NPV
Dendrolimus	*pini*	Lasiocampidae	Lepidoptera	NPV
Dendrolimus	*punctatus*	Lasiocampidae	Lepidoptera	NPV
Dendrolimus	*sibiricus*	Lasiocampidae	Lepidoptera	GV
Dendrolimus	*spectabilis*	Lasiocampidae	Lepidoptera	GV, NPV
Dermestes	*lardarius*	Dermestidae	Coleoptera	NPV
Diachrysia	*orichalcea*	Noctuidae	Lepidoptera	NPV
Diacrisia	*obliqua*	Arctiidae	Lepidoptera	GV, NPV
Diacrisia	*purpurata*	Arctiidae	Lepidoptera	NPV
Diacrisia	*virginica*	Arctiidae	Lepidoptera	GV, NPV

Genus	Species	Family	Order	Virus Type[1]
Diaphora	*mendica*	Arctiidae	Lepidoptera	NPV
Diatraea	*grandiosella*	Pyralidae	Lepidoptera	NPV
Diatraea	*saccharalis*	Pyralidae	Lepidoptera	GV, NPV
Dichocrocis	*punctiferalis*	Pyralidae	Lepidoptera	NPV
Dictyoploca	*japonica*	Saturniidae	Lepidoptera	NPV
Dicycla	*oo*	Noctuidae	Lepidoptera	NPV
Dilta	*hibernica*	Praemachilidae	Thysanura	NPV
Dionychopus	*amasis*	Arctiidae	Lepidoptera	GV
Dioryctria	*abietella*	Pyralidae	Lepidoptera	GV
Dioryctria	*pseudotsugella*	Pyralidae	Lepidoptera	NPV
Diparopsis	*watersi*	Noctuidae	Lepidoptera	NPV
Diprion	*hercyniae*	Diprionidae	Hymenoptera	NPV
Diprion	*leuwanensis*	Diprionidae	Hymenoptera	NPV
Diprion	*nipponica*	Diprionidae	Hymenoptera	NPV
Diprion	*pallida*	Diprionidae	Hymenoptera	NPV
Diprion	*pindrowi*	Diprionidae	Hymenoptera	NPV
Diprion	*pini*	Diprionidae	Hymenoptera	NPV
Diprion	*polytoma*	Diprionidae	Hymenoptera	NPV
Diprion	*similis*	Diprionidae	Hymenoptera	NPV
Dirphia	*gragatus*	Saturniidae	Lepidoptera	NPV
Doratifera	*casta*	Limacodidae	Lepidoptera	NPV
Dryobota	*furva*	Noctuidae	Lepidoptera	GV, NPV
Dryobota	*protea*	Noctuidae	Lepidoptera	NPV
Dryobotodes	*monochroma*	Noctuidae	Lepidoptera	NPV
Earias	*insulana*	Noctuidae	Lepidoptera	NPV
Ecpantheria	*icasia*	Arctiidae	Lepidoptera	GV, NPV
Ectropis	*crepuscularia*	Geometridae	Lepidoptera	NPV
Ectropis	*obliqua*	Geometridae	Lepidoptera	GV, NPV
Ennomos	*quercaria*	Geometridae	Lepidoptera	NPV
Ennomos	*quercinaria*	Geometridae	Lepidoptera	NPV
Ennomos	*subsignarius*	Geometridae	Lepidoptera	NPV
Enypia	*venata*	Geometridae	Lepidoptera	NPV
Epargyreus	*clarus*	Hesperiidae	Lepidoptera	NPV
Ephestia	*elutella*	Pyralidae	Lepidoptera	NPV
Epinotia	*aporema*	Tortricidae	Lepidoptera	GV
Epiphyas	*postvittana*	Tortricidae	Lepidoptera	NPV
Erannis	*ankeraria*	Geometridae	Lepidoptera	NPV
Erannis	*defoliaria*	Geometridae	Lepidoptera	NPV
Erannis	*tiliaria*	Geometridae	Lepidoptera	NPV
Erannis	*vancouverensis*	Geometridae	Lepidoptera	NPV
Eratmapodites	*quinquevittatus*	Culicidae	Diptera	NPV
Erinnyis	*ello*	Sphingidae	Lepidoptera	NPV
Eriogyna	*pyretorum*	Saturniidae	Lepidoptera	NPV
Estigmene	*acrea*	Arctiidae	Lepidoptera	GV,NPV
Eupithecia	*annulata*	Geometridae	Lepidoptera	NPV
Eupithecia	*longipalpata*	Geometridae	Lepidoptera	NPV
Euplexia	*lucipara*	Noctuidae	Lepidoptera	GV
Euproctis	*bipunctapex*	Lymantriidae	Lepidoptera	NPV
Euproctis	*chrysorrhoea*	Lymantriidae	Lepidoptera	NPV
Euproctis	*flava*	Lymantriidae	Lepidoptera	NPV
Euproctis	*flavinata*	Lymantriidae	Lepidoptera	NPV

Genus	Species	Family	Order	Virus Type[1]
Euproctis	*karghalica*	Lymantriidae	Lepidoptera	NPV
Euproctis	*pseudoconspersa*	Lymantriidae	Lepidoptera	NPV
Euproctis	*similis*	Lymantriidae	Lepidoptera	NPV
Euproctis	*subflava*	Lymantriidae	Lepidoptera	NPV
Eupsilia	*satellitia*	Noctuidae	Lepidoptera	GV
Eupsilia	sp.	Noctuidae	Lepidoptera	NPV
Euthyatira	*pudens*	Thyatiridae	Lepidoptera	NPV
Euxoa	*auxiliaris*	Noctuidae	Lepidoptera	GV, NPV
Euxoa	*messoria*	Noctuidae	Lepidoptera	GV, NPV
Euxoa	*ochrogaster*	Noctuidae	Lepidoptera	GV, NPV
Euxoa	*scandens*	Noctuidae	Lepidoptera	NPV
Exartema	*appendiceum*	Tortricidae	Lepidoptera	GV
Feltia	*subterranea*	Noctuidae	Lepidoptera	GV
Feralia	*jacosa*	Noctuidae	Lepidoptera	NPV
Galleria	*mellonella**	Pyralidae	Lepidoptera	NPV*
Gastropacha	*quercifolia*	Lasiocampidae	Lepidoptera	NPV
Glena	*bisulca*	Geometridae	Lepidoptera	GV
Grapholitha	*molesta*	Tortricidae	Lepidoptera	GV
Griselda	*radicana*	Tortricidae	Lepidoptera	GV
Hadena	*basilinea*	Noctuidae	Lepidoptera	GV
Hadena	*sordida*	Noctuidae	Lepidoptera	GV, NPV
Halisidota	*argentata*	Arctiidae	Lepidoptera	NPV
Halisidota	*caryae*	Arctiidae	Lepidoptera	NPV
Harrisina	*brillians*	Zygaenidae	Lepidoptera	GV
Heliothis	*armigera*	Noctuidae	Lepidoptera	GV, NPV
Heliothis	*assulta*	Noctuidae	Lepidoptera	NPV
Heliothis	*obtectus*	Noctuidae	Lepidoptera	NPV
Heliothis	*paradoxa*	Noctuidae	Lepidoptera	NPV
Heliothis	*peltigera*	Noctuidae	Lepidoptera	NPV
Heliothis	*phloxiphaga*	Noctuidae	Lepidoptera	NPV
Heliothis	*punctigera*	Noctuidae	Lepidoptera	GV, NPV
Heliothis	*rubrescens*	Noctuidae	Lepidoptera	NPV
Heliothis	*subflexa*	Noctuidae	Lepidoptera	NPV
Heliothis	*virescens*	Noctuidae	Lepidoptera	NPV
Heliothis	*zea**	Noctuidae	Lepidoptera	GV, NPV*
Hemerobius	*stigma*	Hemerobiidae	Neuroptera	NPV
Hemichroa	*crocea*	Tenthredinidae	Hymenoptera	NPV
Hemileuca	*eglanterina*	Saturniidae	Lepidoptera	GV, NPV
Hemileuca	*maia*	Saturniidae	Lepidoptera	NPV
Hemileuca	*oliviae*	Saturniidae	Lepidoptera	GV, NPV
Hemileuca	sp.	Saturniidae	Lepidoptera	NPV
Hemileuca	*tricolor*	Saturniidae	Lepidoptera	NPV
Hesperumia	*sulphuraria*	Geometridae	Lepidoptera	NPV
Hippotion	*eson*	Sphingidae	Lepidoptera	NPV
Homona	*coffearia*	Tortricidae	Lepidoptera	GV
Homona	*magnanima*	Tortricidae	Lepidoptera	GV, NPV
Hoplodrina	*ambigua*	Noctuidae	Lepidoptera	NPV
Hyalophora	*cecropia*	Saturniidae	Lepidoptera	NPV
Hydria	*prunivora*	Geometridae	Lepidoptera	GV

Genus	Species	Family	Order	Virus Type[1]
Hydriomena	*irata*	Geometridae	Lepidoptera	NPV
Hydriomena	*nubilofasciata*	Geometridae	Lepidoptera	NPV
Hyles	*euphorbiae*	Sphingidae	Lepidoptera	NPV
Hyles	*gallii*	Sphingidae	Lepidoptera	NPV
Hyles	*lineata*	Sphingidae	Lepidoptera	NPV
Hylesia	*nigricans*	Saturniidae	Lepidoptera	NPV
Hylesia	sp.	Saturniidae	Lepidoptera	NPV
Hyloicus	*pinastri*	Sphingidae	Lepidoptera	NPV
Hyperetis	*amicaria*	Geometridae	Lepidoptera	NPV
Hyphantria	*cunea*	Arctiidae	Lepidoptera	GV, NPV
Hyphorma	*minax*	Limacodidae	Lepidoptera	NPV
Hypocrita	*jacobaeae*	Arctiidae	Lepidoptera	NPV
Inachis	*io*	Nymphalidae	Lepidoptera	NPV
Ilragoides	*fasciata*	Limacodidae	Lepidoptera	NPV
Ivela	*auripes*	Lymantriidae	Lepidoptera	NPV
Ivela	*ochropoda*	Lymantriidae	Lepidoptera	NPV
Jankowskia	*athleta*	Geometridae	Lepidoptera	NPV
Junonia	*coenia*	Nymphalidae	Lepidoptera	GV, NPV
Lacanobia	*oleracea*	Noctuidae	Lepidoptera	GV, NPV
Lambdina	*fiscellaria*	Geometridae	Lepidoptera	GV, NPV
Laothoe	*populi*	Sphingidae	Lepidoptera	NPV
Lasiocampa	*quercus*	Lasiocampidae	Lepidoptera	NPV
Lasiocampa	*trifolii*	Lasiocampidae	Lepidoptera	NPV
Lathronympha	*phaseoli*	Tortricidae	Lepidoptera	GV
Lebeda	*nobilis*	Lasiocampidae	Lepidoptera	NPV
Lechriolepis	*basirufa*	Lasiocampidae	Lepidoptera	NPV
Leucoma	*candida*	Lymantriidae	Lepidoptera	NPV
Leucoma	*salicis*	Lymantriidae	Lepidoptera	NPV
Lobesia	*botrana*	Tortricidae	Lepidoptera	GV
Lophopteryx	*camelina*	Notodontidae	Lepidoptera	NPV
Loxostege	*sticticalis*	Pyralidae	Lepidoptera	GV, NPV
Luehdorfia	*japonica*	Papilionidae	Lepidoptera	NPV
Lymantria	*dispar**	Lymantriidae	Lepidoptera	NPV*
Lymantria	*dissoluta*	Lymantriidae	Lepidoptera	NPV
Lymantria	*fumida*	Lymantriidae	Lepidoptera	NPV
Lymantria	*incerta*	Lymantriidae	Lepidoptera	NPV
Lymantria	*mathura*	Lymantriidae	Lepidoptera	NPV
Lymantria	*monacha*	Lymantriidae	Lepidoptera	NPV
Lymantria	*ninayi*	Lymantriidae	Lepidoptera	NPV
Lymantria	*obfuscata*	Lymantriidae	Lepidoptera	NPV
Lymantria	*violaswinhol*	Lymantriidae	Lepidoptera	NPV
Lymantria	*xylina*	Lymantriidae	Lepidoptera	NPV
Macroglossum	*bombylans*	Sphingidae	Lepidoptera	GV
Macrothylacia	*rubi*	Lasiocampidae	Lepidoptera	NPV
Mahasena	*miniscula*	Pyschidae	Lepidoptera	NPV
Malacosoma	*alpicola*	Lasiocampidae	Lepidoptera	NPV
Malacosoma	*americanum*	Lasiocampidae	Lepidoptera	NPV
Malacosoma	*californicum*	Lasiocampidae	Lepidoptera	NPV

Genus	Species	Family	Order	Virus Type[1]
Malacosoma	*constrictum*	Lasiocampidae	Lepidoptera	NPV
Malacosoma	*disstria*	Lasiocampidae	Lepidoptera	NPV
Malacosoma	*fragile*	Lasiocampidae	Lepidoptera	NPV
Malacosoma	*lutescens*	Lasiocampidae	Lepidoptera	NPV
Malacosoma	*neustria*	Lasiocampidae	Lepidoptera	NPV
Malacosoma	*pluviale*	Lasiocampidae	Lepidoptera	GV, NPV
Mamestra	*brassicae**	Noctuidae	Lepidoptera	GV, NPV*
Mamestra	*configurata*	Noctuidae	Lepidoptera	GV, NPV
Mamestra	sp.	Noctuidae	Lepidoptera	GV
Mamestra	*suasa*	Noctuidae	Lepidoptera	NPV
Manduca	*quinquemaculata*	Sphingidae	Lepidoptera	GV
Manduca	*sexta*	Sphingidae	Lepidoptera	GV, NPV
Megalopyge	*opercularis*	Megalopygidae	Lepidoptera	GV
Melanchra	*persicariae*	Noctuidae	Lepidoptera	GV
Melanolophia	*imitata*	Geometridae	Lepidoptera	NPV
Melitaea	*didyma*	Nymphalidae	Lepidoptera	NPV
Merophyas	*divulsana*	Tortricidae	Lepidoptera	NPV
Mesonura	*rufonota*	Tenthredinidae	Hymenoptera	NPV
Moma	*champa*	Noctuidae	Lepidoptera	NPV
Myrteta	*tinagmaria*	Geometridae	Lepidoptera	NPV
Nacoleia	*diemenalis*	Pyralidae	Lepidoptera	GV
Nacoleia	*octosema*	Pyralidae	Lepidoptera	NPV
Nadata	*gibbosa*	Notodontidae	Lepidoptera	NPV
Natada	*nararia*	Limacodidae	Lepidoptera	GV
Nematocampa	*filamentaria*	Geometridae	Lepidoptera	GV
Nematus	*olfaciens*	Tenthredinidae	Hymenoptera	NPV
Neodiprion	*abietis*	Diprionidae	Hymenoptera	NPV
Neodiprion	*excitans*	Diprionidae	Hymenoptera	NPV
Neodiprion	*lecontei*	Diprionidae	Hymenoptera	NPV
Neodiprion	*nanulus*	Diprionidae	Hymenoptera	NPV
Neodiprion	*pratti*	Diprionidae	Hymenoptera	NPV
Neodiprion	*sertifer**	Diprionidae	Hymenoptera	NPV*
Neodiprion	*swainei*	Diprionidae	Hymenoptera	NPV
Neodiprion	*taedae*	Diprionidae	Hymenoptera	NPV
Neodiprion	*tsugae*	Diprionidae	Hymenoptera	NPV
Neodiprion	*virginiana*	Diprionidae	Hymenoptera	NPV
Neophasia	*menapia*	Pieridae	Lepidoptera	NPV
Neopheosia	*excurvata*	Notodontidae	Lepidoptera	NPV
Neophylax	sp.	Limnephilidae	Trichoptera	NPV
Nephelodes	*emmedonia*	Noctuidae	Lepidoptera	GV, NPV
Nepytia	*freemani*	Geometridae	Lepidoptera	NPV
Nepytia	*phantasmaria*	Geometridae	Lepidoptera	NPV
Noctua	*pronuba*	Noctuidae	Lepidoptera	NPV
Nyctobia	*limitaria*	Geometridae	Lepidoptera	NPV
Nymphalis	*antiopa*	Nymphalidae	Lepidoptera	GV, NPV
Nymphalis	*polychloros*	Nymphalidae	Lepidoptera	NPV
Nymphula	*depunctalis*	Pyralidae	Lepidoptera	NPV
Ocinara	*varians*	Bombycidae	Lepidoptera	NPV
Operophtera	*bruceata*	Geometridae	Lepidoptera	NPV
Operophtera	*brumata*	Geometridae	Lepidoptera	NPV

Genus	Species	Family	Order	Virus Type[1]
Opisina	*arenosella*	Cryptophasidae	Lepidoptera	NPV
Opisthograptis	*luteolata*	Geometridae	Lepidoptera	NPV
Oporinia	*autumnata*	Geometridae	Lepidoptera	NPV
Opsiphanes	*cassina*	Brassolidae	Lepidoptera	NPV
Oraesia	*emarginata*	Noctuidae	Lepidoptera	NPV
Orgyia	*anartoides*	Lymantriidae	Lepidoptera	NPV
Orgyia	*antiqua*	Lymantriidae	Lepidoptera	NPV
Orgyia	*australis*	Lymantriidae	Lepidoptera	NPV
Orgyia	*badia*	Lymantriidae	Lepidoptera	NPV
Orgyia	*gonostigma*	Lymantriidae	Lepidoptera	NPV
Orgyia	*leucostigma*	Lymantriidae	Lepidoptera	NPV
Orgyia	*postica*	Lymantriidae	Lepidoptera	NPV
Orgyia	*pseudotsugata**	Lymantriidae	Lepidoptera	NPV*
Orgyia	*turbata*	Lymantriidae	Lepidoptera	NPV
Orgyia	*vetusta*	Lymantriidae	Lepidoptera	NPV
Orthosia	*hibisci*	Noctuidae	Lepidoptera	NPV
Orthosia	*incerta*	Noctuidae	Lepidoptera	NPV
Ostrinia	*nubilalis*	Pyralidae	Lepidoptera	NPV
Pachypasa	*capensis*	Lasiocampidae	Lepidoptera	NPV
Pachypasa	*otus*	Lasiocampidae	Lepidoptera	NPV
Paleacrita	*vernata*	Geometridae	Lepidoptera	NPV
Panaxia	*dominula*	Arctiidae	Lepidoptera	NPV
Pandemis	*heparana*	Tortricidae	Lepidoptera	NPV
Pandemis	*lamprosana*	Tortricidae	Lepidoptera	NPV
Panolis	*flammea*	Noctuidae	Lepidoptera	NPV
Pantana	*phyllostachysae*	Lymantriidae	Lepidoptera	NPV
Panthea	*portlandia*	Noctuidae	Lepidoptera	NPV
Papaipema	*purpurifascia*	Noctuidae	Lepidoptera	GV
Parasa	*bicolor*	Limacodidae	Lepidoptera	GV
Parasa	*consocia*	Limacodidae	Lepidoptera	GV, NPV
Parasa	*lepida*	Limacodidae	Lepidoptera	GV, NPV
Parasa	*sinica*	Limacodidae	Lepidoptera	GV, NPV
Parnara	*guttata*	Hesperiidae	Lepidoptera	NPV
Parnara	*mathias*	Hesperiidae	Lepidoptera	NPV
Papilio	*daunis*	Papilionidae	Lepidoptera	NPV
Papilio	*demoleus*	Papilionidae	Lepidoptera	NPV
Papilio	*podalirius*	Papilionidae	Lepidoptera	NPV
Papilio	*polyxenes*	Papilionidae	Lepidoptera	NPV
Papilio	*xuthus*	Papilionidae	Lepidoptera	NPV
Pectinophora	*gossypiella*	Gelechiidae	Lepidoptera	NPV
Peribatoides	*simpliciaria*	Geometridae	Lepidoptera	NPV
Pericallia	*ricini*	Arctiidae	Lepidoptera	GV, NPV
Peridroma	*saucia*	Noctuidae	Lepidoptera	GV, NPV
Peridroma	sp.	Noctuidae	Lepidoptera	GV, NPV
Pero	*behrensarius*	Geometridae	Lepidoptera	NPV
Pero	*mizon*	Geometridae	Lepidoptera	NPV
Persectania	*ewingii*	Noctuidae	Lepidoptera	GV
Phalera	*assimilis*	Notodontidae	Lepidoptera	NPV
Phalera	*bucephala*	Notodontidae	Lepidoptera	NPV
Phalera	*flavescens*	Notodontidae	Lepidoptera	NPV
Phauda	*flammans*	Zygaenidae	Lepidoptera	NPV

Genus	Species	Family	Order	Virus Type[1]
Phigalia	*titea*	Geometridae	Lepidoptera	NPV
Phlogophora	*meticulosa*	Noctuidae	Lepidoptera	NPV
Phragmatobia	*fuliginosa*	Arctiidae	Lepidoptera	GV
Phryganidia	*californica*	Dioptidae	Lepidoptera	NPV
Phthonosema	*tendinosaria*	Geometridae	Lepidoptera	NPV
Phthorimaea	*operculella**	Gelechiidae	Lepidoptera	GV*, NPV
Pieris	*brassicae**	Pieridae	Lepidoptera	GV*
Pieris	*melete*	Pieridae	Lepidoptera	GV
Pieris	*napi*	Pieridae	Lepidoptera	GV
Pieris	*rapae**	Pieridae	Lepidoptera	GV*, NPV
Pieris	*virginiensis*	Pieridae	Lepidoptera	GV
Pikonema	*dimmockii*	Tenthredinidae	Hymenoptera	NPV
Plathypena	*scabra*	Noctuidae	Lepidoptera	GV, NPV
Platynota	*idaesalis*	Tortricidae	Lepidoptera	NPV
Plodia	*interpunctella**	Pyralidae	Lepidoptera	GV*
Plusia	*argentifera*	Noctuidae	Lepidoptera	NPV
Plusia	*balluca*	Noctuidae	Lepidoptera	NPV
Plusia	*circumflexa*	Noctuidae	Lepidoptera	GV
Plusia	*signata*	Noctuidae	Lepidoptera	NPV
Plutella	*xylostella*	Plutellidae	Lepidoptera	GV, NPV
Polygonia	*c-album*	Nymphalidae	Lepidoptera	NPV
Polygonia	*satyrus*	Nymphalidae	Lepidoptera	NPV
Pontia	*daplidice*	Pieridae	Lepidoptera	GV
Porthesia	*scintillans*	Lymantriidae	Lepidoptera	NPV
Pristophora	*erichsonii*	Tenthredinidae	Hymenoptera	NPV
Pristophora	geniculata	Tenthredinidae	Hymenoptera	NPV
Prodenia	androgea	Noctuidae	Lepidoptera	GV
Prodenia	litosia	Noctuidae	Lepidoptera	NPV
Prodenia	praefica	Noctuidae	Lepidoptera	NPV
Prodenia	terricola	Noctuidae	Lepidoptera	NPV
Protoboarmia	*porcelaria*	Geometridae	Lepidoptera	NPV
Pseudaletia	*convecta*	Noctuidae	Lepidoptera	GV, NPV
Pseudaletia	*separata*	Noctuidae	Lepidoptera	GV, NPV
Pseudaletia	*unipuncta*	Noctuidae	Lepidoptera	GV, NPV
Pseudoplusia	*includens*	Noctuidae	Lepidoptera	NPV
Psilogramma	*menephron*	Sphingidae	Lepidoptera	GV
Psorophora	*confinnis*	Culicidae	Diptera	NPV
Psorophora	*ferox*	Culicidae	Diptera	NPV
Psorophora	*varipes*	Culicidae	Diptera	NPV
Pterolocera	*amplicornis*	Anthelidae	Lepidoptera	NPV
Ptycholomoides	*aeriferana*	Tortricidae	Lepidoptera	NPV
Ptychopoda	*seriata*	Geometridae	Lepidoptera	NPV
Pygaera	*anachoreta*	Notodontidae	Lepidoptera	GV
Pygaera	*anastomosis*	Notodontidae	Lepidoptera	GV, NPV
Pygaera	*fulgurita*	Notodontidae	Lepidoptera	NPV
Pyrausta	*diniasalis*	Pyralidae	Lepidoptera	NPV
Rachiplusia	*nu*	Noctuidae	Lepidoptera	NPV
Rachiplusia	*ou**	Noctuidae	Lepidoptera	NPV*
Rheumaptera	*hastata*	Geometridae	Lepidoptera	GV
Rhyacionia	*buoliana*	Tortricidae	Lepidoptera	GV
Rhyacionia	*duplana*	Tortricidae	Lepidoptera	GV, NPV

Genus	Species	Family	Order	Virus Type[1]
Rhyacionia	*frustrana*	Tortricidae	Lepidoptera	GV
Rhynchosciara	*angelae*	Sciaridae	Diptera	NPV
Rhynchosciara	*hollaenderi*	Sciaridae	Diptera	NPV
Rhynchosciara	*milleri*	Sciaridae	Diptera	NPV
Rondiotia	*menciana*	Bombycidae	Lepidoptera	NPV
Sabulodes	*caberata*	Geometridae	Lepidoptera	GV
Samia	*cynthia*	Saturniidae	Lepidoptera	NPV
Samia	*pryeri*	Saturniidae	Lepidoptera	NPV
Samia	*ricini*	Saturniidae	Lepidoptera	NPV
Saturnia	*pyri*	Saturniidae	Lepidoptera	NPV
Sceliodes	*cordalis*	Pyralidae	Lepidoptera	NPV
Sciaphila	*duplex*	Tortricidae	Lepidoptera	GV
Scirpophaga	*incertulas*	Pyralidae	Lepidoptera	NPV
Scoliopteryx	*libatrix*	Noctuidae	Lepidoptera	NPV
Scopelodes	*contracta*	Limacodidae	Lepidoptera	NPV
Scopelodes	*venosa*	Limacodidae	Lepidoptera	NPV
Scopula	*subpunctaria*	Geometridae	Lepidoptera	NPV
Scotogramma	*trifolii*	Noctuidae	Lepidoptera	GV, NPV
Selenephera	*lunigera*	Lasiocampidae	Lepidoptera	NPV
Selepa	*celtis*	Noctuidae	Lepidoptera	GV
Selidosema	*suavis*	Geometridae	Lepidoptera	NPV
Semidonta	*biloba*	Notodontidae	Lepidoptera	NPV
Semiothisa	*sexmaculata*	Geometridae	Lepidoptera	GV
Sesamia	*calamistis*	Noctuidae	Lepidoptera	NPV
Sesamia	*cretica*	Noctuidae	Lepidoptera	GV
Sesamia	*inferens*	Noctuidae	Lepidoptera	NPV
Sesamia	*nonagrioides*	Noctuidae	Lepidoptera	GV
Smerinthus	*ocellata*	Sphingidae	Lepidoptera	NPV
Sparganothis	*pettitana*	Tortricidae	Lepidoptera	NPV
Sphinx	*ligustri*	Sphingidae	Lepidoptera	NPV
Spilarctia	*subcarnea*	Arctiidae	Lepidoptera	NPV
Spilonota	*ocellana*	Tortricidae	Lepidoptera	NPV
Spilosoma	*lubricipeda*	Arctiidae	Lepidoptera	NPV
Spodoptera	*exempta**	Noctuidae	Lepidoptera	NPV*
Spodoptera	*exigua*	Noctuidae	Lepidoptera	GV, NPV
Spodoptera	*frugiperda**	Noctuidae	Lepidoptera	GV*, NPV*
Spodoptera	*latifascia*	Noctuidae	Lepidoptera	NPV
Spodoptera	*littoralis**	Noctuidae	Lepidoptera	GV, NPV*
Spodoptera	*litura*	Noctuidae	Lepidoptera	GV, NPV
Spodoptera	*mauritia*	Noctuidae	Lepidoptera	NPV
Spodoptera	*ornithogalli*	Noctuidae	Lepidoptera	NPV
Spodoptera	sp.	Noctuidae	Lepidoptera	NPV
Synaxis	*jubararia*	Geometridae	Lepidoptera	NPV
Synaxis	*pallulata*	Geometridae	Lepidoptera	NPV
Syngrapha	*selecta*	Noctuidae	Lepidoptera	NPV
Tetralopha	*scortealis*	Pyralidae	Lepidoptera	NPV
Tetropium	*cinnamopterum*	Cerambycidae	Coleoptera	NPV
Thaumetopoea	*pityocampa*	Thaumetopoeidae	Lepidoptera	GV, NPV
Thaumetopoea	*processionea*	Thaumetopoeidae	Lepidoptera	NPV
Thaumetopoea	*wilkinsoni*	Thaumetopoeidae		

Genus	Species	Family	Order	Virus Type[1]
Theophila	mandarina	Bombycidae	Lepidoptera	NPV
Theretra	japonica	Sphingidae	Lepidoptera	NPV
Thosea	baibarana	Limacodidae	Lepidoptera	NPV
Thosea	sinensis	Limacodidae	Lepidoptera	GV
Thymelicus	lineola	Hesperiidae	Lepidoptera	NPV
Thyridopteryx	ephemeraeformis	Pyschidae	Lepidoptera	NPV
Ticera	castanea	Lasiocampidae	Lepidoptera	NPV
Tinea	pellionella	Tineidae	Lepidoptera	NPV
Tineola	bisselliella	Tineidae	Lepidoptera	NPV
Tipula	paludosa*	Tipulidae	Diptera	NPV*
Tiracola	plagiata	Noctuidae	Lepidoptera	NPV
Tortrix	loeflingiana	Tortricidae	Lepidoptera	NPV
Tortrix	viridana	Tortricidae	Lepidoptera	NPV
Toxorhynchites	brevipalpis	Culicidae	Diptera	NPV
Trabala	vishnou	Lasiocampidae	Lepidoptera	NPV
Trichiocampus	irregularis	Tenthredinidae	Hymenoptera	NPV
Trichiocampus	viminalis	Tenthredinidae	Hymenoptera	NPV
Trichoplusia	ni*	Noctuidae	Lepidoptera	GV*, NPV*
Ugymyia	sericariae	Tachinidae	Diptera	NPV
Uranotaenia	sapphirina	Culicidae	Diptera	NPV
Urbanus	proteus	Hesperiidae	Lepidoptera	NPV
Vanessa	atalanta	Nymphalidae	Lepidoptera	NPV
Vanessa	cardui	Nymphalidae	Lepidoptera	NPV
Vanessa	prorsa	Nymphalidae	Lepidoptera	NPV
Wiseana	cervinata	Hepialidae	Lepidoptera	GV, NPV
Wiseana	signata	Hepialidae	Lepidoptera	NPV
Wiseana	umbraculata	Hepialidae	Lepidoptera	GV, NPV
Wyeomyia	smithii	Culicidae	Diptera	NPV
Xylena	curvimacula	Noctuidae	Lepidoptera	NPV
Yponomeuta	cognatella	Yponomeutidae	Lepidoptera	NPV
Yponomeuta	evonymella	Yponomeutidae	Lepidoptera	NPV
Yponomeuta	malinellus	Yponomeutidae	Lepidoptera	NPV
Yponomeuta	padella	Yponomeutidae	Lepidoptera	NPV
Zeiraphera	diniana	Tortricidae	Lepidoptera	GV, NPV
Zeiraphera	pseudotsugana	Tortricidae	Lepidoptera	NPV
Zeiraphera	sp.	Tortricidae	Lepidoptera	GV

Compiled from Martignoni, M.E. and Iwai, P.J. (1986) A Catalog of Virus Diseases of Insects, Mites and Ticks. Fourth Edition Revised. General Technical Report PNW-195, US Department of Agriculture.

The table does not differentiate between hosts from which nuclear polyhedrosis viruses and granulosis viruses were first isolated and those where infection was observed during virus cross-transmission studies.

[1]NPV = Nuclear polyhedrosis virus; GV = Granulosis virus.
*Viruses which are described in more detail elsewhere in the dictionary.

Appendix B

A list of insect species in which Cytoplasmic Polyhedrosis Virus infections have been recorded

Genus	Species	Family	Order	Virus Type (where known)[1]
*Abraxas	grossulariata	Geometridae	Lepidoptera	8*
Actias	luna	Saturniidae	Lepidoptera	
*Actias	selene	Saturniidae	Lepidoptera	4*
Adoxophyes	fasciata	Tortricidae	Lepidoptera	
Adoxophyes	orana	Tortricidae	Lepidoptera	
Adris	tyrannus	Noctuidae	Lepidoptera	
Aedes	aegypti	Culicidae	Diptera	
Aedes	cantator	Culicidae	Diptera	
Aedes	sierrensis	Culicidae	Diptera	
Aedes	sollicitans	Culicidae	Diptera	
Aedes	sticticus	Culicidae	Diptera	
Aedes	taeniorhynchus	Culicidae	Diptera	
Aedes	thibaulti	Culicidae	Diptera	
Aedes	triseriatus	Culicidae	Diptera	
Aglais	urticae	Nymphalidae	Lepidoptera	2,6
Aglaope	infausta	Zygaenidae	Lepidoptera	
Agraulis	vanillae	Nymphalidae	Lepidoptera	2
Agrochola	helvola	Noctuidae	Lepidoptera	6
Agrochola	lychnidis	Noctuidae	Lepidoptera	6
Agrotis	ipsilon	Noctuidae	Lepidoptera	
*Agrotis	segetum	Noctuidae	Lepidoptera	9*
Alsophila	pometaria	Geometridae	Lepidoptera	
Amathes	c-nigrum	Noctuidae	Lepidoptera	
Anaitis	plagiata	Geometridae	Lepidoptera	3,6
Anopheles	bradleyi	Culicidae	Diptera	
Anopheles	crucians	Culicidae	Diptera	
Anopheles	freeborni	Culicidae	Diptera	
Anopheles	quadrimaculatus	Culicidae	Diptera	
Anoplonyx	destructor	Tenthredinidae	Hymenoptera	
Antheraea	eucalypti	Saturniidae	Lepidoptera	
Antheraea	mylitta	Saturniidae	Lepidoptera	4
Antheraea	paphia	Saturniidae	Lepidoptera	
Antheraea	pernyi	Saturniidae	Lepidoptera	4
Antheraea	polyphemus	Saturniidae	Lepidoptera	
Antitype	xanthomista	Noctuidae	Lepidoptera	6
Aporia	crataegi	Pieridae	Lepidoptera	
*Aporophyla	lutulenta	Noctuidae	Lepidoptera	10*

266

Genus	Species	Family	Order	Virus Type (where known)[1]
Arcte	*coerulea*	Noctuidae	Lepidoptera	
Arctia	*caja*	Arctiidae	Lepidoptera	2,3
Arctia	*villica*	Arctiidae	Lepidoptera	2
Autographa	*californica*	Noctuidae	Lepidoptera	
Autographa	*gamma*	Noctuidae	Lepidoptera	12
Automeris	*aurantiaca*	Saturniidae	Lepidoptera	
Automeris	*io*	Saturniidae	Lepidoptera	
Automeris	*memusae*	Saturniidae	Lepidoptera	
Axylia	*putris*	Noctuidae	Lepidoptera	
*Biston	*betularia*	Geometridae	Lepidoptera	6*
Blepharita	*solieri*	Noctuidae	Lepidoptera	
Boloria	*dia*	Nymphalidae	Lepidoptera	2
*Bombyx	*mori*	Bombycidae	Lepidoptera	1*
Bupalus	*piniarius*	Geometridae	Lepidoptera	
Cactoblastis	*cactorum*	Pyralidae	Lepidoptera	
Cadra	*cautella*	Pyralidae	Lepidoptera	
Calophasia	*lunula*	Noctuidae	Lepidoptera	
Ceramica	*picta*	Noctuidae	Lepidoptera	
Cerura	*vinula*	Notodontidae	Lepidoptera	
Chilo	*sacchariphagus*	Pyralidae	Lepidoptera	
Chironomus	*plumosus*	Chironomidae	Diptera	
Choristoneura	*fumiferana*	Tortricidae	Lepidoptera	
Chrysodeixis	*eriosoma*	Noctuidae	Lepidoptera	
Cnephia	*mutata*	Simuliidae	Diptera	
Colias	*eurytheme*	Pieridae	Lepidoptera	
Crocallis	*elinguaria*	Geometridae	Lepidoptera	
Crymodes	*devastator*	Noctuidae	Lepidoptera	
Cryptophlebia	*leucotreta*	Tortricidae	Lepidoptera	
Culex	*erraticus*	Culicidae	Diptera	
Culex	*peccator*	Culicidae	Diptera	
Culex	*pipiens*	Culicidae	Diptera	
Culex	*restuans*	Culicidae	Diptera	
Culex	*salinarius*	Culicidae	Diptera	
Culex	*tarsalis*	Culicidae	Diptera	
Culex	*territans*	Culicidae	Diptera	
Culicoides	sp.	Ceratopogonidae	Diptera	
Culiseta	*inornata*	Culicidae	Diptera	
Culiseta	*melanura*	Culicidae	Diptera	
Cyclophragma	*undans*	Lasiocampidae	Lepidoptera	
Cyclophragma	*yamadai*	Lasiocampidae	Lepidoptera	
Danaus	*plexippus*	Danaidae	Lepidoptera	3
Dasychira	*axutha*	Lymantriidae	Lepidoptera	
Dasychira	*pudibunda*	Lymantriidae	Lepidoptera	2
Deilephila	*elpenor*	Sphingidae	Lepidoptera	
Dendrolimus	*pini*	Lasiocampidae	Lepidoptera	
Dendrolimus	*punctatus*	Lasiocampidae	Lepidoptera	
Dendrolimus	*spectabilis*	Lasiocampidae	Lepidoptera	1

Genus	Species	Family	Order	Virus Type (where known)[1]
Dendrolimus	*superans*	Lasiocampidae	Lepidoptera	
Dendrolimus	*tabulaeformis*	Lasiocampidae	Lepidoptera	
Diacrisia	*purpurata*	Arctiidae	Lepidoptera	
Diacrisia	*virginica*	Arctiidae	Lepidoptera	
Diaphora	*mendica*	Arctiidae	Lepidoptera	
Dictyoploca	*japonica*	Saturniidae	Lepidoptera	
Dilina	*tiliae*	Sphingidae	Lepidoptera	
Diparopsis	*watersi*	Noctuidae	Lepidoptera	
Dira	*megera*	Satyridae	Lepidoptera	
Drepana	*lacertinaria*	Drepanidae	Lepidoptera	
Earias	*biplaga*	Noctuidae	Lepidoptera	
Earias	*insulana*	Noctuidae	Lepidoptera	
Erannis	*tiliaria*	Geometridae	Lepidoptera	
Eriogaster	*lanestris*	Lasiocampidae	Lepidoptera	2,6
Estigmene	*acrea*	Arctiidae	Lepidoptera	
Euchloe	*cardamines*	Pieridae	Lepidoptera	
Euplagia	*quadripunctaria*	Arctiidae	Lepidoptera	
Euproctis	*chrysorrhoea*	Lymantriidae	Lepidoptera	
Euproctis	*pseudoconspersa*	Lymantriidae	Lepidoptera	
Euproctis	*similis*	Lymantriidae	Lepidoptera	
Euxoa	*messoria*	Noctuidae	Lepidoptera	
Euxoa	*ochrogaster*	Noctuidae	Lepidoptera	
Euxoa	*scandens*	Noctuidae	Lepidoptera	5
Galleria	*mellonella*	Pyralidae	Lepidoptera	
Gastropacha	*quercifolia*	Lasiocampidae	Lepidoptera	
Goeldichironomus	*holoprasinus*	Chironomidae	Diptera	
Gonometa	*rufibrunnea*	Lasiocampidae	Lepidoptera	3
Gonopteryx	*rhamni*	Pieridae	Lepidoptera	
Hada	*nana*	Noctuidae	Lepidoptera	
Hadena	*serena*	Noctuidae	Lepidoptera	
Heliophobus	*albicolon*	Noctuidae	Lepidoptera	
Heliothis	*armigera*	Noctuidae	Lepidoptera	5,8,11
Heliothis	*virescens*	Noctuidae	Lepidoptera	
Heliothis	*zea*	Noctuidae	Lepidoptera	11
Hepialus	*lupulinus*	Hepialidae	Lepidoptera	
Herse	*convolvuli*	Sphingidae	Lepidoptera	
Homona	*magnanima*	Tortricidae	Lepidoptera	
Hyalophora	*cecropia*	Saturniidae	Lepidoptera	
Hyles	*euphorbiae*	Sphingidae	Lepidoptera	
Hyloicus	*pinastri*	Sphingidae	Lepidoptera	2
Hyphantria	*cunea*	Arctiidae	Lepidoptera	
Hypocrita	*jacobaeae*	Arctiidae	Lepidoptera	
*Inachis	*io*	Nymphalidae	Lepidoptera	2*
Isia	*isabella*	Arctiidae	Lepidoptera	
Lacanobia	*oleracea*	Noctuidae	Lepidoptera	2

Genus	Species	Family	Order	Virus Type (where known)[1]
Lampra	*fimbriata*	Noctuidae	Lepidoptera	
Laothoe	*populi*	Sphingidae	Lepidoptera	
Lasiocampa	*quercus*	Lasiocampidae	Lepidoptera	6
Leucoma	*candida*	Lymantriidae	Lepidoptera	
Leucoma	*salicis*	Lymantriidae	Lepidoptera	
Lithophane	*leautieri*	Noctuidae	Lepidoptera	
Lobesia	*botrana*	Tortricidae	Lepidoptera	
Lophopteryx	*capucina*	Notodontidae	Lepidoptera	
Lycaena	*phlaeas*	Lycaenidae	Lepidoptera	
Lymantria	*dispar*	Lymantriidae	Lepidoptera	1, 11
Lymantria	*fumida*	Lymantriidae	Lepidoptera	
Lymantria	*mathura*	Lymantriidae	Lepidoptera	
Lymantria	*monacha*	Lymantriidae	Lepidoptera	
Malacosoma	*americanum*	Lasiocampidae	Lepidoptera	
Malacosoma	*disstria*	Lasiocampidae	Lepidoptera	8
Malacosoma	*neustria*	Lasiocampidae	Lepidoptera	2, 3
Mamestra	*brassicae*	Noctuidae	Lepidoptera	2, 7, 11, 12
Manduca	*sexta*	Sphingidae	Lepidoptera	
Mania	*maura*	Noctuidae	Lepidoptera	
Maruca	*testulalis*	Pyralidae	Lepidoptera	
Megachile	*rotundata*	Megachilidae	Hymenoptera	
Mocis	undata	Noctuidae	Lepidoptera	
Neodiprion	*merkeli*	Diprionidae	Hymenoptera	
Nepytia	*freemani*	Geometridae	Lepidoptera	
*Noctua	*pronuba*	Noctuidae	Lepidoptera	7*
Nudaurelia	*cytherea*	Saturniidae	Lepidoptera	8
Nymphalis	*antiopa*	Nymphalidae	Lepidoptera	
Oiketycus	*kirbyi*	Psychidae	Lepidoptera	
Operophtera	*bruceata*	Geometridae	Lepidoptera	
Operophtera	*brumata*	Geometridae	Lepidoptera	2, 3
Operophtera	*fugata*	Geometridae	Lepidoptera	
Oporinia	*autumnata*	Geometridae	Lepidoptera	
Oraesia	*emarginata*	Noctuidae	Lepidoptera	
Oraesia	*excavata*	Noctuidae	Lepidoptera	
Orgyia	*antiqua*	Lymantriidae	Lepidoptera	
Orgyia	*leucostigma*	Lymantriidae	Lepidoptera	
Orgyia	*pseudotsugata*	Lymantriidae	Lepidoptera	5
Orthopodomyia	*signifera*	Culicidae	Diptera	
Ourapteryx	*sambucaria*	Geometridae	Lepidoptera	
Pachysphinx	*modesta*	Sphingidae	Lepidoptera	
Paleacrita	*vernata*	Geometridae	Lepidoptera	
Panaxia	*dominula*	Arctiidae	Lepidoptera	
Papilio	*machaon*	Papilionidae	Lepidoptera	2
Paradiarsia	*glareosa*	Noctuidae	Lepidoptera	
Pararge	*aegeria*	Satyridae	Lepidoptera	
Parasemia	*plantaginis*	Arctiidae	Lepidoptera	

Genus	Species	Family	Order	Virus Type (where known)[1]
Pectinophora	*gossypiella*	Gelechiidae	Lepidoptera	11
Peridroma	*saucia*	Noctuidae	Lepidoptera	
Phalera	*bucephala*	Notodontidae	Lepidoptera	2
Phlogophora	*meticulosa*	Noctuidae	Lepidoptera	3, 8
Phragmatobia	*fuliginosa*	Arctiidae	Lepidoptera	
Pieris	*brassicae*	Pieridae	Lepidoptera	
Pieris	*rapae*	Pieridae	Lepidoptera	2, 3, 12
Platynota	*idaeusalis*	Tortricidae	Lepidoptera	
Polygonia	*c-album*	Nymphalidae	Lepidoptera	
Porthesia	*xanthocampa*	Lymantriidae	Lepidoptera	
Prosimulium	*mixtum*	Simuliidae	Diptera	
Pseudaletia	*unipuncta*	Noctuidae	Lepidoptera	11
Psorophora	*confinnis*	Culicidae	Diptera	
Psorophora	*ferox*	Culicidae	Diptera	
Pygaera	*anastomosis*	Notodontidae	Lepidoptera	
Samia	*cynthia*	Saturniidae	Lepidoptera	
Samia	*pryeri*	Saturniidae	Lepidoptera	
Samia	*ricini*	Saturniidae	Lepidoptera	
Saturnia	*pyri*	Saturniidae	Lepidoptera	
Schizura	*concinna*	Notodontidae	Lepidoptera	
Scoliopteryx	*libatrix*	Noctuidae	Lepidoptera	
Scopelodes	*contracta*	Limacodidae	Lepidoptera	
Scotogramma	*trifolii*	Noctuidae	Lepidoptera	
Selenia	*lunaria*	Geometridae	Lepidoptera	
Selidosema	*suavis*	Geometridae	Lepidoptera	
Semiothisa	*liturata*	Geometridae	Lepidoptera	
Sesamia	*calamistis*	Noctuidae	Lepidoptera	
Setora	*postornata*	Limacodidae	Lepidoptera	
Sibine	*apicalis*	Limacodidae	Lepidoptera	
Simulium	*argyreatum*	Simuliidae	Diptera	
Simulium	*aureum*	Simuliidae	Diptera	
Simulium	*tuberosum*	Simuliidae	Diptera	
Simulium	*venustum*	Simuliidae	Diptera	
Simulium	*vittatum*	Simuliidae	Diptera	
Sirex	*juvencus*	Siricidae	Hymenoptera	
Sirex	*noctilio*	Siricidae	Hymenoptera	
Smerinthus	*ocellata*	Sphingidae	Lepidoptera	
Sphinx	*ligustri*	Sphingidae	Lepidoptera	6
Spilarctia	*subcarnea*	Arctiidae	Lepidoptera	
Spilosoma	*lubricipeda*	Arctiidae	Lepidoptera	
Spilosoma	*lutea*	Arctiidae	Lepidoptera	
Spilosoma	*punctaria*	Arctiidae	Lepidoptera	
*Spodoptera	*exempta*	Noctuidae	Lepidoptera	3*, 5, 8, 11, 12*
*Spodoptera	*exigua*	Noctuidae	Lepidoptera	11*
Spodoptera	*frugiperda*	Noctuidae	Lepidoptera	
Spodoptera	*litura*	Noctuidae	Lepidoptera	
Thaumetopoea	*pityocampa*	Thaumetopoeidae	Lepidoptera	
Thaumetopoea	*processionea*	Thaumetopoeidae	Lepidoptera	

Genus	Species	Family	Order	Virus Type (where known)[1]
Thaumetopoea	*wilkinsoni*	Thaumetopoeidae	Lepidoptera	
Theophila	*mandarina*	Bombycidae	Lepidoptera	
Tinea	*pellionella*	Tineidae	Lepidoptera	
Tineola	*bisselliella*	Tineidae	Lepidoptera	
Tortrix	*viridana*	Tortricidae	Lepidoptera	
*Trichoplusia	*ni*	Noctuidae	Lepidoptera	5*
Triphaena	*pronuba*	See *Noctua pronuba*		
Uranotaenia	*sapphirina*	Culicidae	Diptera	
Urocerus	*gigas*	Siricidae	Hymenoptera	
Urocerus	*tardigradus*	Siricidae	Hymenoptera	
Vanessa	*cardui*	Nymphalidae	Lepidoptera	
Xeris	*spectrum*	Siricidae	Hymenoptera	
Xylomoges	*conspicillaris*	Noctuidae	Lepidoptera	

Compiled from Martignoni, M.E. and Iwai, P.J. (1986) A Catalog of Virus Diseases of Insects, Mites and Ticks. Fourth Edition Revised. General Technical Report PNW-195, US Department of Agriculture.

The table does not differentiate between hosts from which cytoplasmic polyhedrosis viruses were first isolated and those where infection was observed during virus cross-transmission studies.

[1]Virus "type" as recorded in a provisional classification of cytoplasmic polyhedrosis viruses based on RNA electropherotype; reported in Payne, C.C. and Mertens, P.P.C. (1983) **In** The Reoviridae. p.425. ed. Joklik, W.K. Plenum Press: New York.

*Viruses which are the type member of their electropherotype and which are described in more detail elsewhere in the dictionary.

Appendix C

A list of insect species in which Densovirus infections have been recorded

Genus	Species	Family	Order
Acheta	*domesticus*	Gryllidae	Orthoptera1
Aedes	*aegypti*	Culicidae	Diptera1
Aedes	*cantans*	Culicidae	Diptera
Aedes	*cinereus*	Culicidae	Diptera
Aedes	*geniculatus*	Culicidae	Diptera
Aedes	*vexans*	Culicidae	Diptera
Aglais	*urticae*	Nymphalidae	Lepidoptera1
Agraulis	*vanillae*	Nymphalidae	Lepidoptera1
Apis	*mellifera*	Apidae	Hymenoptera
Bombyx	*mori*	Bombycidae	Lepidoptera1
Casphalia	*extranea*	Limacodidae	Lepidoptera
Catopsilia	*pomona*	Pieridae	Lepidoptera
Culex	*pipiens*	Culicidae	Diptera
Culiseta	*annulata*	Culicidae	Diptera
Diatraea	*saccharalis*	Pyralidae	Lepidoptera1
Euxoa	*auxiliaris*	Noctuidae	Lepidoptera1
Galleria	*mellonella*	Pyralidae	Lepidoptera1
Glyphodes	pyloalis	Pyralidae	Lepidoptera1
Junonia	*coenia*	Nymphalidae	Lepidoptera1
Latoia	*viridissima*	Limacodidae	Lepidoptera
Leucorrhinia	*dubia*	Libellulidae	Odonata1
Lymantria	*dispar*	Lymantriidae	Lepidoptera
Mamestra	*brassicae*	Noctuidae	Lepidoptera
Margaronia	*pyloalis*	Pyralidae	Lepidoptera
Periplaneta	*fuliginosa*	Blattidae	Orthoptera1
Pieris	*rapae*	Pieridae	Lepidoptera1
Pseudoplusia	*includens*	Noctuidae	Lepidoptera
Sibine	*fusca*	Limacodidae	Lepidoptera1
Simulium	*vittatum*	Simuliidae	Diptera1

Compiled from Martignoni, M.E. and Iwai, P.J. (1986). A Catalog of Virus diseases of Insects, Mites and Ticks. Fourth Edition Revised. General Technical Report PNW-195 US Department of Agriculture.

The table does not differentiate between hosts from which densoviruses were first isolated and those where infection was observed during virus cross-transmission studies

[1]Viruses described in more detail by Kawase, S. (1985). **In** Virus Insecticides for Biological Control. p. 197. ed. Maramorosch, K. and Sherman, K.E. New York; Academic Press.

*Type species of the Densovirus genus; described in more detail elsewhere in the Dictionary.

Appendix D

A list of insect species in which Entomopoxvirus infections have been recorded

Genus	Species	Family	Order	Virus Subgenus[2]
Acrobasis	*zelleri*	Pyralidae	Lepidoptera	
Adoxophyes	*fasciata*	Tortricidae	Lepidoptera	
Adoxophyes	*orana*	Tortricidae	Lepidoptera	
Aedes	*aegypti*	Culicidae	Diptera1	C
Aedes	*flavescens*	Culicidae	Diptera	
Amphimallon	*solstitialis*	Scarabaeidae	Coleoptera	
*Amsacta	*moorei*	Arctiidae	Lepidoptera1	B
Anomala	*cuprea*	Scarabaeidae	Coleoptera1	A
Anopheles	*albimanus*	Culicidae	Diptera	
Anoplognathus	*porosus*	Scarabaeidae	Coleoptera	
Anoxia	*villosa*	Scarabaeidae	Coleoptera	
Antitrogus	*morbillosus*	Scarabaeidae	Coleoptera	
Aphodius	*tasmaniae*	Scarabaeidae	Coleoptera1	A
Archippus	*breviplicanus*	Tortricidae	Lepidoptera	
Archippus	*isshikii*	Tortricidae	Lepidoptera	
Archips	*fuscocupreanus*	Tortricidae	Lepidoptera	
Arphia	*conspersa*	Acrididae	Orthoptera	
Bombus	*fervidus*	Apidae	Hymenoptera	
Bombus	*impatiens*	Apidae	Hymenoptera	
Bombus	*pennsylvanicus*	Apidae	Hymenoptera	
Bombyx	*mori*	Bombycidae	Lepidoptera	
Chironomus	*attenuatus*	Chironomidae	Diptera1	C
Chironomus	*decorus*	Chironomidae	Diptera	
*Chironomus	*luridus*	Chironomidae	Diptera1	C
Chironomus	*tentans*	Chironomidae	Diptera1	C
(=Camptochironomus)				
Choristoneura	*biennis*	Tortricidae	Lepidoptera1	B
Choristoneura	*conflictana*	Tortricidae	Lepidoptera1	B
Choristoneura	*diversana*	Tortricidae	Lepidoptera1	B
Choristoneura	*fumiferana*	Tortricidae	Lepidoptera	
Culex	*pipiens*	Culicidae	Diptera	
Dasygnathus	sp.	Scarabaeidae	Coleoptera	
Demodena	*boranensis*	Scarabaeidae	Coleoptera1	A
Dermolepida	*albohirtum*	Scarabaeidae	Coleoptera1	A
Elasmopalpus	*lignosellus*	Pyralidae	Lepidoptera	
Estigmene	*acrea*	Arctiidae	Lepidoptera	

Genus	Species	Family	Order	Virus Subgenus[2]
Euxoa	*auxiliaris*	Noctuidae	Lepidoptera[1]	B
Figulus	*sublaevis*	Lucanidae	Coleoptera	A
Galleria	*mellonella*	Pyralidae	Lepidoptera	
Geotrupes	*silvaticus*	Scarabaeidae	Coleoptera	
Geotrupes	*stercorosus*	Scarabaeidae	Coleoptera[1]	A
Goeldichironomus	*holoprasinus*	Chironomidae	Diptera[1]	C
Gomphocerus	*sibiricus*	Acrididae	Orthoptera	
Homona	*magnanima*	Tortricidae	Lepidoptera	
Hoplia	sp.	Scarabaeidae	Coleoptera	
Locusta	*migratoria*	Acrididae	Orthoptera	
Lymantria	*dispar*	Lymantriidae	Lepidoptera	
Melanoplus	*bivittatus*	Acrididae	Orthoptera	
Melanoplus	*differentialis*	Acrididae	Orthoptera	
Melanoplus	*sanguinipes*	Acrididae	Orthoptera[1]	B
*Melolontha	*melolontha*	Scarabaeidae	Coleoptera[1]	A
Oncopera	*alboguttata*	Hepialidae	Lepidoptera	B
Operophtera	*brumata*	Geometridae	Lepidoptera[1]	B
Oreopsyche	*angustella*	Psychidae	Lepidoptera[1]	B
Othnonius	*batesi*	Scarabaeidae	Coleoptera[1]	A
Phoetaliotes	*nebrascensis*	Acrididae	Orthoptera	
Phyllopertha	*horticola*	Scarabaeidae	Coleoptera[1]	A
Phyllophaga	*pleei*	Scarabaeidae	Coleoptera	
Proagopertha	*lucidula*	Scarabaeidae	Coleoptera	
Pseudaletia	*separata*	Noctuidae	Lepidoptera	
Rhopaea	*verrauxi*	Scarabaeidae	Coleoptera	
Schistocerca	*americana*	Acrididae	Orthoptera	
Wiseana	*cervinata*	Hepialidae	Lepidoptera	
Wiseana	*umbraculata*	Hepialidae	Lepidoptera	
Xanthippus	*corallipes*	Acrididae	Orthoptera	
Zeiraphera	*diniana*	Tortricidae	Lepidoptera	

Compiled from Martignoni, M.E. and Iwai, P.J. (1986). A Catalog of Viral diseases of Insects, Mites, and Ticks. Fourth Edition Revised. General Technical Report PNW-195, US Department of Agriculture. The table does not differentiate between hosts from which entomopoxviruses were first isolated and those where infection was observed during virus cross-transmission studies.

[1]Viruses described in more detail in Arif, B.M. (1984). Adv. Virus Res., **29**, 195.

[2]Viruses are allocated to probable subgenera A, B and C on the basis of morphological and biochemical similarities to the type species of each subgenus.

*Viruses which are the type species for their subgenus and which are described in more detail elsewhere in the Dictionary.

Appendix E

A list of insect species in which Iridescent virus infections have been recorded

Genus	Species	Family	Order	Virus Type[1]
Acrida	*turrita*	Acrididae	Orthoptera	
Aedes	*aegypti*	Culicidae	Diptera	
Aedes	*albopictus*	Culicidae	Diptera	
Aedes	*annulipes*	Culicidae	Diptera2	5 C
Aedes	*cantans*	Culicidae	Diptera2	4 C
Aedes	*caspius*	Culicidae	Diptera2	
Aedes	*cataphylla*	Culicidae	Diptera	
Aedes	*detritus*	Culicidae	Diptera2	14, 15 C
Aedes	*dorsalis*	Culicidae	Diptera2	
Aedes	*excrucians*	Culicidae	Diptera	
Aedes	*flavescens*	Culicidae	Diptera	
Aedes	*fulvus*	Culicidae	Diptera2	
Aedes	*sierrensis*	Culicidae	Diptera	
Aedes	*sollicitans*	Culicidae	Diptera	
Aedes	*sticticus*	Culicidae	Diptera2	
Aedes	*stimulans*	Culicidae	Diptera2	11 C
Aedes	*stramineus*	Culicidae	Diptera	
*Aedes	*taeniorhynchus*	Culicidae	Diptera	3 C
Aedes	*vexans*	Culicidae	Diptera2	
Algedonia	*coclesalis*	Pyralidae	Lepidoptera	
Allomyrina	*dichotomus*	Scarabaeidae	Coleoptera	
Anadevidia	*peponis*	Noctuidae	Lepidoptera	
Anopheles	*albimanus*	Culicidae	Diptera	
Anopheles	*quadrimaculatus*	Culicidae	Diptera	
Antheraea	*eucalypti*	Saturniidae	Lepidoptera	
Antheraea	*yamamai*	Saturniidae	Lepidoptera	
Anthonomus	*grandis*	Curculionidae	Coleoptera	
Anticarsia	*gemmatalis*	Noctuidae	Lepidoptera4	
Apatele	*major*	Noctuidae	Lepidoptera	
Aphania	*geminata*	Tortricidae	Lepidoptera	
Apoda	*dentatus*	Limacodidae	Lepidoptera	
Apis	*cerana*	Apidae	Hymenoptera2	24 I
Apis	*mellifera*	Apidae	Hymenoptera	
Arcte	*coerulea*	Noctuidae	Lepidoptera	
Argyreus	*hyperbius*	Nymphalidae	Lepidoptera	
Artona	*funeralis*	Zygaenidae	Lepidoptera	
Atractomorpha	*bedeli*	Pyrgomorphidae	Orthoptera	
Bezzia	*pygmaea*	Ceratopogonidae	Diptera	
Bibio	*marci*	Bibionidae	Diptera	

Genus	Species	Family	Order	Virus Type[1]
Bombyx	*mori*	Bombycidae	Lepidoptera	
Brachmia	*macroscopa*	Gelechiidae	Lepidoptera	
Calliphora	*vomitoria*	Calliphoridae	Diptera	
Calospilos	*miranda*	Geometridae	Lepidoptera	
Carecomotis	*repulsaria*	Geometridae	Lepidoptera	
Cephonodes	*hylas*	Sphingidae	Lepidoptera	
*Chilo	*suppressalis*	Pyralidae	Lepidoptera[2]	6 I
Chironomus	*plumosus*	Chironomidae	Diptera[2]	
Choristoneura	*fumiferana*	Tortricidae	Lepidoptera	
Chrysomela	*vigintipunctata*	Chrysomelidae	Coleoptera	
Cnaphalocrocis	*medinalis*	Pyralidae	Lepidoptera	
Coccinella	*septempunctata*	Coccinellidae	Coleoptera	
Colladonus	*montanus*	Cicadellidae	Hemiptera	
Corethrella	*appendiculata*	Chaoboridae	Diptera	
Corethrella	*brakeleyi*	Chaoboridae	Diptera[2]	13 C
Costelytra	*zealandica*	Scarabaeidae	Coleoptera	16 I
Culex	*peccator*	Culicidae	Diptera	
Culex	*pipiens*	Culicidae	Diptera	
Culex	*salinarius*	Culicidae	Diptera	
Culex	*territans*	Culicidae	Diptera[2]	
Culicoides	*arboricola*	Ceratopogonidae	Diptera	
Culicoides	*clastrieri*	Ceratopogonidae	Diptera	
Culicoides	*cubitalis*	Ceratopogonidae	Diptera	
Culicoides	*odibilis*	Ceratopogonidae	Diptera	
Culicoides	sp.	Ceratopogonidae	Diptera[2]	8 C
Culiseta	*annulata*	Culicidae	Diptera[2]	
Culiseta	*inornata*	Culicidae	Diptera	
Culiseta	*melanura*	Culicidae	Diptera	
Culiseta	*morsitans*	Culicidae	Diptera	
Curculio	*dentipes*	Curculionidae	Coleoptera	
Cystidia	*stratonice*	Geometridae	Lepidoptera	
Dasychira	*pseudabietis*	Lymantriidae	Lepidoptera	
Diarsia	*canescens*	Noctuidae	Lepidoptera	
Dictyoploca	*japonica*	Saturniidae	Lepidoptera	
Elygma	*narcissus*	Noctuidae	Lepidoptera	
(Ephemeropteran)	-	Ephemeroptera[2]		26 I
Estigmene	*acrea*	Arctiidae	Lepidoptera	
Ethmia	*assamensis*	Ethmiidae	Lepidoptera	
Eucosma	*ancyrota*	Tortricidae	Lepidoptera	
Euproctis	*flava*	Lymantriidae	Lepidoptera	
Euproctis	*pseudoconspersa*	Lymantriidae	Lepidoptera	
Euproctis	*similis*	Lymantriidae	Lepidoptera	
Exorista	*sorbillans*	Tachinidae	Diptera	
Formica	*lugubris*	Formicidae	Hymenoptera[2]	
Galleria	*mellonella*	Pyralidae	Lepidoptera	
Gastropacha	*quercifolia*	Lasiocampidae	Lepidoptera	
Gonodontis	*arida*	Geometridae	Lepidoptera	

Genus	Species	Family	Order	Virus Type[1]
Graphium	*sarpedon*	Papilionidae	Lepidoptera	
Heliothis	*armigera*	Noctuidae	Lepidoptera[2]	21 I
Heliothis	*zea*	Noctuidae	Lepidoptera	
Hestina	*japonica*	Nymphalidae	Lepidoptera	
Heterolocha	*aristonaria*	Geometridae	Lepidoptera	
Heteronychus	*arator*	Scarabaeidae	Coleoptera	23 I
Homona	*coffearia*	Tortricidae	Lepidoptera	
Hydrillodes	*morosa*	Noctuidae	Lepidoptera	
Illiberis	*pruni*	Zygaenidae	Lepidoptera	
Inachis	*io*	Nymphalidae	Lepidoptera	
Iridomyrmex	*itoi*	Formicidae	Hymenoptera	
Laodelphax	*striatella*	Delphacidae	Hemiptera	
Leptinotarsa	*decemlineata*	Chrysomelidae	Coleoptera	
Lethocerus	*columbiae*	Belostomatidae	Hemiptera[2]	28 I
Lycaena	*phlaeas*	Lycaenidae	Lepidoptera	
Lymantria	*dispar*	Lymantriidae	Lepidoptera	
Macrodorcus	*rectus*	Lucanidae	Coleoptera	
Macrodorcus	*rubrofemoratus*	Lucanidae	Coleoptera	
Macroglossum	*pyrrhosticta*	Sphingidae	Lepidoptera	
Mamestra	*brassicae*	Noctuidae	Lepidoptera	
Margaronia	*pyloalis*	Pyralidae	Lepidoptera	
Microleon	*longipalpis*	Limacodidae	Lepidoptera	
Mochlonyx	*velutinus*	Chaoboridae	Diptera	
Nagia	*balteata*	Pyralidae	Lepidoptera	
Nagia	*inferior*	Pyralidae	Lepidoptera	
Nephotettix	*cincticeps*	Cicadellidae	Hemiptera	
Notarcha	*derogata*	Pyralidae	Lepidoptera	
Odagmia	*ornata*	Simuliidae	Diptera	
Odontria	*striata*	Scarabaeidae	Coleoptera[2]	19 I
Opogonia	sp.	Scarabaeidae	Coleoptera[2]	18 I
Papilio	*machaon*	Papilionidae	Lepidoptera	
Paranthrene	*pernix*	Sesiidae	Lepidoptera	
Parectopa	*geometropis*	Gracillariidae	Lepidoptera	
Parnara	*guttata*	Hesperiidae	Lepidoptera	
Phyllophaga	*anxia*	Scarabaeidae	Coleoptera[2]	
Pidorus	*glaucopis*	Zygaenidae	Lepidoptera	
Pieris	*brassicae*	Pieridae	Lepidoptera	
Pieris	*melete*	Pieridae	Lepidoptera	
Plutella	*xylostella*	Plutellidae	Lepidoptera	
Potanthus	*flavum*	Hesperiidae	Lepidoptera	
Prosimulium	sp.	Simuliidae	Diptera[2]	
Pryeria	*sinica*	Zygaenidae	Lepidoptera	
Pseudaletia	*unipuncta*	Noctuidae	Lepidoptera	
Psilogramma	*increta*	Sphingidae	Lepidoptera	
Psorophora	*confinnis*	Culicidae	Diptera	

Genus	Species	Family	Order	Virus Type[1]
Psorophora	*ferox*	Culicidae	Diptera[2]	
Psorophora	*horrida*	Culicidae	Diptera[2]	
Psorophora	*varipes*	Culicidae	Diptera[2]	
Pterostichus	*mandidus*	Carabidae	Coleoptera[2]	17 I
Pygaera	*anachoreta*	Notodontidae	Lepidoptera	
Samia	*pryeri*	Saturniidae	Lepidoptera	
Scapteriscus	*vicinus*	Gryllidae	Orthoptera[3]	
Scopula	sp.	Geometridae	Lepidoptera	
Scythris	*sinensis*	Scythrididae	Lepidoptera	
Sericesthis	*pruinosa*	Scarabaeidae	Coleoptera[2]	2 I
Simulium	*callidum*	Simuliidae	Diptera[2]	
Simulium	*earlei*	Simuliidae	Diptera[2]	
Simulium	*metallicum*	Simuliidae	Diptera	
Simulium	*neornatipes*	Simuliidae	Diptera[5]	
Simulium	*ornatum*	Simuliidae	Diptera[2]	7 C
Simulium	*rubicundulum*	Simuliidae	Diptera[2]	
Simulium	sp.	Simuliidae	Diptera[2]	22 I
Simulium	*virgatum*	Simuliidae	Diptera	
Spilarctia	*flammeolus*	Arctiidae	Lepidoptera	
Spilosoma	*lubricipeda*	Arctiidae	Lepidoptera	
Spilosoma	*punctaria*	Arctiidae	Lepidoptera	
Spodoptera	*frugiperda*	Noctuidae	Lepidoptera	
Spodoptera	*litura*	Noctuidae	Lepidoptera	
Stenodryas	*clavigera*	Cerambycidae	Coleoptera	
Sybrida	*misakiensis*	Pyralidae	Lepidoptera	
Tenebrio	*molitor*	Tenebrionidae	Coleoptera	29 I
Theretra	*nessus*	Sphingidae	Lepidoptera	
Theretra	*oldenlandiae*	Sphingidae	Lepidoptera	
Tinea	*pellionella*	Tineidae	Lepidoptera	
Tipula	*livida*	Tipulidae	Diptera	
Tipula	*oleracea*	Tipulidae	Diptera	
Tipula	*paludosa*	Tipulidae	Diptera[2]	1 I
Tipula	sp.	Tipulidae	Diptera[2]	25 I
Trachys	*auricollis*	Buprestidae	Coleoptera	
Trichoplusia	*ni*	Noctuidae	Lepidoptera	
Vanessa	*cardui*	Nymphalidae	Lepidoptera	
Wiseana	*cervinata*	Hepialidae	Lepidoptera[2]	9 I
Witlesia	*sabulosella*	Pyralidae	Lepidoptera[2]	10 I
Yponomeuta	*mayumivorellus*	Yponomeutidae	Lepidoptera	

Compiled from Martignoni, M.E. and Iwai, P.J. (1986). A catalog of virus diseases of Insects, Mites and Ticks. Fourth Edition Revised. General Technical Report PNW-195, US Department of Agriculture.

The table does not differentiate between hosts from which iridoviruses were first isolated and those where infection was observed during virus cross-transmission studies.

[1] An interim classification system for iridescent viruses assigned each new isolate a type number. Although now discontinued, the 'type' numbers already allocated have been included in the table. 'C' refers to viruses classified within the Chloriridovirus genus. 'I' refers to viruses classified within the Iridovirus genus.

[2] Viruses described in more detail in the review by Hall, D.W. (1985). **In** Viral Insecticides for Biological Control. p. 163. ed.Maramorosch, K. and Sherman, K.E. NewYork: Academic Press.

[3] Boucias, D.G. *et al.* (1987) J. Invertebr. Pathol. **50**, 238.

[4] Sieburth, P.J. and Carner, G.R. (1987) J. Invertebr. Pathol. **49**, 49.

[5] Batson, B.S. (1986) J. Invertebr. Pathol. **48**, 384.

*Viruses which are the type species for their genus and which are described in more detail elsewhere in the Dictionary.

Appendix F

Properties of Phage Isolates

The following table includes phage isolates from bacteria, cyanobacteria and mycoplasmas. While attempts have been made to make the table as comprehensive as possible, the complexity of phage nomenclature and classification has made it necessary to concentrate on those isolates that have been studied in detail and/or listed in recent attempts to classify the viruses. Defective phages and phage-like bacteriocins have been excluded. The morphotype (*see* PHAGE) and host of each isolate is recorded as well as the 'species group', genus or family to which the virus has been attributed, largely on the basis of morphology (possible 'species group' relationships are shown in parentheses).

Phage names marked with an asterisk (*) are the type species for their genus or family and/or are described in more detail elsewhere in the Dictionary.

A list of references used in the construction of the table is given at the end.

Phages are listed in alphabetic order. Phage names beginning with a Greek letter are listed within the alphabetic order according to their phonetic pronunciation (e.g. δ (delta) is listed under 'D'; and λ (lambda) under 'L'). Phage names beginning with Arabic or Roman numerals are listed in numerical order at the end.

Phage name	Morpho-type	Taxonomy 'Species group'	Family (Genus)	Host	Reference
A	B1	A	*Siphoviridae*	*Corynebacterium*	1,4,10,19
a	A1	CM₁	*Myoviridae*	*Rhizobium*	17,19
a	B2	3ML	*Siphoviridae*	*Streptococcus*	3,19
a	C1	(H39)	*Podoviridae*	*Streptococcus*	19
A-1	A1		*Myoviridae*	*Gluconobacter*	19
A-1	C1	(A-4(L))	*Podoviridae*	*Anabaena*	19
A-1(L)	A1	N-1	*Myoviridae*	*Anabaena*	1,12,19
A1-Dat	B1	A1-Dat	*Siphoviridae*	*Dactylosporangium*	1,4,10,19
A1	A1		*Siphoviridae*	*Acinetobacter*	1,10,19
A1-1	B1		*Siphoviridae*	*Azospirillum*	19
A1/A79	A1		*Myoviridae*	*Pseudomonas*	8
A-2	A1	AS-1	*Myoviridae*	*Anabaena*	1,12,19
A2	B		*Siphoviridae*	*Streptomyces*	18
A2	A1	(N-1)	*Myoviridae*	*Anabaena*	19
A2P	A1	(ViI)	*Myoviridae*	Enterobacteria	19
A-4(L)	C1	A-4(L)	*Podoviridae*	*Anabaena*	1,12,19
A5	B1	A25	*Siphoviridae*	*Streptococcus*	19
A5/A6	B1	A5/A6	*Siphoviridae*	*Alcaligenes*	8,10,19
A5/415	B1	(A5/A6)	*Siphoviridae*	*Alcaligenes*	1,8,19
A6	A1	A6	*Myoviridae*	*Alcaligenes*	1,8,10,19
A8	B2	3A	*Siphoviridae*	*Staphylococcus*	19
A9-A10	A1		*Myoviridae*	*Acinetobacter*	1,19

Phage name	Morpho-type	Taxonomy 'Species group'	Family (Genus)	Host	Reference
A10	B2	3A	*Siphoviridae*	*Staphylococcus*	19
A11-A79	A1		*Myoviridae*	*Alcaligenes*	1
A12	C1		*Podoviridae*	*Azotobacter*	1,19
A13	B1		*Siphoviridae*	*Azotobacter*	19
A13	B2	3A	*Siphoviridae*	*Staphylococcus*	19
A14	A1		*Myoviridae*	*Azotobacter*	1,19
A16	C1	(T7)	*Podoviridae*	Enterobacteria	39
A19	A1	A19	*Myoviridae*	*Brocothrix*	19
A20/415	B1	(A5/A6)	*Siphoviridae*	*Alcaligenes*	8,19
A21	C1		*Podoviridae*	*Azotobacter*	1,19
A22	C1		*Podoviridae*	*Azotobacter*	1,19
A25	B1	A25	*Siphoviridae*	*Streptococcus*	19
A31	B1		*Siphoviridae*	*Azotobacter*	1
A31	C1		*Podoviridae*	*Acinetobacter*	1,19
A41	C1		*Podoviridae*	*Azotobacter*	19
A64/A62	B1	(A5/A6)	*Siphoviridae*	*Alcaligenes*	1,8,19
A74/A3	B1	(A5/A6)	*Siphoviridae*	*Alcaligenes*	8,19
A86/A88	B1	(A5/A6)	*Siphoviridae*	*Alcaligenes*	1,8,19
A422	C1	Tb	*Podoviridae*	*Brucella*	1,13,19
A586	B1	F116	*Siphoviridae*	*Pseudomonas*	8,19
AA-1	C1	AA-1	*Podoviridae*	*Aeromonas*	5,19
AB	E1	(serogroup I)	*Leviviridae (Levivirus)*	Enterobacteria	19
Ab-1	B1		*Siphoviridae*	*Azospirillum*	19
AC-1	C1 ·	AC-1	*Podoviridae*	*Anacystis, Chroococcus*	1,12,19
AC3	B2	3A	*Siphoviridae*	*Staphylococcus*	19
ACP13	B		*Siphoviridae*	*Streptomyces*	18
AE2	F1		*Inoviridae (Inovirus)*	Enterobacteria	1,10,19
Aeh1	A2	65	*Myoviridae*	*Aeromonas*	5,19
Aeh2	A1	Aeh2	*Myoviridae*	*Aeromonas*	5,19
AG	E1	(serogroup III)	*Leviviridae (Levivirus)*	Enterobacteria	19
AG1	B1	Leo	*Siphoviridae*	*Mycobacterium*	1,4,19
AIO-1	A2	9266	*Myoviridae*	Enterobacteria	19
AN-10	A1		*Myoviridae*	*Anabaena*	34
AN-13	C1		*Podoviridae*	*Anabaena*	34
AN-23	C1		*Podoviridae*	*Anabaena*	34
AN25S-1	C1	AN25S-1	*Podoviridae*	*Arthrobacter*	4,19
AN29R-2	C1	(AN25S-1)	*Podoviridae*	*Arthrobacter*	4,19
AO-1	A2	T4	*Myoviridae*	Enterobacteria	19
AO-2	A2	T4	*Myoviridae*	Enterobacteria	19
AO-3	A2	T4	*Myoviridae*	Enterobacteria	19
AP50	D4	PRD1	*Tectiviridae (Tectivirus)*	*Bacillus*	1,10,11
AR1	A1	SP8	*Myoviridae*	*Bacillus*	1,11,19
AR2	A1	(SP50)	*Myoviridae*	*Bacillus*	1,11,19
AR3	A1	(SP50)	*Myoviridae*	*Bacillus*	1,11,19
AR9	A1	PBS1	*Myoviridae*	*Bacillus*	1,11,19

Phage name	Morpho-type	Taxonomy 'Species group'	Family (Genus)	Host	Reference
AR13	C2	φ29	*Podoviridae*	*Bacillus*	11,19
Arp	B1	Arp	*Siphoviridae*	*Arthrobacter*	4,19
AS-1	A1	AS-1	*Myoviridae*	*Anacystis, Synechococcus*	1,12,19
AS-1M	A1	AS-1	*Myoviridae*	*Anacystis, Synechococcus*	19
AT298	B1		*Siphoviridae*	*Streptococcus*	33
AU	A1	AU	*Myoviridae*	*Pasteurella*	19
Av-1	C1	Av-1	*Podoviridae*	*Actinomyces*	4,19
α	B1	α	*Siphoviridae*	*Bacillus*	1,10,11,19
α1	A2	T4	*Myoviridae*	Enterobacteria	1,10,19
α1a	B1	α3a	*Siphoviridae*	*Vibrio*	2,19
α2	B1	α3a	*Siphoviridae*	*Vibrio*	2,19
α3	B2	α3a	*Siphoviridae*	*Vibrio*	19
α3	D1		*Microviridae (Microvirus)*	Enterobacteria	1,10,19
α3a	B2	α3a	*Siphoviridae*	*Vibrio*	2,19
α3b	B1	α3a	*Siphoviridae*	*Vibrio*	2
α10	D1		*Microviridae (Microvirus)*	Enterobacteria	1,10,19
α15	E1		*Leviviridae (Levivirus)*	Enterobacteria	1,10,19
B	B1	A	*Siphoviridae*	*Corynebacterium*	1,4,19
B	B		*Siphoviridae*	*Streptomyces*	18
b	B2	3ML	*Siphoviridae*	*Streptococcus*	3,19
B1	B1	Leo	*Siphoviridae*	*Mycobacterium*	1,4,19
B₁	C2	GA-1	*Podoviridae*	*Bacillus*	11,19
B₂	C2	(GA-1)	*Podoviridae*	*Bacillus*	11,19
b2	A1	hw	*Myoviridae*	*Lactobacillus*	19
B3	B1	B33	*Siphoviridae*	*Mycobacterium*	1,8
B3	B1	(D3)	*Siphoviridae*	*Pseudomonas*	19
B6	E1	serogroup R	*Leviviridae (Levivirus)*	Enterobacteria	1,10,19
B7	E1	serogroup R	*Leviviridae (Levivirus)*	Enterobacteria	1,10,19
B₉ GP	C1		*Podoviridae*	*Acinetobacter*	1,19
B₉ PP	B2		*Siphoviridae*	*Acinetobacter*	1,19
B11	B1	β4	*Siphoviridae*	Enterobacteria	1,19
B₂₄	B2	R1-Myb	*Siphoviridae*	*Mycobacterium*	1,19
B26	B1	F116	*Siphoviridae*	*Pseudomonas*	8,19
B33	B1	(D3)	*Siphoviridae*	*Pseudomonas*	8,10,19
b ⁵⁸¹	B1	77	*Siphoviridae*	*Staphylococcus*	1,3,19
b594n	B2	3A	*Siphoviridae*	*Staphylococcus*	1,3,19
B932a	B1	(C-2)	*Siphoviridae*	*Pasteurella*	19
Bam35	D4	φNS11	*Tectiviridae (Tectivirus)*	*Bacillus*	10,11
BAOR	C1	(P22)	*Podoviridae*	Enterobacteria	1,19
B · C₃	B1	Leo	*Siphoviridae*	*Mycobacterium*	1,4,19
BE/1	D1		*Microviridae (Microvirus)*	Enterobacteria	19

Phage name	Morpho-type	Taxonomy 'Species group'	Family (Genus)	Host	Reference
Beccles	A1	Beccles	*Myoviridae*	Enterobacteria	[1,19]
Berkeley	Synonym: phage Bk				[19]
Berne 6/29	C2	Berne 6/29	*Podoviridae*	*Vibrio*	[2,19]
BF23	B1	T5	*Siphoviridae*	Enterobacteria	[20]
BG2	B1	(lacticola)	*Siphoviridae*	*Mycobacterium*	[4,19]
Bir	B1	Bir	*Siphoviridae*	*Bifidobacterium*	[4,19]
Bk	C1	Tb	*Podoviridae*	*Brucella*	[13,19]
BL3	B1	BL3	*Siphoviridae*	*Brocothrix*	[19]
BLE	B3	BLE	*Siphoviridae*	*Bacillus*	[10,11,19]
BM$_{29}$	C1	Tb	*Podoviridae*	*Brucella*	[13,19]
BN1	C1	MV-L3	*Podoviridae*	*Acholeplasma*	[9,19]
BO1	E1	(serogroup I)	*Leviviridae* (*Levivirus*)	Enterobacteria	[19]
Bo1	B2	R1-Myb	*Siphoviridae*	*Mycobacterium*	[1,4,19]
BP1	C1		*Podoviridae*	*Acinetobacter*	[1,19]
BPP-10	C2	GA-1	*Podoviridae*	*Bacillus*	[11,19]
BPV-G	C1	I	*Podoviridae*	*Vibrio*	[2,19]
BPV-J	C1	I	*Podoviridae*	*Vibrio*	[2,19]
BPV-K	C1	I	*Podoviridae*	*Vibrio*	[2,19]
BPV-P	C1	I	*Podoviridae*	*Vibrio*	[2,19]
BPV-Q	C1	I	*Podoviridae*	*Vibrio*	[2,19]
BPV-S	C1	I	*Podoviridae*	*Vibrio*	[2,19]
BPV-T	C1	I	*Podoviridae*	*Vibrio*	[2,19]
Br1	A1	A1	*Myoviridae*	*Mycoplasma*	[19]
BR2	D1	φX174	*Microviridae* (*Microvirus*)	Enterobacteria	[14]
BT11	B1		*Siphoviridae*	*Streptococcus*	[33]
BZ13	E1	(serogroup II)	*Leviviridae* (*Levivirus*)	Enterobacteria	[19]
β	B1	β	*Siphoviridae*	*Corynebacterium*	[1,4,19]
β	A1	kappa	*Myoviridae*	*Vibrio*	[2,19]
β	E1	(serogroup I)	*Leviviridae* (*Levivirus*)	Enterobacteria	[1,7,10,19]
β4	B1	β4	*Siphoviridae*	Enterobacteria	[1,10,19]
β22	A1	SP50	*Myoviridae*	*Bacillus*	[11,19]
C	B1	A	*Siphoviridae*	*Corynebacterium*	[1,4,19]
C	B1	IV	*Siphoviridae*	*Vibrio*	[2,19]
c	A1	CM$_1$	*Myoviridae*	*Rhizobium*	[1,19]
C-1	E1		*Leviviridae* (*Levivirus*)	Enterobacteria	[19]
C1	C1	C1	*Podoviridae*	*Micrococcus*	[1,10]
C1	C1	C$_{vir}$	*Podoviridae*	*Streptococcus*	[19]
C1	B1	(T5)	*Siphoviridae*	Enterobacteria	[1,19]
c1	B2	c2	*Siphoviridae*	*Streptococcus*	[3,19]
C-2	B1	C-2	*Siphoviridae*	*Pasteurella*	[19]
C-2	F1		*Inoviridae* (*Inovirus*)	Enterobacteria	[19]
C2	E1		*Leviviridae* (*Levivirus*)	Enterobacteria	[19]
C2	A1		*Myoviridae*	*Flexibacter*	[1,19]

Phage name	Morpho-type	Taxonomy 'Species group'	Family (Genus)	Host	Reference
C2	B1	Leo	*Siphoviridae*	*Mycobacterium*	1,4,19
C2	C3	7-11	*Podoviridae*	Enterobacteria	1,19
c2	B2	c2	*Siphoviridae*	*Streptococcus*	1,3,19
C2F	C3	7-11	*Podoviridae*	Enterobacteria	19
C3	B1	(χ)	*Siphoviridae*	Enterobacteria	19
C3	C1	C3	*Podoviridae*	*Spiroplasma*	19
c3	B1	(936)	*Siphoviridae*	*Streptococcus*	3,19
C4	A2	T4	*Myoviridae*	Enterobacteria	1,19
C-5	C1		*Podoviridae*	*Lactobacillus*	1
C5	C1		*Podoviridae*	*Pseudomonas*	1,8
c6A	B		*Siphoviridae*	*Streptococcus*	41
c10I	B2	3ML	*Siphoviridae*	*Streptococcus*	1,3,19
c10II	B2	3ML	*Siphoviridae*	*Streptococcus*	1,3,19
c10III	A1	(RZh)	*Myoviridae*	*Streptococcus*	1,3,19
c11	B1	(936)	*Siphoviridae*	*Streptococcus*	3,19
c13	B1	(936)	*Siphoviridae*	*Streptococcus*	3,19
C16	A2	T4	*Myoviridae*	Enterobacteria	1,10,19
c20-1	B1	24	*Siphoviridae*	*Streptococcus*	3,19
C22	B1	(SD1)	*Siphoviridae*	*Pseudomonas*	1,8,19
C31	Synonym: phage φC31				19
C131	B		*Siphoviridae*	*Streptomyces*	18
C163	A1	PS17	*Myoviridae*	*Pseudomonas*	8,19
C557	B1	β4	*Siphoviridae*	Enterobacteria	1,19
CA1	B1	A25	*Siphoviridae*	*Streptococcus*	19
CA1	C1	HM2	*Podoviridae*	*Clostridium*	19
CB3	A1	(PB-1)	*Myoviridae*	*Pseudomonas*	8,19
CC	E1	(serogroup I)	*Leviviridae* (*Levivirus*)	Enterobacteria	19
CEβ	A1	CEβ	*Myoviridae*	*Clostridium*	19
CF	E1	(serogroup III)	*Leviviridae* (*Levivirus*)	Enterobacteria	7,19
Cf	F1		*Inoviridae* (*Inovirus*)	*Xanthomonas*	10,19
chZ	B1	(y5)	*Siphoviridae*	*Lactobacillus*	19
CL31	B		*Siphoviridae*	*Corynebacterium*	23
CK235	C1	(T7)	*Siphoviridae*	Enterobacteria	40
CM$_1$	A1	CM$_1$	*Myoviridae*	*Rhizobium*	1,19
CM$_2$	A1		*Myoviridae*	*Rhizobium*	1
cmf1	B1		*Siphoviridae*	*Flavobacterium*	19
c-n71	A1	CEβ	*Myoviridae*	*Clostridium*	19
CN8	Synonym: phage CON8				19
CN11	Synonym: phage CON11				19
CON8	B1	CONX	*Siphoviridae*	*Corynebacterium*	19
CON11	B1	CONX	*Siphoviridae*	*Corynebacterium*	19
CON77	B1	CONX	*Siphoviridae*	*Corynebacterium*	19
CONRH	B1	CONX	*Siphoviridae*	*Corynebacterium*	19
CONX	B1	CONX	*Siphoviridae*	*Corynebacterium*	19
CP-1	C2	CP-1	*Podoviridae*	*Streptococcus*	3,19
CP1	A1	PS17	*Myoviridae*	*Pseudomonas*	8,10,19
CP-7	C2	CP-1	*Podoviridae*	*Streptococcus*	3,19
CP10	C2	CP-1	*Podoviridae*	*Streptococcus*	3,19

Phage name	Morpho-type	Taxonomy 'Species group'	Family (Genus)	Host	Reference
CP51	A1	SP8	*Myoviridae*	*Bacillus*	[1,11,19]
CP-T1	A1	II	*Myoviridae*	*Vibrio*	[2,19]
CS$_1$	B1	(SPP1)	*Siphoviridae*	*Bacillus*	[19]
Cslx13a	B1		*Siphoviridae*	*Caryophanon*	[1,19]
Cslx13b	A1		*Myoviridae*	*Caryophanon*	[1,19]
CT1	A1		*Myoviridae*	*Rhizobium*	[1]
CT2	A1		*Myoviridae*	*Rhizobium*	[1]
CT3	A1	(m)	*Myoviridae*	*Rhizobium*	[1,17,19]
CT4	A1	CT4	*Myoviridae*	*Rhizobium*	[1,10,17,19]
CT5	A1		*Myoviridae*	*Rhizobium*	[1]
CT6	A1		*Myoviridae*	*Rhizobium*	[1]
Ct^{kas}	B1		*Siphoviridae*	*Caryophanon*	[1,19]
C$_{vir}$	C1	C$_{vir}$	*Podoviridae*	*Streptococcus*	[19]
χ	B1	χ	*Siphoviridae*	Enterobacteria	[1,10,19]
D		Synonym: phage P (from *Brucella*)			[19]
D	B1	A	*Siphoviridae*	*Corynebacterium*	[1,4,19]
D	B2	R1-Myb	*Siphoviridae*	*Mycobacterium*	[1,4,19]
D	B2	3A	*Siphoviridae*	*Staphylococcus*	[3,19]
D	C1		*Podoviridae*	*Pseudomonas*	[8]
d	B1		*Siphoviridae*	*Rhizobium*	[1]
D$_1$	C1	182	*Podoviridae*	*Streptococcus*	[3,19]
D3	B1	D3	*Siphoviridae*	*Pseudomonas*	[1,8,10,19]
D-5	A1	SP50	*Myoviridae*	*Bacillus*	[11,19]
D20	B1	(T1)	*Siphoviridae*	Enterobacteria	[43]
D108	A1	Mu	*Myoviridae*	Enterobacteria	[1,19]
D3112	B1	(SD1)	*Siphoviridae*	*Pseudomonas*	[8,19]
*dar*1	A1	SP50	*Myoviridae*	*Bacillus*	[11]
DdVi	A2	T4	*Myoviridae*	Enterobacteria	[1,10,19]
DdVII	B1	β	*Siphoviridae*	Enterobacteria	[1,19]
*den*1	A1	(SP50)	*Myoviridae*	*Bacillus*	[11,19]
DF$_{78}$	A1	(RZh)	*Myoviridae*	*Streptococcus*	[3,19]
DM-11	A1	(P2)	*Myoviridae*	Enterobacteria	[1]
DM-31	C3	7-11	*Podoviridae*	Enterobacteria	[1,19]
DM-51	A2	T4	*Myoviridae*	Enterobacteria	[19]
dφ3	D1		*Microviridae* (*Microvirus*)	Enterobacteria	[1,10,19]
dφ4	D1		*Microviridae* (*Microvirus*)	Enterobacteria	[1,10,19]
dφ5	D1		*Microviridae* (*Microvirus*)	Enterobacteria	[19]
drc1	B2	c2	*Siphoviridae*	*Streptococcus*	[3,19]
drc3	B2	3ML	*Siphoviridae*	*Streptococcus*	[1]
DSP$_1$	B1	Leo	*Siphoviridae*	*Mycobacterium*	[1,4,19]
Dundee	A1	01	*Myoviridae*	Enterobacteria	[19]
δ	B1	β	*Siphoviridae*	*Corynebacterium*	[1,4,19]
δ1	D1		*Microviridae* (*Microvirus*)	Enterobacteria	[1,10,19]
δA	F1		*Inoviridae* (*Inovirus*)	Enterobacteria	[1,10,19]
δ^H	C1	(T7)	*Podoviridae*	Enterobacteria	[1,19]

Phage name	Morpho-type	Taxonomy 'Species group'	Family (Genus)	Host	Reference
E	B1	A	*Siphoviridae*	*Corynebacterium*	4,19
E	A1	(m)	*Myoviridae*	*Rhizobium*	17,19
e	A1	CM_1	*Myoviridae*	*Rhizobium*	1,10,17,19
e	B2	3ML	*Siphoviridae*	*Streptococcus*	3,19
E1	A1	(01)	*Myoviridae*	Enterobacteria	1,19
e1	C1	III	*Podoviridae*	*Vibrio*	2,19
e2	C1	III	*Podoviridae*	*Vibrio*	2,19
E3	C3	7-11	*Podoviridae*	Enterobacteria	19
e3	C1	I	*Podoviridae*	*Vibrio*	2,19
e4	C1	I	*Podoviridae*	*Vibrio*	2,19
e5	C1	III	*Podoviridae*	*Vibrio*	2,19
E15P	A2	(01)	*Myoviridae*	Enterobacteria	19
E16B	A3	16-19	*Myoviridae*	Enterobacteria	1,19
E16P	C3	7-11	*Podoviridae*	Enterobacteria	1,19
E62	C3	7-11	*Podoviridae*	Enterobacteria	19
E79	A1	(PB-1)	*Myoviridae*	*Pseudomonas*	1,8,19
eb4	B1	(936)	*Siphoviridae*	*Streptococcus*	3,19
eb7	B2	3ML	*Siphoviridae*	*Streptococcus*	3,19
Ec9	F1		*Inoviridae* (*Inovirus*)	Enterobacteria	1,10,19
EI	E1	(serogroup II)	*Leviviridae* (*Levivirus*)	Enterobacteria	7,19
Erh1	C3	7-11	*Podoviridae*	Enterobacteria	19
Esc-7-11	C3	Esc-7-11	*Podoviridae*	Enterobacteria	19
ET25	A1	kappa	*Myoviridae*	*Vibrio*	2,19
ET42	B1		*Siphoviridae*	*Streptococcus*	33
EV	B1		*Siphoviridae*	*Ancalomicrobium*	19
EV	C1	(T7)	*Podoviridae*	Enterobacteria	1,19
ε15	C1	(N4)	*Podoviridae*	Enterobacteria	1,19
ε34	C1	(P22)	*Podoviridae*	Enterobacteria	1,19
η8	D1		*Microviridae* (*Microvirus*)	Enterobacteria	1,10,19
F	B1	A	*Siphoviridae*	*Corynebacterium*	4,19
F	A2	(44RR2.8t)	*Myoviridae*	*Aeromonas*	5,19
F-1	B1	1b6	*Siphoviridae*	*Lactobacillus*	19
F1	B2	F1	*Siphoviridae*	*Clostridium*	1,10,19
F_1	B1	T5	*Siphoviridae*	Enterobacteria	1,19
F_1	C1	Tb	*Podoviridae*	*Brucella*	1,13,19
F_1	A1	(RZh)	*Myoviridae*	*Streptococcus*	3,19
f1	F1		*Inoviridae* (*Inovirus*)	Enterobacteria	1,10,19
F_{1m}	C1	Tb	*Podoviridae*	*Brucella*	13,19
F_1U	C1	Tb	*Podoviridae*	*Brucella*	13,19
F_2	C1	Tb	*Podoviridae*	*Brucella*	1,13,19
F_2	A1	(RZh)	*Myoviridae*	*Streptococcus*	3,19
f2	E1	serogroup I	*Leviviridae* (*Levivirus*)	Enterobacteria	1,7,10,19
F_3	C1	Tb	*Podoviridae*	*Brucella*	13,19
F_4	C1	Tb	*Podoviridae*	*Brucella*	13,19

Phage name	Morpho-type	Taxonomy 'Species group'	Family (Genus)	Host	Reference
f4	E1	(serogroup I)	*Leviviridae (Levivirus)*	Enterobacteria	7,19
F4/L425	B2		*Siphoviridae*	*Rhizobium*	19
F4/L425I	B2	φ2037/1	*Siphoviridae*	*Rhizobium*	17,19
F5	B2	φ2037/1	*Siphoviridae*	*Rhizobium*	1,17,19
F$_5$	C1	Tb	*Podoviridae*	*Brucella*	13,19
F5/L422	B2	φ2037/1	*Siphoviridae*	*Rhizobium*	17,19
F6	A1	(P2)	*Myoviridae*	Enterobacteria	1,19
F$_6$	C1	Tb	*Podoviridae*	*Brucella*	13,19
F7	B		*Siphoviridae*	*Pseudomonas*	21
F$_7$	C1	Tb	*Podoviridae*	*Brucella*	13,19
F8	A1	PB-1	*Myoviridae*	*Pseudomonas*	21
F9	C1		*Podoviridae*	*Rhizobium*	1
F10	B1	D3	*Siphoviridae*	*Pseudomonas*	21
F11	B1	β4	Siphoviridae	Enterobacteria	1,19
F13	A2	T4	*Myoviridae*	Enterobacteria	1,19
F$_{25}$	C1	Tb	*Podoviridae*	*Brucella*	13,19
F$_{25}$U	C1	Tb	*Podoviridae*	*Brucella*	13,19
F$_{44}$	C1	Tb	*Podoviridae*	*Brucella*	13,19
F$_{45}$	C1	Tb	*Podoviridae*	*Brucella*	13,19
F46/3039	C3	7-11	*Podoviridae*	Enterobacteria	1,19
F$_{48}$	C1	Tb	*Podoviridae*	*Brucella*	13,19
F116	B1	F116	*Siphoviridae*	*Pseudomonas*	8,10,19
F119	A1		*Myoviridae*	*Pseudomonas*	8
FC3-1	A2	T4	*Myoviridae*	Enterobacteria	19
FC3-3	A2	T4	*Myoviridae*	Enterobacteria	19
FC3-4	A2	T4	*Myoviridae*	Enterobacteria	19
FC3-6	A2	T4	*Myoviridae*	Enterobacteria	19
FC3-7	A2	T4	*Myoviridae*	Enterobacteria	19
FC3-9	A1	FC3-9	*Myoviridae*	Enterobacteria	19
fcan1	E1	(serogroup I)	*Leviviridae (Levivirus)*	Enterobacteria	1,10,19
fd*	F1	fd	*Inoviridae (Inovirus)*	Enterobacteria	1,10,19
FF4	B		*Siphoviridae*	*Streptomyces*	18
FH5	E1	serogroup I	*Leviviridae*	Enterobacteria	1,7,10,19
FIA	A1	(CT4)	*Myoviridae*	*Rhizobium*	17,19
FI	E1	(serogroup IV)	*Leviviridae (Levivirus)*	Enterobacteria	7,19
Fi		Synonym: phage Fz			19
FI1	A2	T4	*Myoviridae*	Enterobacteria	19
FI-5	A1	(01)	*Myoviridae*	Enterobacteria	19
FIA	A1	(CT4)	*Myoviridae*	*Rhizobium*	17,19
Firenze 75/13		Synonym: phage Fz			19
F-K	B2	R1-Myb	*Siphoviridae*	*Mycobacterium*	1,4,19
Fk	C		*Podoviridae*	*Vibrio*	2
FM	E1	(serogroup II)	*Leviviridae (Levivirus)*	Enterobacteria	19
FO$_1$	C1	Tb	*Podoviridae*	*Brucella*	13
F$_o$*lac*	E1		*Leviviridae (Levivirus)*	Enterobacteria	19

Phage name	Morpho-type	Taxonomy 'Species group'	Family (Genus)	Host	Reference
FoS₁	A1	(SP8)	*Myoviridae*	*Bacillus*	[19]
FOS₂	A1	(SP50)	*Myoviridae*	*Bacillus*	[19]
FP₁	A1	(SP50)	*Myoviridae*	*Bacillus*	[19]
FP₂	C1	Tb	*Podoviridae*	*Brucella*	[13,19]
FP4	B		*Podoviridae*	*Streptomyces*	[18]
fr	E1	serogroup I	*Leviviridae* (*Levivirus*)	Enterobacteria	[1,7,10,19]
FRI	E1	(serogroup I)	*Leviviridae* (*Levivirus*)	Enterobacteria	[19]
fri	A1	fri	*Myoviridae*	*Lactobacillus*	[19]
F-S	B2	R1-Myb	*Siphoviridae*	*Mycobacterium*	[1,4,19]
FS1	C1	(T7)	*Podoviridae*	Enterobacteria	[1,19]
FS₁	B1	(SPP1)	*Siphoviridae*	*Bacillus*	[19]
FS2	B1	β4	*Siphoviridae*	Enterobacteria	[1,19]
FS₂	B1	(SPP1)	*Siphoviridae*	*Bacillus*	[19]
FS₃	B1	(SPP1)	*Siphoviridae*	*Bacillus*	[19]
FS₄	A1	(SP50)	*Myoviridae*	*Bacillus*	[19]
FS₅	B1	(SPP1)	*Siphoviridae*	*Bacillus*	[19]
FS7	A2	T4	*Myoviridae*	Enterobacteria	[1,19]
FS₈	B1	(SPP1)	*Siphoviridae*	*Bacillus*	[19]
FS₉	B1	(SPP1)	*Siphoviridae*	*Bacillus*	[19]
FS10	A2	T4	*Myoviridae*	Enterobacteria	[1,19]
FSa	A1	(P2)	*Myoviridae*	Enterobacteria	[1]
Fsα	A2	T4	*Myoviridae*	Enterobacteria	[1,19]
FSD2a	A2	T4	*Myoviridae*	Enterobacteria	[1,19]
FSD2e	A1	(01)	*Myoviridae*	Enterobacteria	[1,19]
FSD8	A2	T4	*Myoviridae*	Enterobacteria	[1,19]
FYc	A1	222a	*Myoviridae*	*Lactobacillus*	[19]
Fz	C1	Tb	*Podoviridae*	*Brucella*	[13,19]
Fz75/13	Synonym: phage Fz				[19]
G	B1	A	*Siphoviridae*	*Corynebacterium*	[4,19]
G	A1	G	*Myoviridae*	*Bacillus*	[1,10,11,19]
G	B1	PL-1	*Siphoviridae*	*Lactobacillus*	[19]
GIII	Synonym: phage LPP-1G				[19]
G4	A1		*Myoviridae*	*Acinetobacter*	[1]
G4	C1	(P22)	*Podoviridae*	Enterobacteria	[19]
G4	D1		*Microviridae* (*Microvirus*)	Enterobacteria	[1,10,19]
G5	B1	(T5)	*Siphoviridae*	Enterobacteria	[19]
G6	D1	φX 174	*Microviridae* (*Microvirus*)	Enterobacteria	[1,10,19]
g8	C2	GA-1	*Podoviridae*	*Bacillus*	[11,19]
G₁₀	B1	PL-1	*Siphoviridae*	*Lactobacillus*	[19]
G13	D1	φX 174	*Microviridae* (*Microvirus*)	Enterobacteria	[1,10,19]
G14	D1		*Microviridae*	Enterobacteria	[19]
g18	B2	*mor*1	*Siphoviridae*	*Bacillus*	[11,19]
G101	B2	M6	*Siphoviridae*	*Pseudomonas*	[1,8,19]
GA	E1	(serogroup II)	*Leviviridae* (*Levivirus*)	Enterobacteria	[7,19]

Phage name	Morpho-type	Taxonomy 'Species group'	Family (Genus)	Host	Reference
ga	A1		*Myoviridae*	*Pseudomonas*	8
GA-1	C2	GA-1	*Podoviridae*	*Bacillus*	1,10,11,19
*gal*1	A1	SP50	*Myoviridae*	*Bacillus*	11,19
gb	A1		*Myoviridae*	*Pseudomonas*	8
gb-4	C1		*Podoviridae*	*Methylosinus*	19
gc	B1		*Siphoviridae*	*Pseudomonas*	8
gd	B1		*Siphoviridae*	*Pseudomonas*	8
ge	B1		*Siphoviridae*	*Pseudomonas*	8
gf	B1		*Siphoviridae*	*Pseudomonas*	8
GF-2	A1	(SP3)	*Myoviridae*	*Bacillus*	1,11,19
gh-1	C1	gh-1	*Podoviridae*	*Pseudomonas*	1,10,19
GM	E1	(serogroup III)	*Leviviridae* (*Levivirus*)	Enterobacteria	19
GM	C1		*Podoviridae*	*Plectonema*	1
GP1	B2	PT11	*Siphoviridae*	*Agrobacterium*	17,19
GR	E1		*Leviviridae* (*Levivirus*)	Enterobacteria	7
Group 1	A1		*Myoviridae*	*Neisseria*	1
Group 2	A1		*Myoviridae*	*Neisseria*	1
GS$_1$	A1	(SP8)	*Myoviridae*	*Bacillus*	19
GS2	A1	PIIBNV6	*Myoviridae*	*Agrobacterium*	19
GS4E	B1	Leo	*Siphoviridae*	*Mycobacterium*	1,14,19
GS6	A1	PIIBNV6	*Myoviridae*	*Agrobacterium*	19
GS$_7$	B1	Leo	*Siphoviridae*	*Mycobacterium*	1,4,19
GT-6	A1	(SP50)	*Myoviridae*	*Bacillus*	1,11,19
GT-234	B1	A25	*Siphoviridae*	*Streptococcus*	19
GV-3	C2	GA-1	*Podoviridae*	*Bacillus*	1,11,19
GV-5	C2	GA-1	*Podoviridae*	*Bacillus*	1,11,19
GV-6	A1	(SP50)	*Myoviridae*	*Bacillus*	1,11,19
GW6210	A1		*Myoviridae*	*Gluconobacter*	19
γ	B1	β	*Siphoviridae*	*Corynebacterium*	4,19
γ	A1	SP50	*Myoviridae*	*Bacillus*	11,19
γ2	B1	(β4)	*Siphoviridae*	Enterobacteria	1,19
H	A2	T4	*Myoviridae*	Enterobacteria	19
H	C1	T7	*Podoviridae*	Enterobacteria	1,10,19
H	B1	A	*Siphoviridae*	*Corynebacterium*	4,19
h1	B2	3ML	*Siphoviridae*	*Streptococcus*	3,19
H8/73	B1	2389	*Siphoviridae*	*Listeria*	19
H10G	B1	H387	*Siphoviridae*	*Listeria*	19
H-19J	B3	H-19J	*Siphoviridae*	Enterobacteria	19
H19	B1	H387	*Siphoviridae*	*Listeria*	19
H20	A1	(A6)	*Myoviridae*	*Alcaligenes*	1,8,19
H21	B1	H387	*Siphoviridae*	*Listeria*	19
H39	C1	H39	*Podoviridae*	*Streptococcus*	19
H43	B1	2671	*Siphoviridae*	*Listeria*	19
H46	B1	H387	*Siphoviridae*	*Listeria*	19
H107	B1	2389	*Siphoviridae*	*Listeria*	19
H108	B1	2389	*Siphoviridae*	*Listeria*	19
H110	B1	2671	*Siphoviridae*	*Listeria*	19

Phage name	Morpho-type	Taxonomy 'Species group'	Family (Genus)	Host	Reference
H163/74	B1	2389	*Siphoviridae*	*Listeria*	19
H312	B1	2671	*Siphoviridae*	*Listeria*	19
H340	B1	2389	*Siphoviridae*	*Listeria*	19
H391/73	B1	2389	*Siphoviridae*	*Listeria*	19
H648/74	B1	2389	*Siphoviridae*	*Listeria*	19
H924A	B1	2685	*Siphoviridae*	*Listeria*	19
Hanoi III	A1	kappa	*Myoviridae*	*Vibrio*	2,19
hb	A1	hw	*Myoviridae*	*Lactobacillus*	19
HDC-1	C or D2		Unclassified	*Bdellovibrio*	1,19
HDC-2	A1		*Myoviridae*	*Bdellovibrio*	1
HF	C1		*Podoviridae*	*Pseudomonas*	1
Hh-1	B1		*Siphoviridae*	*Halobacterium*	19
Hh-3	B1		*Siphoviridae*	*Halobacterium*	19
HI	E1	(serogroup III)	*Leviviridae* (*Levivirus*)	Enterobacteria	7,19
Hi	B1	(T1)	*Siphoviridae*	Enterobacteria	43
HK022	B1	λ	*Siphoviridae*	Enterobacteria	19
HK2	B2	3A	*Siphoviridae*	*Staphylococcus*	3,19
HK97	B1	λ	*Siphoviridae*	Enterobacteria	19
HK122	A1	(P2)	*Myoviridae*	Enterobacteria	42
HK124	A1	(P2)	*Myoviridae*	Enterobacteria	42
HK162	A1	(P2)	*Myoviridae*	Enterobacteria	42
HK166	A1	(P2)	*Myoviridae*	Enterobacteria	42
HK167	A1	(P2)	*Myoviridae*	Enterobacteria	42
HK175	B1	(λ)	*Siphoviridae*	Enterobacteria	38
HK180	A1	(P2)	*Myoviridae*	Enterobacteria	42
HK183	B1	(λ)	*Siphoviridae*	Enterobacteria	38
HK187	B1	(λ)	*Siphoviridae*	Enterobacteria	38
HK189	B1	(λ)	*Siphoviridae*	Enterobacteria	38
HK239	A1	(P2)	*Myoviridae*	Enterobacteria	42
HK243	B2	ZG/3A	*Siphoviridae*	Enterobacteria	19
HK250	B1	(λ)	*Siphoviridae*	Enterobacteria	38
HK253	B1	(λ)	*Siphoviridae*	Enterobacteria	26
HK321	A1	(P2)	*Myoviridae*	Enterobacteria	42
HK332	B1	(λ)	*Siphoviridae*	Enterobacteria	38
HM2	C1	HM2	*Podoviridae*	*Clostridium*	1,10,19
HM3	A1	HM3	*Myoviridae*	*Clostridium*	1,10,19
HM7	B1	HM7	*Siphoviridae*	*Clostridium*	1,10,19
HMT	C1	HM2	*Podoviridae*	*Clostridium*	1,19
HP	B2	R1-Myb	*Siphoviridae*	*Mycobacterium*	1,4,19
HP1	A1		*Myoviridae*	*Haemophilus*	1,19
HP1	A2		*Myoviridae*	*Acinetobacter*	1,19
HP2 to HP4	A1		*Myoviridae*	*Acinetobacter*	1,19
HR	F1		*Inoviridae* (*Inovirus*)	Enterobacteria	1,10,19
H-Sh	A1	(P2)	*Myoviridae*	Enterobacteria	1
HSO47	B1	2389	*Siphoviridae*	*Listeria*	19
HT-2	A1		*Myoviridaae*	*Thiobacillus*	1,19
hv	A1	hv	*Myoviridae*	*Lactobacillus*	19
hv-1	B1	hv-1	*Siphoviridae*	*Vibrio*	2,19

Phage name	Morpho-type	Taxonomy 'Species group'	Family (Genus)	Host	Reference
hw	A1	hw	*Myoviridae*	*Lactobacillus*	19
hw1	A1	hw	*Myoviridae*	*Lactobacillus*	19
HXX	B1	HXX	*Siphoviridae*	*Xanthomonas*	19
Hyφ1a	C1		*Podoviridae*	*Hyphomicrobium*	19
Hyφ22a	B1		*Siphoviridae*	*Hyphomicrobium*	19
Hyφ30	C1		*Myoviridae*	*Hyphomicrobium*	1,19
Hyφ32a	B1		*Siphoviridae*	*Hyphomicrobium*	1,19
I	A1	CM$_1$	*Myoviridae*	*Rhizobium*	19
I	C1	Tb	*Podoviridae*	*Brucella*	1,13,19
I-1	C		*Podoviridae*	*Clostridium*	1
I1	A1	(I3)	*Myoviridae*	*Mycobacterium*	37
I$_1$/H$_{21}$	B1	A	*Siphoviridae*	*Corynebacterium*	4,19
I$_1$/H$_{36}$	B1	A	*Siphoviridae*	*Corynebacterium*	4,19
I-2	A2		*Myoviridae*	*Rhodopseudomonas*	1,19
I2	A2	T4	*Myoviridae*	Enterobacteria	1,19
I$_2$-2	F1		*Inoviridae* (*Inovirus*)	Enterobacteria	19
I3	A1	I3	*Myoviridae*	*Mycobacterium*	4,10,19
I5	A1	(I3)	*Myoviridae*	*Mycobacterium*	37
I8	?		?	*Mycobacterium*	37
I9	A1	SP8	*Myoviridae*	*Bacillus*	1,11,19
I10	A1	PBS1	*Myoviridae*	*Bacillus*	1,11,19
ID2	E1	(serogroup IV)	*Leviviridae* (*Levivirus*)	Enterobacteria	19
If1	F1		*Inoviridae* (*Inovirus*)	Enterobacteria	1,10,19
If2	F1		*Inoviridae* (*Inovirus*)	Enterobacteria	1,10,19
IKe	F1		*Inoviridae* (*Inovirus*)	Enterobacteria	1,10,19
Im	C1	Tb	*Podoviridae*	*Brucella*	19
IPy-1	B1	IPy-1	*Siphoviridae*	*Bacillus*	19
IS$_1$	A1	(PBS1)	*Myoviridae*	*Bacillus*	19
Iα	E1		*Leviviridae* (*Levivirus*)	Enterobacteria	19
J	C1	7480b	*Podoviridae*	Enterobacteria	19
J	A1	(CM1)	*Myoviridae*	*Rhizobium*	17,19
J	B1	A	*Siphoviridae*	*Corynebacterium*	4,19
j	A1	(m)	*Myoviridae*	*Rhizobium*	17,19
J1	B1	PL-1	*Siphoviridae*	*Lactobacillus*	19
Ja.1	A1		*Myoviridae*	*Halobacterium*	19
Jersey	B1	Jersey	*Siphoviridae*	Enterobacteria	1,10,19
JP34	E1	(serogroup II)	*Leviviridae* (*Levivirus*)	Enterobacteria	19
JP500	E1	(serogroup II)	*Leviviridae* (*Levivirus*)	Enterobacteria	19
JP501	E1	(serogroup I)	*Leviviridae* (*Levivirus*)	Enterobacteria	19

Phage name	Morpho-type	Taxonomy 'Species group'	Family (Genus)	Host	Reference
JW2040	C1		*Podoviridae*	*Gluconobacter* [19]	
K	A1	Twort	*Myoviridae*	*Staphylococcus*	[3,19]
K	A1	(01)	*Myoviridae*	Enterobacteria	[1,19]
K	B1	β	*Siphoviridae*	*Corynebacterium*	[1,4,19]
K	B1	A	*Siphoviridae*	*Corynebacterium*	[4,19]
K	B2	F1	*Siphoviridae*	*Clostridium*	[1,19]
K₁	A1		*Myoviridae*	*Pseudomonas*	[1]
K3	A2	T4	*Myoviridae*	Enterobacteria	[19]
K₅	C1		*Podoviridae*	*Pseudomonas*	[1]
K₆	C1		*Podoviridae*	*Pseudomonas*	[1]
K11	C1	(T7)	*Podoviridae*	Enterobacteria	[39]
K31	C1	(T7)	*Podoviridae*	Enterobacteria	[40]
K32	A1		*Myoviridae*	Enterobacteria	[1]
K104	B1	F116	*Siphoviridae*	*Pseudomonas*	[8,19]
K170/11	A2	T2	*Myoviridae*	Enterobacteria	[1]
K328	C1	(N4)	*Podoviridae*	Enterobacteria	[1,19]
kappa	A1	kappa	*Myoviridae*	*Vibrio*	[2,19]
KD9	B1	(T1)	*Siphoviridae*	Enterobacteria	[43]
KF	E1	(serogroup III)	*Leviviridae* (*Levivirus*)	Enterobacteria	[19]
KF1	B1	KF1	*Siphoviridae*	*Pseudomonas*	[19]
KJ	E1	(serogroup II)	*Leviviridae* (*Levivirus*)	Enterobacteria	[7,19]
Kl1 to Kl4	A2	T4	*Myoviridae*	Enterobacteria	[1,19]
Kl4B	A1	(01)	*Myoviridae*	Enterobacteria	[1,19]
Kl4B	C1	(N4)	*Podoviridae*	Enterobacteria	[1,19]
Kl5 to Kl8	A2	T4	*Myoviridae*	Enterobacteria	[1,19]
Kl6B	A1	FC3-9	*Myoviridae*	Enterobacteria	[19]
Kl9	A1	K19	*Myoviridae*	Enterobacteria	[1,10,19]
Kl112	C1	(N4)	*Podoviridae*	Enterobacteria	[1,19]
Kl115	A1	ViI	*Myoviridae*	Enterobacteria	[1,19]
Kl116	C1	(N4)	*Podoviridae*	Enterobacteria	[1,19]
Kl117	C1	(N4)	*Podoviridae*	Enterobacteria	[19]
Kl119	A2	T2	*Myoviridae*	Enterobacteria	[1]
Kl23	C1	(T7)	*Podoviridae*	Enterobacteria	[1,19]
Kl24	C1	(N4)	*Podoviridae*	Enterobacteria	[1,19]
Kl27	A2	T2	*Myoviridae*	Enterobacteria	[1]
Kl29	A1	ViI	*Myoviridae*	Enterobacteria	[1,19]
Kl31	A2	T2	*Myoviridae*	Enterobacteria	[1]
Kl32	A1	(01)	*Myoviridae*	Enterobacteria	[1,19]
Kl35	A1	(01)	*Myoviridae*	Enterobacteria	[1,19]
Kl42B	B1	β4	*Siphoviridae*	Enterobacteria	[1,19]
Kl106B	C1	(sd)	*Podoviridae*	Enterobacteria	[19]
Kl141B	B1	β4	*Siphoviridae*	Enterobacteria	[1,19]
Kl171B	A2	T2	*Myoviridae*	Enterobacteria	[1,19]
Kl181B	A1	(01)	*Myoviridae*	Enterobacteria	[1,19]
Kl832B	A1	(01)	*Myoviridae*	Enterobacteria	[1,19]
KLXI	C1	(N4)	*Podoviridae*	Enterobacteria	[1,19]

Phage name	Morpho-type	Taxonomy 'Species group'	Family (Genus)	Host	Reference
KM	E1	(serogroup I)	*Leviviridae* (*Levivirus*)	Enterobacteria	1 9
KSY1	C3	KSY1	*Podoviridae*	*Streptococcus*	3,19
KT	A1	HM3	*Myoviridae*	*Clostridium*	1,19
KU1	E1	(serogroup II)	*Leviviridae* (*Levivirus*)	Enterobacteria	19
L	C1	P22	*Podoviridae*	Enterobacteria	19
L	B1	β	*Siphoviridae*	*Corynebacterium*	1,4,19
L	B1	A	*Siphoviridae*	*Corynebacterium*	4,19
L1	F2	MVL51	*Inoviridae* (*Plectrovirus*)	*Acholeplasma*	19
L₁	B2	M6	*Siphoviridae*	*Pseudomonas*	19
L2	G	L2	*Plasmaviridae* (*Plasmavirus*)	*Acholeplasma*	1,6,9,10
L3	C1	L3	*Podoviridae*	*Acholeplasma*	19
L17	D4	PRD1	*Tectiviridae* (*Tectivirus*)	Gram-negatives	10
L34	B1	χ	*Siphoviridae*	Enterobacteria	19
L39x35	B1	(77)	*Siphoviridae*	*Staphylococcus*	1,3,19
L172*	G		Unclassified	*Acholeplasma*	1,6,9,10
L.228	B1	(χ)	*Siphoviridae*	Enterobacteria	19
L.232	B1	(χ)	*Siphoviridae*	Enterobacteria	19
L.359	B2	ZG/3A	*Siphoviridae*	Enterobacteria	19
L412a	B1		*Siphoviridae*	*Rhizobium*	1
L419	B2	φ2037/1	*Siphoviridae*	*Rhizobium*	1,19
L426a	B1		*Siphoviridae*	*Rhizobium*	1
L434a	B1	φ2037/1	*Siphoviridae*	*Rhizobium*	17,19
La₆K₁	B1	(PS8)	*Siphoviridae*	*Rhizobium*	17,19
Labol	A1	X29	*Myoviridae*	*Vibrio*	2,19
lacticola	B1	lacticola	*Siphoviridae*	*Mycobacterium*	4,19
Lc-58	B1	(PS8)	*Siphoviridae*	*Rhizobium*	17,19
Lcg	B1	(PS8)	*Siphoviridae*	*Rhizobium*	17,19
Leo	B1	Leo	*Siphoviridae*	*Mycobacterium*	1,4,10,19
LHIIBNV6-2	B1	PS8	*Siphoviridae*	*Agrobacterium*	1,17,19
LHIIBV7-2	B1	PS8	*Siphoviridae*	*Agrobacterium*	1,17,19
LIIBNV6-1	B1	PS8	*Siphoviridae*	*Agrobacterium*	1,17,19
LIIBV7-1	B1	PS8	*Siphoviridae*	*Agrobacterium*	1,17,19
LL-H	B1		*Siphoviridae*	*Lactobacillus*	27
LP7	C1	P22	*Podoviridae*	Enterobacteria	1,19
LPP-1	C1	LPP-1	*Podoviridae*	*Plectonema*	1,12,19
LPP-1G	C1	LPP-1	*Podoviridae*	*Plectonema*	19
LPP-2	C1	LPP-1	*Podoviridae*	*Plectonema*	12,19
Lr-4	B1	(PS8)	*Siphoviridae*	*Rhizobium*	17,19
LV-1	B1	PS8	*Siphoviridae*	*Agrobacterium*	1,17,19
λ*	B1	λ	*Siphoviridae*	Enterobacteria	1,10,19
M	A2	T4	*Myoviridae*	Enterobacteria	19
M	B1	A	*Siphoviridae*	*Corynebacterium*	4,19
M	C1	Tb	*Podoviridae*	*Brucella*	1,13,19
M	E1		*Leviviridae* (*Levivirus*)	Enterobacteria	19

Phage name	Morpho-type	Taxonomy 'Species group'	Family (Genus)	Host	Reference
m	A1	m	*Myoviridae*	*Rhizobium*	1,10,17,19
M1	C1		*Podoviridae*	*Methylosinus*	19
M₁	B2	M1	*Siphoviridae*	*Thermoactinomyces*	1,4,10,19
M₃	B2	M1	*Siphoviridae*	*Thermoactinomyces*	1,4,19
M4	A1	PB-1	*Myoviridae*	*Pseudomonas*	21
M6	B2	M6	*Siphoviridae*	*Pseudomonas*	1,8,19
M6a	B2	M6	*Siphoviridae*	*Pseudomonas*	8,19
M12	E1	(serogroup I)	*Leviviridae (Levivirus)*	Enterobacteria	1,7,10,19
M13*	F1		*Inoviridae (Inovirus)*	Enterobacteria	1,10,19
m13	B2	3ML	*Siphoviridae*	*Streptococcus*	3,19
M20	D1		*Microviridae (Microvirus)*	Enterobacteria	1,10,19
M51	C1	Tb	*Podoviridae*	*Brucella*	13,19
m59	C1	(N4)	*Podoviridae*	Enterobacteria	1,19
M85	Synonym: phage M51				19
MAC-1	E1		*Leviviridae (Levivirus)*	*Bdellovibrio*	1,10,19
MAC-1'	E1		*Leviviridae (Levivirus)*	*Bdellovibrio*	1,10,19
MAC-2	E1		*Leviviridae (Levivirus)*	*Bdellovibrio*	1,10,19
MAC-3	A2		*Myoviridae*	*Bdellovibrio*	1,19
MAC-4	E1		*Leviviridae (Levivirus)*	*Bdellovibrio*	1,10,19
MAC-4'	E1		*Leviviridae (Levivirus)*	*Bdellovibrio*	1,10,19
MAC-5	E1		*Leviviridae (Levivirus)*	*Bdellovibrio*	1,10,19
MAC-6	A1		*Myoviridae*	*Bdellovibrio*	1,19
MAC-7	E1		*Leviviridae (Levivirus)*	*Bdellovibrio*	1,10,19
MB	E1	(serogroup I)	*Leviviridae (Levivirus)*	Enterobacteria	19
MB78	B1	(Jersey)	*Siphoviridae*	Enterobacteria	19
MC	E1	(serogroup I)	*Leviviridae (Levivirus)*	Enterobacteria	7,19
MC-1	B1	(Leo)	*Siphoviridae*	*Mycobacterium*	19
MC-3	B1	(Leo)	*Siphoviridae*	*Mycobacterium*	19
MC/75	C1	Tb	*Podoviridae*	*Brucella*	13,19
Me1	A2	(T4)	*Myoviridae*	Enterobacteria	36
Mex	B2	(MSP8)	*Siphoviridae*	*Streptomyces*	4,19
MG40	C1	P22	*Podoviridae*	Enterobacteria	1,19
minetti	B1	Leo	*Siphoviridae*	*Mycobacterium*	1,4,19
MJ-1	B1	(MP15)	*Siphoviridae*	*Bacillus*	19
MJ-4	B2	BLE	*Siphoviridae*	*Bacillus*	19
MLSP1	B2	MSP8	*Siphoviridae*	*Streptomyces*	4,19
MLSP2	B2	MSP8	*Siphoviridae*	*Streptomyces*	4,19
MLSP3	B2	MSP8	*Siphoviridae*	*Streptomyces*	4,19

Phage name	Morpho-type	Taxonomy 'Species group'	Family (Genus)	Host	Reference
MNP8	B1	(R1)	*Siphoviridae*	*Nocardia*	4,19
MM₁	C1		*Podoviridae*	*Rhizobium*	1
MO1	A1		*Myoviridae*	*Acetobacter*	19
*mor*1	B2	*mor*1	*Siphoviridae*	*Bacillus*	1,10,11,19
MP3	A1		*Myoviridae*	*Methylomonas*	1,19
MP9	A1	MP13	*Myoviridae*	*Bacillus*	19
MP10	B1	MP15	*Siphoviridae*	*Bacillus*	19
MP12	B1	MP15	*Siphoviridae*	*Bacillus*	19
MP13	A1	MP13	*Myoviridae*	*Bacillus*	19
MP14	B1	MP15	*Siphoviridae*	*Bacillus*	19
MP15	B1	MP15	*Siphoviridae*	*Bacillus*	19
MP20	C2	(φ29)	*Podoviridae*	*Bacillus*	19
MP21	A1	MP13	*Myoviridae*	*Bacillus*	19
MP25	A1	MP13	*Myoviridae*	*Bacillus*	19
MP27	C2	(φ29)	*Podoviridae*	*Bacillus*	19
MP28 to MP39	A1	MP13	*Myoviridae*	*Bacillus*	19
MP49	C2	(φ29)	*Podoviridae*	*Bacillus*	19
MP50	A1	MP13	*Myoviridae*	*Bacillus*	19
MS2*	E1	(serogroup I)	*Leviviridae* (*Levivirus*)	Enterobacteria	1,7,10,19
MSP1	B2	MSP8	*Siphoviridae*	*Streptomyces*	4,19
MSP2	B2	MSP8	*Siphoviridae*	*Streptomyces*	1,4,19
MSP3	B2	(MSP8)	*Siphoviridae*	*Streptomyces*	4,19
MSP5	B2	(MSP8)	*Siphoviridae*	*Streptomyces*	4,19
MSP8	B2	MSP8	*Siphoviridae*	*Streptomyces*	1,4,10,18,19
MSP9	B2	(MSP8)	*Siphoviridae*	*Streptomyces*	4,19
MSP12	B2	(MSP8)	*Siphoviridae*	*Streptomyces*	4,19
MSP13	B2	(MSP8)	*Siphoviridae*	*Streptomyces*	4,19
MSP14	B2	(MSP8)	*Siphoviridae*	*Streptomyces*	4,19
MT	B1	MT	*Siphoviridae*	*Brocothrix*	19
Mu	A1	Mu	*Myoviridae*	Enterobacteria	1,19
MVBr1	Synonym: phage Br				19
MVG51	F2	MVL51	*Inoviridae* (*Plectrovirus*)	*Acholeplasma*	1,9,10
MVL1	Synonym: phage L1				19
MVL2*	Synonym: phage L2				19
MVL3	Synonym: phage L3				19
MVL51*	F2	MVL51	*Inoviridae* (*Plectrovirus*)	*Acholeplasma*	6,9,10
MVL52	F2	MVL51	*Inoviridae* (*Plectrovirus*)	*Acholeplasma*	9
MV-Lg-pS2 -L172	Synonym: phage L172				19
MX-1	A1		*Myoviridae*	*Myxococcus*	1
MX1	E1	(serogroup IV)	*Leviviridae* (*Levivirus*)	Enterobacteria	19
MX2	E1	(serogroup IV)	*Leviviridae* (*Levivirus*)	Enterobacteria	19
MX4	A1		*Myoviridae*	*Myxococcus*	19
Mx8	C1		*Podoviridae*	*Myxococcus*	19

Phage name	Morpho-type	Taxonomy 'Species group'	Family (Genus)	Host	Reference
MX41	A1		*Podoviridae*	*Myxococcus*	[19]
MY	E1	(serogroup I)	*Leviviridae (Levivirus)*	Enterobacteria	[7,19]
μ2	E1	(serogroup I)	*Leviviridae (Levivirus)*	Enterobacteria	[1,7,10,19]
N	A2	T4	*Myoviridae*	Enterobacteria	[19]
N	B1	A	*Siphoviridae*	*Corynebacterium*	[4,19]
N-1	A2	T4	*Myoviridae*	Enterobacteria	[19]
N-1	A1	N-1	*Myoviridae*	*Anabaena*	[1,12,19]
N1	B1	N1	*Siphoviridae*	*Micrococcus*	[1,3,10,19]
N2	B1	N1	*Siphoviridae*	*Micrococcus*	[3,19]
N2	C1		*Podoviridae*	*Veillonella*	[19]
N₂	A1		*Myoviridae*	*Pseudomonas*	[1]
N3	B1	N1	*Siphoviridae*	*Micrococcus*	[3,19]
N3	B1		*Siphoviridae*	*Haemophilus*	[1]
N4	B1	N1	*Siphoviridae*	*Micrococcus*	[3,19]
N4	B2	ZG/3A	*Siphoviridae*	Enterobacteria	[19]
N4	C1	N4	*Podoviridae*	Enterobacteria	[1,10,19]
N₄	C1		*Podoviridae*	*Pseudomonas*	[1]
N-5	A2	T4	*Myoviridae*	Enterobacteria	[19]
N5	B1	II	*Siphoviridae*	*Bacillus*	[1,11,19]
N5	B1	N5	*Siphoviridae*	*Micrococcus*	[1,10,19]
N₅	C1		*Podoviridae*	*Pseudomonas*	[1,19]
N6	B1	N1	*Siphoviridae*	*Micrococcus*	[1,19]
N7	B1	N1	*Siphoviridae*	*Micrococcus*	[1,19]
N8	B1	N1	*Siphoviridae*	*Micrococcus*	[1,19]
N9	B2	3A	*Siphoviridae*	*Staphylococcus*	[19]
N-10	A2	T4	*Myoviridae*	Enterobacteria	[19]
N15	B2	3A	*Siphoviridae*	*Staphylococcus*	[19]
N-17	A2	T4	*Myoviridae*	Enterobacteria	[19]
N17	A1	(SP50)	*Myoviridae*	*Bacillus*	[1,11,19]
N20	B1		*Siphoviridae*	*Veillonella*	[19]
N-22	A2	T4	*Myoviridae*	Enterobacteria	[19]
Nb1			Unclassified	*Nitrobacter*	[1,19]
NCMB384	B1		*Siphoviridae*	*Cytophaga*	[1,19]
ND-2 to ND-7	A2	T4	*Myoviridae*	Enterobacteria	[1,19]
Nf	C2	φ29	*Podoviridae*	*Bacillus*	[1,11,19]
NH	E1	(serogroup III)	*Leviviridae (Levivirus)*	Enterobacteria	[7,19]
NHc	A1	222a	*Myoviridae*	*Lactobacillus*	[19]
NHs	B1	PL-1	*Siphoviridae*	*Lactobacillus*	[19]
NLP-1	A1	SP8	*Myoviridae*	*Bacillus*	[19]
NM	E1	(serogroup III)	*Leviviridae (Levivirus)*	Enterobacteria	[7,19]
NM1	B1	NM1	*Siphoviridae*	*Rhizobium*	[10,17,19]
NM2 to NM4	B1		*Siphoviridae*	*Rhizobium*	[1]
NM5	B2	(7-7-7)	*Siphoviridae*	*Rhizobium*	[1,17,19]
N°1	A1	SP50	*Myoviridae*	*Bacillus*	[1,11,19]
nt-1	A2	nt-1	*Myoviridae*	*Vibrio*	[2,19]
NT1	B1	(16-6-12)	*Siphoviridae*	*Rhizobium*	[1,19]

Phage name	Morpho-type	Taxonomy 'Species group'	Family (Genus)	Host	Reference
NT2	B1	NT2	*Siphoviridae*	*Rhizobium*	[1,10,17,19]
NT3-to-NT4	B1		*Siphoviridae*	*Rhizobium*	[1]
nt-6	C1	I	*Podoviridae*	*Vibrio*	[2,19]
NTc	A1	222a	*Myoviridae*	*Lactobacillus*	[19]
O	B1	A	*Siphoviridae*	*Corynebacterium*	[4,19]
O$_{06}$	C1	HM2	*Podoviridae*	*Clostridium*	[1]
O1	A1		*Myoviridae*	*Vibrio*	[10]
O6N-12P	B1		*Siphoviridae*	*Flavobacterium*	[1,19]
O6N-21P	A1		*Myoviridae*	*Vibrio*	[1,2]
O6N-22P	A2		*Myoviridae*	*Vibrio*	[1]
O6N-25P	C1		*Podoviridae*	*Pseudomonas*	[1]
O6N-34P	C1		*Podoviridae*	*Vibrio*	[1,2]
O6N-52P	B1		*Siphoviridae*	*Pseudomonas*	[1]
O6N-58P	D3	(PM2)	*Corticoviridae* (*Corticovirus*)	*Vibrio*	[1,10]
O111	A2	T4	*Myoviridae*	Enterobacteria	[1,19]
ON	E1	(serogroup III)	*Leviviridae* (*Levivirus*)	Enterobacteria	[19]
Ox-1	A2	T4	*Myoviridae*	Enterobacteria	[19]
Ox2	A2	T4	*Myoviridae*	Enterobacteria	[19]
Ox5	A2	T4	*Myoviridae*	Enterobacteria	[19]
Ox-6	A2	T4	*Myoviridae*	Enterobacteria	[19]
OXN-32P	B1		*Siphoviridae*	*Pseudomonas*	[1]
OXN-36P	C1		*Podoviridae*	*Vibrio*	[1]
OXN-52P	B1	OXN-52P	*Siphoviridae*	*Vibrio*	[1,2,19]
OXN-69P	A1	(II)	*Myoviridae*	*Vibrio*	[1,2,19]
OXN-72P	C1		*Podoviridae*	*Vibrio*	[1,2]
OXN-85P	C1		*Podoviridae*	*Vibrio*	[1,2]
OXN-86P	A1	(II)	*Myoviridae*	*Vibrio*	[2,19]
OXN-100P	C1	OXN-100P	*Podoviridae*	*Vibrio*	[1,2,19]
o6	D1		*Microviridae* (*Microvirus*)	Enterobacteria	[1,10,19]
Ω	B1	PS8	*Siphoviridae*	*Agrobacterium*	[1,19]
ω	B1	β	*Siphoviridae*	*Corynebacterium*	[1,4,19]
ω	Synonym: phage Ω				[19]
ω1	B1	(ω3)	*Siphoviridae*	*Streptococcus*	[3,19]
ω2	B1	ω3	*Siphoviridae*	*Streptococcus*	[3,19]
ω3	B1	ω3	*Siphoviridae*	*Streptococcus*	[3,19]
ω4	B1	(ω3)	*Siphoviridae*	*Streptococcus*	[3,19]
ω5	B1	(ω3)	*Siphoviridae*	*Streptococcus*	[3,19]
ω6	B1	(ω3)	*Siphoviridae*	*Streptococcus*	[3,19]
Ω8	C1	Ω8	*Podoviridae*	Enterobacteria	[1,10,19]
ω8	B1	ω3	*Siphoviridae*	*Streptococcus*	[3,19]
ω10	B1	(ω3)	*Siphoviridae*	*Streptococcus*	[3,19]
P	C1	Tb	*Podoviridae*	*Brucella*	[13,19]
P	B1	A	*Siphoviridae*	*Corynebacterium*	[4,19]
P	B1	949	*Siphoviridae*	*Streptococcus*	[3,19]
P	B1	(β)	*Siphoviridae*	*Corynebacterium*	[4]
P001	B2	3ML	*Siphoviridae*	*Streptococcus*	[3,19]

Phage name	Morpho-type	Taxonomy 'Species group'	Family (Genus)	Host	Reference
P008	B1	(936)	*Siphoviridae*	*Streptococcus*	3,19
P04	B1	(SD1)	*Siphoviridae*	*Pseudomonas*	8,19
P013	B1	(P047)	*Siphoviridae*	*Streptococcus*	3,19
P013	B1	(P059)	*Siphoviridae*	*Streptococcus*	3,19
P026	B1	949	*Siphoviridae*	*Streptococcus*	3,19
P034	C2	P034	*Podoviridae*	*Streptococcus*	3,19
P047	B1	P047	*Siphoviridae*	*Streptococcus*	3,19
P059	B1	P059	*Siphoviridae*	*Streptococcus*	3,19
P087	A1	(RZh)	*Myoviridae*	*Streptococcus*	3,19
P1	A1	P1	*Myoviridae*	Enterobacteria	1,10,19
P2	A1	(Twort)	*Myoviridae*	*Staphylococcus*	1,3,19
P2	A1	P2	*Myoviridae*	Enterobacteria	1,10,19
P2	B1		*Siphoviridae*	*Pseudomonas*	1
P3	A1	(Twort)	*Myoviridae*	*Staphylococcus*	1,3,19
P3	A1	(P2)	*Myoviridae*	Enterobacteria	1
P4	A1	(Twort)	*Myoviridae*	*Staphylococcus*	1,3,19
P6	A1		*Myoviridae*	*Azotobacter*	19
P7	A1	(P1)	*Myoviridae*	Enterobacteria	32
P8	A1	(Twort)	*Myoviridae*	*Staphylococcus*	1,3,19
P9	A1	(Twort)	*Myoviridae*	*Staphylococcus*	1,3,19
P9a	A1	(P2)	*Myoviridae*	Enterobacteria	1
P10	A1	(Twort)	*Myoviridae*	*Staphylococcus*	1,3,19
P10	A1	(01)	*Myoviridae*	Enterobacteria	1,19
P11-M15	B1	P11-M15	*Siphoviridae*	*Staphylococcus*	1,3,10,19
P22	C1	P22	*Podoviridae*	Enterobacteria	1,10,19
P22-4	C1	P22	*Podoviridae*	Enterobacteria	1,19
P22-7	C1	P22	*Podoviridae*	Enterobacteria	1,19
P22-11	C1	P22	*Podoviridae*	Enterobacteria	1,19
P22a1	C1	P22	*Podoviridae*	Enterobacteria	1,19
P$_{23}$	C1	HM2	*Podoviridae*	*Clostridium*	19
P42D	B1	77	*Siphoviridae*	*Staphylococcus*	1,3,19
P$_{46}$	C1	HM2	*Podoviridae*	*Clostridium*	1,19
P52	B2	3A	*Siphoviridae*	*Staphylococcus*	19
P52A	B1	P11-M15	*Siphoviridae*	*Staphylococcus*	1,3,19
P68	C1	44AHJD	*Podoviridae*	*Staphylococcus*	3,19
P78	C1		*Podoviridae*	*Acinetobacter*	19
P87	B2	3A	*Siphoviridae*	*Staphylococcus*	19
P107	B1	(ω3)	*Siphoviridae*	*Streptococcus*	3,19
P109	B2	c2	*Siphoviridae*	*Streptococcus*	3,19
P142	B1	(936)	*Siphoviridae*	*Streptococcus*	3,19
P191	B1	(936)	*Siphoviridae*	*Streptococcus*	3,19
P219	B1	(936)	*Siphoviridae*	*Streptococcus*	3,19
P-a-1 to P-a-9	B1	P-a-1	*Siphoviridae*	*Propionibacterium*	1,410,19
PA2	B1	λ	*Siphoviridae*	Enterobacteria	1,10
PA6	B1	(PS8)	*Siphoviridae*	*Agrobacterium*	1,17,19
Pal6	B		*Siphoviridae*	*Streptomyces*	18
PB	B1	(PL-1)	*Siphoviridae*	*Lactobacillus*	19
PB-1	A1	PB-1	*Myoviridae*	*Pseudomonas*	1,8,10,19
PB2	B2	M6	*Siphoviridae*	*Pseudomonas*	1,8,19
PB2A	B1	PS8	*Siphoviridae*	*Agrobacterium*	1,17,19

Phage name	Morpho-type	Taxonomy 'Species group'	Family (Genus)	Host	Reference
PB-6	Synonym: phage Ω				[19]
PBA12	B2	BLE	*Siphoviridae*	*Bacillus*	[1,11,19]
PBL1	B2	*mor*1	*Siphoviridae*	*Bacillus*	[19]
PBM12	A1	PBS1	*Myoviridae*	*Bacillus*	[11]
PBP1	B1	PBP1	*Siphoviridae*	*Bacillus*	[1,10,11,19]
PBS1	A1	PBS1	*Myoviridae*	*Bacillus*	[1,10,11,19]
PC	B1		*Siphoviridae*	*Pseudomonas*	[1]
Pc	B1		*Siphoviridae*	*Pseudomonas*	[1]
PE1	B2	PE1	*Siphoviridae*	*Streptococcus*	[3,19]
PE69	C1		*Podoviridae*	*Pseudomonas*	[1]
Pf	C1		*Podoviridae*	*Pseudomonas*	[1]
PF1	C1	HM2	*Podoviridae*	*Clostridium*	[1,19]
Pf1	F1		*Inoviridae* (*Inovirus*)	*Pseudomonas*	[1,10,19]
Pf2	F1		*Inoviridae* (*Inovirus*)	*Pseudomonas*	[1,10,19]
Pf3	F1		*Inoviridae* (*Inovirus*)	Gram-negatives	[1,10,19]
PF64FS	F1		*Inoviridae* (*Inovirus*)	Enterobacteria	[19]
Pg2	B		*Siphoviridae*	*Streptomyces*	[18]
PG60	B1		*Siphoviridae*	*Pseudomonas*	[8]
Pg81	B		*Siphoviridae*	*Streptomyces*	[18]
Pg100	B		*Siphoviridae*	*Streptomyces*	[18]
PH	B1	(PL-1)	*Siphoviridae*	*Lactobacillus*	[19]
Ph-1	C1	III	*Podoviridae*	*Vibrio*	[1,2,19]
Ph5	A1	Twort	*Myoviridae*	*Staphylococcus*	[3,19]
Ph9	A1	Twort	*Myoviridae*	*Staphylococcus*	[3,19]
Ph10	A1	Twort	*Myoviridae*	*Staphylococcus*	[3,19]
Ph13	A1	Twort	*Myoviridae*	*Staphylococcus*	[3,19]
phlei	B	Leo	*Siphoviridae*	*Mycobacterium*	[1,4,19]
PIIBNV6	A1	PIIBNV6	*Myoviridae*	*Agrobacterium*	[1,10,17,19]
PIIBNV6-C	C1	PIIBNV6-C	*Podoviridae*	*Agrobacterium*	[1,10,17,19]
pilHα	E1		*Leviviridae* (*Levivirus*)	Enterobacteria	[19]
Pk	A1	P2	*Myoviridae*	Enterobacteria	[1,19]
PL-1	B1	PL-1	*Siphoviridae*	*Lactobacillus*	[1,19]
PL25	C1	(P22)	*Podoviridae*	Enterobacteria	[1,19]
PL26	C1	(P22)	*Podoviridae*	Enterobacteria	[1,19]
PL37	C1	(P22)	*Podoviridae*	Enterobacteria	[1,19]
PL163/10	C1	I	*Podoviridae*	*Vibrio*	[2,19]
PLS-1	B1		*Siphoviridae*	*Lactobacillus*	[1]
PLT22	Synonym: phage P22				[19]
Pm1	C1	(N4)	*Podoviridae*	Enterobacteria	[1,19]
PM2*	D3	PM2	*Corticoviridae* (*Corticovirus*)	*Alteromonas*	[1,10]
PM5006M	C1		*Podoviridae*	*Proteus*	[35]
PMB1	A1	SP10	*Myoviridae*	*Bacillus*	[19]
PMB12	A1	PBS1	*Myoviridae*	*Bacillus*	[19]
PMJ1	A1	SP50	*Myoviridae*	*Bacillus*	[11,19]

Phage name	Morpho-type	Taxonomy 'Species group'	Family (Genus)	Host	Reference
PO4	B1		*Siphoviridae*	*Pseudomonas*	1
Polonus I	B2	R1-Myb	*Siphoviridae*	*Mycobacterium*	1,4,19
Polonus II	B1	(lacticola)	*Siphoviridae*	*Mycobacterium*	4,19
PP1	A1	(PB-1)	*Myoviridae*	*Pseudomonas*	8,19
PP4	B1	(SD1)	*Siphoviridae*	*Pseudomonas*	1,19
PP7	E1		*Leviviridae* (*Levivirus*)	*Pseudomonas*	1,10,19
PP8	A1	PP8	*Myoviridae*	*Pseudomonas*	1,8,10,19
PPR1	E1		*Leviviridae* (*Levivirus*)	*Pseudomonas*	10
PR3	D4	PRD1	*Tectiviridae* (*Tectivirus*)	Gram-negatives	1,10
PR4	D4	PRD1	*Tectiviridae* (*Tectivirus*)	Gram-negatives	1,10
PR5	D4	PRD1	*Tectiviridae* (*Tectivirus*)	Gram-negatives	1,10
PR10	A1		*Myoviridae*	*Azotobacter*	19
PR-590a	C1	PIIBNV6-C	*Podoviridae*	*Agrobacterium*	19
PR772	D4	PRD1	*Tectiviridae* (*Tectivirus*)	Gram-negatives	1,10
PR-1001	C1	(PIIBNV6-C)	*Podoviridae*	*Agrobacterium*	1,19
PRD1*	D4	PRD1	*Tectiviridae* (*Tectivirus*)	*Pseudomonas*	1,10
pro2	B1	pro2	*Siphoviridae*	*Leuconostoc*	3,19
PRR1	E1		*Leviviridae* (*Levivirus*)	Gram negatives	1,19
PS3	A1	PS17	*Myoviridae*	*Pseudomonas*	1,8,19
PS4	B1	PS4	*Siphoviridae*	*Pseudomonas*	1,8,10,19
PS8	B1	PS8	*Siphoviridae*	*Agrobacterium*	1,10,17,19
PS10	B1		*Siphoviridae*	*Pseudomonas*	1
PS17	A1	PS17	*Myoviridae*	*Pseudomonas*	19
PS20	A1	(01)	*Myoviridae*	Enterobacteria	1,19
PS-192	C1	PIIBNV6-C	*Podoviridae*	*Agrobacterium*	19
Psa1	A1	(PB-1)	*Myoviridae*	*Pseudomonas*	1,8,19
PSA68	C1	P22	*Podoviridae*	Enterobacteria	1,19
PsR-1012	C1	PIIBNV6-C	*Podoviridae*	*Agrobacterium*	19
Pssy15	B2	M6	*Siphoviridae*	*Pseudomonas*	19
Pssy41	B1	SD1	*Siphoviridae*	*Pseudomonas*	19
Pssy42	B1	SD1	*Siphoviridae*	*Pseudomonas*	19
Pssy403	B1	SD1	*Siphoviridae*	*Pseudomonas*	19
Pssy404	B1	SD1	*Siphoviridae*	*Pseudomonas*	19
Pssy420	B1	SD1	*Siphoviridae*	*Pseudomonas*	19
Pssy923	B1	SD1	*Siphoviridae*	*Pseudomonas*	19
Pssy4210	B1	M6	*Siphoviridae*	*Pseudomonas*	19
Pssy4220	B1	M6	*Siphoviridae*	*Pseudomonas*	19
PST	A2	T4	*Myoviridae*	Enterobacteria	1,10,19
PT11	B2	PT11	*Siphoviridae*	*Agrobacterium*	1,10,17,19
PTB	C1	T7	*Podoviridae*	Enterobacteria	10,19
PV-1	Synonym: phage LV-1				19
PX1	C1		*Podoviridae*	*Pseudomonas*	1

Phage name	Morpho-type	Taxonomy 'Species group'	Family (Genus)	Host	Reference
PX2	B2	M6	*Siphoviridae*	*Pseudomonas*	[1,8,19]
PX3	C1	(gh-1)	*Podoviridae*	*Pseudomonas*	[8,19]
PX4	A1	(PB-1)	*Myoviridae*	*Pseudomonas*	[1,8,19]
PX5	B2	M6	*Siphoviridae*	*Pseudomonas*	[1,8,19]
PX7	A1	(PB-1)	*Myoviridae*	*Pseudomonas*	[1,8,19]
PX10	C1	(gh-1)	*Podoviridae*	*Pseudomonas*	[1,8,19]
PX12	C1	(gh-1)	*Podoviridae*	*Pseudomonas*	[8,19]
PX14	C1		*Podoviridae*	*Pseudomonas*	[1]
Pz	B1		*Siphoviridae*	*Pseudomonas*	[1]
PZA	C2	φ29	*Podoviridae*	*Bacillus*	[19]
PZE	C2	φ29	*Podoviridae*	*Bacillus*	[19]
φ1	A2	φ1	*Myoviridae*	*Pseudomonas*	[19]
φ1.2.	C1	(T7)	*Podoviridae*	Enterobacteria	[40]
φ1Sv	B2	3ML	*Siphoviridae*	*Streptococcus*	[3,19]
φI	C1	T7	*Podoviridae*	Enterobacteria	[1,10,19]
φIA	A1		*Myoviridae*	*Rhizobium*	[1]
φ2	A1	(01)	*Myoviridae*	Enterobacteria	[1,19]
φ-2	A1		*Myoviridae*	*Pseudomonas*	[1]
φ2SV	B2	3ML	*Siphoviridae*	*Streptococcus*	[3,19]
φII	C1	T7	*Podoviridae*	Enterobacteria	[1,10]
Φ3Sv	B2	3ML	*Siphoviridae*	*Streptococcus*	[3,19]
φ3T	B1	(SPβ)	*Siphoviridae*	*Bacillus*	[11,19]
Φ4St	B2	3ML	*Siphoviridae*	*Streptococcus*	[3,19]
φ6*	E2	φ6	*Cystoviridae* (*Cystovirus*)	*Pseudomonas*	[1,10]
φ6	B2		*Siphoviridae*	*Caulobacter*	[1,19]
Φ6St	B2	3ML	*Siphoviridae*	*Streptococcus*	[3,19]
Φ12Sv	B2	3ML	*Siphoviridae*	*Streptococcus*	[3]
φ13Sv	B2	3ML	*Siphoviridae*	*Streptococcus*	[3]
φ15	C2	φ29	*Podoviridae*	*Bacillus*	[11,19]
φ17	C1	φ17	*Podoviridae*	*Streptomyces*	[1,4,10,19]
φ18	A1		*Myoviridae*	*Rhizobium*	[1]
φ25	A1	SP8	*Myoviridae*	*Bacillus*	[1,11,19]
φ29	C2	φ29	*Podoviridae*	*Bacillus*	[1,10,11,19]
φ41k	B1	PL-1	*Siphoviridae*	*Lactobacillus*	[19]
φ42	B1		*Siphoviridae*	*Streptococcus*	[33]
φ56	B1	XP12	*Siphoviridae*	*Pseudomonas*	[8,19]
φ63	A1	(SP50)	*Myoviridae*	*Bacillus*	[19]
φ80	B1	λ	*Siphoviridae*	Enterobacteria	[1,10,19]
φ92	A1	φ92	*Myoviridae*	Enterobacteria	[19]
φ101	B1		*Siphoviridae*	*Caulobacter*	[1,19]
φ102	B1		*Siphoviridae*	*Caulobacter*	[1,19]
φ105	B1	φ105	*Siphoviridae*	*Bacillus*	[1,10,11,19]
φ112	B1	XP12	*Siphoviridae*	*Pseudomonas*	[8,19]
φ-115A	B1	φ115A	*Siphoviridae*	*Thermoactinomyces*	[4,19]
φ118	B1		*Siphoviridae*	*Caulobacter*	[1,19]
φ138	A1		*Myoviridae*	*Vibrio*	[29]
φ149	B1		*Siphoviridae*	*Rhizobium*	[1]
φ-150A	B1	φ150A	*Siphoviridae*	*Micropolyspora*	[4]

Phage name	Morpho-type	Taxonomy 'Species group'	Family (Genus)	Host	Reference
φ151	B1		*Siphoviridae*	*Caulobacter*	1,19
φ218	A1	222a	*Myoviridae*	*Lactobacillus*	19
φ219	B1	(PL-1)	*Siphoviridae*	*Lactobacillus*	19
φ393	B1	(PL-1)	*Siphoviridae*	*Lactobacillus*	19
φ448	B		*Siphoviridae*	*Streptomyces*	18
φ630A	B1	(lacticola)	*Siphoviridae*	*Mycobacterium*	4,19
φ786	B1	(PL-1)	*Siphoviridae*	*Lactobacillus*	19
φ853	B1	(936)	*Siphoviridae*	*Streptococcus*	3,19
φ923	B2	3ML	*Siphoviridae*	*Streptococcus*	3,19
φ1178	B2	c2	*Siphoviridae*	*Streptococcus*	3,19
φ2037/1	B2	φ2037/1	*Siphoviridae*	*Rhizobium*	1,17,19
φ2037/2 to φ2037/7	B2	φ2037/1	*Siphoviridae*	*Rhizobium*	17,19
φ2042	C1	φ2042	*Podoviridae*	*Rhizobium*	1,10,17,19
φ2048	B1		*Siphoviridae*	*Rhizobium*	1
φ2193/2	A1	(m)	*Myoviridae*	*Rhizobium*	1,17,19
φ2205	B2	φ2037/1	*Siphoviridae*	*Rhizobium*	17
φA	D1	φX174	*Microviridae* (*Microvirus*)	Enterobacteria	1,10,19
φA1	B1		*Siphoviridae*	*Bacteroides*	1,19
φAa17	A1		*Myoviridae*	*Actinobacillus*	19
φC20	B1		*Siphoviridae*	*Asticcaulis*	1,19
φAc31	B1		*Siphoviridae*	*Asticcaulis*	19
φAc41	B3		*Siphoviridae*	*Asticcaulis*	19
φAcM$_2$	B2		*Siphoviridae*	*Asticcaulis*	19
φAcS$_1$	B3		*Siphoviridae*	*Asticcaulis*	19
φAcS$_2$	B3		*Siphoviridae*	*Asticcaulis*	19
φAG8010	B1	φAG8010	*Siphoviridae*	*Arthrobacter*	1,4,10,19
φBA1	C		*Podoviridae*	*Bacillus*	28
φC	B1	φC	*Siphoviridae*	*Rhodococcus*	1,4,10,19
φC8	Synonym: phage CON8				19
φC11	Synonym: phage CON11				19
φC31	B1	φC31	*Siphoviridae*	*Streptomyces*	18,19
φC43	B		*Siphoviridae*	*Streptomyces*	18
φC62	B		*Siphoviridae*	*Streptomyces*	18
φCb2	E1		*Leviviridae* (*Levivirus*)	*Caulobacter*	19
φCb4	E1		*Leviviridae* (*Levivirus*)	*Caulobacter*	19
φCb5	E1		*Leviviridae* (*Levivirus*)	*Caulobacter*	1,10,19
φCb8r	E1		*Leviviridae* (*Levivirus*)	*Caulobacter*	1,10,19
φCb9	E1		*Leviviridae* (*Levivirus*)	*Caulobacter*	19
φCb12r	E1		*Leviviridae* (*Levivirus*)	*Caulobacter*	1,10,19
φCb13	B3		*Siphoviridae*	*Caulobacter*	1
φCb23r	E1		*Leviviridae* (*Levivirus*)	*Caulobacter*	1,10,19

Phage name	Morpho-type	Taxonomy 'Species group'	Family (Genus)	Host	Reference
φCbK	B3		*Siphoviridae*	*Caulobacter*	1,19
φCC814/1	C1		*Podoviridae*	*Rhizobium*	1
φCd1	C1		*Podoviridae*	*Caulobacter*	1,19
φCj1	A1		*Myoviridae*	*Cytophaga*	19
φCj7	A1		*Myoviridae*	*Cytophaga*	19
φCj27	A1		*Myoviridae*	*Cytophaga*	19
φCLV-29	B2		*Siphoviridae*	*Caulobacter*	1,19
φCp2	E1		*Leviviridae* (*Levivirus*)	*Caulobacter*	10,19
φCp18	E1		*Leviviridae* (*Levivirus*)	*Caulobacter*	10,19
φCp34	B3		*Siphoviridae*	*Caulobacter*	19
φCR24	A1		*Myoviridae*	*Caulobacter*	19
φCR26	A1		*Myoviridae*	*Caulobacter*	19
φCR30	A1		*Myoviridae*	*Caulobacter*	19
φCR40	C1		*Podoviridae*	*Caulobacter*	19
φCr1	B1		*Siphoviridae*	*Caulobacter*	19
φCr2	B3		*Siphoviridae*	*Caulobacter*	19
φCr14	E1		*Leviviridae* (*Levivirus*)	*Caulobacter*	10,19
φCr22	B1		*Siphoviridae*	*Caulobacter*	19
φCr28	E1		*Leviviridae* (*Levivirus*)	*Caulobacter*	10,19
φD	B1	(77)	*Siphoviridae*	*Staphylococcus*	1,3,19
φD326	B1		*Siphoviridae*	Enterobacteria	1,10,19
φe	A1	SP8	*Myoviridae*	*Bacillus*	1,11,19
φEC	B1	(φC)	*Siphoviridae*	*Rhodococcus*	1,4,19
φFSV	B1	φFSW	hiphoviridae	*Lactobacillus*	19
φFSW	B1	φFSW	*Siphoviridae*	*Lactobacillus*	19
φg	B1	β4	*Siphoviridae*	Enterobacteria	19
φgal-1/0W	A1	φgal-1/R	*Myoviridae*	*Rhizobium*	19
φgal-1/R	A1	φgal-1/R	*Myoviridae*	*Rhizobium*	19
φgal-3/0W	A1	CM$_1$	*Myoviridae*	*Rhizobium*	19
φgal-3/R	A1	CM$_1$	*Myoviridae*	*Rhizobium*	19
φH	A1		*Myoviridae*	*Halobacterium*	19
φK	D1	φX174	*Microviridae* (*Microvirus*)	Enterobacteria	14
φKZ	A1	φKZ	*Myoviridae*	*Pseudomonas*	8,10,19
φ-LT	A1	PS17	*Myoviridae*	*Pseudomonas*	8,19
φm	C1	(P22)	*Podoviridae*	Enterobacteria	19
φM12	A1	CM$_1$	*Myoviridae*	*Rhizobium*	19
φMC	C1		*Podoviridae*	*Pseudomonas*	1
φMU1	B1	(936)	*Siphoviridae*	*Streptococcus*	3,19
φNS11	D4	PRD1	*Tectiviridae* (*Tectivirus*)	*Bacillus*	10,11
φPLS-1	A1	(PB-1)	*Myoviridae*	*Pseudomonas*	8,19
φPS	B1	XP12	*Siphoviridae*	*Xanthomonas*	1,8,19
φR	D1	φX174	*Microviridae* (*Microvirus*)	Enterobacteria	1,10,19
φRE	B2	3A	*Siphoviridae*	*Staphylococcus*	1,3,19

Phage name	Morpho-type	Taxonomy 'Species group'	Family (Genus)	Host	Reference
φRS	B1	XP12	*Siphoviridae*	*Pseudomonas*	8,19
φRsG1	B2		*Siphoviridae*	*Rhodopseudomonas*	19
φS1	C1		*Podoviridae*	*Pseudomonas*	1
φSD	B1	XP12	*Siphoviridae*	*Pseudomonas*	8,19
φSK311	A1	Twort	*Myoviridae*	*Staphylococcus*	19
φSL	B1	XP12	*Siphoviridae*	*Pseudomonas*	8,19
φUW21	B1	φUW21	*Siphoviridae*	*Micromonospora*	4
φW	Synonym: phage φII				19
φW-14	A1	φW-14	*Myoviridae*	*Pseudomonas*	1,8,10,19
φX174*	D1	φX174	*Microviridae* (*Microvirus*)	Enterobacteria	1,10,19
φYS40	A1		*Myoviridae*	*Thermus*	1
φX7	B1	χ	*Siphoviridae*	Enterobacteria	1,19
π1	C1	(P22)	*Podoviridae*	Enterobacteria	1,19
ψ	B1	ψ	*Siphoviridae*	*Agrobacterium*	18
Q	B1	A	*Siphoviridae*	*Corynebacterium*	4,19
q	B1	T5	*Myoviridae*	Enterobacteria	1,19
Q_{16}	C1	HM2	*Podoviridae*	*Clostridium*	19
Q_{21}	C1	HM2	*Podoviridae*	*Clostridium*	19
Q_{26}	C1	HM2	*Podoviridae*	*Clostridium*	19
Q_{46}	C1	HM2	*Podoviridae*	*Clostridium*	19
Qβ*	E1	serogroup III	*Leviviridae* (*Levivirus*)	Enterobacteria	1,7,10,19
Q_{04}	C1	HM2	*Podoviridae*	*Clostridium*	19
Q_{05}	C1	HM2	*Podoviridae*	*Clostridium*	19
Q_{06}	C1	HM2	*Podoviridae*	*Clostridium*	19
R	B1	A	*Siphoviridae*	*Corynebacterium*	4,19
R	C1	T7	*Podoviridae*	Enterobacteria	1,10,19
R	C1		*Podoviridae*	*Pseudomonas*	8
R	C1	Tb	*Podoviridae*	*Brucella*	13,19
R1	B1	R1	*Siphoviridae*	*Nocardia*	1,4,10,19
R1	Synonym: phage R1-Myb				
R1-Myb	B2	R1-Myb	*Siphoviridae*	*Mycobacterium*	4,19
R_2	B2	R_2	*Siphoviridae*	*Streptomyces*	1,4,10,19
R4	B1		*Siphoviridae*	*Agrobacterium*	1
$R4c_1$	B		*Siphoviridae*	*Streptomyces*	18
R17	E1	serogroup I	*Leviviridae* (*Levivirus*)	Enterobacteria	1,7,10,19
R23	E1	(serogroup I)	*Leviviridae* (*Levivirus*)	Enterobacteria	1,7,10,19
R34	E1	(serogroup I)	*Leviviridae* (*Levivirus*)	Enterobacteria	1,10,19
R40	E1	(serogroupI)	*Leviviridae* (*Levivirus*)	Enterobacteria	1,7,10,19
rabinovitchi	B1	Leo	*Siphoviridae*	*Mycobacterium*	1,4,19
RG	A1	Twort	*Myoviridae*	*Staphylococcus*	19

Phage name	Morpho-type	Taxonomy 'Species group'	Family (Genus)	Host	Reference
RL1	B1	NT2	*Siphoviridae*	*Rhizobium*	19
Roy	B2	R1-Myb	*Siphoviridae*	*Mycobacterium*	1,4,19
Rp1.Rp1	C1		*Podoviridae*	*Rhodopseudomonas*	1,19
RP2	B		*Siphoviridae*	*Streptomyces*	18
Rφ-1	B1		*Podoviridae*	*Rhodopseudomonas*	1,19
Rφ6P	B1		*Siphoviridae*	*Rhodopseudomonas*	19
RR66	C1	RR66	*Podoviridae*	*Xanthomonas*	19
RS1	C1		*Podoviridae*	*Rhodopseudomonas*	1,19,30
RZh	A1	RZh	*Myoviridae*	*Streptococcus*	1,3,10,19
ρ	B1	(β)	*Siphoviridae*	*Corynebacterium*	19
ρ11	B1	(SPβ)	*Siphoviridae*	*Bacillus*	11
ρ15	B1	(SPP1)	*Siphoviridae*	*Bacillus*	25
S	A1	SP50	*Myoviridae*	*Bacillus*	11,19
S	A1	(PB-1)	*Myoviridae*	*Pseudomonas*	19
S	A2	T4	*Myoviridae*	*Enterobacteria*	19
S	B1	A	*Siphoviridae*	*Corynebacterium*	4,19
S	C1	(2)	*Podoviridae*	*Rhizobium*	1,17,19
s	C1		*Podoviridae*	Enterobacteria	10
S-1	B1	S-2L	*Siphoviridae*	*Microcystis, Synechococcus*	12,19
S1	B2	3A	*Siphoviridae*	*Staphylococcus*	3,19
S1	B		*Siphoviridae*	*Streptomyces*	18
S_1	C1	(N4)	*Podoviridae*	Enterobacteria	19
S1BL	B1	(β4)	*Siphoviridae*	Enterobacteria	1,19
S-2	Synonym: phage S-2L				19
S2	B1	P11-M15	*Siphoviridae*	*Staphylococcus*	3,19
S2	A1		*Myoviridae*	*Haemophilus*	1,19
S2	B2	F1	*Siphoviridae*	*Clostridium*	1,19
S_2	A1	(12S)	*Myoviridae*	*Pseudomonas*	19
S-2L	B1	S-2L	*Siphoviridae*	*Microcystis, Synechococcus*	12,19
S3K	A1	Twort	*Myoviridae*	*Staphylococcus*	1,3,19
S-4L	B1	S-4L	*Siphoviridae*	*Synechococcus*	19
S-5L	C1	(SM-1)	*Podoviridae*	*Synechococcus*	19
S6	B2	3A	*Siphoviridae*	*Staphylococcus*	19
S-6(L)	A1	S-6(L)	*Myoviridae*	*Synechococcus*	19
S-9	B1		*Siphoviridae*	*Lactobacillus*	1
S13	D1	φX174	*Microviridae* (*Microvirus*)	Enterobacteria	1,10,19
S14	B		*Siphoviridae*	*Streptomyces*	18
S45	B1		*Siphoviridae*	*Halobacterium*	19
S66	A1	(P2)	*Myoviridae*	Enterobacteria	1
S70	B1	T5	*Myoviridae*	Enterobacteria	1,19
S_{111}	C1	HM2	*Podoviridae*	*Clostridium*	19
S171	B1		*Siphoviridae*	*Lactobacillus*	1
S206	B1	T5	*Myoviridae*	Enterobacteria	1,19
S708	C1	Tb	*Podoviridae*	*Brucella*	13,19
SA_{02}	C1	HM2	*Podoviridae*	*Clostridium*	19

Phage name	Morpho-type	Taxonomy 'Species group'	Family (Genus)	Host	Reference
SA1 to SA5	B2	MSP8	*Siphoviridae*	*Streptomyces*	[19]
Sa-S	B1		*Siphoviridae*	*Lactobacillus*	[1]
SB	E1	(serogroup II)	*Leviviridae* (*Levivirus*)	Enterobacteria	[7,19]
S$_B$-1	A1	Twort	*Myoviridae*	*Staphylococcus*	[3,19]
SB24	C1	2BV	*Siphoviridae*	*Streptococcus*	[3,19]
SBX-1	A1		*Myoviridae*	*Xanthomonas*	[1]
SD	E1	(serogroup II)	*Leviviridae* (*Levivirus*)	Enterobacteria	[7,19]
sd	C1	sd	*Podoviridae*	Enterobacteria	[10,19]
SD1	B1	SD1	*Siphoviridae*	*Pseudomonas*	[1,8,10,19]
SE	A1		*Myoviridae*	*Vibrio*	[1]
SE3	A1	kappa	*Myoviridae*	*Vibrio*	[2,19]
'Serotype'V	A1		*Myoviridae*	*Vibrio*	[2]
'Serotype'VI	C		*Podoviridae*	*Vibrio*	[2]
'Serotype'VII	C		*Podoviridae*	*Vibrio*	[2]
'Serotype'VIII	A1	II	*Myoviridae*	*Vibrio*	[2,19]
'Serotype'IX	C		*Podoviridae*	*Vibrio*	[2]
'Serotype'X	C		*Podoviridae*	*Vibrio*	[2]
SF-5	C2	GA-1	*Podoviridae*	*Bacillus*	[11,19]
SF6	B1	SPP1	*Siphoviridae*	*Bacillus*	[19]
SF6	C1	(P22)	*Podoviridae*	Enterobacteria	[19]
SG	E1	(serogroup III)	*Leviviridae* (*Levivirus*)	Enterobacteria	[7,19]
SG35	A2	T4	*Myoviridae*	Enterobacteria	[1,19]
SG55	A2	T4	*Myoviridae*	Enterobacteria	[1,19]
SG3201	A2	T4	*Myoviridae*	Enterobacteria	[1,19]
SHII	A2	T4	*Myoviridae*	Enterobacteria	[1,19]
SH3	B		*Siphoviridae*	*Streptomyces*	[18]
SHIII	B2	ZG/3A	*Siphoviridae*	Enterobacteria	[1,19]
SHIV	A1	(01)	*Myoviridae*	Enterobacteria	[1,19]
SHV	A2	T4	*Myoviridae*	Enterobacteria	[1,19]
SH6	C1		*Podoviridae*	*Pseudomonas*	[8]
SHVI	A1	(01)	*Myoviridae*	Enterobacteria	[1,19]
SHVIII	A1	(01)	*Myoviridae*	Enterobacteria	[1,19]
SH10	B		*Siphoviridae*	*Streptomyces*	[18]
SHX	A2	T4	*Myoviridae*	Enterobacteria	[1,19]
SH11	B		*Siphoviridae*	*Streptomyces*	[18]
SHXI	B2	ZG/3A	*Siphoviridae*	Enterobacteria	[1,19]
SH12	B		*Siphoviridae*	*Streptomyces*	[18]
SH133	C1		*Podoviridae*	*Pseudomonas*	[8]
SiI*	D2		Unclassified	*Aquaspirillum*	[1]
SK	E1		*Leviviridae* (*Levivirus*)	Enterobacteria	[7]
SKI	B1	β4	*Siphoviridae*	Enterobacteria	[1,19]
SKII	A2	T4	*Myoviridae*	Enterobacteria	[1,19]
SKIII	A1	(01)	*Myoviridae*	Enterobacteria	[1,19]
SKIV	A1	(01)	*Myoviridae*	Enterobacteria	[1,19]
SKIVa	A1	(01)	*Myoviridae*	Enterobacteria	[1,19]
SKV	A2	T4	*Myoviridae*	Enterobacteria	[1,19]

Phage name	Morpho- type	Taxonomy 'Species group'	Family (Genus)	Host	Reference
SKVI	A1	(01)	*Myoviridae*	Enterobacteria	1,19
SKVIII	A1	(01)	*Myoviridae*	Enterobacteria	1,19
SKVIIIa	A1	(01)	*Myoviridae*	Enterobacteria	1,19
SKX	A2	T4	*Myoviridae*	Enterobacteria	1,19
SKXI	B2	ZG/3A	*Siphoviridae*	Enterobacteria	1,19
SKXII	B1	(T5)	*Myoviridae*	Enterobacteria	1,19
SK66	A1	(P2)	*Myoviridae*	Enterobacteria	1
SKF12	C1	(T7)	*Podoviridae*	Enterobacteria	1,19
SL4	B2	PT11	*Siphoviridae*	Agrobacterium	17,19
SL6	B2	PT11	*Siphoviridae*	Agrobacterium	17,19
SL7	B2	PT11	*Siphoviridae*	Agrobacterium	17,19
SLE111	C1	(φ17)	*Podoviridae*	*Streptomyces*	19
SM-1	C1	SM-1	*Podoviridae*	*Synechococcus*	1,12,19
SM-2	B1	(S-2L)	*Siphoviridae*	*Synechococcus, Microcystis*	12,19
sm26	B1	(N1)	*Siphoviridae*	*Micrococcus*	3,19
sm59	B1	(N1)	*Siphoviridae*	*Micrococcus*	3,19
SMB	A2	T2	*Myoviridae*	Enterobacteria	10
SMB1	B2	(ZG/3A)	*Siphoviridae*	Enterobacteria	1,19
SMB2	A2	T4	*Myoviridae*	Enterobacteria	1,19
SMB3	A1	(P2)	*Myoviridae*	Enterobacteria	1
SMP2	A2	T4	*Myoviridae*	Enterobacteria	1,10,19
SMP5	A1	(P2)	*Myoviridae*	Enterobacteria	1
SN	E1		*Leviviridae (Levivirus)*	Enterobacteria	7
SN₁	B2		*Siphoviridae*	*Sphaerotiles*	19
SO	E1	(serogroup III)	*Leviviridae (Levivirus)*	Enterobacteria	7,19
SP	E1	(serogroup IV)	*Leviviridae (Levivirus)*	Enterobacteria	7,19
Sp	B1		*Siphoviridae*	*Ancalomicrobium*	19
SP01	A1	SP8	*Myoviridae*	*Bacillus*	1,11,19
SP02	B1	(φ105)	*Siphoviridae*	*Bacillus*	1,11,19
SP3	A1	SP3	*Myoviridae*	*Bacillus*	1,10,11,19
SP5	A1	SP8	*Myoviridae*	*Bacillus*	1,11,19
SP5	C2	GA-1	*Podoviridae*	*Bacillus*	1,19
SP6	A1	SP8	*Myoviridae*	*Bacillus*	1,11,19
SP7	A1	SP8	*Myoviridae*	*Bacillus*	1,11,19
SP8	A1	SP8	*Myoviridae*	*Bacillus*	1,10,11,19
SP9	A1	SP8	*Myoviridae*	*Bacillus*	1,11
SP10	A1	SP10	*Myoviridae*	*Bacillus*	19
SP13	A1	SP8	*Myoviridae*	*Bacillus*	1,11,19
SP15	A1	SP15	*Myoviridae*	*Bacillus*	10,11,19
SP50	A1	SP50	*Myoviridae*	*Bacillus*	1,10,11,19
SP82	A1	SP8	*Myoviridae*	*Bacillus*	1,11,19
SPβ	B1	SPβ	*Siphoviridae*	*Bacillus*	10,11,19
SPP1	B1	SPP1	*Siphoviridae*	*Bacillus*	1,10,11,19
SPV1/KC3		SPV1/KC3	Unclassified	*Spiroplasma*	9
spv-1	C	SVC3	*Podoviridae*	*Drosophila SRO (Spiroplasma)*	6,9

Phage name	Morpho-type	Taxonomy 'Species group'	Family (Genus)	Host	Reference
spv-2	C	SVC3	*Podoviridae*	*Drosophila SRO (Spiroplasma)*	6,9
SpV4*	D1		Unclassified	*Spiroplasma*	19
SPy-1	A1	(SP8)	*Myoviridae*	*Bacillus*	19
SPy-2	A1	SPy-2	*Myoviridae*	*Bacillus*	19
SS	E1	(serogroup II)	*Leviviridae (Levivirus)*	Enterobacteria	7,19
SST	A1	SST	*Myoviridae*	*Bacillus*	19
SSV1*			Unclassified	*Sulfolobus*	19
ST	E1	(serogroup III)	*Leviviridae (Levivirus)*	Enterobacteria	7,19
ST-1	D1		*Microviridae (Microvirus)*	Enterobacteria	1,10,19
ST1	C1		*Podoviridae*	*Rhizobium*	1
STV	A2	T4	*Myoviridae*	Enterobacteria	1,19
STVIII	A2	T4	*Myoviridae*	Enterobacteria	1,19
STXI	A1	(P2)	*Myoviridae*	Enterobacteria	1
*Sub*1	A1	SP50	*Myoviridae*	*Bacillus*	11,19
SV2	B1	SV2	*Siphoviridae*	*Streptomyces*	4,19
SV4*	Synonym: phage SpV4				19
SV-C1	F2	(MVL51)	*Inoviridae (Plectrovirus)*	*Spiroplasma*	6,10
SVC2	B		*Siphoviridae*	*Spiroplasma*	6,9
SVC3	Synonym: phage C3			*Spiroplasma*	19
SW	A1	SP8	*Myoviridae*	*Bacillus*	1,11,19
SW	E1	(serogroup II)	*Leviviridae (Levivirus)*	Enterobacteria	7,19
SW7	B2	PT11	*Siphoviridae*	*Agrobacterium*	17,19
SW12	B2	PT11	*Siphoviridae*	*Agrobacterium*	17,19
SW14	B2	PT11	*Siphoviridae*	*Agrobacterium*	17,19
SZ	E1	(serogroup I)	*Leviviridae (Levivirus)*	Enterobacteria	19
T	B1	A	*Siphoviridae*	*Corynebacterium*	4,19
t	E1		*Leviviridae (Levivirus)*	*Enterobacteria*	19
T1	B1	(λ)	*Siphoviridae*	Enterobacteria	1,10,19
T2*	A2	T4	*Myoviridae*	Enterobacteria	1,10,19
T3	C1	T7	*Podoviridae*	Enterobacteria	1,10,19
T3C	B1	β4	*Siphoviridae*	Enterobacteria	1,19
T4*	A2	T4	*Myoviridae*	Enterobacteria	1,19
T5	B1	T5	*Siphoviridae*	Enterobacteria	1,10,19
T6	A2	T4	*Myoviridae*	Enterobacteria	1,10,19
T7*	C1	T7	*Podoviridae*	Enterobacteria	1,10,19
Ta₁	C1	Ta₁	*Podoviridae*	*Thermoactinomyces*	4,19
Taunton	A1	Beccles	*Myoviridae*	Enterobacteria	1,19
TB	Synonym: phage Tb				19
Tb	C1	Tb	*Podoviridae*	*Brucella*	1,13,19
Tbilisi	Synonym: phage Tb				19
Td8	B1	(φ105)	*Siphoviridae*	*Bacillus*	19

Phage name	Morpho-type	Taxonomy 'Species group'	Family (Genus)	Host	Reference
Tg7	B1	(II)	*Siphoviridae*	*Bacillus*	11,19
Tg8	A1	SP50	*Myoviridae*	*Bacillus*	11,19
Tg12	A1	SP50	*Myoviridae*	*Bacillus*	11,19
Tg14	A1	SP50	*Myoviridae*	*Bacillus*	11,19
Tg21	B1	(φ105)	*Siphoviridae*	*Bacillus*	19
TH1	E1	(serogroup II)	*Leviviridae* (*Levivirus*)	Enterobacteria	19
*thu*1	A1	SP8	*Myoviridae*	*Bacillus*	11,19
thu2-thu3	B2	*mor*1	*Siphoviridae*	*Bacillus*	1,11,19
thu4	A1	SP50	*Myoviridae*	*Bacillus*	11,19
Tm1	B2	*mor*1	*Siphoviridae*	*Bacillus*	19
Tm2	B2	type F	*Siphoviridae*	*Bacillus*	19
Tm3	B1	(φ105)	*Siphoviridae*	*Bacillus*	19
TP-13	A1	SP15	*Myoviridae*	*Bacillus*	19
TP50	A1	SP50	*Myoviridae*	*Bacillus*	1,11,19
TP446	A1	(43)	*Myoviridae*	*Aeromonas*,	19
TSP-1	A1	SP50	*Myoviridae*	*Bacillus*	1,11,19
TTV1* to TTV4	F3		Unclassified	*Thermoproteus*	19
Tt4	B1	(φ105)	*Siphoviridae*	*Bacillus*	19
TuII*-6	A2	T4	*Myoviridae*	Enterobacteria	19
TuII*-24	A2	T4	*Myoviridae*	Enterobacteria	19
TuII*-46	A2	T4	*Myoviridae*	Enterobacteria	19
TuII*-60	A2	T4	*Myoviridae*	Enterobacteria	19
TW18	E1	(serogroup III)	*Leviviridae* (*Levivirus*)	Enterobacteria	19
TW19	E1	(serogroup IV)	*Leviviridae* (*Levivirus*)	Enterobacteria	19
TW28	E1	(serogroup IV)	*Leviviridae* (*Levivirus*)	Enterobacteria	19
Twort	A1	Twort	*Myoviridae*	*Staphylococcus*	1,3,10,19
TX	A1	AU	*Myoviridae*	*Pasteurella*	19
type D	B1	(φ105)	*Siphoviridae*	*Bacillus*	11,19
type F	B2	type F	*Siphoviridae*	*Bacillus*	10,11,19
type G	B2	*mor*1	*Siphoviridae*	*Bacillus*	11,19
type Ia (*S. spheroides*)	B2	MSP8	*Siphoviridae*	*Streptomyces*	1,4,19
type II (*S. streptomycini*)	B2	MSP8	*Siphoviridae*	*Streptomyces*	1,4,19
U	B1	A	*Siphoviridae*	*Corynebacterium*	4,19
U3	D1		*Microviridae* (*Microvirus*)	Enterobacteria	1,10,19
UC-1	B1	(β4)	*Siphoviridae*	Enterobacteria	19
UC-18	B1	P11-M15	*Siphoviridae*	*Staphylococcus*	1,3,19
UW	E1	(serogroup II)	*Leviviridae* (*Levivirus*)	Enterobacteria	19
UZ	B1	PL-1	*Siphoviridae*	*Lactobacillus*	19
V	A1	SP50	*Myoviridae*	*Bacillus*	11,19
v1	G	(L2)	*Plasmaviridae* (*Plasmavirus*)	*Acholeplasma*	10

Phage name	Morpho-type	Taxonomy 'Species group'	Family (Genus)	Host	Reference
v2	G	(L2)	*Plasmaviridae (Plasmavirus)*	*Acholeplasma*	10
V3	B1		*Siphoviridae*	*Campylobacter*	1
v4	G	(L2)	*Plasmaviridae (Plasmavirus)*	*Acholeplasma*	10
v5	G	(L2)	*Plasmaviridae*	*Acholeplasma*	10
v6	F1		*Inoviridae (Inovirus)*	*Vibrio*	1,2,10,19
v7	G	(L2)	*Plasmaviridae (Plasmavirus)*	*Acholeplasma*	10
V16	B1		*Siphoviridae*	*Campylobacter*	1
V19		Synonym: phage Vfi-6			19
V40	A1	(ViI)	*Myoviridae*	Enterobacteria	19
V.43	B1	(χ)	*Siphoviridae*	Enterobacteria	19
V45	B1		*Siphoviridae*	*Campylobacter*	1
V56	A1	(Kl9)	*Myoviridae*	Enterobacteria	19
Va	B1		*Siphoviridae*	*Ancalomicrobium*	19
VcA-1	A1		*Myoviridae*	*Vibrio*	2
VcA-2	A1	kappa	*Myoviridae*	*Vibrio*	2,19
VD13	B3	VD13	*Siphoviridae*	*Streptococcus*	1,3,10,19
Versailles	C2	GA-1	*Podoviridae*	*Bacillus*	11,19
Vf12	F1		*Inoviridae (Inovirus)*	*Vibrio*	19
Vf33	F1		*Inoviridae (Inovirus)*	*Vibrio*	19
Vfi-3	B1		*Siphoviridae*	*Campylobacter*	1,19
Vfi-6	B1		*Siphoviridae*	*Campylobacter*	1,19
VI	C1	(N4)	*Podoviridae*	Enterobacteria	1,19
ViI	A1	ViI	*Myoviridae*	Enterobacteria	1,10,19
ViII	B1	ViII	*Siphoviridae*	Enterobacteria	1,10,19
ViIII to ViVII	C1	(N4)	*Podoviridae*	Enterobacteria	1,19
VK	E1	(serogroup III)	*Leviviridae (Levivirus)*	Enterobacteria	19
VL	A1	AU	*Myoviridae*	*Pasteurella*	19
VL-1	A1		*Siphoviridae*	*Bdellovibrio*	1
VP1	A1	VP1	*Myoviridae*	*Vibrio*	2,19
VP2	A		*Myoviridae*	*Vibrio*	2
VP3	B1	VP3	*Siphoviridae*	*Vibrio*	2,19
VP4	A		*Myoviridae*	*Vibrio*	2
VP5	B1	VP5	*Siphoviridae*	*Streptomyces*	1,4,10,18,19
VP5	B2	VP5	*Siphoviridae*	*Vibrio*	2,19
VP6	B1	VP3	*Siphoviridae*	*Vibrio*	2,19
VP7	A1	(VP1)	*Myoviridae*	*Vibrio*	2,19
VP7	B1	VP5	*Siphoviridae*	*Streptomyces*	1,4,18,19
VP8	A		*Myoviridae*	*Vibrio*	2
VP9	A		*Myoviridae*	*Vibrio*	2
VP10	A		*Myoviridae*	*Vibrio*	2
VP11	B		*Siphoviridae*	*Streptomyces*	18
VP11	B2	VP11	*Siphoviridae*	*Vibrio*	2,19
VP12	B		*Siphoviridae*	*Streptomyces*	18
VP12	B1	VP12	*Siphoviridae*	*Vibrio*	2

Phage name	Morpho-type	Taxonomy 'Species group'	Family (Genus)	Host	Reference
VP13	B1	(VP12)	*Siphoviridae*	*Vibrio*	2
VP14	B		*Siphoviridae*	*Streptomyces*	18
VP15	B2	VP5	*Siphoviridae*	*Vibrio*	2,19
VP16	B2	VP5	*Siphoviridae*	*Vibrio*	2,19
VP17	A		*Myoviridae*	*Vibrio*	2
VP18	A1	(VP1)	*Myoviridae*	*Vibrio*	2,19
VP19	A		*Myoviridae*	*Vibrio*	2
Vx	A1	SP8	*Myoviridae*	*Bacillus*	2,11,19
W	B1	N1	*Siphoviridae*	*Micrococcus*	3,19
w	B1	T5	*Siphoviridae*	Enterobacteria	1,19
W1	B		*Siphoviridae*	*Streptomyces*	18
W-1a	B		*Siphoviridae*	*Streptomyces*	18
W-2a	B		*Siphoviridae*	*Streptomyces*	18
W2c	A1	(01)	*Myoviridae*	Enterobacteria	1,19
W3	B		*Siphoviridae*	*Streptomyces*	18
W5	B		*Siphoviridae*	*Streptomyces*	18
W31	C1	T7	*Podoviridae*	Enterobacteria	1,10,19
w40x1	B1	949	*Siphoviridae*	*Streptococcus*	3,19
W407x1	B1	949	*Siphoviridae*	*Streptococcus*	3,19
W_{111}	C1	HM2	*Podoviridae*	*Clostridium*	19
W_{523}	C1	HM2	*Podoviridae*	*Clostridium*	19
WA_{01}	C1	HM2	*Podoviridae*	*Clostridium*	19
WA_{03}	C1	HM2	*Podoviridae*	*Clostridium*	19
WA/1	D1		*Microviridae* (*Microvirus*)	Enterobacteria	1,10,19
Wb	C1	Tb	*Podoviridae*	*Brucella*	1,11,19
Weybridge	Synonym: phage Wb				19
WF/1	D1		*Microviridae* (*Microvirus*)	Enterobacteria	1,10,19
Worksop	C1	(P22)	*Podoviridae*	Enterobacteria	1,19
Wφ	A1	P2	*Myoviridae*	Enterobacteria	1,19
WSP1	B2	(MSP8)	*Siphoviridae*	*Streptomyces*	4,19
WSP4	B2	MSP8	*Siphoviridae*	*Streptomyces*	4,19
WSP5	B2	MSP8	*Siphoviridae*	*Streptomyces*	4,19
WT1	A1	WT1	*Myoviridae*	*Rhizobium*	1,10,17,19
WW/1	D1		*Microviridae* (*Microvirus*)	Enterobacteria	19
X	B1	N1	*Siphoviridae*	*Micrococcus*	3,19
X	C1	Tb	*Podoviridae*	*Brucella*	19
X	F1		*Inoviridae* (*Inovirus*)	Enterobacteria	19
X29	A1	X29	*Myoviridae*	*Vibrio*	2,19
Xf	F1		*Inoviridae* (*Inovirus*)	*Xanthomonas*	1,10,19
Xf2	F1		*Inoviridae* (*Inovirus*)	*Xanthomonas*	10,19
XO1	A1	XP5	*Myoviridae*	*Xanthomonas*	19
XP5	A1	XP5	*Myoviridae*	*Xanthomonas*	1,8,10,19
Xp10	?		?	*Xanthomonas*	24

Phage name	Morpho-type	Taxonomy 'Species group'	Family (Genus)	Host	Reference
XP12	B2	XP12	*Siphoviridae*	*Xanthomonas*	[1,8,19]
ξ11	B1	SPβ	*Siphoviridae*	*Bacillus*	[19]
Y	C1	T7	*Podoviridae*	Enterobacteria	[1,10,19]
y5	B1	y5	*Siphoviridae*	*Lactobacillus*	[19]
Y7	B1	Leo	*Siphoviridae*	*Mycobacterium*	[1,4,19]
Y10	B1	Leo	*Siphoviridae*	*Mycobacterium*	[1,4,19]
Y46(E2)	A1	(01)	*Myoviridae*	Enterobacteria	[19]
Ya1	A1		*Myoviridae*	*Pseudomonas*	[1]
Ya4	A1		*Myoviridae*	*Pseudomonas*	[1]
Ya5	C1		*Podoviridae*	*Pseudomonas*	[1]
Ya7	C1		*Podoviridae*	*Pseudomonas*	[1]
Ya11	A1		*Myoviridae*	*Pseudomonas*	[1]
YG	E1	(serogroup II)	*Leviviridae* (*Levivirus*)	Enterobacteria	[19]
Z_1	B1	P11-M15	*Siphoviridae*	*Staphylococcus*	[1,3,19]
Z-1/H-16	C1		*Podoviridae*	*Alcaligenes*	[1]
Z1K/1	E1	(serogroup II)	*Leviviridae* (*Levivirus*)	Enterobacteria	[1,1,19]
Z4	B2	3A	*Siphoviridae*	*Staphylococcus*	[1,3,19]
Z-33	A1		*Myoviridae*	*Hydrogenomonas*	[1]
ZD/13	D1		*Microviridae* (*Microvirus*)	Enterobacteria	[19]
ZG/1	E1	(serogroup R)	*Leviviridae* (*Levivirus*)	Enterobacteria	[1,10,19]
ZG/2	F1		*Inoviridae* (*Inovirus*)	Enterobacteria	[1,10,19]
ZG/3A	B2	ZG/3A	*Siphoviridae*	Enterobacteria	[1,10,19]
ZJ/1	E1	(serogroup II)	*Leviviridae* (*Levivirus*)	Enterobacteria	[1,10,19]
ZJ/2	F1	fd	*Inoviridae* (*Inovirus*)	Enterobacteria	[1,10,19]
ZL/3	E1	(serogroup R)	*Leviviridae* (*Levivirus*)	Enterobacteria	[1,10,19]
ZR	E1	(serogroup I)	*Leviviridae* (*Levivirus*)	Enterobacteria	[7,19]
ZS/3	E1	(serogroup R)	*Leviviridae* (*Levivirus*)	Enterobacteria	[1,10,19]
ζ3	D1		*Microviridae* (*Microvirus*)	Enterobacteria	[10,19]
01	A1	01	*Myoviridae*	Enterobacteria	[1,19]
02	A1	01	*Myoviridae*	Enterobacteria	[1,19]
03	A1	01	*Myoviridae*	Enterobacteria	[1,19]
03c1R	F2	MVL51	*Inoviridae* (*Plectrovirus*)	*Acholeplasma*	[1,10]
06	A1	Twort	*Myoviridae*	*Staphylococcus*	[1,3,19]
011c1r	F2		*Inoviridae* (*Plectrovirus*)	*Acholeplasma*	[1,10]
025	A1	121	*Myoviridae*	Enterobacteria	[19]

Phage name	Morpho- type	Taxonomy 'Species group'	Family (Genus)	Host	Reference
1	A1	AU	*Myoviridae*	*Pasteurella*	19
1	B3	VD13	*Siphoviridae*	*Streptococcus*	3,19
1	B2	φ2037/1	*Siphoviridae*	*Rhizobium*	1,17,19
1	B1	XP12	*Siphoviridae*	*Pseudomonas*	8,19
1	C1	sd	*Podoviridae*	Enterobacteria	1,19
1	B1	Jersey	*Siphoviridae*	Enterobacteria	1,19
1	A1	CM₁	*Myoviridae*	*Rhizobium*	17,19
1	B2	F1	*Siphoviridae*	*Clostridium*	1,19
1	C1	(P22)	*Podoviridae*	Enterobacteria	1,19
1	C1	φ17	*Podoviridae*	*Streptomyces*	4,19
I	C1	I	*Podoviridae*	*Vibrio*	2,19
1.2	C1	(N4)	*Podoviridae*	Enterobacteria	19
1(40)	C1	(P22)	*Podoviridae*	Enterobacteria	1,19
1/72	A1	(PB-1)	*Myoviridae*	*Pseudomonas*	1,8,19
1-97	C2	GA-1	*Podoviridae*	*Bacillus*	11,19
1A	A2	T4	*Myoviridae*	Enterobacteria	19
1A	B1	1A	*Siphoviridae*	*Bacillus*	19
1b6	B1	1b6	*Siphoviridae*	*Lactobacillus*	19
1C	A1	CEβ	*Myoviridae*	*Clostridium*	19
1D	A1	CEβ	*Myoviridae*	*Clostridium*	19
1φ1	D1	φX174	*Microviridae* (*Microvirus*)	Enterobacteria	1,10,19
1φ3	D1	φX174	*Microviridae* (*Microvirus*)	Enterobacteria	1,10,19
1φ7	D1	φX174	*Microviridae* (*Microvirus*)	Enterobacteria	1,10,19
1φ9	D1	φX174	*Microviridae* (*Microvirus*)	Enterobacteria	1,10,19
2	A1	AU	*Myoviridae*	*Pasteurella*	19
2	A1	RZh	*Myoviridae*	*Streptococcus*	3,19
2	C1	2	*Podoviridae*	*Rhizobium*	1,10,17,19
2	B1	1A	*Siphoviridae*	*Bacillus*	19
2	B1		*Siphoviridae*	*Pseudomonas*	1
2	C1	gh-1	*Podoviridae*	*Pseudomonas*	21
2	B1	Jersey	*Siphoviridae*	Enterobacteria	1,10,19
II	A1	II	*Myoviridae*	*Vibrio*	2,19
II	A2	T4	*Myoviridae*	Enterobacteria	19
II	B1	II	*Siphoviridae*	*Bacillus*	1,10,11,19
2.5	C1	sd	*Podoviridae*	Enterobacteria	19
2/79	A1	(PB-1)	*Myoviridae*	*Pseudomonas*	8,19
2b	B2	(MSP8)	*Siphoviridae*	*Streptomyces*	4,19
2BV	C1	2BV	*Podoviridae*	*Streptococcus*	3,19
2C	A1	SP8	*Myoviridae*	*Bacillus*	1,11,19
2D	A1	CEβ	*Myoviridae*	*Clostridium*	19
2F	B1		*Siphoviridae*	*Pseudomonas*	1
3	A1	37	*Myoviridae*	*Aeromonas,*	5,19
3	A1	PS17	*Myoviridae*	*Pseudomonas*	1,8,19
3	A2	T4	*Myoviridae*	Enterobacteria	1,10,19

Phage name	Morpho-type	Taxonomy 'Species group'	Family (Genus)	Host	Reference
3	B1	1A	*Siphoviridae*	*Bacillus*	[19]
3	C1	22	*Podoviridae*	*Pasteurella*	[19]
3	C1	Tb	*Podoviridae*	*Brucella*	[13,19]
III	A1	(SP50)	*Myoviridae*	*Bacillus*	[1,11,19]
III	C1	III	*Podoviridae*	*Vibrio*	[2,19]
3/10	A1	AU	*Myoviridae*	*Pasteurella*	[19]
3/26	C2	7/26	*Podoviridae*	*Kurthia*	[4]
3-793	B1		*Siphoviridae*	*Lactobacillus*	[1]
3A	A2	T4	*Myoviridae*	Enterobacteria	[19]
3A	B2	3A	*Siphoviridae*	*Staphylococcus*	[1,3,10,19]
3a	B1	Jersey	*Siphoviridae*	Enterobacteria	[1,19]
3aI	B1	Jersey	*Siphoviridae*	Enterobacteria	[1,19]
3B	B2	3A	*Siphoviridae*	*Staphylococcus*	[1,3,19]
3b	C1	(P22)	*Podoviridae*	Enterobacteria	[1,19]
3C	A1	CEβ	*Myoviridae*	*Clostridium*	[19]
3C	B2	3A	*Siphoviridae*	*Staphylococcus*	[1,3,19]
3/DO	A1	(PB-1)	*Myoviridae*	*Pseudomonas*	[8,19]
3ML	B2	3ML	*Siphoviridae*	*Streptococcus*	[1,3,10,19]
3NT	A1	PBS1	*Myoviridae*	*Bacillus*	[1,11,19]
3T+	A2	T4	*Myoviridae*	Enterobacteria	[1,10,19]
4	A1	RZh	*Myoviridae*	*Streptococcus*	[3,19]
4	A1	37	*Myoviridae*	*Aeromonas*	[5,19]
4	A1	(CT4)	*Myoviridae*	*Rhizobium*	[1,17,19]
4	B2	MSP8	*Siphoviridae*	*Streptomyces*	[1,4,19]
IV	B1	IV	*Siphoviridae*	*Vibrio*	[2,19]
4/10	A1	AU	*Myoviridae*	*Pasteurella*	[19]
4/273	A1	(PB-1)	*Myoviridae*	*Pseudomonas*	[8,19]
4C	A1	CEβ	*Myoviridae*	*Clostridium*	[19]
5	B1	5	*Siphoviridae*	*Rhizobium*	[1,19]
5	B1	(χ)	*Siphoviridae*	Enterobacteria	[19]
5	B2	ZG/3A	*Siphoviridae*	Enterobacteria	[1,19]
5	B2	MSP8	*Siphoviridae*	*Streptomyces*	[19]
5/27	C2	7/26	*Podoviridae*	*Kurthia*	[4]
5/4O6	A1	(PB-1)	*Myoviridae*	*Pseudomonas*	[8,19]
5φ	A2	T4	*Myoviridae*	Enterobacteria	[19]
6	C1	(φ2042)	*Podoviridae*	*Rhizobium*	[1,17,19]
6	B2	3A	*Siphoviridae*	*Staphylococcus*	[1,3,19]
6	C1	Tb	*Podoviridae*	*Brucella*	[13,19]
6	C1	(P22)	*Podoviridae*	Enterobacteria	[1,19]
VI	A2	T4	*Myoviridae*	Enterobacteria	[19]
6/14(18)	C1	(P22)	*Podoviridae*	Enterobacteria	[1,19]
6/27	C2	7/26	*Podoviridae*	*Kurthia*	[4]
6/6660	A1	(PB-1)	*Myoviridae*	*Pseudomonas*	[8,19]
6C	A1	PS17	*Myoviridae*	*Pseudomonas*	[1,8,19]
6SR	D1	φX174	*Microviridae* (*Microvirus*)	Enterobacteria	[14]

Phage name	Morpho-type	Taxonomy 'Species group'	Family (Genus)	Host	Reference
7	B1	(ViII)	*Siphoviridae*	Enterobacteria	19
7	C1	(N4)	*Podoviridae*	Enterobacteria	19
7	B2	3A	*Siphoviridae*	*Staphylococcus*	19
7	C1	Tb	*Podoviridae*	*Brucella*	13,19
7	C1	gh-1	*Podoviridae*	*Pseudomonas*	21
7-7-1	A1	(WT1)	*Myoviridae*	*Rhizobium*	1,19
7-7-7	B3	7-7-7	*Siphoviridae*	*Rhizobium*	1,10,17,19
7-11	C3	7-11	*Podoviridae*	Enterobacteria	1,10,19
7/26	C2	7/26	*Podoviridae*	*Kurthia*	4
7/184	A1	(PB-1)	*Myoviridae*	*Pseudomonas*	8,19
7-*E. coli*	C1	(N4)	*Podoviridae*	Enterobacteria	1
7-*Klebsiella*	B1	(ViII)	*Siphoviridae*	Enterobacteria	1
7m	B1		*Siphoviridae*	*Pseudomonas*	1
7s	E1		*Leviviridae* (*Levivirus*)	*Pseudomonas*	1,10,19
7v	A1	(PB-1)	*Myoviridae*	*Pseudomonas*	1,8,19
8	B1	(χ)	*Siphoviridae*	Enterobacteria	19
8	C1	(N4)	*Podoviridae*	Enterobacteria	1,19
8/27	C2	7/26	*Podoviridae*	*Kurthia*	4
8/280	A1	(PB-1)	*Myoviridae*	*Pseudomonas*	8,19
9	A1	01	*Myoviridae*	Enterobacteria	1,19
9	B1	T5	*Siphoviridae*	Enterobacteria	1,19
9	C1	φ17	*Podoviridae*	*Streptomyces*	4,19
IX	A2	T4	*Myoviridae*	Enterobacteria	19
9/0	A2	T4	*Myoviridae*	Enterobacteria	1,10,19
9/41	B1		*Siphoviridae*	*Dactylosporangium*	19
9/95	A1	(PB-1)	*Myoviridae*	*Pseudomonas*	8,19
9F	B1	(χ)	*Siphoviridae*	Enterobacteria	19
10	A1	AU	*Myoviridae*	*Pasteurella*	19
X	A2	T4	*Myoviridae*	Enterobacteria	19
10/1	B1	(χ)	*Siphoviridae*	Enterobacteria	19
10/I	C1	Tb	*Podoviridae*	*Brucella*	1,11,19
10/1A	A1	(Beccles)	*Myoviridae*	Enterobacteria	19
10/27	C2	7/26	*Podoviridae*	*Kurthia*	4
10/502	A1	(PB-1)	*Myoviridae*	*Pseudomonas*	8,19
10tur	F2	MVL51	*Inoviridae* (*Plectrovirus*)	*Acholeplasma*	1,10
10w	B2	3ML	*Siphoviridae*	*Streptococcus*	1,3,19
11	A1	37	*Siphoviridae*	*Aeromonas,*	5,19
11	B1	P11-M15	*Siphoviridae*	*Staphylococcus*	19
11	A1	(OKZ)	*Myoviridae*	*Pseudomonas*	19
11	B2	R1-Myb	*Siphoviridae*	*Mycobacterium*	1,4,19
11	C1	(N4)	*Podoviridae*	Enterobacteria	1,19
11/DE	A1	(φKZ)	*Siphoviridae*	*Pseudomonas*	1,19
11F	Synonym: phage Ox-1				19

Phage name	Morpho-type	Taxonomy 'Species group'	Family (Genus)	Host	Reference
12/27	C2	7/26	*Siphoviridae*	*Kurthia*	4
12/100	A1	(PB-1)	*Myoviridae*	*Pseudomonas*	8,19
12B	C1		*Podoviridae*	*Pseudomonas*	1
12m	C1	Tb	*Podoviridae*	*Brucella*	13,19
12S	A1	12S	*Myoviridae*	*Pseudomonas*	1,8,10,19
13	A1	II	*Myoviridae*	*Vibrio*	2,19
13	A1	13	*Myoviridae*	*Mycobacterium*	1
13/3a	C3	7-11	*Podoviridae*	Enterobacteria	1,19
13/441	B1		*Siphoviridae*	*Pseudomonas*	1
13*vir*	A1	(01)	*Myoviridae*	Enterobacteria	1,19
14	A1	(RZh)	*Myoviridae*	*Streptococcus*	3,19
14	C1	φ17	*Podoviridae*	*Streptomyces*	4,19
14	A1	kappa	*Myoviridae*	*Vibrio*	2,19
14	C1	(gh-1)	*Podoviridae*	*Pseudomonas*	1,8,19
14(6,7)	C1	(P22)	*Podoviridae*	Enterobacteria	1,19
15	C1		*Podoviridae*	*Pseudomonas*	1
15	B2	(MSP8)	*Siphoviridae*	*Streptomyces*	1,4,19
16	A1	PB-1	*Myoviridae*	*Pseudomonas*	21
16	A1	II	*Myoviridae*	*Vibrio*	2,19
16	B2	3A	*Siphoviridae*	*Staphylococcus*	1,3,19
16-2-4	B1		*Siphoviridae*	*Rhizobium*	1
16-3-2	C1	(φ2042)	*Podoviridae*	*Rhizobium*	1,19
16-6-12	B1	16-6-12	Siphoviridae	Rhizobium	1,10,17,19
16-6-14	C1		*Podoviridae*	*Rhizobium*	1
16-7-1	A1	(WT1)	*Myoviridae*	*Rhizobium*	1,17,19
16-12-1	B1		*Siphoviridae*	*Rhizobium*	1
16-19	A3	16-19	*Myoviridae*	Enterobacteria	1,10,19
16-22-2	B1		*Siphoviridae*	*Rhizobium*	1
16-35-5	A1		*Myoviridae*	*Rhizobium*	1
17	A1	(SP8)	*Myoviridae*	*Bacillus*	11,19
18	A1	(P2)	*Myoviridae*	Enterobacteria	31
18	A1	(SP8)	*Myoviridae*	*Bacillus*	11,19
19	A1	(SP8)	*Myoviridae*	*Bacillus*	11,19
20	C1	(gh-1)	*Podoviridae*	*Pseudomonas*	1,8,19
20	B1		*Siphoviridae*	*Xanthomonas*	1
20.2	A3	16-19	*Myoviridae*	Enterobacteria	1,19
20A	A1	(01)	*Myoviridae*	Enterobacteria	19
20E	B1	χ	*Siphoviridae*	Enterobacteria	19
21	A1	φKZ	*Myoviridae*	*Pseudomonas*	1,19
21	B1	(T1)	*Siphoviridae*	Enterobacteria	43
21	B2	3A	*Siphoviridae*	*Staphylococcus*	1,3,19
22	B1		*Siphoviridae*	*Pseudomonas*	8
22	B1	χ	*Siphoviridae*	Enterobacteria	19

Phage name	Morpho-type	Taxonomy 'Species group'	Family (Genus)	Host	Reference
22	C1	22	*Podoviridae*	*Pasteurella*	19
23	C1		*Podoviridae*	*Pseudomonas*	1
23	B1	(T1)	*Siphoviridae*	Enterobacteria	43
24	A1	PB-1	*Myoviridae*	*Pseudomonas*	21
24	B1	24	*Siphoviridae*	*Pseudomonas*	1,3
24	B1	24	*Siphoviridae*	*Streptococcus*	1,10,19
24	C1	φ17	*Podoviridae*	*Streptomyces*	4,19
24	A1	II	*Myoviridae*	*Vibrio*	2,19
24/II	C1	Tb	*Podoviridae*	*Brucella*	13,19
24/25, 14	B2	ZG/3A	*Siphoviridae*	Enterobacteria	1,19
25	A2	44RR2.8t	*Myoviridae*	*Aeromonas*	5,19
25F	A1		*Myoviridae*	*Pseudomonas*	1
26	C1	(sd)	*Podoviridae*	Enterobacteria	1,19
26	B2	(MSP8)	*Siphoviridae*	*Streptomyces*	1,19
27	A1		*Myoviridae*	*Pseudomonas*	1
27	C1	φ17	*Podoviridae*	*Streptomyces*	4,19
27	C1	(P22)	*Podoviridae*	Enterobacteria	1
28-1	A1	ViI	*Myoviridae*	Enterobacteria	1,19
28.2	C1	(sd)	*Podoviridae*	Enterobacteria	1,19
29	A1	29	*Myoviridae*	*Aeromonas*	5,19
29	C1	(N4)	*Podoviridae*	Enterobacteria	1,19
29	B1	(77)	*Siphoviridae*	*Staphylococcus*	1
30	C1	(N4)	*Podoviridae*	Enterobacteria	1,19
31	A2	44RR2.8t	*Myoviridae*	*Aeromonas*	5,19
31	C1	(N4)	*Podoviridae*	Enterobacteria	1,19
31	B2	R1-Myb	*Siphoviridae*	*Mycobacterium*	1,4,19
31	C1	gh-1	*Podoviridae*	*Pseudomonas*	21
31B	B1	P11-M15	*Siphoviridae*	*Staphylococcus*	3,19
32	B1	32	*Siphoviridae*	*Pasteurella*	19
32	A1	37	*Myoviridae*	*Aeromonas*	5,19
32	A1	kappa	*Myoviridae*	*Vibrio*	2,19
32	C1	(N4)	*Podoviridae*	Enterobacteria	1,19
32/6	B1	(χ)	*Siphoviridae*	Enterobacteria	19
32f	A1		*Myoviridae*	*Campylobacter*	1,19
34	C1	(P22)	*Podoviridae*	Enterobacteria	19
34/13	C1	(P22)	*Podoviridae*	Enterobacteria	1,19
34B	B1	χ	*Siphoviridae*	Enterobacteria	19
34Ct	B1	χ	*Siphoviridae*	Enterobacteria	19
34D	C1	(sd)	*Podoviridae*	Enterobacteria	19
34H	A1	(ViI)	*Myoviridae*	Enterobacteria	19
34P	B1	χ	*Siphoviridae*	Enterobacteria	19
35	C1	I	*Podoviridae*	*Vibrio*	2,19
36	A3	16-19	*Myoviridae*	Enterobacteria	1,19
37	A1	37	*Myoviridae*	*Aeromonas*	5,19
37	B1	(χ)	*Siphoviridae*	Enterobacteria	19
37	C1	I	*Podoviridae*	*Vibrio*	19
37	C1	(P22)	*Podoviridae*	Enterobacteria	1,19
38	C1	(N4)	*Podoviridae*	Enterobacteria	1,19
38T	A1	(Kl19)	*Myoviridae*	Enterobacteria	19
39	C1	(N4)	*Podoviridae*	Enterobacteria	1,19

Phage name	Morpho-type	Taxonomy 'Species group'	Family (Genus)	Host	Reference
40	A1	Twort	*Myoviridae*	*Staphylococcus*	1,3,19
40	B2	R1-Myb	*Siphoviridae*	*Mycobacterium*	1,4,19
40.3	C3	7-11	*Podoviridae*	Enterobacteria	1,19
41	A1	RZh	*Myoviridae*	*Streptococcus*	3,19
41	B1	χ	*Siphoviridae*	Enterobacteria	19
41c	B1	SPP1	*Siphoviridae*	*Bacillus*	19
42	C1	(N4)	*Podoviridae*	Enterobacteria	1,19
42₂	C1	(P22)	*Podoviridae*	Enterobacteria	1
42B	B2	3A	*Siphoviridae*	*Staphylococcus*	3,19
42C	B2	3A	*Siphoviridae*	*Staphylococcus*	3,19
42D	B1	77	*Siphoviridae*	*Staphylococcus*	1
42E	B2	3A	*Siphoviridae*	*Staphylococcus*	1,3,19
43	A1	43	*Myoviridae*	*Aeromonas*	5,19
44	A1	PB-1	*Myoviridae*	*Pseudomonas*	21
44	B1	(T1)	*Siphoviridae*	Enterobacteria	43
44	C1	Tb	*Podoviridae*	*Brucella*	13,19
44A	B1	P11-M15	*Siphoviridae*	*Staphylococcus*	3,19
44AHJD	C1	44AHJD	*Podoviridae*	*Staphylococcus*	3,19
44RR2.8t	A	44RR2.8t	*Myoviridae*	*Aeromonas*	5,19
45/III	C1	Tb	*Podoviridae*	*Brucella*	1,13,19
47	B2	3A	*Siphoviridae*	*Staphylococcus*	1,3,19
47A	B2	3A	*Siphoviridae*	*Staphylococcus*	3,19
49	B1		*Siphoviridae*	*Pseudomonas*	1
50	B2	R1-Myb	*Siphoviridae*	*Mycobacterium*	1,4,19
50	A2	T4	*Myoviridae*	Enterobacteria	1,10,19
51	C1	III	*Podoviridae*	*Vibrio*	2,19
51	A1	51	*Myoviridae*	*Aeromonas*	5,19
51	B1	(T1)	*Siphoviridae*	Enterobacteria	43
52	B1	P11-M15	*Siphoviridae*	*Staphylococcus*	3,19
52	B1	(T1)	*Siphoviridae*	Enterobacteria	43
52A	B1	P11-M15	*Siphoviridae*	*Staphylococcus*	1,3,19
52B	B1	P11-M15	*Siphoviridae*	*Staphylococcus*	3,19
53	B1	P11-M15	*Siphoviridae*	*Staphylococcus*	1,3,19
53	B1	32	*Siphoviridae*	*Pasteurella*	19
54	B2	3A	*Siphoviridae*	*Staphylococcus*	1,3,19
54x1	B2	3A	*Siphoviridae*	*Staphylococcus*	1,3,19
55	B1	(77)	*Siphoviridae*	*Staphylococcus*	1
55	C1	22	*Podoviridae*	*Pasteurella*	19
55R.1	A19.1		*Myoviridae*	*Aeromonas*	5,19
56	A11		*Myoviridae*	*Aeromonas*	5,19
56	B1	(χ)	*Siphoviridae*	Enterobacteria	19
56P	B1	χ	*Siphoviridae*	Enterobacteria	19
56RR.2	A1	59.1	*Myoviridae*	*Aeromonas*	5,19
57	A1	51	*Myoviridae*	*Aeromonas*	5,19
57	C1	I	*Podoviridae*	*Vibrio*	2
57	B2	*mor*1	*Siphoviridae*	*Bacillus*	1
58	A1	Twort	*Myoviridae*	*Staphylococcus*	1,3,19
59.1	A1	59.1	*Myoviridae*	*Aeromonas*	5,19
60	B2	*mor*1	*Siphoviridae*	*Bacillus*	1

Phage name	Morpho-type	Taxonomy 'Species group'	Family (Genus)	Host	Reference
60P	B1	χ	*Siphoviridae*	Enterobacteria	[19]
61	B1		*Siphoviridae*	*Pseudomonas*	[1]
61/6	B1	(χ)	*Siphoviridae*	Enterobacteria	[19]
63	B1	(1A)	*Siphoviridae*	*Bacillus*	[19]
63-F	C1		*Podoviridae*	*Methylocystis*	[19]
64	B1	(1A)	*Siphoviridae*	*Bacillus*	[19]
65	A2	65	*Myoviridae*	*Aeromonas*	[5,19]
66F	A2	T2	*Myoviridae*	Enterobacteria	[1,10]
67.1	B1		*Siphoviridae*	*Bacteroides*	[1]
68	A1	φkz	*Myoviridae*	*Pseudomonas*	[21]
70	B2	3A	*Siphoviridae*	*Staphylococcus*	[1,3,19]
70.1 to 70.7	B1	PS8	*Siphoviridae*	*Agrobacterium*	[1,17,19]
71	B1	P11-M15	*Siphoviridae*	*Staphylococcus*	[1,3,19]
73	B		*Siphoviridae*	*Pseudomonas*	[21]
73	B2	3A	*Siphoviridae*	*Staphylococcus*	[1,3,19]
74/6	B1	(χ)	*Siphoviridae*	Enterobacteria	[19]
75	B2	3A	*Siphoviridae*	*Staphylococcus*	[1,19]
75	C1	Tb	*Podoviridae*	*Brucella*	[13,19]
76	B2		*Siphoviridae*	*Caulobacter*	[1,19]
76/4	B1	(χ)	*Siphoviridae*	Enterobacteria	[19]
77	B1	77	*Siphoviridae*	*Staphylococcus*	[1,3,10,19]
78	B2	3A	*Siphoviridae*	*Staphylococcus*	[3,19]
79	B1	P11-M15	*Siphoviridae*	*Staphylococcus*	[3,19]
80	C1	HM2	*Podoviridae*	*Clostridium*	[1,19]
80	B1	P11-M15	*Siphoviridae*	*Staphylococcus*	[1,3,19]
80α	B1	P11-M15	*Siphoviridae*	Staphylococcus	[19]
81	B2	3A	*Siphoviridae*	*Staphylococcus*	[3,19]
82	B2	3A	*Siphoviridae*	*Staphylococcus*	[3,19]
83A	B1	P11-M15	*Siphoviridae*	*Staphylococcus*	[19]
84	B1	77	*Siphoviridae*	*Staphylococcus*	[19]
84	C1	Tb	*Podoviridae*	*Brucella*	[13,19]
85	B1	P11-M15	*Siphoviridae*	*Staphylococcus*	[19]
88	B2	3A	*Siphoviridae*	*Staphylococcus*	[1,3,19]
88A	B1	107	*Siphoviridae*	*Staphylococcus*	[1,3,19]
93	B2	3A	*Siphoviridae*	*Staphylococcus*	[19]
93	B1	(77)	*Siphoviridae*	*Staphylococcus*	[1,3,19]
94	B2	3A	*Siphoviridae*	*Staphylococcus*	[1,3,19]
95	A1	(PB-1)	*Myoviridae*	*Pseudomonas*	[8]
101	B2	3A	*Siphoviridae*	*Staphylococcus*	[1,19]
101/8900	B1	(χ)	*Siphoviridae*	Enterobacteria	[19]
102	B1	P11-M15	*Siphoviridae*	*Staphylococcus*	[1,3,19]
102	B1	(T1)	*Siphoviridae*	Enterobacteria	[43]
103	B1	(T1)	*Siphoviridae*	Enterobacteria	[43]
103a	B2	R1-Myb	*Siphoviridae*	*Mycobacterium*	[1,4,19]
103b	B2	R1-Myb	*Siphoviridae*	*Mycobacterium*	[1,4,19]
105	B2	3A	*Siphoviridae*	*Staphylococcus*	[19]
106	B2	3ML	*Siphoviridae*	*Streptococcus*	[1,3,19]

Phage name	Morpho-type	Taxonomy 'Species group'	Family (Genus)	Host	Reference
107	B1	107	*Siphoviridae*	*Staphylococcus*	1,3,10,19
107/69	A1	(P2)	*Myoviridae*	Enterobacteria	1
108/106	A2	108/106	*Myoviridae*	*Thermomonospora*	1,4,10,19
109	A1	PB-1	*Myoviridae*	*Pseudomonas*	21
110	B2	3A	*Siphoviridae*	*Staphylococcus*	19
114	C1	114	*Podoviridae*	*Thermomonospora*	1,4,10,19
115	B2	3A	*Siphoviridae*	*Staphylococcus*	19
115	C1	22	*Podoviridae*	*Pasteurella*	19
115/10	A1	AU	*Myoviridae*	*Pasteurella*	19
119	B1	119	*Siphoviridae*	*Thermomonospora*	1,4,10,19
119	A1	Twort	*Myoviridae*	*Staphylococcus*	1,3,19
119x	C1	gh-1	*Podoviridae*	*Pseudomonas*	21
121	A1	121	*Myoviridae*	Enterobacteria	1,10,19
121Q	A1	121	*Myoviridae*	Enterobacteria	19
123	B1	(T1)	*Siphoviridae*	Enterobacteria	43
124	B1	(T1)	*Siphoviridae*	Enterobacteria	43
126	B1	(T1)	*Siphoviridae*	Enterobacteria	43
126-16	B1	107	*Siphoviridae*	*Staphylococcus*	1,3,19
128	B2	R1-Myb	*Siphoviridae*	*Mycobacterium*	1,4,19
129-16	B2	3A	*Siphoviridae*	*Staphylococcus*	9
130	A1	Twort	*Myoviridae*	*Staphylococcus*	1,3,19
131	A1	Twort	*Myoviridae*	*Staphylococcus*	1,3,19
140	B1	(T1)	*Siphoviridae*	*Enterobacteria*	43
143tur	F2	MVL51	*Inoviridae (Plectrovirus)*	*Acholeplasma*	1
150	B1	(T1)	*Siphoviridae*	*Enterobacteria*	43
168	B1	(T1)	*Siphoviridae*	*Enterobacteria*	43
170	C1	(gh-1)	*Podoviridae*	*Pseudomonas*	1,8,19
171	B1	(T1)	*Siphoviridae*	Enterobacteria	43
172	B1	(T1)	*Siphoviridae*	Enterobacteria	43
174	B1	(T1)	*Siphoviridae*	Enterobacteria	43
174	B2	3A	*Siphoviridae*	*Staphylococcus*	1,3,19
175	B1	(T1)	*Siphoviridae*	Enterobacteria	43
179tur	F2	MVL51	*Inoviridae (Plectrovirus)*	Acholeplasma	1
182	C2	182	*Podoviridae*	*Streptococcus*	19
182tur	F2	MVL51	*Inoviridae (Plectrovirus)*	*Acholeplasma*	1
184	B1	2671	*Siphoviridae*	*Listeria*	16,19
186	A1	P2	*Myoviridae*	Enterobacteria	1,19
187	B1	187	*Siphoviridae*	*Staphylococcus*	1,3,10,19
187	B1	(936)	*Siphoviridae*	*Streptococcus*	3,19
188	B1	(936)	*Siphoviridae*	*Streptococcus*	3,19
200	A1	Twort	*Myoviridae*	*Staphylococcus*	1,3,19
200	B31	VD13	*Siphoviridae*	*Streptococcus*	3,19
206	A1	222a	*Myoviridae*	*Lactobacillus*	1,19
212/XV	C1	Tb	*Podoviridae*	*Brucella*	13,19
222a	A1	222a	*Myoviridae*	*Lactobacillus*	1,19
223	B2	223	*Siphoviridae*	*Lactobacillus*	1,19
225	C1	182	*Podoviridae*	*Streptococcus*	3,19

Phage name	Morpho-type	Taxonomy 'Species group'	Family (Genus)	Host	Reference
227	B1	(χ)	*Siphoviridae*	Enterobacteria	[1] [9]
227	B1		*Siphoviridae*	*Lactobacillus*	[1]
235	B3	VD13	*Siphoviridae*	*Streptococcus*	[3,19]
277	C1		*Podoviridae*	*Fusobacterium*	[1,19]
289	B1	χ	*Siphoviridae*	Enterobacteria	[19]
293	B1	(T1)	*Siphoviridae*	Enterobacteria	[43]
299	A1	P2	*Myoviridae*	Enterobacteria	[1,19]
300	A1	(222a)	*Myoviridae*	*Lactobacillus*	[1,19]
315	A1	222a	*Myoviridae*	*Lactobacillus*	[1,19]
317	B1	317	*Siphoviridae*	*Rhizobium*	[1,10,17,19]
341	B3	VD13	*Siphoviridae*	*Streptococcus*	[3,19]
345G	A1	(ViI)	*Myoviridae*	Enterobacteria	[19]
345P	A1	(01)	*Myoviridae*	Enterobacteria	[19]
352	A1	PB-1	*Myoviridae*	*Pseudomonas*	[21]
356	A1	(222a)	*Myoviridae*	*Lactobacillus*	[1,19]
371	Synonym: phage 371/XXIX				[19]
371/XXIX	C1	Tb	*Podoviridae*	*Brucella*	[1,13,19]
380	C1	(N4)	*Podoviridae*	Enterobacteria	[1,19]
449C	A3	16-19	*Myoviridae*	Enterobacteria	[1,19]
501B	A1	(01)	*Myoviridae*	Enterobacteria	[19]
512	B1	(χ)	*Siphoviridae*	Enterobacteria	[19]
513	C1	Tb	*Podoviridae*	*Brucella*	[13,19]
514	A1	222a	*Myoviridae*	*Lactobacillus*	[19]
514-3	B		*Siphoviridae*	*Streptomyces*	[18]
525	C1		*Podoviridae*	*Pseudomonas*	[8]
531	B2		*Siphoviridae*	*Acinetobacter*	[1,19]
535/222a	B1		*Siphoviridae*	*Lactobacillus*	[1]
575	B1	2671	*Siphoviridae*	*Listeria*	[16,19]
581	B1	77	*Siphoviridae*	*Staphylococcus*	[1,3,19]
594	B2	3A	*Siphoviridae*	*Staphylococcus*	[1,3,19]
633	B1	2671	*Siphoviridae*	*Listeria*	[16,19]
643	B2	3ML	*Siphoviridae*	*Streptococcus*	[3,19]
744	B1	2671	*Siphoviridae*	*Listeria*	[16,19]
836/IV	C2	GA-1	*Podoviridae*	*Bacillus*	[11,19]
837/IV	C2	GA-1	*Podoviridae*	*Bacillus*	[1]
853	B1	936	*Siphoviridae*	*Streptococcus*	[3,19]
864/100	B2	(ZG/3A)	*Siphoviridae*	Enterobacteria	[1]
867	A1	RZh	*Siphoviridae*	*Streptococcus*	[3,19]
877	B1	949	*Siphoviridae*	*Streptococcus*	[3,19]
895	A1	AU	*Myoviridae*	*Pasteurella*	[19]
896	C1	22	*Podoviridae*	*Pasteurella*	[19]
900	B1	2671	*Siphoviridae*	*Listeria*	[16,19]
917	B2	3ML	*Siphoviridae*	*Streptococcus*	[3,19]
923	B2	3ML	*Siphoviridae*	*Streptococcus*	[3]

Phage name	Morpho-type	Taxonomy 'Species group'	Family (Genus)	Host	Reference
929	B1	936	*Siphoviridae*	*Streptococcus*	3,19
933J	B3	H-19J	*Siphoviridae*	Enterobacteria	19
936	B1	936	*Siphoviridae*	*Streptococcus*	3,19
949	B1	949	*Myoviridae*	*Streptococcus*	3,19
963	A1	(RZh)	*Myoviridae*	*Streptococcus*	3,19
966A	A3	16-19	*Myoviridae*	Enterobacteria	1,19
967	B1	32	*Siphoviridae*	*Pasteurella*	19
971	B1	949	*Siphoviridae*	*Streptococcus*	3,19
994	C1	22	*Podoviridae*	*Pasteurella*	19
995	C1	22	*Podoviridae*	*Pasteurella*	19
1009	B1	936	*Siphoviridae*	*Streptococcus*	3,19
1010	B1	Jersey	*Siphoviridae*	Enterobacteria	1,19
1039	B2	3ML	*Siphoviridae*	*Streptococcus*	3,19
1075	B1	32	*Siphoviridae*	*Pasteurella*	19
1090	B1	2671	*Siphoviridae*	*Listeria*	16,19
1204	B1	936	*Siphoviridae*	*Streptococcus*	3,19
1214	A1	PB-1	*Myoviridae*	*Pseudomonas*	21
1214	B2	1214	*Siphoviridae*	*Pseudomonas*	19
1250	B1	936	*Siphoviridae*	*Streptococcus*	3,19
1277	B1	936	*Siphoviridae*	*Streptococcus*	3,19
1280	B2	3ML	*Siphoviridae*	*Streptococcus*	3,19
1283	B1	949	*Siphoviridae*	*Streptococcus*	3,19
1299	B1	949	*Siphoviridae*	*Streptococcus*	3,19
1304clr	F2	MVL51	*Inoviridae* (*Plectrovirus*)	*Acholeplasma*	1
1307	G	L2	*Plasmaviridae* (*Plasmavirus*)	*Acholeplasma*	1,10
1310	B2	3ML	*Siphoviridae*	*Streptococcus*	3,19
1317	B1	2671	*Siphoviridae*	*Listeria*	16,19
1337	B2	3ML	*Siphoviridae*	*Streptococcus*	3,19
1358	B1	1358	*Siphoviridae*	*Streptococcus*	3,19
1363/14	B2	3A	*Siphoviridae*	*Staphylococcus*	19
1364	B1	936	*Siphoviridae*	*Streptococcus*	3,19
1367	B1	936	*Siphoviridae*	*Streptococcus*	3,19
1370	B1	936	*Siphoviridae*	*Streptococcus*	3,19
1374	B1	936	*Siphoviridae*	*Streptococcus*	3,19
1378	B2	3ML	*Siphoviridae*	*Streptococcus*	3,19
1404	B1	1358	*Siphoviridae*	*Streptococcus*	3,19
1405	B1	936	*Siphoviridae*	*Streptococcus*	3,19
1412	C1	1412	*Podoviridae*	Enterobacteria	1,19
1444	B1	2671	*Siphoviridae*	*Listeria*	16,19
1623	A1	(Twort)	*Myoviridae*	*Staphylococcus*	1,3,19
1652	B1	2671	*Siphoviridae*	*Listeria*	16,19
1806	B1	2685	*Siphoviridae*	*Listeria*	16,19
1807	B1	2685	*Siphoviridae*	*Listeria*	16,19

Phage name	Morpho-type	Taxonomy 'Species group'	Family (Genus)	Host	Reference
1967	B1	2671	*Siphoviridae*	*Listeria*	19
2389	B1	2389	*Siphoviridae*	*Listeria*	16
2425	B1	2671	*Siphoviridae*	*Listeria*	16,19
2460	B2	3A	*Siphoviridae*	*Staphylococcus*	19
2671	B1	2671	*Siphoviridae*	*Listeria*	16,19
2685	B1	2685	*Siphoviridae*	*Listeria*	16
2792A	B1	2848A	*Siphoviridae*	*Staphylococcus*	19
2847/10	B1	χ	*Siphoviridae*	Enterobacteria	19
2847/10b	C1	7480b	*Podoviridae*	Enterobacteria	19
2848A	B1	2848A	*Siphoviridae*	*Staphylococcus*	19
3000	B1	β4	*Siphoviridae*	Enterobacteria	1,19
3111-D	B2	R1-Myb	*Siphoviridae*	*Mycobacterium*	1,4,19
3215-D	B2	R1-Myb	*Siphoviridae*	*Mycobacterium*	1,4,19
3274	B1	2671	*Siphoviridae*	*Listeria*	16,19
3550	B1	2389	*Siphoviridae*	*Listeria*	16,19
3551	B1	2389	*Siphoviridae*	*Listeria*	16,19
3552	B1	2389	*Siphoviridae*	*Listeria*	16,19
4211	A1	4211	*Myoviridae*	*Listeria*	15,19
4276	B1	2671	*Siphoviridae*	*Listeria*	15,19
4277	B1	2671	*Siphoviridae*	*Listeria*	15,19
4286	A1	4211	*Myoviridae*	*Listeria*	15,19
4292	B1	2685	*Siphoviridae*	*Listeria*	15,19
4477	B1	2679	*Siphoviridae*	*Listeria*	16,19
4996	C1	4996	*Podoviridae*	*Vibrio*	2,19
5337	B1	2671	*Siphoviridae*	*Listeria*	15,19
5845	A2	T4	*Myoviridae*	Enterobacteria	1,10,19
6214	A1	kappa	*Myoviridae*	*Vibrio*	2,19
7050	A1	X29	*Myoviridae*	*Vibrio*	2,19
7479	B1	T5	*Siphoviridae*	Enterobacteria	1,19
7480b	C1	7480b	*Podoviridae*	Enterobacteria	1,10,19
8238	B2	MSP8	*Siphoviridae*	*Streptomyces*	1,4,19
8764	B1	8764	*Siphoviridae*	*Alcaligenes*	1,8,10,19
8893	A2	T4	*Myoviridae*	Enterobacteria	1,10,19
9211/9295	C1	(P22)	*Podoviridae*	Enterobacteria	1,19
9213/9211b	C1	(P22)	*Podoviridae*	Enterobacteria	1,19
9248	C1	(T7)	*Podoviridae*	Enterobacteria	1,19
9266	A2	9266	*Myoviridae*	Enterobacteria	1,10,19
9266Q	A2	9266	*Myoviridae*	Enterobacteria	19
10041/2815	C1	(P22)	*Podoviridae*	Enterobacteria	1,19
12029	B1	2671	*Siphoviridae*	*Listeria*	16,19